Practical Skills in Biology

Practical Skills in Biology

Sixth Edition

Allan Jones

Rob Reed

Jonathan Weyers

Harlow, England • London • New York • Boston • San Francisco • Toronto • Sydney • Auckland • Singapore • Hong Kong
Tokyo • Seoul • Taipei • New Delhi • Cape Town • São Paulo • Mexico City • Madrid • Amsterdam • Munich • Paris • Milan

Pearson Education Limited
Edinburgh Gate
Harlow CM20 2JE
United Kingdom
Tel: +44 (0)1279 623623
Web: www.pearson.com/uk

First published 1994 (print)
Second edition 1998 (print)
Third edition 2003 (print)
Fourth edition 2007 (print)
Fifth edition 2012 (print and electronic)
Sixth edition published 2016 (print and electronic)

ISBN: 978-1-292-09432-8 (print)
    978-1-292-09436-6 (PDF)
    978-1-292-09434-2 (eText)

British Library Cataloguing-in-Publication Data
A catalogue record for the print edition is available from the British Library

Library of Congress Cataloging-in-Publication Data
A catalog record for the print edition is available from the Library of Congress

10 9 8 7 6 5 4 3 2 1
20 19 18 17 16

Print edition typeset in Times NR MT Pro 10/12 by Lumina Datamatics
Print edition printed in Malaysia (CTP-PJB)

NOTE THAT ANY PAGE CROSS REFERENCES REFER TO THE PRINT EDITION

Contents

Contents

Companion Website

For open-access **student resources** specifically written to complement this textbook and support your learning, please visit **www.pearsoned.co.uk/practicalskills**

Lecturer Resources

For password-protected online resources tailored to support the use of this textbook in teaching, please visit **www.pearsoned.co.uk/practicalskills**

List of boxes

Preface to the sixth edition

The biosciences exist in an environment of current hypotheses rather than certainty, where natural variation occurs and can confuse empirical data. Knowledge of research design and the appropriate use of statistical analysis to enable a valid interpretation of experimental results is required. The biosciences are essentially practical and experimental subjects. Students are required to undertake appropriate practical education throughout their programme, which is progressive in nature and designed to supplement other academic learning.

> *Draft Subject Benchmark Statement for Biosciences* (2015)
> Quality Assurance Agency for Higher Education.
> Available: http://www.qaa.ac.uk/en/Publications/
> Documents/SBS-Biosciences-consultation-15.pdf.
> Last accessed 18/07/15.

All knowledge and theory in biology have originated from observation and experiment. As a result, laboratory and field work are important components of undergraduate training, and successful students develop a number of practical skills, ranging from those required to observe, measure and record accurately to those associated with operating up-to-date analytical equipment, alongside broader skills involved in teamwork and effective study. In creating the sixth edition of *Practical Skills in Biology*, we have maintained the approach of the previous editions, with the aim of supporting students (and lecturers) over a wide range of practical topics. As before, this is provided in a concise but user-friendly manner, with key points and definitions, illustrations and worked examples, tips and hints, 'How to' boxes and checklists and 'Safety note' boxes where appropriate.

This new edition consolidates the changes made for the fifth edition, which included additional material covering environmental investigations, chemical analysis, photosynthesis and respiration, stable isotopes and electrophoresis. We have completely revised and updated the text references and sources for further study, and incorporated over 50 new tips, figures, tables and boxes. Throughout the book we have changed the guidance regarding Microsoft *Office* software to apply in a generic sense rather than to any specific version. This may mean that readers may need to adjust commands if these are not appropriate. Guidance on specific commands and their syntax can usually be found using the software's help facility. Boxes giving details of approaches based on *Office 2003* and *Office 2007* that appeared in previous additions will be made available via the book's online resource at www.pearsoned. co.uk/practicalskills. This website continues to host the answers to the study exercises as well as text references and sources for further study – with live web links where applicable.

We should like to take this opportunity to thank our wives and families for their continued support, and to recognise the following colleagues and friends who have provided assistance, comment and food for thought at various points during the production of all six editions: Richard A'Brook, Margaret Adamson, Gail Alexander, Steve Atkins, Janet Aucock, Chris Baldwin, Abdellah Barekate, Gary Black, Geoff Bosson, Olivia Bragg, Sally Brown, Eldridge Buultjens, Richard Campbell, Cathy Caudwell, Bob Cherry, Mirela Cuculescu, John Dean, Charlie Dixon, Jackie Eager, Brian Eddy, Charmain Elder, Neil Fleming, Jennifer Gallacher, Karen Gowlett-Holmes, Alan Grant, Margaret Gruver, Mhairi Haggerty, Bryan Harrison, Rod Herbert, Dave Holmes, Helen Hooper, David Hopkins, Steve Hubbard, Jane Illés, Hugh Ingram, Wendy James, Andy Johnston, Alan Jones, Lorraine Kay, Ian Kill, Rhonda Knox, Lisa Lee-Jones, Phil Manning, Pete Maskrey, Fiona McKie-Bell, Steve Millam, Kirsty Millar, Stephen Moore, Rachel Morris, Fiona O'Donnell, Roy Oliver, Neil Paterson, John Raven, Steve Reed, Pete Rowell, David Sillars, J. Andrew C. Smith, Philip Smith, Susan Smith, Peter Sprent, Rob Sunley, Bill Tomlinson, Ruth Valentine, Lorraine Walsh, Dave Wealleans, Will Whitfield, Ian Winship, Bob Young and Hilary-Kay Young.

A special word of thanks is extended to Kathleen McMillan, whose work with one of us (JW) on Pearson Education's *Smarter Student* series has influenced much of our writing in the first three sections of this book. We should also like to thank the staff of Pearson Education for their friendly support over the years, and would wish to acknowledge Pat Bond, Rufus Curnow, Pauline Gillett, Owen Knight, Simon Lake, Dawn Phillips, Alex Seabrook and Richelle Zakrzewski for their encouragement and commitment to the *Practical Skills* series. Our thanks are also extended to Louise Attwood, Lisa Blackwell, Gary Hall and Mary Lince for their excellent work at the editing and proofreading stages. As with the previous editions, we should be grateful to hear of any errors you might notice, so that these can be put right at the earliest opportunity.

ALLAN JONES (allan.jones9@btinternet.com)
ROB REED (r.reed@cqu.edu.au)
JONATHAN WEYERS (j.d.b.weyers@dundee.ac.uk)

For the student

This book aims to provide guidance and support over the broad range of your undergraduate course, including laboratory classes, project work, lectures, tutorials, seminars and examinations, as outlined below.

Chapters 1-8 cover general skills

These include a number of transferable skills that you will develop during your course: for example, self-evaluation; time management; teamwork; preparing for examinations; creating a CV. They also provide guidance on how to study effectively and how to approach examinations and assessments.

Chapters 9-19 deal with IT, learning resources and communicating information

These chapters will help you get the most out of the resources and information available in your library, and on the Web, as well as providing helpful guidance on the use of software packages for data analysis, preparing assignments, essays and laboratory reports, etc. The ability to evaluate information is an increasingly important skill in contemporary society, and practical advice is provided here.

Chapters 20-70 cover a wide range of specific practical skills required in biology

These are based on the authors' experience of the questions students often ask in practical classes, and the support that is needed in order to get the most out of particular exercises. The text includes tips, hints, definitions, worked examples and 'How to' boxes that set out the key procedures in a step-by-step manner, with appropriate comments on safe working practice. The material ranges from basic laboratory procedures, such as preparing solutions, through specimen collection, identification and manipulation to the more advanced practical procedures that you might use during a final year project, e.g. radioisotope work.

Chapters 71-76 explain data analysis and presentation

This will be an important component of your course and you will find that these chapters guide you through the skills and techniques required, ranging from the presentation of results as graphs or tables through to the application of statistical tests. Worked examples are used to reinforce the numerical aspects wherever possible.

Study exercises and problems

We added these as a new feature in the third edition, following comments from students and staff at UK universities, who felt that it would provide a useful opportunity to practise some of the skills covered in the book and a check on the understanding of the material. We hope that the exercises will continue be useful both to learners and to their tutors.

Most of the problems assume students are working on their own, using the information supplied; however, tutors might wish to provide alternative starting material (e.g. a set of data from a practical class). We have also assumed that students will have access to a scientific calculator and, sometimes, to a networked PC with typical 'office' programs (especially word processor and spreadsheet), plus Internet access via a modem and browser. Where a library is mentioned, this is assumed to include access to standard reference works and a selection of scientific journals.

We recommend that students work together for some exercises – this is a valuable means of learning and, where there is no single correct answer to a problem, teamwork provides a mechanism for checking and discussing different approaches.

Answers are provided on the book's website at www.pearsoned.co.uk/practicalskills. For numerical problems, the working out is shown with the final answer, while for non-numerical exercises, we provide 'answers' in the form of tips, general guidance or illustrative examples, etc.

We hope that you will find this book a helpful guide throughout your course, and beyond.

Acknowledgements

We are grateful to the following for permission to reproduce copyright material:

Figures

Figure 40.2 adapted from *New Flora of the British Isles*, 2nd edn, Cambridge University Press (Stace, C.A. 1997) Cambridge University Press, Cambridge; Figure 54.1 from www.ukgbc.org/site/document/download/?document_id=556, http://jncc.defra.gov.uk/, Joint Nature Conservation Committee; Figure 56.2 from *Soil Taxonomy: A basic system of soil classification for making and interpreting soil surveys*, 2nd edn, ftp://ftp-fc.sc.egov.usda.gov/NSSC/Soil_Taxonomy/tax.pdf, US Government Printing Office, Washington. USA; Figure 58.1 from *Methods for Physical and Chemical Analysis of Fresh Waters*, 2nd edn, Blackwell Scientific (Golterman, H.L., Clymo, R.S. and Ohnstad, M.A.M. 1978) No 8, Wiley-Blackwell; Figure 58.3 adapted from *Methods for Physical and Chemical Analysis of Fresh Waters*, 1st edn, Blackwell Scientific, Oxford (Golterman, H.L.,Clymo, R.S. and Ohnstad, M.A.M. 1978) Wiley-Blackwell; Figure 66.2 from http://ppsystems.com/co2-gas-analyzers/, www.ppsystems.com, PP Systems; Figure 67.1 from http://www.fisher.co.uk/index.php/en/component/searchenhanced/, www.fisher.co.uk, Fisher Scientific UK Ltd, Thermo Fisher Scientific.

Maps

Figure 54.2 from NERC 100017897, http//data.nbn.org.uk, National Biodiversity Network, Crown Copyright. https://www.nationalarchives.gov.uk/doc/open-government-licence/version/3/

Screenshots

Screenshot 42.4 from http://www.whitman.edu/content/virtualpig, Whitman College

Tables

Table 54.1 after *National Vegetation Classification: Users' Handbook*, Joint Nature Conservation Committee (Rodwell, J.C. 2006) Joint Nature Conservation Committee; Tables 54.3 and 54.4 from Age specific survivorship and reproduction in *Phlox drummondii*, *The American Naturalist*, 113, pp. 881–903 (Leverlich, W.J. and Levin, D.A. 1979), University of Chicago Press; Table 58.2 from *The Biology of Seaweeds*, University of California Press (Luning, K.J. Editors: C.S. Lobban and .J. Wynne 1981) pp. 326–55, University of California Press; Table 58.2 from *The Biology of Seaweeds* by Lobban, Christopher S.; Wynne, Michael J. Reproduced with permission of Blackwell Scientific via Copyright Clearance Center.

Text

Box 5.1 adapted from A Guide to Learning Preferences, vark-learn.com, Version 7.1 (2011) held by Neil D. Fleming, Christchurch, New Zealand.

Photographs

(Key: b-bottom; c-centre; l-left; r-right; t-top)

123RF.com: petervick167 355; American Physiological Society: 420; David Bryson/Clandonia Resources: 346; Allan Jones: 240; KNAUER, Germany. www.knauer.net: 161; Supplied by Microscopy an operating division of KeyMed (Medical & industrial Equipment) Ltd: 272, 275; with permission Oregon Scientific Global Distribution Ltd. www.oregonscientific.com: 365l, 365r, 365l, 365r; Courtesy of PP Systems, Amesbury, MA, USA (http://www.ppsystems.com): 432; Property of bioMerieux S.A.: 251; Rob Reed: 284, 294; Skye Instruments Ltd: 373; Courtesy of Heinz Walz GmbH, Effeltrich, Germany (http://www.walz.com): 434; Jonathan Weyers: 262, 267, 272b, 274, 277, 311.

List of abbreviations

A	absorbance
A_r	relative atomic mass
ACDP	Advisory Committee on Dangerous Pathogens
ADP	adenosine diphosphate
ANOVA	analysis of variance
ATP	adenosine triphosphate
ASA	American Standards Association
BOD	biological (or biochemical) oxygen demand
BSA	bovine serum albumin
CD	compact disc
CE	capillary electrophoresis
CFU	colony-forming unit
CGE	capillary gel electrophoresis
CIEF	capillary isoelectric focusing
COD	chemical oxygen demand
COSHH	Control of Substances Hazardous to Health
CoV	coefficient of variance
CZE	capillary zone electrophoresis
DCMU	3-(3',4'-dichlorphenyl)-1,1-dimethylurea
DCPIP	2,6-dichlorophenol-indophenol
DEFT	direct epifluorescence technique
d.f.	degrees of freedom
DIN	Deutsches Institut fur Normung
DMO	5,5-dimethyl-2,4-oxazolidinedione
DNA	deoxyribonucleic acid
DNP	dinitrophenol
DO	dissolved oxygen
d.p.m.	disintegrations per minute
DTT	dithiothreitol
EC	electron capture
EDTA	ethylenediaminetetraacetic acid
EI	electron impact ionisation
EIA	enzyme immunoassay
ELISA	enzyme-linked immunosorbent assay
EM	electron microscopy
EMI	extended matching items
EMR	electromagnetic radiation
ESR	electron spin resonance
F	Faraday constant
FAA	formalin acetic acid
FTP	file transfer protocol
g	acceleration due to gravity
GA	graduate attributes
GC	gas chromatography
GFP	green fluorescent protein
GMO	genetically modified organism
GPS	global positioning system
HCG	human chorionic gonadotrophin
HEPA	high-efficiency particulate air
HEPES	N-[2-hydroxyethyl]piperazine-N'-[2-ethanesulphonic acid]
HPLC	high-performance liquid chromatography
HRP	horseradish peroxidise
IAPSO	International Association for Physical Sciences of the Ocean
IEF	isoelectric focusing
Ig	immunoglobulin
IPTG	isothiopropylgalactoside
IR	infrared (radiation)
IRGA	infrared gas analyser
IRMA	immunoradiometric assay
IRMS	isotope ratio mass spectrometry
ISO	International Organization for Standardization
Km	Michaelis constant
Kw	ionisation constant of water
LAN	local area network
LM	light microscopy
LOI	loss on ignition
LSD	least significant difference
M	molar (mol l^{-1})
MCQ	multiple-choice question
MECC	micellar electrokinetic capillary chromatography
MEL	maximum exposure limits
MMR	measles-mumps-rubella
MPN	most probable number
MRI	magnetic resonance imaging
M_r	relative molecular mass
MS	mass spectrometry
NAD^+	nicotinamide adenine dinucleotide (oxidised form)
NADH	nicotinamide adenine dinucleotide (reduced form)
$NADP^+$	nicotinamide adenine dinucleotide phosphate (oxidised form)
NADPH	nicotinamide adenine dinucleotide phosphate (reduced form)
NH	null hypothesis
NMR	nuclear magnetic resonance
NPQ	non-photochemical quenching
NTU	national turbidity unit
OC	organic carbon
OD	optical density
OES	occupational exposure standards
OM	organic matter
PAGE	polyacrylamide gel electrophoresis
PAM	pulse-amplified modulated
PAR	photosynthetically active radiation
PBL	problem-based learning
PCR	polymerase chain reaction
PDP	personal development planning

PEG	polyethylene glycol
PFD	photon flux density
PFU	plaque-forming unit
pH	$-\log_{10}$ proton concentration (activity), in mol m^{-1}
PI	photosynthetic irradiance
pK_a	$\log_{10}$ acid dissociation constant
PM	particulate materials
PMF	proton-motive force
PPFD	photosynthetic photon flux density
PQ	photosynthetic quotient
PS II	photosystem II
PTS	personal transferable skills
QIP	quench indication parameter
R	universal gas constant
RCF	relative centrifugal field
rDNA	recombinant deoxyribonucleic acid
R_F	relative frontal mobility
RIA	radioimmunoassay
RID	radioimmunodiffusion
RNA	ribonucleic acid
RP-HPLC	reverse phase high-performance liquid chromatography
r.p.m.	revolutions per minute
SAQ	short-answer question
SD	standard deviation

SDS	sodium dodecyl sulphate
SE	standard error (of the sample mean)
SEM	scanning electron microscopy
SI	Systeme Internationale d'Unités
SLR	single lens reflex
SPM	selectively permeable membrane
STP	standard temperature and pressure
T	absolute temperature (in kelvin)
TCA	tricarboxylic acid
TEM	transmission electron microscopy
TEMED	tetramethylethylenediamine
TLC	thin-layer chromatography
TMB	tetramethylbenzidine
TPMD	tetramethylphenylenediamine
TPP$^+$	tetraphenylphosphonium
TRIS	tris(hydroxymethyl)aminomethane *or* 2-amino-2-hydroxymethyl-1,3-propanediol
TS	transverse section
TTL	through the lens
URL	uniform resource locator
USB	Universal Serial Bus
UV	ultraviolet (radiation)
V_{max}	maximum velocity
WWW	World Wide Web
XGAL	5-bromo-4-chloro-3-indolyl-β-D-galactoside (62)

Study and examination skills

1 The importance of transferable skills

Skills terminology – different phrases may be used to describe transferable skills and associated personal qualities, depending on place or context. These include: 'graduate attributes', 'personal transferable skills' (PTS), 'key skills', 'core skills' and 'competences'.

Using course materials – study your course handbook and the schedules for each practical session to find out what skills you are expected to develop at each point in the curriculum. Usually the learning objectives/outcomes (p. 30) will outline the skills involved.

Example The skills involved in teamwork cannot be developed fully without a deeper understanding of the interrelationships involved in successful groups. The context will be different for every group and a flexible approach will always be required, according to the individuals involved and the nature of the task.

This chapter outlines the range of transferable skills and their significance to biologists. It also indicates where practical skills fit into this scheme. Having a good understanding of this topic will help you place your studies at university in a wider context. You will also gain an insight into the qualities that employers expect you to have developed by the time you graduate. Awareness of these matters will help when carrying out personal development planning (PDP) as part of your studies.

The range of transferable skills

Table 1.1 provides a comprehensive listing of university-level transferable skills under six skill categories. There are many possible classifications – and a different one may be used in your institution or field of study. Note particularly that 'study skills', while important, and rightly emphasised at the start of many courses, constitute only a subset of the skills acquired by most university students.

The phrase '*Practical Skills*' in the title of this book indicates that there is a special area of transferable skills related to work in the laboratory or field. However, although this text deals primarily with skills and techniques required for laboratory practicals, fieldwork and associated studies, a broader range of material is included. This is because the skills concerned are important, not only in the biosciences but also in the wider world. Examples include time management, evaluating information and communicating effectively.

> **KEY POINT** Biology is essentially a practical subject, and therefore involves highly developed laboratory and field skills. The importance that your lecturers place on practical skills will probably be evident from the large proportion of curriculum time you will spend on practical work in your course.

The word 'skill' implies much more than the robotic learning of, for example, a laboratory routine. Of course, some of the tasks you will be asked to carry out in practical classes *will* be repetitive. Certain techniques require manual dexterity and attention to detail if accuracy and precision are to be attained, and the necessary competence often requires practice to make perfect. However, a deeper understanding of the context of a technique is important if the skill is to be appreciated fully and then transferred to a new situation. That is why this text is not simply a 'recipe book' of methods and why it includes background information, tips and worked examples, as well as study exercises to test your understanding.

Transferability of skills

'Transferability' implies that someone with knowledge, understanding or ability gained in one situation can adapt or extend this for application in a different context. In some cases, the transfer of a skill is immediately obvious. Take, for example, the ability to use a spreadsheet to summarise biological data and create a graph to illustrate results. Once the key concepts and commands are learned (Chapter 13), they can be applied to

Table 1.1 Transferable skills identified as important in the biosciences. The list is based on the UK Quality Assurance Agency for Higher Education *Subject Benchmark Statement for the Biosciences* (QAA, 2007). Particularly relevant chapters are shown for the skills covered by this book (numbers in **bold** blue text indicate a deeper, or more extensive, treatment)

Skill category	Examples of skills and competences	Relevant chapters in this textbook
Generic skills for biologists	Having an appreciation of the complexity and diversity of life and life processes	36–8; **39**; 40–1; 61–2
	Reading and evaluating biological literature with a full and critical understanding	4; 9; **10**
	Capacity to communicate a clear and accurate account of a biological topic, both verbally and in writing	**14–16**; 17–19
	Applying critical and analytical skills to evaluate evidence regarding theories and hypotheses in biology	10; **33**
	Using a variety of methods for studying biology	36–70
	Having the ability to think independently, set personal tasks and solve problems	33; 35; **74**
Intellectual skills	Recognising and applying biological theories, concepts and principles	10; 33
	Analysing, synthesising and summarising information critically	10; 71–6
	Obtaining evidence to formulate and test hypotheses; applying knowledge to address familiar and unfamiliar problems	27–32; **33**; 34–5
	Recognising and explaining moral, ethical and legal issues in biological research	**21**; 22; 31; 36; 40
Experimental (practical) and observational skills	Carrying out basic laboratory and field techniques and understanding the principles that underlie them	**20**; 22–31; 34; 40–1; 43–5
	Working in lab or field safely, responsibly and legally, with due attention to ethical aspects	20–1; **22**; 31; 35–6; 38; 42
	Designing, planning, conducting and reporting on biological investigations and data arising from them	15; 18; 31; **33**; 35; 52
	Obtaining, recording, collating and analysing data in the field and laboratory	27; 29; **31**; 32–3; 36–41; **51**; **52**; 53–8; **71**; 72–6
Numeracy, communication and IT skills	Understanding and using data in several forms (e.g. numerical, textual, verbal and graphical)	4; 10; 29; 71–4
	Communicating in written, verbal, graphical and visual forms	**14, 15, 16**; 17–19; **72**; **73**; 74
	Citing and referencing the work of others in an appropriate manner	9
	Obtaining data, including the concepts behind sampling and sampling errors, calibration and types of error	29; **32**; 33; **34**; **53**; 71; 74; **75**; 76
	Processing, interpreting and presenting data, and applying appropriate statistical methods for summarising and analysing data	71–4; **75**; 76
	Solving problems with calculators and computers, including the use of tools such as spreadsheets	11; **12**; **20**; 74
	Using computer technology to communicate and as a source of information in biology	**11**; 12; 13
Interpersonal and teamwork skills	Working individually or in teams as appropriate; identifying individual and group goals and acting responsibly and appropriately to achieve them	3
	Recognising and respecting the views and opinions of others	3
	Evaluating your own performance and that of others	3; 8
	Appreciating the interdisciplinary nature of contemporary biology	1; 19
Self-management and professional development skills	Working independently, managing time and organising activities	2; 31; 33; 35
	Identifying and working towards targets for personal, academic and career development	1; 8
	Developing an adaptable and effective approach to study and work (including revision and exam technique)	5; 6; 7

many instances outside the biosciences where this type of output is used. This is not only true for similar data sets, but also in unrelated situations, such as making up a financial balance sheet and creating a pie chart to show sources of expenditure. Similarly, knowing the requirements for good graph drawing and tabulation (Chapters 72 and 73), perhaps practised by hand in earlier work, might help you use spreadsheet commands to make the output suit your needs.

Other cases may be less clear but equally valid. For example, towards the end of your undergraduate studies you may be involved in designing experiments as part of your project work. This task will draw on several skills gained at earlier stages in your course, such as preparing solutions (Chapters 23–26), deciding about numbers of replicates and experimental layout (Chapters 32 and 33) and perhaps carrying out some particular method of observation, measurement or analysis (Chapters 42–70). How and when might you transfer this complex set of skills? In the workplace, it is unlikely that you would be asked to repeat the same process, but in critically evaluating a problem or in planning a complex project for a new employer, you will need to use many of the time-management, organisational and analytical skills developed when designing and carrying out experiments. The same applies to information retrieval and evaluation and writing essays and dissertations, when transferred to the task of analysing or writing a business report.

Opportunities to develop and practise skills in your private or social life – you could, for example, practise spreadsheet skills by organising personal or club finances using Microsoft *Excel*, or teamwork skills within any university clubs or societies you may join (see Chapter 7).

Types of PDP portfolio and their benefits – some PDP schemes are centred on academic and learning skills, while others are more focused on career planning. They may be carried out independently or possibly in tandem with a personal tutor or advisory system. Certain PDP schemes involve creating an online portfolio, while others are primarily paper-based. Each method has specific goals and advantages, but whichever way your scheme operates, maximum benefit will be gained from fully involving yourself in the process.

Personal development planning

Many universities have schemes for PDP, which may go under slightly different names such as progress file or professional development plan. You will usually be expected to create a portfolio of evidence on your progress, then reflect on this, and subsequently set yourself plans for the future, including targets and action points. Analysis of your transferable skills profile will probably form part of your PDP (Box 1.1). Other aspects commonly included are:

- **your aspirations, goals, interests and motivations;**
- **your learning style or preference** (see p. 23);
- **your assessment transcript or academic profile information** (e.g. record of grades in your modules);
- **your developing CV** (see p. 45).

Taking part in PDP can help you to focus your thoughts about your university studies and future career. This is important in biology, because most biological sciences degrees do not lead only to a specific occupation. The PDP process will introduce you to some new terms and will help you to describe your personality and abilities. This will be useful when constructing your CV and when applying for jobs.

Definition

Employability – the 'combination of in-depth subject knowledge, work aware-ness, subject-specific, generic and career management skills, and personal attributes and attitudes that enable a student to secure suitable employment and perform excellently throughout a career spanning a range of employers and occupations.' (Anon, 2015: Higher Education Academy Centre for Bioscience, *Define Employability in the Context of Teaching Bioscience*).

Graduate attributes and employability

The skills emphasised in biology courses (Table 1.1) are sometimes considered alongside a university-wide framework of graduate attributes that are intended to summarise the qualities and skills that an employer might expect in those with qualifications from your institution. The associated notion of 'graduateness' summarises the effect of degree-level

Box 1.1 How to carry out a personal skills audit

1. **Create a list of appropriate skills.** As noted on p. 3, there are many systems for categorising skills. If your university publishes a specific skill set, e.g. as part of its framework for PDP or graduate attributes (GA), then you should use that. If not, you could adapt the listing in Table 1.1 or consult a text such as McMillan and Weyers (2013). Your list should relate to you personally, your intended career and any specific skills associated with your intended qualification.

2. **Lay out your list in table format.** You will need to create a table using a word processor or spreadsheet program. Your table should have four columns, as shown in Table 1.2.

3. **Rate your skills.** This may be challenging for many students as it is difficult to be objective and tough to gauge employer expectations. A confident student may rate a certain skill strongly, while a self-critical person may consider the same level of skill to be deficient. However, this does not matter too much as you will effectively be comparing yourself at different stages in your learning, rather than judging yourself against an outside standard. The suggested method is to use a scale of 1 to 10, with low values indicating that the skill 'needs lots of development' and high values indicating that, for the time being, you feel your competence is 'well above average'.

4. **Note actions.** This especially applies to skills with low scores in the previous column – and you may wish to prioritise certain ones. You will need to think about ways in which you could improve, and this may require some research on your part. Is there a book you could read? Is there a training workshop you could attend? Could an extracurricular activity help you to develop? Should you sign up to speak to a skills adviser? It is important that you recognise that the solution to any deficiencies you perceive lies in your own hands. At university, no one will do the work for you.

5. **Add comments and progress notes.** Here is where you can add any comments to amplify or assist with the action points. The addition of progress notes implies that you will revisit the list from time to time. If your university PDP system allows you to add the list to a portfolio, then do this.

Inevitably, your skills audit will become out of date after a period. It will still be useful, however, to look back at so you can see how you have progressed. This will give a sense of achievement and self-awareness that could be valuable when speaking to academic tutors or careers advisers and potential employers. You may wish to set up a new list at intervals, perhaps at the start of each academic year.

Table 1.2 One possible way of creating a personal skills audit. The second row provides guidance about the content of each column. The third row provides an example of possible content

Skill	Rating at [date] with notes	Proposed actions	Comments and notes on progress
You should be quite specific. It may be a good idea to subdivide complex skills like 'communication'	Provide a realistic evaluation of your competence in the skill at a specific point in time	This column will note what you intend to do to try to improve the skill. You might tick these off as completed	This column will summarise your progress. You may wish to add a revised rating
Giving spoken presentations	4/10 [3rd March 2012] Wasn't satisfied with presentation to tutorial group – nervous, a little disorganised and ppt too 'wordy'	1. Read Ch 15 in Practical Skills in Biology ✓ 2. Learn how to use advanced features of PowerPoint ✓ 3. Ask more questions in tutorials ✓	Gave second presentation to tutorial group; went well, although quite nervous at start. Slides much better. Make sure not to rush the introduction next time. 7/10

experience and learning on an individual. This in turn is connected with the concept of 'employability' which encompasses those skills and qualities required to gain and maintain employment. An understanding of these notions is important for every student, as this not only leads to a better understanding of the value of certain activities and assessments, but also provides a specialised vocabulary and gives insights about personal and career development.

At the end of your course, which may seem some time away, you will aim to get a job and start on your chosen career path. You will need to sell yourself to your future employer, firstly in your application form and curriculum vitae (Chapter 8), and perhaps later at interview. Companies rarely employ bioscience graduates simply because they know how to carry out a particular lab routine or because they can recall specific facts about their chosen degree subject. Instead, they will be looking for a range of graduate-level skills and attributes. Typically, for example, they will seek employees who can demonstrate the ability to work in a team, to speak effectively and write clearly about their work. All of these skills and attributes can be developed at different stages during your university studies.

> **KEY POINT** Factual knowledge can be important in degrees with a strong vocational element, but understanding how to find and evaluate information is usually rated more highly by employers than the ability to memorise facts.

Most likely, your future employer(s) will seek someone with an organised yet flexible mind, capable of demonstrating a logical approach to problems – someone who has a range of skills and who can transfer these skills to new situations. Many competing applicants will probably have similar qualifications. If you want the job, you will have to show that your additional skills and personal attributes place you above the other candidates.

Text references

Anon. *Define employability in the context of teaching bioscience.* Available at http://www.bioscience. heacademy.ac.uk/ftp/events/empforum/definition.pdf Last accessed 12/02/15.
[Part of the HE Academy Centre for Bioscience website.]

McMillan, K.M. and Weyers, J.D.B. (2013) *The Study Skills Book,* 3rd edn. Pearson Education, Harlow.

QAA (2007) *Subject benchmark statement – Biosciences 2007.* Available: http://www.qaa.ac.uk/academicinfra structure/benchmark/statements/biosciences07.asp Last accessed 12/02/15.
[Part of the Quality Assurance Agency Academic Infrastructure.]

Sources for further study

Drew, S. and Bingham, R. (2010) *The Guide to Learning and Study Skills.* Gower Publishing Ltd, Aldershot.

Race, P. (2007) *How to Get a Good Degree: Making the Most of Your Time at University*, 2nd edn. Open University Press, Buckingham.

STUDY EXERCISES

1.1 Evaluate your skills. Examine the list of skill topics shown in Table 1.1. Now create a new table with two columns, like the one on the right-hand side. The first half of this table should indicate *five* skills you feel confident about in column 1 and show where you demonstrated this skill in column 2 (for example, 'working in a team' and 'in a first year group project on marine biology'). The second half of the table should show *five* skills you do not feel confident about, or you recognise need development (e.g. communicating in verbal form). List these in column 1 and in column 2 list ways in which you think the course material and activities in your current modules will provide you with the opportunity to develop these skills.

1.2 Find skills resources. For at least one of the skills in the second half of the table in Study exercise 1.1, check your university's library database to see if there are any texts on that subject. Borrow an appropriate book and read the relevant sections. Alternatively, carry out a search for relevant websites (there are many); decide which are useful and bookmark them for future use (see Chapter 11).

Skills I feel confident about	Where demonstrated
1.	
2.	
3.	
4.	
5.	
Skills that I could develop	Opportunities for development
6.	
7.	
8.	
9.	
10.	

1.3 Analyse your goals and aspirations. Spend a little time thinking what you hope to gain from university. See if your friends have the same aspirations. Think about and/or discuss how these goals can be achieved, while keeping the necessary balance between university work, paid employment and your social life.

Answers to these study exercises are available at www.pearsoned.co.uk/practicalskills

2   Managing your time

One of the most important activities that you can do is to organise your personal and working time effectively. There is a lot to do at university and a common complaint is that there is just not enough time to accomplish everything. In fact, research shows that most people use up a lot of their time without realising it through ineffective study or activities such as extended coffee breaks. Developing your time-management skills will help you achieve more in work, rest and play, but it is important to remember that putting time-management techniques into practice is an individual matter, requiring a level of self-discipline not unlike that required for dieting. A new system will not always work perfectly straight away, but through time you can evolve a system that is effective for you. An inability to organise your time effectively, of course, results in feelings of failure, frustration, guilt and being out of control in your life.

Setting your goals

The first step is to identify clearly what you want to achieve, both in work and in your personal life. We all have a general idea of what we are aiming for, but to be effective, your goals must be clearly identified and priorities allocated. Clear, concise objectives can provide you with a framework in which to make these choices. Try using the 'SMART' approach, in which objectives should be:

- **Specific** – clear and unambiguous, including what, when, where, how and why.
- **Measurable** – having quantified targets and benefits to provide an understanding of progress.
- **Achievable** – being attainable within your resources.
- **Realistic** – being within your abilities and expectations.
- **Timed** – stating the time period for completion.

Having identified your goals, you can now move on to answer four very important questions:

1. **Where does your time go?**
2. **Where should your time go?**
3. **What are your time-wasting activities?**
4. **What strategies can help you?**

Analysing your current activities

The key to successful development of time management is a realistic knowledge of how you currently spend your time. Start by keeping a detailed time log for a typical week (Fig. 2.1), but you will need to be truthful in this process. Once you have completed the log, consider the following questions:

- **How many hours do I work in total and how many hours do I use for 'relaxation'?**
- **What range of activities do I do?**

Time slots	Activity										Notes
7.00–7.15											
7.15–7.30											
7.30–7.45											
7.45–8.00											
8.00–8.15											
8.15–8.30											
8.30–8.45											
8.45–9.00											
9.00–9.15											

Fig. 2.1 Example of how to lay out a time log. Write activities along the top of the page, and divide the day into 15-min segments as shown. Think beforehand how you will categorise the different things you do, from the mundane (laundry, having a shower, drinking coffee, etc.) to the well-timetabled (tutorial meeting, sports club meeting), and add supplementary notes if required. At the end of each day, place a dot in the relevant column for each activity and sum the dots to give a total at the bottom of the page. You will need to keep a diary like this for at least a week before you see patterns emerging.

- **How long do I spend on each activity?**
- **What do I spend most of my time doing?**
- **What do I spend the least amount of my time doing?**
- **Are my allocations of time in proportion to the importance of my activities?**
- **How much of my time is ineffectively used**, e.g. for uncontrolled socialising or interruptions?

If you wish, you could use a spreadsheet (Chapter 13) to produce graphical summaries of time allocations in different categories as an aid to analysis and management. Divide your time into:

- **Committed time** – timetabled activities involving your main objectives/ goals.
- **Maintenance time** – time spent supporting your general life activities (shopping, cleaning, laundry, etc.).
- **Discretionary time** – time for you to use as you wish, e.g. recreation, sport, hobbies, socialising.

Avoiding time-wasting activities

Look carefully at those tasks that could be identified as time-wasting activities. They include gossiping, over-long breaks, uninvited interruptions and even ineffective study periods. Try to reduce these to a minimum, but do not count on eliminating them entirely. Remember also that some relaxation *should* be programmed into your daily schedule.

Organising your tasks

Having analysed your time usage, you can now use this information, together with your objectives and prioritised goals, to organise your activities, on both a short-term and a long-term basis. Consider using a diary-based system (such as those produced by Filofax, TMI and Day-Timer) that will help you plan ahead and analyse your progress.

Quality in time management – avoid spending a lot of time doing unproductive studying, e.g. reading a textbook without specific objectives for that reading.

Being assertive – if friends and colleagues continually interrupt you, find a way of controlling them before they control you. Indicate clearly on your door that you do not wish to be disturbed and explain why. Otherwise, try to work away from disturbance.

WEEKLY DIARY Week beginning: _____

	Sunday	Monday	Tuesday	Wednesday	Thursday	Friday	Saturday
DATE							

	Sunday	Monday	Tuesday	Wednesday	Thursday	Friday	Saturday
7–8 am		Breakfast	Breakfast	Breakfast	Breakfast	Breakfast	
8–9		Preparation	Preparation	Preparation	Preparation	Preparation	Breakfast
9–10	Breakfast	PE112(L)	PE112(L)	PE112(L)	PE112(L)	BIOL(P)	Travel
10–11	FREE	CHEM(L)	CHEM(L)	CHEM(L)	CHEM(L)	BIOL(P)	WORK
11–12	STUDY	STUDY	STUDY	STUDY	STUDY	BIOL(P)	WORK
12–1 pm	STUDY	BIOL(L)	BIOL(L)	BIOL(L)	BIOL(L)	TUTORIAL	WORK
1–2	Lunch	Lunch	Lunch	Lunch	Lunch	Lunch	Lunch
2–3	(VOLLEY-	CHEM(P)	STUDY	SPORT	PE112(P)	STUDY	WORK
3–4	BALL	CHEM(P)	STUDY	(VOLLEY-	PE112(P)	STUDY	WORK
4–5	MATCH)	CHEM(P)	STUDY	BALL	PE112(P)	SHOPPING	WORK
5–6	FREE	STUDY	STUDY	CLUB)	STUDY	TEA ROTA	WORK
6–7	Tea	Tea	Tea	Tea	Tea	Tea	Tea
7–8	FREE*	STUDY	STUDY	FREE*	STUDY	FREE*	FREE
8–9	FREE*	STUDY	STUDY	FREE*	STUDY	FREE*	FREE
9–10	FREE*	FREE*	STUDY	FREE*	STUDY	FREE*	FREE

	Sunday	Monday	Tuesday	Wednesday	Thursday	Friday	Saturday
Study (h)	2	10	11	4	11	6	0
Other (h)	13	5	4	11	4	9	15

Total study time = 44 h

Fig. 2.2 A weekly diary with an example of entries for a first-year science student with a Saturday job and active membership of a volleyball club. Note that 'free time' can change to 'study time', in sessions marked with an asterisk when work or exam revision requires the extra effort. Study time (including attendance at lectures, practicals and tutorials) thus represents between 42 and 50% of the total time.

Divide your tasks into several categories, such as:

- **Urgent** – must be done as a top priority and at short notice (e.g. doctor's appointment).
- **Routine** – predictable and regular and therefore easily scheduled (e.g. preparation, lectures or playing sport).
- **One-off activities** – usually with rather shorter deadlines and which may be of high priority (e.g. a tutorial assignment or seeking advice).
- **Long-term tasks** – sometimes referred to as 'elephant tasks' that are too large to 'eat' in one go (e.g. learning a language). These are best managed by scheduling frequent small 'bites' to achieve the task over a longer timescale.

You should make a weekly plan (Fig. 2.2) for the routine activities, with gaps for less predictable tasks. This should be supplemented by individual daily checklists, preferably written at the end of the previous working day. Such plans and checklists should be flexible, forming the basis for most of your activities except when exceptional circumstances intervene. The planning must be kept brief, however, and should be scheduled into your activities. Box 2.1 provides tips for effective planning and working.

Matching your work to your body's rhythm – everyone has times of day when they feel more alert and able to work. Decide when these times are for you and programme your work accordingly. Plan relaxation events for periods when you tend to be less alert.

Use checklists as often as possible – post your lists in places where they are easily and frequently visible, such as in front of your desk. Ticking things off as they are completed gives you a feeling of accomplishment and progress, increasing motivation.

KEY POINT Review each day's plan at the end of the previous day, making such modifications as are required by circumstances, e.g. adding an uncompleted task from the previous day or a new and urgent task.

Box 2.1 Tips for effective planning and working

- **Set guidelines and review expectations regularly.**

- **Do not procrastinate:** do not keep putting off doing things you know are important – they will not go away but they will increase to crisis point.

- **Do not be a perfectionist** – perfection is paralysing.

- **Learn from past experience** – review your management system regularly.

- **Do not set yourself unrealistic goals and objectives** – this will lead to procrastination and feelings of failure.

- **Avoid recurring crises** – they are telling you something is not working properly and needs to be changed.

- **Learn to concentrate effectively** and do not let yourself be distracted by casual interruptions.

- **Learn to say 'no' firmly** but graciously when appropriate.

- **Know your own body rhythms:** e.g. are you a morning person or an evening person?

- **Learn to recognise the benefits of rest and relaxation** at appropriate times.

- **Take short but complete breaks** from your tasks – come back feeling refreshed in mind and body.

- **Work in suitable study areas** and keep your own workspace organised.

- **Avoid clutter** (physical and mental).

- **Learn to access and use information effectively** (Chapter 10).

- **Learn to read and write accurately and quickly** (Chapters 4 and 16).

Sources for further study

Anon. *Day-Timer*. Available: http://www.daytimer.co.uk Last accessed 06/03/15.
[Website for products of Day-Timer Europe Ltd, Fairfax Road, Newton Abbot, TQ12 6UD, UK.]

Anon. *Filofax*. Available: http://www.filofax.co.uk Last accessed 06/03/15.
[Website for products of Filofax UK, the Broadgate Tower, Primrose Street, London, EC2A 2EW, UK.]

Evans, C. (2008) *Time Management for Dummies,* UK edn. John Wiley & Sons, Chichester.

Study Guides and Strategies Website Time Management Series. Available: http://www.studygs.net/timman.htm. Last accessed 06/03/15.

STUDY EXERCISES

2.1 **Evaluate your time usage.** Compile a spreadsheet to keep a record of your daily activities in 15-min segments for a week. Analyse this graphically and identify areas for improvement.

2.2 **List your short-, medium- and long-term tasks and allocate priorities.** Produce several lists, one for each of the three timescales and prioritise each item. Use this list to plan your time management by scheduling high-priority tasks and leave low-priority activities to 'fill in' the spare time that you may identify. This task should be done on a regular (monthly) basis to allow for changing situations.

2.3 **Plan an 'elephant' task.** Spend some time planning how to carry out a large or difficult task (learning a language or learning to use a complex computer program) by breaking it down into achievable segments ('bites').

Answers to these study exercises are available at www.pearsoned.co.uk/practicalskills

It is highly likely that you will be expected to work with fellow students during practicals, tutorials and special study exercises. This might take the form of sharing tasks or casual collaboration through discussion, or it might be formally directed teamwork such as problem-based learning (Box 7.1) or preparing a poster (Chapter 14). Interacting with others in this way can be extremely rewarding and realistically represents the professional world, where teamworking is common. Advantages of such collaborative work include:

- **Synergistic interactions among team members** – these often result in better ideas, produced by the interchange of views, and better output, due to the complementary skills of team members.
- **Support offered to individuals by other members of the team.**
- **Enhanced levels of personal commitment** – because individuals will not want to let the team down.
- **Sharing responsibilities for complex or difficult tasks.**

However, some people approach teamwork with negative feelings. Reasons for this include:

- **Initial reservations about working with strangers** – not knowing whether you will be able to form a friendly and productive relationship.
- **Lack of personal commitment to the specified task** – perhaps due to unfamiliarity or perceived dislike concerning the topic.
- **Apprehension that others may not pull their weight** – especially if you hope to do really well in the course.
- **Lack of previous experience** – worries about the kinds of personal interactions likely to occur and the team role likely to suit you best.
- **Concerns about the outcomes of peer assessment** – in particular, whether others will give you a fair mark for your efforts.

In most cases, these feelings can be transformed through positive participation and a common desire to perform well, so that the team as a whole achieves its target.

Peer assessment – this term applies to marking schemes in which all or a proportion of the marks for a teamwork exercise are allocated by the team members themselves. Read the instructions carefully before embarking on the exercise, so you know which aspects of your work your fellow team members will be assessing. When deciding what marks to allocate yourself, try to be as fair as possible with your marking.

Teamwork skills

Some of the key skills you will need to develop to maximise the success of your teamworking activities include:

- **Interpersonal skills.** These include the way in which you react to new people, how you listen to and communicate with others, and how you deal with conflicts and disagreements.
- **Delegation and sharing of tasks.** This implies that you are willing to share effort and responsibility, are willing to trust your fellow team members, and have mechanisms to deal with unexpected outcomes or even failure to contribute.
- **Effective listening.** This involves giving others the opportunity and time to express their views and to take these seriously, however expressed.

Gaining confidence through experience – the more you take part in teamwork, the more you know how teams operate and how to make teamwork effective for you.

- **Speaking clearly and concisely.** Effective communication is a vital part of teamwork, not only between team members, but also when presenting team outcomes to others (see Chapter 15).
- **Providing constructive criticism.** It is all too easy to be negative, but only constructive criticism will have a positive effect on interactions with others.

The dynamics of teamworking

It is important that team activities are properly structured so that each member knows what is expected of them. Several studies of groups have identified different team roles that derive from differences in personality. You should be aware of such categorisations, both in terms of your own predispositions and those of your fellow team members, as it will help the group to interact more productively. Belbin (2010) identified eight such roles, recently extended to nine, as shown in Table 3.1.

In formal team situations, your course organiser should deal with these issues; even if they do not, it is important that you are aware of these roles and their potential impact on the success or failure of teamwork. You should try to identify your own 'natural' role.

Allocation of responsibilities usually requires the clear identification of a leader. If asked to form a team, bear the different roles in mind during your selection of colleagues and your interactions with them. The ideal team should contain members capable of adopting most of these roles. However, you should also note the following points:

- **People will probably fit one of these roles naturally** as a function of their personality and skills.
- **Group members may be suited to more than one of these roles.**
- **In some circumstances, team members may be required to adapt** and take a different role from the one that they feel suits them.
- **No one role is 'better' than any other.** For good teamwork, the group should have a balance of personality types present.
- **People may have to adopt multiple roles**, especially if the team size is small.

> **KEY POINT** In formal teamwork situations, be clear as to how individual contributions are to be identified and recognised. This might require discussion with the course organiser. Make sure that recognition, including assessment marks, is truly reflective of effort. Failure to ensure that this is the case can lead to disputes and feelings of unfairness.

Your lab partner

Many laboratory sessions in the life sciences involve working in pairs. In some cases, you may work with the same partner for a series of practicals or for a complete module. The relationship you develop as a team is important to your progress, and can enhance your understanding of the material and the grades you obtain. Tips for building a constructive partnership include:

- **Introduce yourselves at the first session** and take a continuing interest in each other's interests and progress at university.

Web-based resources and support for teamwork – websites such as www.belbin.com give further information and practical advice for teamworking.

Recording group discussions – make sure you structure meetings (including writing agendas) and note their outcomes (taking minutes and noting action points).

Table 3.1 A summary of the team roles described by Belbin (2010). No one role should be considered 'better' than any other, and a good team requires members who are able to undertake appropriate roles at different times. Each role provides important strengths to a team, and its compensatory weaknesses should be accepted within the group framework

Team role	Personality characteristics	Typical function in a team	Strengths	Allowable weaknesses
Coordinator	Self-confident, calm and controlled	Leading: causing others to work towards goals	Good at spotting others' talents and delegating activities	Often less creative or intellectual than others in the group
Shaper	Strong need for achievement; outgoing; dynamic; highly strung	Leading: generating action within a team; imposing shape and pattern to work	Providing drive and realism to group activities	Can be headstrong, emotional and less patient than others
Innovator*	Individualistic, serious-minded; often unorthodox	Generating new proposals and solving problems	Creative, innovative and knowledgeable	Tendency to work in isolation; ideas may not always be practical
Monitor–evaluator	Sober, unemotional and prudent	Analysing problems and evaluating ideas	Shrewd judgement	May work slowly; not usually a good motivator
Implementer	Well-organised and self-disciplined, with practical common sense	Doing what needs to be done	Organising abilities and common sense	Lack of flexibility and tendency to resist new ideas
Teamworker	Sociable, mild and sensitive	Being supportive, perceptive and diplomatic; keeping the team going	Good listeners; reliable and flexible; promote team spirit	Not comfortable when leading; may be indecisive
Resource investigator	Extrovert, enthusiastic, curious and communicative	Exploiting opportunities; finding resources; external relations	Quick thinking; good at developing others' ideas	May lose interest rapidly
Completer–finisher	Introvert and anxious; painstaking, orderly and conscientious	Ensuring completion of activity to high standard	Good focus on fulfilling objectives and goals	Obsessive about details; may wish to do all the work to control quality
Specialist	Professional, self-motivated and dedicated	Providing essential skills	Commitment and technical knowledge	Contribute on a narrow aspect of project; tend to be single-minded

* May also be called 'plant' in some texts.

- **At appropriate points, discuss the practical** (both theory and tasks) and your understanding of what is expected of you.
- **Work jointly to complete the practical effectively**, avoiding the situation where one partner dominates the activities and gains most from the practical experience.
- **Share tasks according to your strengths**, but do this in such a way that one partner can learn new skills and knowledge from the other.
- **Make sure you ask questions of each other** and communicate any doubts about what you have to do.
- **Discuss other aspects of your course**, e.g. by comparing notes from lectures or ideas about in-course assessments.
- **Consider meeting up outside the practical sessions** to study, revise and discuss exams.

Tutorial groups

Tutorials represent a group learning environment. Discussions in tutorials can help you develop your ideas by observing others' approaches, views and ideas and assimilating the facts and concepts they introduce (see p. 27). This depends on all present participating fully and enthusiastically, while at the same time respecting the right of others to make a contribution and express their opinions.

If you feel you might like to team up with someone for study, but lack a suitable 'study buddy', then remember that your fellow tutees might be feeling the same. Tutorial meetings are therefore chances to set up further informal meetings for study or exam revision (see 'Collaboration for learning' below).

Collaboration for learning

Much collaboration is informal and consists of pairs or groups of individuals getting together to exchange materials and ideas while studying. It may consist of a 'brainstorming' session for a topic or piece of work, or sharing efforts to research a topic. This has much to commend it and is generally encouraged. However, it is vital that this collaborative learning is distinguished from the collaborative writing of assessed documents: the latter is not usually acceptable and, in its most extreme form, is plagiarism, usually with a heavy potential punishment in university assessment systems. Make sure you know what plagiarism is, what unacceptable collaboration is, and how they are treated within your institution (see p. 61).

Studying with others – teaming up with someone else on your course for revision ('study buddying') is a potentially valuable activity and may especially suit some types of learners (p. 25). It can help keep your morale high when things get tough. You might consider:

- sharing notes, textbooks and other information;
- going through past papers together, dissecting the questions and planning answers;
- talking to each other about a topic (good for aural learners: see Box 5.1);
- giving tutorials to each other about parts of the course that have not been fully grasped.

KEY POINT Collaboration is inappropriate during the final phase of an assessed piece of work unless you have been directed to produce a group report. Collaboration is encouraged during research and learning activities but the final write-up must normally be your own work. The extreme of producing copycat write-ups is regarded as plagiarism and will be punished accordingly.

Text reference

Belbin, R.M. (2010) *Team Roles at Work,* 2nd edn. Butterworth-Heinemann, Oxford.

Sources for further study

Belbin, R.M. *Belbin® Team Roles.* Available. http://www.belbin.com/ Last accessed 12/2/15.

STUDY EXERCISES

3.1 Evaluate your 'natural' team role(s). Using Table 3.1 as a source, decide which team role best fits your personality.

3.2 Keep a journal during a group activity. Record your feelings and observations about experiences of working with other students. After the event, review the journal, then draw up a strategy for developing aspects where you feel you might have done better.

3.3 Reflect upon your teamwork abilities. Draw up a list of your reactions to previous efforts at collaboration or teamwork and analyse your strengths and weaknesses. How could these interactions have been improved or supported more effectively?

Answers to these study exercises are available at www.pearsoned.co.uk/practicalskills

4 Taking notes from lectures and texts

Choose note-taking methods
appropriately – the method you choose
to take notes might depend on the subject;
the lecturer and their style of delivery; and
your own preference.

Compare lecture notes with a colleague –
looking at your notes for the same lecture
may reveal interesting differences in
approach, depth and detail.

Try to adjust to the different styles of
your lecturers – recognise that different
approaches to lecture delivery demand
different approaches to note-taking. For
example, if a lecturer seems to tell lots of
anecdotes or spends much of the time on
examples during a lecture, do not switch
off – you still need to be listening carefully
to recognise the important take-home
messages. Similarly, if a lecture includes a
section consisting mainly of images, you
should still try to take notes – names of
organisms, locations, key features, even
quick sketches. These will help prompt
your memory when revising. Do not be
deterred by lecturers' idiosyncrasies; in
every case you still need to focus and take
useful notes (see also Box 5.2).

Note-taking is an essential skill that you will require in many different
situations, such as:

- **listening to lectures and seminars;**
- **attending meetings and tutorials;**
- **reading texts and research papers;**
- **finding information on the Web.**

> **KEY POINT** Good performance in assessments and exams is built on effective
> learning and revision (Chapters 5, 6 and 7). However, both ultimately depend
> on the quality of your notes.

Taking notes from lectures

Taking legible and meaningful lecture notes is essential if you are to make
sense of them later, but many students find it difficult when starting their
university studies. Begin by noting the date, course, topic and lecturer
on the first page of each day's notes. Number every page in case they get
mixed up later. The most popular way of taking notes is to write in a linear
sequence down the page, emphasising the underlying structure via outline
headings, as in Fig. 16.3. However, the 'pattern' and 'Mind Map' methods
(Figs 4.1 and 4.2) have their advocates: experiment, to see which method
you prefer.

Whatever technique you use, do not try to take down all the lecturer's
words, except when an important definition or example is being given, or
when the lecturer has made it clear that he/she is dictating. Listen first,
then write. Your goal should be to take down the structure and reasoning
behind the lecturer's approach in as few words and phrases as possible.
At this stage, follow the lecturer's sequence of delivery. Use headings and
leave plenty of space, but do not worry too much about being tidy – it

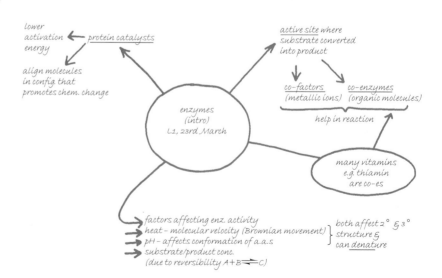

Fig. 4.1 An example of 'pattern' notes, an alternative
to the more commonly used 'linear' format. Note
the similarity to the 'spider diagram' method of
brainstorming ideas (Fig. 16.2).

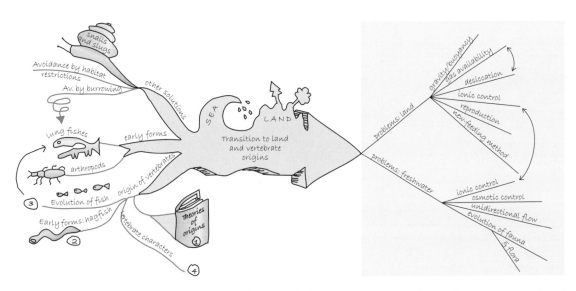

Fig. 4.2 Example of the 'Mind Map' approach to note-taking and 'brainstorming'. Start at the centre with the overall topic title, adding branches and sub-branches for themes and subsidiary topics. 'Basic' maps consist of a branched hierarchy overwritten with key words (e.g. shaded portion above). Connections should be indicated with arrows: numbering and abbreviations are encouraged. To aid recall and creativity, Buzan and Buzan (2009) recommend use of colour, different fonts, three-dimensional doodles and other forms of emphasis (e.g. non-shaded portion above).

is more important that you get down the appropriate information in a readable form.

Abbreviations can be useful to help you to save time when note-taking. As well as standard and well-known symbols (Box 4.1), you can also make up your own contractions relevant to the context. For example, if a lecturer is repeatedly mentioning photosynthesis, you might write 'PS' instead of the full term.

Make sure you note down references to texts and take special care to ensure accuracy of definitions and numerical examples. If the lecturer repeats or otherwise emphasises a point, highlight (e.g. by underlining) or make a margin note of this – it could come in useful when revising. If there is something you do not understand, ask at the end of the lecture, and make an appointment to discuss the matter if there is no time to deal with it then. Tutorials may provide an additional forum for discussing course topics.

Printing *PowerPoint* slides – use the 'Black and White' option on the Print menu to avoid wasting ink on printing coloured backgrounds. If you wish to use colour, remember that slides can be difficult to read if printed in small format. Always print a sample page before printing the whole lecture.

Box 4.1 Some abbreviations commonly used in note-taking

The following symbols are commonly used in the sciences.

$\because$	because	$\equiv, \not\equiv$	equivalent, not equivalent to
$\propto$	is proportional to	$<, >$	smaller than, bigger than
$\rightarrow$	leads to, into	$\gg$	much bigger than
$\leftarrow$	comes from, from	$[X]$	concentration of X
$\rightarrow\rightarrow$	involves several processes in a sequence	$\sum$	sum
$1°, 2°$	primary, secondary (etc.)	Δ	change
$\approx, \cong$	approximately, roughly equal to	f	function
$=, \neq$	equals, not equal to	$\#$	number
		∞	infinity, infinite

Further general abbreviations can be found at http://public.oed.com/how-to-use-the-oed/abbreviations/ and some relevant to biology at http://www.allacronyms.com/biology/topic

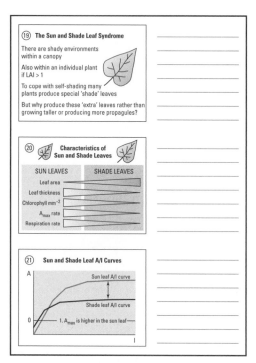

Fig. 4.3 An example of a printout from PowerPoint in 'Handouts (3 slides per page)' format.

Lectures delivered by PowerPoint *or similar presentation programs*

Some students make the mistake of thinking that lectures delivered as computer-based presentations with an accompanying handout or Web resource require little or no effort by way of note-taking. While it is true that you may be freed from the need to copy out large diagrams and the basic text may provide structure, you will still need to adapt and add to the lecturer's pre-prepared points. Much of the important detail and crucial emphasis will still be delivered verbally. Furthermore, if you simply listen passively to the lecture, or worse, try to work from the handout alone, it will be far more difficult to understand and remember the content.

If you are not supplied with handouts, you may be able to print out the presentation beforehand, perhaps in the '3 slides per page' format that allows space for notes alongside each slide (Fig. 4.3). Scan through this before the lecture if you can; then, during the presentation, focus on listening to what the lecturer has to say. Note down any extra details, points of emphasis and examples. After lectures, you could also add notes from supplementary reading. The text in presentations can be converted to word processor format if you have access to the electronic file. In *PowerPoint,* this can be achieved from the *Outline* option in the *'Normal' Presentation View.* You can copy and paste text between programs in the normal fashion, then modify font size and colour as appropriate.

'Making up' your notes

As soon as possible after each lecture, work through your notes, tidying them up and adding detail where necessary. Add emphasis to any headings you have made, so that the structure is clearer. If you feel it would be more logical for your purposes, change the order. Compare your notes with material in a textbook and correct any inconsistencies. Make notes from, or photocopy, any useful material you see in textbooks, ready for revision.

Taking notes from books and journal papers

Scanning

Good reading skills are vital when taking notes from written sources. When consulting a new source for specific information, the first thing you need to do is to orientate yourself by understanding the text's scope and structure. This is often called 'scanning'.

- **For books**, this might involve a quick inspection of the contents section, a check on how each chapter is constructed, and noting, for example, whether the book has a glossary that might be useful. Once you are familiar with the structure and layout, then you might either go directly to the appropriate chapter or section, or consult the index.
- **For research publications**, the best idea is to read the Abstract before consulting specific text, figures or tables, as this should detail the essential methods and findings. The paper's layout thereafter will probably involve the same component sections ('IMRaD', p. 114), but occasionally in a journal-specific order. Referencing systems may also differ (see p. 55) and you may wish to note this.

At the next stage, when you are scanning through the text to find material relevant to your goal, it is useful to have a good understanding of how most writing is structured. Both paragraphs and larger pieces of writing generally

Identifying a study goal – reading around your subject is worthwhile and rewarding, but if your time is limited, it is important to identify a purpose for your reading and note-making. Ideally, the goal should relate to specific learning outcomes or assessments.

Scanning effectively – you need to stay focused on your key words, otherwise you may be distracted by apparently interesting but irrelevant material.

Spotting sequences – writers often number their points (firstly, secondly, thirdly, etc.) and looking for these words in the text can help you skim it quickly.

Making sure you have all the details – when taking notes from a text or journal paper: (a) always take full details of the source (Chapter 9); (b) if copying word-for-word make sure you indicate this using quotes and take special care to ensure you do not alter the original.

Definitions

Quoting – this involves lifting text directly from your source and demonstrating this clearly by using quotation marks (and sometimes italics). If you wish to miss words out, use ellipsis punctuation marks (...), as shown in this example: *Wilson and Brand (2010) concluded: 'this indicates that sample size was small in this study... but that this was acceptable in view of the low variability encountered.'* Here, the words *'because of financial constraints'* have been omitted as irrelevant to the author's main point. For longer quotes (say 40 words or more) create a separate paragraph of text with a citation (normally at the end).

Summarising – this is taking the essence from a text and expressing it briefly in your own words. For example: *Other researchers defend the use of small sample sizes (e.g. Wilson and Brand, 2010).*

Paraphrasing – this is to quote others' ideas or points by restating them in different words. It is different from summary in that you may include more detail. For example: *Wilson and Brand (2010), based on their studies of the snail population in Arran, take the view that small sample size may be irrelevant when variability is low.*

In all cases above, note the citation of a relevant reference, in this case using Harvard style (author names and date, p. 55). Full details of the source would be provided in a reference or bibliography section (see p. 56). See also Chapter 10 (especially Box 10.1) in relation to plagiarism.

include an introduction, a main body and a conclusion. The introductory or 'topic' sentence/paragraph should orientate you to the material that will follow; the main body expands on the topic sentence, with explanation and examples; and the final or 'terminator' sentence/paragraph provides a conclusion and/or a link to the next part.

> **KEY POINT** Working out how a piece of writing is structured will help you to scan quickly for relevance, digest the content faster and get to the material of interest as soon as possible.

Depending on your goal, you may have a specific key word or phase you are searching for, or you may wish to read paragraphs on a specific topic. As well as an index, most biosciences textbooks use systems of headings and sub-headings which facilitate such searches. With digital and online resources, you can press the [Control] and [F] keys together to access a 'find' dialog box that will speed this process – however, take care to enter correctly spelled and relevant words, using alternatives if an initial search draws a blank.

Finally, you need to absorb the content and decide what the salient points are (Chapter 9). These will obviously depend on the task you are carrying out. When taking notes, it is vital that you distinguish between your own paraphrasing or summarising of the text and situations where your notes are direct quotes, or you may be guilty of plagiarism (see p. 61). If transcribing text word-for-word, always use quotation marks and always note the full citation details of the source (p. 56) at the same time.

Skimming

This is a valuable way to gain the maximum amount of information in the minimum amount of time, by reading as little of a text as is required. Essentially, the technique (also termed 'surveying') requires you to look at the structure of the text, rather than the detail. In a sense, you are trying to see the writer's original plan and the purpose behind each part of the text. Look through the whole of the piece first, to gain an overview of its scope and organisation. Headings provide an obvious clue to structure, if present. Next, look for the 'topic sentence' in each paragraph (p. 107), which is often the first. You might then decide that the paragraph contains a definition that is important to note, or it may contain examples, so may not be worth reading in detail for your purpose.

When you have found relevant material, note-taking fulfils the vital purpose of helping you understand and remember the information. If you simply read it, either directly or from a photocopy, you are at risk of accomplishing neither. The act of paraphrasing (using different words to give the same meaning) makes you think about the meaning and express this for yourself. It is an important active learning technique. A popular method of describing skimming and note-taking is called the SQ3R technique (Box 4.2).

> **KEY POINT** Obtaining information and understanding it are distinct parts of the process of learning. As discussed in Chapter 5 (Table 5.1), you must be able to do more than recall facts to succeed.

Methods for finding and evaluating texts and articles are discussed further in Chapters 9 and 10.

Box 4.2 The SQ3R technique for skimming texts

Survey Gain a quick overview of the contents of the book or chapter, perhaps by rapidly reading the contents page or headings.

Question Ask yourself what the material covers and how precisely it relates to your study objectives.

Read Now read the text, paying attention to the ways it addresses your key questions.

Recall Recite to yourself what has been stated every few paragraphs. Write notes of this if appropriate, paraphrasing the text rather than copying it.

Review Think about what you have read and/or review your notes as a whole. Consider where it all fits in.

Text reference

Buzan, T. and Buzan, B. (2009) *The Mind Map Book: Unlock Your Creativity, Boost Your Memory, Change Your Life.* Pearson, Harlow.

Sources for further study

Anon. Mind Map. Available: http://en.wikipedia.org/wiki/Mind_map
Last accessed 13/02/15.

STUDY EXERCISES

4.1 **Experiment with a new note-taking technique.** If you have not tried the pattern or mindmapping methods (Figs 4.1 and 4.2), carry out a trial to see how they might work for you. Research the methods first by consulting appropriate books or websites.

4.2 **Carry out a 'spring clean' of your desk area and notes.** Make a concerted effort to organise your notes and handouts, investing if necessary in files and folders. This will be especially valuable at the start of a revision period.

4.3 **Try out the SQ3R technique.** The next time you need to obtain information from a text, compare this method (Box 4.2) with others you may have adopted in the past. Is it faster, and does it aid your ability to recall the information?

Answers to these study exercises are available at www.pearsoned.co.uk/practicalskills

There are many different ways of learning and at university you have the freedom to choose which approach to study suits you best. You should tackle this responsibility with an open mind, and be prepared to consider new options. Understanding how you learn best and how you are expected to think about your discipline will help you to improve your approach to study and understand your branch of biology at a deeper level.

> **KEY POINT**   At university, you are expected to set your own agenda for learning. There will be timetabled activities, assessments and exam deadlines, but it is your responsibility to decide how you will study and learn, how you will manage your time, and, ultimately, what you will gain from the experience. You should be willing to challenge yourself academically to discover your full potential.

> **Understanding your learning preferences** – although the notion that we have 'hard-wired' individual learning 'styles' is contested by academics, in practice it can be helpful to reflect on your personal preferences, and how they can be used to enhance your learning.

Thinking about thinking

The thinking processes that students are expected to carry out can be presented in a sequence, starting with shallower thought processes and ending with deeper processes, each of which builds on the previous level (see Table 5.1). The first two categories in this ladder apply to gaining basic knowledge and understanding, important when you first encounter a topic. Processes 3 to 6 are those additionally carried out by high-performing university students, with the latter two being especially relevant to final-year students, researchers and professionals. Naturally, the tutors assessing you will want to reward the deepest thinking appropriate for your level of study. This is often signified by the words they use in assessment tasks and marking criteria (column four, Table 5.1, and p. 30), and while this is not an exact relationship, being more aware of this agenda can help you to gain more from your studies and appreciate what is being demanded of you.

Identifying your personal learning preferences

We do not all learn in the same way. Your preferred learning method is simply the one that suits you best for receiving, communicating and understanding information. It therefore involves approaches that will help you learn and perform most effectively. There are many different ways of describing such learning strategies (sometimes called 'styles'), and you may be introduced to specific schemes during your studies. Although methods and terminology may differ among these approaches, it is important to realise that it is the *process* of analysing your learning preferences that is critical, together with the way you use the information to modify your approach to studying, rather than the specific type of learner that you may identify yourself to be.

> **Aligning your learning with a specific approach** – it is important not to read too much into any formal 'diagnosis' of your preferred method of learning. You need to maintain a flexible approach; no matter how you learn, many of the outputs assessed at university (for example, exam answers, essays and reports) are in written formats and you will need to apply what you have learned to these formats. Also, different aspects of your study will require different approaches. For example, practical skills are best learned by 'hands-on' activities – learning by doing.

A useful scheme for describing learning preferences is the VARK system devised by Fleming (2001). By answering a short online questionnaire,

> **Being adaptable** – you should be prepared to adjust your study methods according to the topic, the learning environment (p. 26) and the approach of your lecturers (p. 27).

Table 5.1 A ladder of thinking processes, moving from 'shallower' thought processes (top of table) to 'deeper' levels of thinking (bottom of table). This table is derived from research by Benjamin Bloom and colleagues (Bloom *et al.*, 1956; Anderson *et al.*, 2001). When considering the cue words in typical question instructions, bear in mind that the precise meaning will always depend on the context. For example, while 'describe' is often associated with relatively simple processes of recall, an instruction like 'describe how the human brain works' demands that you exhibit higher-level understanding. Note also that while a 'cue word' is often given at the start of a question/instruction, this is not universally so.

Thinking processes and description (in approximate order of increasing 'depth')	Example in life sciences	Example of typical question structure, with *cue word* highlighted	Other cue words used in question instructions
1. **Remembering (knowing facts).** If you know information, you can *recall* or *recognise* it. This does not always mean you understand it at a higher level.	You might know the order of bases in a piece of DNA but not understand what this means.	*Describe* the main components of a biological membrane.	• define • list • state • identify
2. **Understanding.** To understand information means you can *construct meaning* from it.	You might know the order of bases in a piece of DNA and understand that they code in triplets for specific amino acids.	*Explain* how membrane components are involved in the accumulation of solutes within living cells.	• contrast • compare • distinguish • interpret
3. **Applying.** To apply a fact means that you can *put it to use* in a particular context.	You might be able to take the DNA base sequence and work out the amino acid sequence of the protein for which they code.	Using the Nernst equation, and realistic values for the membrane potential and solute concentrations, *demonstrate* how Na^+ ions must be actively transported out of the cells of marine organisms.	• calculate • illustrate • solve • show
4. **Analysing.** To analyse information means that you are able to *break it down into parts* and show how these components *fit together*.	You might be able to construct a three-dimensional model of a protein derived from the base sequence.	Drawing on information about membrane structure, *defend* the endosymbiotic theory of eukaryote evolution.	• compare • explain • consider • infer
5. **Evaluating.** If you evaluate information, you *arrive at a judgement* based on its importance relative to the topic being addressed.	You might be able to comment on theories about how a protein has evolved, by considering the structure of related proteins and relating this to the taxonomic position of their source species.	*Evaluate* the relative importance of passive and active transport in the accumulation of heavy metal salts by the main groups of soil fungi.	• review • assess • consider • justify
6. **Creating.** To create or synthesise ideas, you need to be able to *extract relevant facts* from a body of knowledge and use these to *address an issue in a novel way* or *create something new*.	You might be able to work out the function of a protein for which you know the sequence of bases, based on a comparison with other like proteins.	*Devise* an experiment to test the hypothesis that a specific membrane fraction contains a functional ATPase involved in glucose transport.	• design • integrate • test • create

you can 'diagnose' yourself as one of the types shown in Box 5.1, which also summarises important outcomes relating to how information and concepts can be assimilated, learned and expressed. People show different degrees of alignment with these categories, and Fleming's research indicates that the majority of students are multi-modal learners – that is, falling into more than one category – rather than being only in one grouping. By carrying out an analysis like this, you can become more aware of your personal characteristics and think about whether the methods of studying you currently use are those that are best suited to your needs.

Box 5.1 How to diagnose your learning preferences using the VARK scheme

Visit www.vark-learn.com to carry out the online diagnostic test, reflect on whether it is a fair description of your preferences, and think about whether you might change the way you study to improve your performance. None of the outcomes should be regarded as prescriptive – you should mix techniques as you see fit and only use methods that you feel comfortable adopting. Adapted with permission from material produced by Fleming (2001).

Optimal learning approach *Description of learning preferences*	Outcomes for your learning, studying and exam technique		
	Advice for taking in information and understanding it	Best methods of studying for effective learning	Ways to cope with exams so you perform better
Visual: *You are interested in colour, layout and design. You probably prefer to learn from visual media or books with diagrams and charts. You tend to add doodles and use highlighters on lecture and revision notes and express ideas and concepts as images.*	Use media incorporating images, diagrams, flowcharts, etc. When constructing notes, employ underlining, different colours and highlighters. Use symbols as much as you can, rather than words. Leave plenty of white space in your notes. Experiment with the 'Mind Map' style of note-taking (p. 19).	Use similar methods to those described in column two. Reduce lecture notes to pictures. Try to construct your own images to aid understanding, then test your learning by redrawing these from memory.	Plan answers diagrammatically. Recall the images and doodles you used in your notes. Use diagrams in your answers (making sure they are numbered and fully labelled). As part of your revision, turn images into words.
Aural: *You prefer discussing subjects and probably like to attend tutorials and listen to lecturers, rather than read textbooks. Your lecture notes may be poor because you would rather listen than take notes.*	Make sure you attend classes, discussions and tutorials. Note and remember the interesting examples, stories, jokes. Leave spaces in your notes for later recall and 'filling'. Discuss topics with a 'study buddy'. Record lectures (with lecturer's permission) and replay to reinforce your learning.	Expand your notes by talking with others and making additional notes from the textbook. Ask others to hear you talk about topics. Read your summarised notes aloud to yourself. Record your vocalised notes and listen to them later.	When writing answers, imagine you are talking to an unseen examiner. Speak your answers inside your head. Listen to your voice and write them down. Practise writing answers to previous exam questions.
Read–Write: *You prefer using text in all formats. Your lecture notes are probably good. You tend to like lecturers who use words well and provide lots of information in sentences and notes. In note-taking, you may convert diagrams to text and text to bullet points.*	Focus on note-taking. You may prefer the 'linear' style of notetaking (p. 18). Use the following in your notes: lists; headings; glossaries; and definitions. Expand your notes by adding further information from handouts, textbooks and library readings.	Reduce your notes to lists or headings. Write out and read the lists again and again (silently). Turn actions, diagrams, charts and flowcharts into words. Rewrite the ideas and principles into other words. Organise diagrams and graphs into statements, e.g. 'The trend is...'.	Plan and write out exam answers using remembered lists. Arrange your words into hierarchies and bullet points.
Kinesthetic: *You tend to recall by remembering real events and lecturers' 'stories'. You probably prefer field excursions and lab work to theory and like lecturers who give real-life examples. Your lecture notes may be weak because the topics did not seem 'concrete' or 'relevant'.*	Focus on examples that illustrate principles. Concentrate on applied aspects and hands-on approaches, but try to understand the theoretical principles that underpin them. When taking in information, use all your senses – sight, touch, taste, smell, hearing.	Put plenty of examples, pictures and photographs into your notes. Use case studies and applications to help with principles and concepts. Talk through your notes with others. Recall your experience of lectures, tutorials, experiments or field trips.	Write practice answers and paragraphs. Recall examples and things you did in the lab or field trip. Role-play the exam situation in your own room.
Multi-modal: *Your preferences fall into two or more of the above categories. You are able to use these different modes as appropriate.*	If you are diagnosed as having two dominant preferences or several equally dominant preferences, read the study strategies above that apply to each of these. You may find it necessary to use more than one strategy for learning and exam-sitting, feeling less secure with only one.		

> **KEY POINT** Having a particular learning preference or style does not mean that you are automatically skilled in using methods generally suited to that type of learner. You must work at developing your ability to take in information, study and cope with assessment.

Be active in your learning – a common theme in all recommended approaches to effective learning is *activity* on the learner's part. Thus, note-making during lectures requires a greater level of engagement with the material than the more passive action of simply listening; discussing concepts with others involves a deeper degree of thinking than solo thought processes when reading text (see also Box 6.3).

Definition

Take-home message(s) – the main point or points that a presenter wishes you to remember from what they have said (perhaps a distillation of facts or a set of conclusions). If a lecture or seminar has been particularly detailed or complex, noting these messages will help you to make sense of it at a later time – and they may be closely related to what is eventually examined.

Lay the foundations for learning – unless you are gifted with photographic recall, it is not easy to commit information to your memory unless you have a conceptual framework of understanding in which to 'place' it. This 'bigger picture' comes from reading, listening and thinking at an early stage in the process.

Learning effectively in different settings

How you learn best depends on your personal characteristics as noted above, but the approaches you use must also suit the different modes of teaching you encounter.

Lectures

These are designed to impart knowledge and understanding efficiently on the part of the staff, who can teach large classes using this method. However, in some instances they can simply be note-taking exercises for you as a student. This may tempt you to approach such teaching in a 'passive' frame of mind, but this is almost certainly not the best way to learn. To get the most from lectures, you need to:

- **Prepare beforehand** by finding out what the topic will be, what the learning outcomes are, and doing some preliminary reading.
- **Arrive in good time**, sitting where you can see and hear the lecturer, with the right materials for taking notes.
- **Listen attentively**, particularly when the lecturer is stating what he or she plans to cover; when they indicate that you are expected to write things down word-for-word; when a definition is being presented; when key facts are stated (often emphasised by repetition); and especially at the end, when the take-home messages are usually covered.
- **Take *useful* notes** – these will not be a transcription of the lecturer's words, but rather a personal digest of the key points that have been made (see Chapter 4). Even if the material is already provided as a *PowerPoint* or *Word* file (or equivalent), your aim is to add value by explaining and interpreting for yourself, and thereby to lay the foundations for memorising the content.
- **Ask questions if you do not understand something**. This might not be at the point of delivery, because things may become clearer when you go over your notes, read a textbook, or talk about the subject with a fellow student. However, it should be as soon as possible after those options are exhausted. There will be opportunities to enquire at tutorials, via email or online discussion boards or at later lectures or meetings.
- **Above all, do attend lectures**, even if you feel their value may be limited in some cases – do not underestimate the way that looking and listening contribute to understanding and memorisation. Also, cold facts on the page rarely have the additional context that a lecture can provide.

Box 5.2 provides some tips for dealing with some of the different teaching styles of lecturers.

Box 5.2 How to accommodate different lecturers' teaching styles

One challenging aspect of lectures is that no two staff members deliver them in the same way. This means you will need to adjust your listening and note-making approaches to suit. For example:

- **It may be hard to extract facts and meaning** from a lecturer who tells meandering stories or keeps the class amused with anecdotes. You must not lose focus when listening and need to be alert when the key points are mentioned, or use the textbook to fill in the missing material when making up your notes later on.

- **You may be distracted** if the lecturer uses technology in a way that makes you think about the medium rather than the message. By all means enjoy the videos, images and electronic wizardry, but do remember to listen out for the underlying messages and focus your notes on these.

- **It may be difficult to stay attentive** if the lecturer's delivery is quiet, monotonous or mumbled (or all three). Here, you will need to ensure that you sit where you can hear, and listen very carefully for key words and phrases, rather than changes in intonation, that might signal importance or change of emphasis. Do not be afraid to ask the lecturer to speak up if you cannot hear.

- **You may find it difficult to abstract general principles** if the lecturer provides great detail about their life in science or current research work. Moreover, the latter may be hard to follow if complex jargon and undefined acronyms are used frequently. The remedy here may be to use the textbook before and after the delivery to create an overarching framework of understanding.

Practicals and field excursions

In a subject such as biology, practical work is essential, so that your understanding is grounded in real examples of specimens, organisms, habitats, equipment, processes and reactions. Many practicals follow the 'scientific method' (Chapter 33) and illustrate how to design and carry out observations and experiments. They may help you develop manipulative skills, as well as those associated with data recording, analysis and reporting. To maximise learning from practicals and field excursions, you should:

- **prepare as well as you can** by reading through the schedule in advance;
- **make sure you have all the equipment and clothing you may need;**
- **arrive at the lab or assembly point in good time;**
- **understand and observe safe practice** (Chapter 22);
- **follow the advised procedures;**
- **ask questions when you do not understand;**
- **work speedily** and use time when waiting for something to happen to start your write-up;
- **take meaningful notes;**
- **draw up informal figures and tables as you go along;**
- **complete your write-up as soon as possible after the event.**

These suggestions are discussed in greater detail in later chapters.

Tutorials

These may be designed to fulfil a number of purposes, including:

- **providing guidance**, as part of an advising system;
- **reviewing and testing your understanding** of material covered elsewhere;

Making the most of demonstrators – these staff or postgraduate students are paid to help you, so ensure you get good value from them by listening to their advice and asking them questions. However, do not expect them to hand you answers on a plate – they will probably be under instruction to make you think for yourself.

- **introducing new material** that is better taught in small groups;
- **developing and assessing essay-writing and numerical skills** through appropriate tasks;
- **promoting interactions among students and with the tutor.**

The face-to-face discussions that can occur during tutorials are rarely possible in larger classes and they add value by allowing views and understanding to develop through an exchange of opinion and knowledge. Think of your tutor as a potential ally. He or she may be able to explain things differently from the lecturer, for example, and thereby help you understand any concepts than you find difficult. Never feel embarrassed about asking questions – the tutor is there to provide answers and the chances are that if you find a subject tricky, so will the rest of the group. Your tutor may also have or assume responsibilities beyond teaching, and be able to help you with subject choices or non-academic issues.

e-Learning and distance learning

These modes of learning can be a mixed blessing. On the one hand, they allow easy access to learning materials at a time of your choosing, while on the other, they assume a personal discipline in your approach to study that may be hard to maintain. Effective learning in this context requires the following:

- **making regular and frequent visits to online modules and portals;**
- **allocating time to study;**
- **paying special attention to announcements, message pages or email systems;**
- **participating in online discussions;**
- **organising materials provided in digital form** – such as lists of 'live' links to website addresses;
- **reading the advised materials and more**, whether online or as texts;
- **making notes as appropriate;**
- **using active learning techniques** (Chapter 6), rather than passively reading the materials provided;
- **observing staff-set milestones** in study, assessment and submission of work;
- **creating personal milestones** in study, self-assessment and preparation of work.

> **Getting the most from tutorials** – this requires preparation beforehand, for example, by:
>
> - carrying out the recommended reading or tackling the set problems;
> - submitting expected work in good time;
> - thinking about the issues involved in the subject matter;
> - preparing any questions you might have about the topic you have been asked to study;
> - reflecting on the potential relevance of the tutorial to your programme of study.

> **Taking account of tutorial assessment** – the tasks prepared for tutorials may be marked and in some cases your participation will be graded. Consult the course handbook to find out what proportion of the total mark derives from tutorials and what the marking criteria are. Ensure that you maximise the marks obtained from this source, as they may be relatively easy to achieve if you put in the necessary work.

> **Alternative modes of distance learning** – the discussion opposite assumes that e-learning facilities are being used (i.e. a virtual learning environment, or VLE), but in some courses the materials are primarily paper-based. If this is the case, then finding the best way to learn may be difficult. You will need to find a method of making the material 'come alive', perhaps by purchasing or borrowing well-illustrated texts or video materials (such as BBC nature documentaries).

> **KEY POINT** The only person who can judge the effectiveness of your learning at university is you. Only you will know how much effort you put in to your studies and what you are expecting, in terms of a mark or grade; only you will know how comfortable you feel with a particular approach; and only you can respond to the feedback you have been given, in terms of how you will learn for future assessments. Be prepared to change your learning methods if you find that they are unsuccessful or difficult to apply in particular circumstances.

Text references

Anderson, L.W., Krathwohl, D.R. and Bloom, B.S. (2001) *A Taxonomy for Learning Teaching, and Assessing. A Revision of Bloom's Taxonomy of Educational Objectives*. Allyn & Bacon, London.

Bloom, B., Englehart, M., Furst, E., Hill, W. and Krathwohl, D. (1956) *Taxonomy of Educational Objectives: The Classification of Educational Goals. Handbook I: Cognitive Domain*. Longmans, Green, New York and Toronto.

Sources for further study

McMillan, K.M. and Weyers, J.D.B. (2013) *The Study Skills Book*, 3rd edn. Pearson Education, Harlow.

Fleming, N.D. (2001) *Teaching and Learning Styles: VARK Strategies*. Neil Fleming, Christchurch.

Fleming, N.D. *VARK: A Guide to Learning Styles*. Available: http://www.vark-learn.com/ Last accessed 13/02/15.

STUDY EXERCISES

5.1 **Visit the vark-learn website (see Box 5.1).** Carry out the short online test there to diagnose your learning preference. Next, review the guidance given on the website for study practices, including taking in information and using information for effective learning. Do you feel that any of the suggestions might be useful to you?

5.2 **Review your note-taking methods.** How well suited are they to your needs? How well suited are they to the lecture styles of the academic staff? Have you captured the important points, or are you merely transcribing exactly what the lecturer says? Where lecturers use *PowerPoint* slides, are you listening for, and capturing in note form, the additional spoken points and examples that the lecturer is adding during the live presentation?

5.3 **Think about your thinking.** Read through Table 5.1 and consider different thinking processes in relation to (a) feedback you have received from a previous assignment or (b) your next assignment and your intended approach to it. Does this reveal that your marks could be suffering because you are not thinking deeply enough or are not capturing your thoughts clearly enough in your writing?

Answers to these study exercises are available at www.pearsoned.co.uk/practicalskills

Where your studies are assessed through an exam, you will need to prepare for this by revising and, because you may have a number of exams in a short period, you will need to be well organised. You will need to align your study effort with the expectations of your lecturers, as explained in the course materials and through the feedback you have been given on previous assignments and exams. Adopting active methods of revision that are suited to your learning personality (Chapter 5) can also make a significant difference to your performance.

The role of assessment and feedback in your learning

Your starting point when preparing for assessment should always be the learning outcomes or objectives for each module, topic or learning activity. You will usually find them in your module handbook. They state in clear terms what your tutors expect you to be able to accomplish after participating in each element and reading around the topic. Also of value will be marking/assessment criteria or grade descriptors, which state in general terms what level of attainment is required for your work to reach specific grades. These are more likely to be defined at faculty/college/school/department level and consequently published in appropriate handbooks or websites. Reading learning outcomes and grade descriptors will give you a good idea of what to expect and the level of performance required to reach your personal goals. Relate them to both the material covered (e.g. in lectures and practicals, or online) and past exam papers. Doing this as you study and revise will indicate whether further reading and independent studying is required, and of what type. You will also have a much clearer picture of how you are likely to be assessed.

> **KEY POINT** Use the learning objectives for your course (normally published in the handbook) as a fundamental part of your revision planning. These indicate what you will be expected to be able to do after taking part in the course, so exam questions are often closely based on them. Check this by reference to past papers.

There are essentially two types of assessment – formative and summative, although the distinction may not always be clear-cut (see margin). The first way you can learn from formative assessment is to consider the grade you obtained in relation to the work you put in. If this is a disappointment to you, then there must be a mismatch between your understanding of the topic and the marking scheme with that of the marker, or a problem in the writing or presentation of your assignment. If you do not understand the reason for your grade, consult your tutor.

The second way to learn from formative assessment is through the written feedback and notes on your work. These comments may be cryptic, or scribbled hastily, so if you do not understand or cannot read them, ask the tutor who marked the work. Most tutors will be pleased to explain how you could have improved your mark. If you find that the same comments appear frequently, it may be a good idea to seek

---

Definitions

Learning objectives/outcomes – statements of the knowledge, understanding or skills that a learner will be able to demonstrate on successful completion of a module, topic or learning activity.

Formative assessments – these may be mid-term or mid-semester tests and are often in the same format as later exams. They are intended to give you feedback on your performance.

Summative assessments – these include end-of-year or end-of-module exams. They inform others about the standard of your work. In continuous or 'in-course' assessment, the summative elements are spread out over the course. Sometimes these assessments may involve a formative aspect, if feedback is given.

Example A set of learning objectives taken from an introductory lecture on *The Ocean as a Habitat*.

'After this lecture, you should be able to:

- Define the following:
  - Benthos
  - Chemolithotroph
  - Photolithotroph
  - Plankton
  - Upper mixed layer
  - Upwelling
- Draw a diagram illustrating changes in light, plant nutrients, oxygen, temperature and density with depth in the ocean
- Discuss the role of seasonal mixing and upwellings, and inputs from rivers and the atmosphere, in the supply of nutrients to the surface of the ocean
- Demonstrate knowledge of magnitudes of ocean area and marine primary productivity relative to land and freshwater habitats.'

(With acknowledgements to Emeritus Professor J.A. Raven, University of Dundee.)

Using tutors' feedback – it is always worth reading any comments on your work as soon as it is returned. If you do not understand the comments, or are unsure about why you might have lost marks in an assignment, ask for an explanation while the topic is fresh in your mind.

help from your university's academic support unit. Take along examples of your work and representative feedback comments so they can give you the best possible advice. Another suggestion is to ask to see the work of another student who obtained a good mark, and compare it with your own. This will help you judge the standard you should be aiming towards.

Preparing for revision and examinations

Before you start revising, find out as much as you can about each exam, including:

- **its format and duration;**
- **the date and location;**
- **the types of questions;**
- **whether any questions/sections are compulsory;**
- **whether the questions are internally or externally set or assessed;**
- **whether the exam is 'open book'**, and if so, which texts or notes are allowed.

Time management when revising – this is vital to success and is best achieved by creating a revision timetable (Box 6.2).

Your course tutor is likely to give you details of exam structure and timing well beforehand, so that you can plan your revision; the course handbook and past papers (if available) can provide further useful details (Box 6.1). Always check that the nature of the exam has not changed before you make plans based on past papers.

Organising and using lecture notes, assignments and practical reports

Filing lecture notes – make sure your notes are kept neatly and in sequence by using a ring binder system. File the notes in lecture or practical sequence, adding any supplementary notes or photocopies alongside.

Given their importance as a source of material for revision, you should have sorted out any deficiencies or omissions in your lecture notes and practical reports at an early stage. For example, you may have missed a lecture or practical due to illness, etc., but the exam is likely to assume attendance throughout the year. Ask others in the class for copies of any notes you may have missed.

Your practical reports and any assignment work will contain specific comments from the teaching staff, indicating where marks were lost, corrections, mistakes, inadequacies, and so on. Most lecturers are quite happy to discuss such details with students on a one-to-one basis and this information may provide you with 'clues' to the expectations of individual lecturers that may be useful in exams set by the same members of staff. However, you should never 'fish' for specific information on possible exam questions, as this will not be well received.

Revision

Recognise when your concentration powers are dwindling – take a short break from revision when this happens and return to work refreshed and ready to learn. Remember that 20 min is often quoted as a typical limit to a spell of full concentration effort.

Begin early, to avoid last-minute panic. Start in earnest several weeks beforehand, and plan your work carefully:

- **Prepare a revision timetable** – an 'action plan' that gives details of specific topics to be covered (Box 6.2). Find out at an early stage when (and where) your examinations are to be held, and plan your revision around this. Try to keep to your timetable. Time management during this period is as important as keeping to time during the exam itself.

Box 6.1 How to use past exam papers in your revision

Past exam papers are a valuable resource for targeting your revision.

1. **Find out where the past (or sample) exam papers are kept.** Copies may be lodged in your department or the library; or they may be accessible online.

2. **Locate and copy relevant papers for your module(s).** Check with your tutor or course handbook that the style of paper will not change for the next set of examinations.

3. **Analyse the design of the exam paper.** Taking into account the length in weeks of your module, and the different lecturers and/or topics for those weeks, note any patterns that emerge. For example, can you translate weeks of lectures/practicals into numbers of questions or sections of the paper? Consider how this might affect your revision plans and exam tactics, taking into account (a) any choices or restrictions offered in the paper, and (b) the different types of questions asked (i.e. multiple choice, short-answer or essay).

4. **Examine carefully the style of questions.** Can you identify the expectations of your lecturers? Can you relate the questions to the learning objectives? How much extra reading do they seem to expect? Are the questions fact-based? Do they require a synthesis based on other knowledge? Can you identify different styles for different lecturers? Consider how the answers to these questions might affect your revision effort and exam strategy.

5. **Practise answering questions.** Perhaps with friends, set up your own mock exam when you have done a fair amount of revision, but not too close to the exams. Use a relevant past exam paper; and try not to study it beforehand. You need not attempt all of the paper at one sitting. You will need a quiet room in a place where you will not be interrupted (e.g. a library). Keep close track of time during the mock exam and try to do each question in the length of time you would normally assign to it (see p. 111) – this gives you a feel for the speed of thought and writing required and the scope of answer possible. Mark each other's papers and discuss how each of you interpreted the question and laid out your answers and your individual marking schemes.

6. **Practise writing answer plans and starting answers.** This can save time compared with the 'mock exam' approach. Practise in starting answers can help you get over stalling at the start and wasting valuable time. Writing essay plans gets you used to organising your thoughts quickly and putting your thoughts into a logical sequence.

Box 6.2 How to prepare and use a revision timetable

1. **Make up a grid showing the number of days until your exams are finished.** Divide each day into several sections. If you like revising in large blocks of time, use a.m., p.m. and evening slots, but if you prefer shorter periods, divide each of these in two, or use hourly divisions (see also the table in study exercise 6.2).

2. **Write in your non-revision commitments,** including any time off you plan to allocate and physical activity at frequent intervals. Try to have about one-third or a quarter of the time off in any one day. Plan this in relation to your best times for useful work – for example, some people work best in the mornings, while others prefer evenings. If you wish, use a system where your relaxation time is a bonus to be worked for; this may help you motivate yourself.

3. **Decide on how you wish to subdivide your subjects** for revision purposes. This might be among subjects, according to difficulty (with the hardest getting the most time), or within subjects, according to topics. Make sure there is an adequate balance of time among topics and especially that you do not avoid working on the subject(s) you find least interesting or most difficult.

4. **Allocate the work to the different slots available on your timetable.** You should work backwards from the exams, making sure that you cover every exam topic adequately in the period just before each exam. You may wish to colour-code the subjects.

5. **As you revise, mark off the slots completed** – this has a positive psychological effect and will boost your self-confidence.

6. **After the exams, revisit your timetable** and decide whether you would do anything differently next time.

Question-spotting – avoid adopting this risky strategy in an attempt to reduce the amount of time you spend revising. Lecturers are aware that this approach may be taken and try to ask questions in an unpredictable manner. You may find that you are unable to answer on unexpected topics that you failed to revise. Moreover, if you have a preconceived idea about what will be asked, you may also fail to grasp the nuances of the exact question set, and provide a response lacking in relevance.

- **Study the learning objectives/outcomes** for each topic (usually published in the course handbook) to get an idea of what lecturers expect from you.
- **Use past papers as a guide** to the form of exam and the type of question likely to be asked (Box 6.1).
- **Remember to have several short (five-minute) breaks during each hour of revision** and a longer break every few hours. In any day, try to work for a maximum of three-quarters of the time.
- **Include recreation within your schedule:** there is little point in tiring yourself with too much revision, as this is unlikely to be profitable.
- **Make your revision as active and interesting as possible** (see Box 6.3): the least productive approach is simply to read and reread your notes.
- **Ease back on the revision near the exam:** plan your revision to avoid last-minute cramming and overload fatigue.

Box 6.3 How to revise actively

The following techniques may prove useful in devising an active revision strategy:

- **'Distil' your lecture notes** to show the main headings and examples. Prepare revision sheets with details for a particular topic on a single sheet of paper, arranged as a numbered checklist. Wall posters are another useful revision aid.
- **Aid recall through effective note-taking:** for example, the Mind Map technique (p. 19), when used to organise ideas, is claimed to enhance recall for some by connecting the material to visual images or linking it to the physical senses.
- **Carry out revision checks:** it is important to test yourself frequently to ensure that you have retained the information you are revising. Assess what you have learned by taking a blank sheet of paper and writing down all you know on a selected topic. Check your full notes to see if you missed anything out. If you did, go back immediately to a fresh blank sheet and redo the example. Repeat, as required.
- **Make up lists of key facts and definitions:** these can be a useful starting point for many exam answers. Test yourself repeatedly on these lists, or get a friend to do this. Try to remember *how many* points you need to know in each case – this will help you recall them all during the exam.
- **Use mnemonics and acronyms** to commit specific factual information to memory. Sometimes, the dafter they are, the better they seem to work.

- **Draw diagrams from memory:** make sure you can label them fully.
- **Try recitation as an alternative to written recall.** Talk about your topic to another person, preferably someone in your class. Talk to yourself (in private) if necessary. Explaining something out loud is an excellent test of your understanding.
- **Associate facts with images or journeys** if you find this method works.
- **Use a wide variety of approaches** to avoid boredom during revision (e.g. record information on audio tape, use cartoons, or any other method, as long as it is not just reading).
- **Form a revision group** to share ideas and discuss topics with other students.
- **Prepare answers to past papers,** e.g. write essays or, if time is limited, write essay plans (see Box 6.1).
- **If your subject involves numerical calculations, work through representative problems.**
- **Make up your own questions:** the act of putting yourself in the examiner's mindset by inventing questions can help revision. However, you should not rely on 'question-spotting': this is a risky practice!
- **Write key points on 'Post-it®' notes or similar:** arrange and rearrange these in clusters or as lists around your study space or room.

KEY POINT When considering assessment questions, look carefully at words used in the instructions. These cues can help you identify what depth is expected in your answer (see Table 5.1). Take special care in multi-part questions, because the first part may require lower-level thinking, while in later parts marks may be awarded for evidence of deeper thinking.

Final preparations – try to get a good night's sleep before an exam. Last-minute cramming will be counter-productive if you are too tired during the exam.

The evening before your exam should be spent in consolidating your material, and checking through summary lists and plans. Avoid introducing new material at this late stage: your aim should be to boost your confidence, putting yourself in the right frame of mind for the exam itself.

Sources for further study

Burns, R. (1997) *The Student's Guide to Passing Exams.* Kogan Page, London.

Hamilton, D. (2003) *Passing Exams: A Guide for Maximum Success and Minimum Stress.* Cengage Learning, Boston, USA.

McMillan, K.M. and Weyers, J.D.B. (2011) *How to Succeed in Exams and Assessments*, 2nd edn. Pearson Education, London.

McMillan, K.M. and Weyers, J.D.B. (2013) *The Study Skills Book,* 3rd edn. Pearson Education, London.

Many universities host study skills websites; these can be found using 'study skills', 'revision' or 'exams' as key words in a search engine.

STUDY EXERCISES

6.1 Make use of past exam papers. Use the techniques discussed in Box 6.1 to improve your revision strategy.

6.2 Draw up a revision timetable. Use the techniques discussed in Box 6.2 to create a revision timetable for your forthcoming exams. You may wish to use or adapt the arrangement below, either on paper or within a spreadsheet.

6.3 Try out a new active revision technique. Try any or all of the methods mentioned in Box 6.3 when revising. Compare notes with a colleague – which seems to be the most successful technique for you and for the topic you are revising?

A revison timetable planner

Date	Morning		Lunch	Afternoon		Tea/Dinner	Evening	
	Session 1	Session 2		Session 1	Session 2		Session 1	Session 2

Answers to these study exercises are available at www.pearsoned.co.uk/practicalskills

Definition

Transcript – this is your record of achievement at university. Normally it will consist of details of each module or course you have taken, and an indication of the grade or mark achieved. It will also show your final (honours) classification, that is: first class, upper second class (2.1), lower second class (2.2), third class or unclassified (some UK universities do not differentiate second class degrees).

Aiming high – your goal should be to perform at your highest possible level and not simply to fulfil the minimum criteria for progression. This will lay sound foundations for your later studies. Remember too that a future employer might ask to see your academic transcript, which will detail *all* your module grades including any fails/resits, and will not just state your final degree classification.

Seeking advice on module options – most universities provide students with access to a staff member who acts as an impartial adviser on curriculum matters. Careers service staff may also offer valuable guidance. Make sure you take advantage of this by setting up and attending meetings with relevant personnel. Senior students can also be a useful source of advice but always compare their opinions with those of staff.

Avoiding plagiarism – this is a key issue for assessed coursework – see p. 60 for a definition and Chapter 9 for appropriate methods of referring to the ideas and results of others using citation.

Many universities adopt a modular system for their biosciences degree courses. This allows greater flexibility in subject choice and accommodates students studying on different degree paths. Modules also break a subject into discrete, easily assimilated elements. They have the advantage of spreading assessment over the academic year, but they can also tempt you to avoid certain difficult subjects or to feel that you can forget about a topic once the module is finished.

> **KEY POINT** You should select your modules with care, mindful of potential degree options and how your transcript and CV will appear to a prospective employer. If you feel you need advice, consult your personal tutor or study adviser.

As you move between levels of the university system, you will be expected to have passed a certain number of modules, as detailed in the progression criteria. These may be expressed using a credit-point system. Students are normally allowed two attempts at each module exam and the resits often take place at the end of the summer vacation. If a student does not pass at the second attempt, they may be asked to 'carry' the subject in a subsequent year, and in severe cases of multiple failure, they may be asked to retake the whole year or even leave the course. Consequently, it is worth finding out about these aspects of your degree. They are usually published in relevant course handbooks.

You are unlikely to have reached this stage in your education without being exposed to the examination process. You may not enjoy being assessed, but you probably want to do well in your course. It is therefore important to understand why and how you are being tested. Identifying and improving the skills required for exam success will allow you to perform to the best of your ability.

Assessed coursework

There is a component of assessed coursework in many modules. This often tests specific skills, and may require you to demonstrate thinking at deeper levels (see Table 5.1). The common types of coursework assessment are covered at various points in this book:

- **Practical exercises** (throughout).
- **Essays** (Chapters 16 and 17).
- **Numerical problems** (Chapter 74).
- **Data analysis** (Chapters 71, 75 and 76).
- **Poster and spoken presentations** (Chapters 14 and 15).
- **Fieldwork** (Chapter 31).
- **Literature surveys and reviews** (Chapter 19).
- **Project work** (Chapters 18 and 35).
- **Problem-based learning** (Box 7.1).

At the start of each year or module, read the course handbook or module guide carefully to find out when any assessed work needs to be

---

Box 7.1 Problem-based learning (PBL)

In this teaching method, you are presented with a 'real-world' problem or issue, often working within a team. As you tackle the problem, you are expected to gain factual knowledge, develop skills and exercise critical thinking (Chapter 10). Because there is a direct and relevant context for your work, and because you have to employ active learning techniques, the knowledge and skills you gain are likely to be more readily assimilated and remembered. This approach also more closely mimics workplace practices. PBL usually proceeds as follows:

1. **You are presented with a problem** (e.g. a case study, a hypothetical patient, a topical issue).

2. **You consider what issues and topics you need to research**, by discussion with others if necessary. You may need to identify where relevant resources can be found (Chapters 9 and 11).

3. **You then need to rank the issues and topics in importance**, allocating tasks to group members, if appropriate.

4. **Having carried out the necessary research, you should review the information that has been obtained.** As a result, new issues may need to be explored and, where appropriate, allocated to group members.

5. **You will be asked to produce an outcome, such as a report, diagnosis, seminar presentation or poster.** An outline structure will be required, and for groups, further allocation of tasks to accomplish this end-point.

If asked to carry out PBL as part of your course, it is important to get off to a good start. At first, the problem may seem unfamiliar. However, once you become involved in the work, you will quickly gain confidence. If working as part of a group, make sure that your group meets as early as possible, that you attend all sessions and that you do the necessary background reading. When working in a team, a degree of self-awareness is necessary regarding your 'natural' role in group situations (Table 3.1). Various methods are used for grading PBL, and this may involve peer assessment.

Definitions

Some terms are used with specialised meaning in university exam contexts.

Invigilator – a person whose role is to ensure an exam is conducted according to the regulations.

Diet – a period during which a number of exams take place, according to a published timetable.

Paper – usually, a set of questions for a single subject.

Question – an element of an exam paper, often couched in terms of an instruction rather than a true question, and often with the marks available stated at the end.

Rubric – the set of instructions at the start of an exam paper.

submitted. Note relevant dates in your diary, and use this information to plan your work. Take special note if deadlines for different modules clash, or if they coincide with social or sporting commitments.

> **KEY POINT** If, for some valid reason (e.g. illness), you will be late with an assessment, speak to your tutors as soon as possible. They may be able to take extenuating circumstances into account by not applying a marking penalty. They will let you know what paperwork you may require to submit to support your claim.

Summative exams – general points

Summative exams (p. 30) normally involve you answering questions without being able to consult other students or your notes. Invigilators are present to ensure appropriate conduct, but departmental representatives may be present for some of the exam. Their role is to sort out any subject-related problems, so if you think something is wrong, ask at the earliest opportunity. It is not unknown for parts of questions to be omitted in error, or for double meanings to arise, for example.

Planning

When preparing for an exam, make a checklist of the items you will need (see p. 42). On the day of the exam, give yourself sufficient time to arrive at the correct room, without the risk of being late. Double-check the times and places of your exams, both well before the exam and also on arrival. If you arrive at the exam venue early, you can always rectify a mistake if you find you have gone to the wrong place.

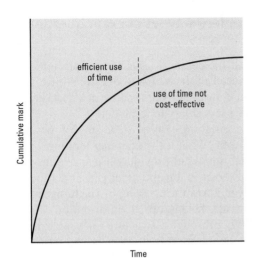

Cumulative mark

efficient use of time

use of time not cost-effective

Time

Fig. 7.1 Potential exam marks as a function of time. The marks awarded in a single answer will follow the law of diminishing returns – it will be far more difficult to achieve the final 25% of the available marks than the initial 25%. Do not spend too long on any one question.

Using the question paper – unless this is specifically forbidden, you *should* write on the question paper to plan your strategy, keep to time and organise answers.

Tackling the paper

Begin by reading the instructions at the top of the exam paper carefully, so that you do not make any errors based on lack of understanding of the exam structure. Make sure that you confirm:

- **how many questions are set;**
- **how many must be answered;**
- **whether the paper is divided into sections;**
- **whether any parts are compulsory;**
- **what each question/section is worth**, as a proportion of the total mark;
- **whether different questions should be answered in different books.**

Do not be tempted to spend too long on any one question or section: the return in terms of marks will not justify the loss of time from other questions (see Fig. 7.1). Take the first few minutes or so to read the paper and plan your strategy, before you begin writing. Do not be put off by those who begin immediately; it is almost certain they are producing unplanned work of a poor standard.

Underline the key phrases in the instructions, to reinforce their message. Next, read through the set of questions. If there is a choice, decide on those questions to be answered and decide on the order in which you will tackle them. Quickly prepare a timetable which takes into account the amount of time required to complete each section and which reflects the allocation of marks – there is little point in spending one-quarter of the exam period on a question worth only 5% of the total marks. Use the exam paper to mark the sequence in which the questions will be answered and write the finishing times alongside; refer to this timetable during the exam to keep yourself on course.

Reviewing your answers

At the end of the exam, you should allow some time to read through your script, to check for:

- **errors of fact;**
- **missing information;**
- **grammatical and spelling errors;**
- **errors of scale and units;**
- **errors in calculations.**

Make sure your name and/or ID number is on each exam book as required and on all other sheets of paper, including graph paper, even if securely attached to your script, as it is in your interest to ensure that your work does not go astray.

> **KEY POINT** *Never* leave any exam early. Most exams assess work carried out over several months in a time period of 2–3 h and there is always something constructive you can do with the remaining time to improve your script.

Adopting different tactics according to the exam – you should adjust your exam strategy (and revision methods) to allow for the differences in question types used in each exam paper.

Special considerations for different types of exam question

Essay questions

Essay questions let examiners test the depth of your comprehension and understanding as well as your recall of facts. Essay questions give you plenty of scope to show what you know. They suit those with a good grasp of principles but who perhaps have less ability to recall details.

Before you tackle a particular question, you must be sure of what is required in your answer. Ask yourself 'What is the examiner looking for in this particular question?' and then set about providing a *relevant* answer. Consider each individual word in the question and highlight, underline or circle the key words. Make sure you know the meaning of the terms given in Table 17.1 so that you can provide the appropriate information, where necessary. Spend some time thinking through a structure for your writing (see Chapter 16). Refer back to the question frequently as you write, to confirm that you are keeping to the subject matter. Box 7.2 gives advice on writing essays under exam conditions.

It is usually a good idea to begin with the question that you are most confident about. This will reassure you before tackling more difficult parts of the paper. If you run out of time, write in note form. Examiners are usually understanding, as long as the main components of the question have been addressed and the intended structure of the answer is clear. Common reasons for poor exam answers in essay-style questions are listed in Box 7.3; try not to fall into any of these traps.

Box 7.2 · Writing under exam conditions

Always go into an exam with a strategy for managing the available time.

- **Allocate some time (say 5% of the total) to consider which questions to answer and in which order.**

- **Share the rest of the time among the questions, according to the marks available.** Aim to optimise the marks obtained. A potentially good answer should be allocated slightly more time than one you do not feel so happy about. However, do not concentrate on any one answer (see Fig. 7.1).

- **For each question divide the time into planning, writing and revision phases** (see p. 111).

Employ time-saving techniques as much as possible.

- **Use spider diagrams** (Fig. 16.2) **or Mind Maps** (Fig. 4.2) to organise and plan your answer.

- **Use diagrams and tables** to save time in making difficult and lengthy explanations, but make sure you refer to each one in the text.

- **Use standard abbreviations** to save time repeating text but always explain them at the first point of use.

- **Consider speed of writing and neatness** – especially when selecting the type of pen to use – ballpoint pens are fastest, but they tend to smudge. You can only gain marks if the examiner can read your script.

- **Keep your answer simple and to the point**, with clear explanations of your reasoning.

Make sure your answer is relevant.

- **Do not include irrelevant facts** just because you memorised them during revision, as this may do you more harm than good. You must answer the specific question that has been set.

- **Remember that time taken to write irrelevant material is time lost from another question.**

Box 7.3 Reasons for poor exam answers to essay-style questions

The following are reasons that lecturers cite when they give low marks for essay answers:

- **Not answering the exact question set.** Either failing to recognise the specialist terms used in the question, or failing to demonstrate an understanding of the terms by not providing definitions, or failing to carry out the precise instruction in a question, or failing to address all aspects of the question.

- **Running out of time.** Failing to match the time allocated to the extent of the answer. Frequently, this results in spending too long on one question and not enough on the others, or even failing to complete the paper.

- **Failing to answer all parts** of a multiple-part question, or to recognise that one part (perhaps involving more complex ideas) may carry more marks than another.

- **Failing to provide evidence** to support an answer. Forgetting to state the 'obvious' – either basic facts or definitions.

- **Failing to illustrate an answer appropriately**, either by not including a relevant diagram, or by providing a diagram that does not aid communication, or by not including relevant examples.

- **Incomplete answer(s).** Failing to answer appropriately due to lack of knowledge.

- **Providing irrelevant evidence** to support an answer. 'Waffling' to fill space.

- **Illegible handwriting.**

- **Poor English**, such that facts and ideas are not expressed clearly.

- **Lack of logic** or structure to the answer.

- **Factual errors**, indicating poor note-taking or poor revision or poor recall.

- **Failing to correct obvious mistakes** by rereading an answer before submitting the script.

At higher levels, the following aspects are especially important:

- **Not providing enough in-depth information.**

- **Providing a descriptive rather than an analytical answer** – focusing on facts, rather than deeper aspects of a topic (Table 5.1).

- **Not setting a problem in context**, or not demonstrating a wider understanding of the topic. (However, make sure you do not overdo this, or you may risk not answering the question set.)

- **Not giving enough evidence of reading around the subject.** This can be demonstrated, for example, by quoting relevant papers and reviews and by giving author names and dates of publication.

- **Not considering both sides of a topic/debate, or not arriving at a conclusion if you have done so.**

Multiple-choice and short-answer questions

Multiple-choice questions (MCQs) and short-answer questions (SAQs) are generally used to test the breadth and detail of your knowledge. The various styles that can be encompassed within the SAQ format allow for more demanding questions than MCQs, which may emphasise specific factual knowledge.

A good approach for MCQ papers is as follows:

1. **First trawl:** read through all the questions fairly rapidly, noting the 'correct' answer for those you can attempt immediately.

2. **Second trawl:** go through the question set again, now dealing with the straightforward but more difficult questions, but leaving those that require a lot of thought or time.

3. **Third trawl:** now tackle the difficult questions and those that require longer to answer (e.g. those based on numerical problems).

4. **Final review:** look again at all your answers, checking for obvious mistakes.

This approach is not only the most efficient, but it also gives scope for other questions to prompt you to recall forgotten facts (which happens surprisingly frequently).

Penalties for guessing – if there is a penalty for incorrect answers in a multiple-choice test, the best strategy is not to answer questions when you know your answer is a complete guess. Depending on the penalty, it may be beneficial to guess if you can narrow the choice down to two options (but beware false or irrelevant alternatives). However, if there are no such penalties, then you should provide an answer to all questions.

Answer the question as requested – this is true for all questions, but especially important for SAQs. If the question asks for a diagram, make sure you provide one; if it asks for *n* aspects of a topic, try to list this number of points; if there are two or more parts, provide appropriate answers to all aspects. This may seem obvious, but many marks are lost for not following instructions.

Typical layout for an extended matching items (EMI) question

1. Description of topic or theme – use this heading to orientate your thoughts when answering.

2. Instructions – this part will explain how you should answer and the weighting of marks among the questions. If all the questions are similar, this may appear at the head of the exam paper; if not, read each instruction carefully, as the rules and marks given may change.

3. A series of options – you will have to select among these to answer the questions which follow. Some options may be correct more than once, some not at all, and you may or may not be told how many are correct in each case. Each time you answer a question you will have to go through the list carefully and see which apply – slow readers beware!

4. A scenario (or 'vignette') – this is a short paragraph that describes the background to the questions. Details here will have a big effect on the answers you give, so you will need to consider the wording very carefully before you answer. It might be a good idea to highlight key words to keep them in your mind.

5. The questions – these may be grouped, and may develop in complexity from those demanding a simple factual answer to those requiring a fair amount of reflection and judgement.

When unsure of an answer, the first stage is to rule out options that are clearly absurd or have obviously been placed there to distract you. Next, looking at the remaining options, can you judge between contrasting pairs with alternative answers? Logically, both cannot be correct, so you should see if you can rule one of the pair out. Watch out, however, in case *both* may be irrelevant to the answer. If the question involves a calculation, try to work this out independently from the answers, so you are not influenced by them.

In SAQ papers, there may be a choice of questions. Choose your options carefully – it may be better to gain half marks for a correct answer to half a question, than to provide a largely irrelevant answer that apparently covers the whole question but lacks the necessary detail. For this form of question, few if any marks are given for writing style. Think in 'bullet point' mode and list the crucial points only. The time for answering SAQ questions may be tight, so get down to work fast, starting with answers that demand remembered facts. Stick to your timetable by moving on to the next question as soon as possible. Strategically, it is probably better to get part-marks for the full number of questions than good marks for only a few.

Extended matching items

Extended matching items (EMIs), sometimes called extended matching sets or extended matching questions, are a more complex and challenging version of the multiple-choice format, where a pool of answers is offered that may be correct in several cases or none at all. Formats differ, but the questions often relate to a complex scenario outlined as a 'vignette'. EMIs are designed to test not only your detailed knowledge but also your deeper understanding and ability to apply this, and are commonly used in disciplines where professional judgements on complex issues must be made.

Exams with an EMI component require you to read carefully, quickly and with understanding, distinguish between relevant and irrelevant information and make accurate connections between statements. As with MCQs, you will require a broad knowledge across your subject, but also a deeper and sometimes quite detailed understanding. An idea of the sorts of topics that might be covered should be apparent from the learning outcomes and case studies covered in the course. Look carefully at any example EMIs or past papers that are offered, as this will help you develop a strategy for answering.

When answering an EMI paper, first look quickly through the topics and select those that most suit your knowledge and revision. Plan to answer those you are most confident about first. Then, for each individual EMI:

1. **Scan-read the whole question quickly** to get a feel for its scope.

2. **Read both the scenario and the questions more carefully,** and highlight key terms. Some people find it helpful next to try to answer the questions without reference to the options as they will not be put off by distracters.

3. **Now run through the list of options ruling them 'in' or 'out'** for each specific question – you should be left with a short list of possible answers.

4. **Select the final answers you wish to give,** with reference to the instructions or rules provided.

5. **Make sure you are keeping up with the necessary pace** by setting a target time for completion of each topic. Try to leave some time in your planning to allow you to review and check your answers (p. 37).

You will need to be particularly well focused and work quickly and decisively – time will often be a constraint in EMI exams. If there are no penalty marks, ensure you give a response to each question.

Practical and information-processing exams

The prospect of a practical or information-processing exam in biology may cause you more concern than a theory exam. This may be due to a limited experience of practical examinations, or to the fact that practical and observational skills are tested, as well as recall, description and analysis of factual information. Your first thoughts may be that it is not possible to prepare for such exams but, in fact, you can improve your performance by mastering the various practical techniques described in this book.

You may be allowed to take your laboratory reports and other texts into the practical exam. Do not assume that this is a soft option, or that revision is unnecessary: you will not have time to read large sections of your reports or to familiarise yourself with basic principles, etc. The main advantage of 'open book' exams is that you can check specific details of methodology, reducing your reliance on memory, provided you know your way around your practical manual. In all other respects, your revision and preparation for such exams should be similar to theory exams. Make sure you are familiar with all of the practical exercises, including any work carried out in class by your partner (since exams are assessed on individual performance). If necessary, check with the teaching staff to see whether you can be given access to the laboratory, to complete any exercises that you have missed.

At the outset of the practical exam, determine or decide on the order in which you will tackle the questions. A question in the latter half of the paper may need to be started early on in the exam period (e.g. an enzyme assay requiring 2-h incubation in a 3-h exam). Such questions are included to test your forward-planning and time-management skills. You may need to make additional decisions on the allocation of material, e.g. if you are given 30 sterile test tubes, there is little value in designing an experiment that uses 25 of these to answer question 1, only to find that you need at least 15 tubes for subsequent questions.

Make sure you explain your choice of apparatus and experimental design. Calculations should be set out in a stepwise manner, so that credit can be given, even if the final answer is incorrect (see p. 494). If there are any questions that rely on recall of factual information and you are unable to remember specific details, e.g. you cannot identify a particular specimen or slide, make sure that you describe the item fully, so that you gain credit for observational skills. Alternatively, leave a gap and return to the question at a later stage.

Oral exams and interviews

An oral interview is sometimes a part of final degree exams, representing a chance for the external examiner(s) to get to know the students personally and to test their abilities directly and interactively. In some departments,

---

Example   These are the principal types of question you are likely to encounter in a practical or information-processing exam:

Manipulative exercises   Often based on work carried out during your practical course. Tests dexterity, specific techniques (e.g. dissection; sterile technique).

'Spot' tests   Short questions requiring an identification, or brief descriptive notes on a specific item (e.g. a prepared slide). Tests knowledge of seen material or ability to transfer this to a new example.

Calculations   For example, the preparation of aqueous solutions at particular concentrations (Chapter 25) or statistical exercises (Chapter 76). Tests numeracy.

Data analyses   May include the preparation and interpretation of graphs (Chapter 72) and numerical information, from data either obtained during the exam or provided by the examiner. Tests problem-solving skills.

Drawing specimens   Accurate representation and labelling will be important. Tests drawing and interpretation abilities.

Preparing specimens for examination with a microscope   Tests staining technique and light microscopy technique (Chapters 44 and 45).

Interpreting images   Sometimes used when it is not possible to provide living specimens, e.g. in relation to fieldwork, or electron microscopy. Can test a variety of skills.

Terminology – the oral exams are sometimes known simply as 'orals' or, borrowing Latin, as 'viva voces' (by or with the living voice) or 'vivas'.

orals are used to validate the exam standard, or to test students on the borderline between exam grades. Sometimes an interview may form part of an assessment, as with project work or posters. This type of exam is often intimidating – many students say they do not know how to revise for an oral – and many candidates worry that they will be so nervous they will not be able to do themselves justice.

Preparation is just as important for orals as it is for written exams:

- **Think about your earlier performances** – if the oral follows written papers, it may be that you will be asked about questions you did not do so well on. These topics should be revised thoroughly. Be prepared to say how you would approach the questions if given a second chance.

- **Read up a little about the examiner** – he or she may focus their questions in their area of expertise.

- **Get used to giving spoken answers** – it is often difficult to transfer between written and spoken modes. Write down a few questions and get a friend to ask you them, possibly with unscripted follow-up queries.

- **Research and think about topical issues in your subject area** – some examiners will feel this reflects how interested you are in your subject.

Your conduct during the oral exam is important too:

- **Arrive promptly and wear reasonably smart clothing.** Not to do either might be considered disrespectful by the examiner.

- **Take your time before answering questions.** Even if you think you know the answer immediately, take a while to check mentally whether you have considered all angles. A considered, logical approach will be more impressive than a quick but ill-considered response.

- **Start answers with the basics, then develop into deeper aspects.** There may be both surface and deeper aspects to a topic and more credit will be given to students who put the latter in context.

- **When your answer is finished, stop speaking.** A short, crisp answer is better than a rambling one.

- **If you do not know the answer, say so.** To waffle and talk about irrelevant material is more damaging than admitting that you do not know.

- **Make sure your answer is balanced.** Talk about the evidence and opinions on both sides of a contentious issue.

- **Do not disagree violently with the examiner.** Politely put your point of view, detailing the evidence behind it. Examiners will be impressed by students who know their own mind and subject area. However, they will expect you to support a position at odds with the conventional viewpoint.

- **Finally, be positive and enthusiastic about your topic.**

Counteracting anxiety before and during exams

Adverse effects of anxiety need to be overcome by anticipation and preparation well in advance (Box 7.4). Exams, with their tight time limits,

---

Allow yourself to relax in oral exams – external examiners are experienced at putting students at ease. They will start by asking 'simple-to-answer' questions, such as what modules you did, how your project research went and what your career aspirations are. Imagine the external examiner as a friend rather than a foe.

Creating an exam action list – knowing that you have prepared well, checked everything on your list and gathered together all you need for an exam will improve your confidence and reduce anxiety. Your list of actions might include:

- Time, date and place of the exam.
- Travel arrangements to exam hall.
- Double-check module handbooks and past papers for exam structure.
- Think through use of time and exam strategy.
- Identify a quiet place near the exam hall to carry out a last-minute check on key knowledge (e.g. formulae, definitions, diagram labels).
- Ensure you have all the items you wish to take to the exam, e.g.
  - pens, pencils (with sharpener and eraser);
  - ruler;
  - correction fluid;
  - calculator (allowable type);
  - sweets and drink, if allowed;
  - tissues;
  - watch or clock;
  - ID card;
  - texts and/or notes, if an open-book exam;
  - mascot.
- Lay out clothes (if exam is early in the morning).
- Set alarm(s) and/or ask a friend or family member to check you are awake on time.

> ### Box 7.4 Strategies for combating the symptoms of exam anxiety
>
> **Sleeplessness** – this is commonplace and does little harm in the short term. Get up, have a snack, do some light reading or other activity, then return to bed. Avoid caffeine (e.g. tea, coffee and cola) for several hours before going to bed.
>
> **Lack of appetite** – again commonplace. Eat what you can, but take sugary sweets into the exam to keep energy levels up in case you become tired.
>
> **Fear of the unknown** – it can be a good idea to visit the exam room beforehand, so you can become familiar with the location. By working through the points given in the exam action list on p. 42 you will be confident that nothing has been left out.
>
> **Worries about timekeeping** – double-check your alarm is functioning and use more than one system if possible. Arrange for an alarm phone call. Ask a friend or relative to make sure you are awake on time. Make reliable travel arrangements, to arrive on time. If your exam is early in the morning, it may be a good idea to get up early for a few days beforehand.
>
> **Blind panic during an exam** – explain how you feel to an invigilator. Ask to go for a supervised walk outside. Do some relaxation exercises (see below), then return to your work. If you are having problems with a specific question, it may be appropriate to speak to the departmental representative at the exam.
>
> **Feeling tense** – shut your eyes, take several slow, deep breaths, do some stretching and relaxing muscle movements. During exams, it may be a good idea to do this between questions, and possibly to have a complete rest for half a minute or so. Prior to exams, try some exercise activity or escape temporarily from your worries by watching TV or a movie.
>
> **Running out of time** – do not panic when the invigilator says 'five minutes left'. It is amazing how much you can write in this time. Write note-style answers or state the areas you would have covered: you may get some credit.
>
> **Lack of preparation** – the remedy, of course, is to adopt a better approach to your revision.

are especially stressful for perfectionists. To counteract this tendency, focus on the following points during the exam:

- **Do not expect to produce a perfect essay** – this will not be possible in the time available.
- **Do not spend too long planning your answer** – once you have an outline plan, get started.
- **Do not spend too much time on the initial parts of an answer**, at the expense of the main message.
- **Concentrate on getting all of the basic points across** – examiners usually reserve some marks for the main points, before allocating extra marks for detail and depth.
- **Do not be obsessed with neatness**, either in handwriting, or in the diagrams you draw, but make sure your answers are legible.
- **Do not worry if you forget something**. You cannot be expected to know everything. Most marking schemes give a first class grade to work that misses out on up to 30% of the marks available.

After the exam – try to avoid becoming involved in prolonged analyses with other students over the 'ideal' answers to the questions; after all, it is too late to change anything at this stage. Go for a walk, watch TV for a while, or do something else that helps you relax, so that you are ready to face the next exam with confidence.

> **KEY POINT** Everyone worries about exams. Anxiety Is a perfectly natural feeling. It works to your advantage, as it helps provide motivation and the adrenaline that can help you 'raise your game' on the day.

There is a lot to be said for treating exams as a game. After all, they are artificial situations contrived to ensure that large numbers of candidates can be assessed together, with little risk of cheating. They have conventions and rules, just like games. If you understand the rationale behind them and follow the rules, this will aid your performance.

Sources for further study

Acres, D. (1998) *How to Pass Exams Without Anxiety: How to get Organised, Be Prepared and Feel Confident of Success.* How To Books, London.

McMillan, K.W. and Weyers, J.D.B. (2011) *How to Succeed in Exams and Assessments.* Pearson Education, London.

McMillan, K.M. and Weyers, J.D.B. (2013) *The Study Skills Book,* 3rd edn. Pearson Education, London.

O'Brien, D. (2007) *How to Pass Exams: Accelerate Your Learning – Memorise Key Facts – Revise Effectively,* 2nd edn. Duncan Baird Publishers, Winchester.

Many universities host study skills websites that cover exam technique; these can be found using 'study skills', 'revision' or 'exams' as key words in a search engine.

STUDY EXERCISES

7.1 **Analyse your past performances.** Think back to past exams and any feedback you received from them. How might you improve your performance? Consider ways in which you might approach the forthcoming exam differently. If you have kept past papers and answers to continuous assessment exercises, look at any specific comments your lecturers may have made.

7.2 **Share revision notes with other students.** Make a revision plan (see pp. 11 and 31) and then allocate some time to discussing your revision notes with a colleague. Try to learn from his or her approach. Discuss any issues you do not agree upon.

7.3 **Plan your exam tactics.** Find out from your module handbook or past papers what the format of each paper will be. Confirm this if necessary with staff. Decide how you will tackle each paper, allocating time to each section and to each question within the sections (see p. 37). Write a personal checklist of requirements for the exam (see p. 42).

Answers to these study exercises are available at www.pearsoned.co.uk/practicalskills

Many students only think about their curriculum vitae immediately before applying for a job. Reflecting this, chapters on preparing a CV are usually placed near the end of texts of this type. Putting this chapter near the beginning of this book emphasises the importance of focusing your thoughts on your CV at an early stage in your studies. There are four main reasons why this can be valuable:

1. **Considering your CV and how it will look to a future employer will help you think more deeply about the direction and value of your academic studies.**
2. **Creating a draft CV will prompt you to assess your skills and personal qualities** and how these fit into your career aspirations.
3. **Your CV can be used as a record of all the relevant things you have done at university** and then, later, will help you communicate these to a potential employer.
4. **Your developing CV can be used when you apply for vacation or part-time employment.**

Personal development planning (PDP) and your CV – many PDP schemes (p. 5) also include an element of career planning that may involve creating a draft or generic CV. The PDP process can help you improve the structure and content of your CV, and the language you use within it.

> **KEY POINT** Developing your skills and qualities needs to be treated as a long-term project. It makes sense to think early about your career aspirations so that you can make the most of opportunities to build up relevant experience. A good focus for such thoughts is your developing curriculum vitae, so it is useful to work on this from a very early stage.

Understanding skills and qualities – it may be helpful to think about how the skills and qualities in Tables 1.1 and 8.1 apply to particular activities during your studies, since this will give them a greater relevance.

Skills and personal qualities

Skills (sometimes called competences) are generally what you have learned to do and have improved with practice. Table 1.1 summarises some important skills for biologists. This list might seem quite daunting, but your tutors will have designed your courses to give you plenty of opportunities for developing your expertise. Personal qualities, on the other hand, are predominately innate. Examples include honesty, determination and thoroughness (Table 8.1). These qualities need not remain static, however, and can be developed or changed according to your experiences. By consciously deciding to take on new challenges and responsibilities, not only can you develop your personal qualities, but you can also provide supporting evidence for your CV.

Personal qualities and skills are interrelated because your personal qualities can influence the skills you gain. For example, you may become highly proficient at a skill requiring manual dexterity if you are particularly adept with your hands. Being able to transfer your skills is highly important (Chapter 1) – many employers take a long-term view and look for evidence of the adaptability that will allow you to be a flexible employee and one who will continue to develop skills.

Focusing on evidence – it is important to be able to provide specific concrete information that will back up the claims you make under the 'skills and personal qualities' and other sections of your CV. A potential employer will be interested in your level of competence (what you can actually do) and in situations where you have used a skill or demonstrated a particular quality. These aspects can also be mentioned in your covering letter or at interview.

Developing your curriculum vitae

The initial stage involves making an audit of the skills and qualities you already have and thinking about those you might need to develop.

Table 8.1 Some positive personal qualities

Adaptability
Conscientiousness
Curiosity
Determination
Drive
Energy
Enthusiasm
Fitness and health
Flexible approach
Honesty
Innovation
Integrity
Leadership
Logical approach
Motivation
Patience
Performance under stress
Perseverance
Prudence
Quickness of thought
Seeing others' viewpoints
Self-confidence
Self-discipline
Sense of purpose
Shrewd judgement
Social skills (sociability)
Taking initiative
Tenacity
Tidiness
Thoroughness
Tolerance
Unemotional approach
Willingness to take on challenges

> **Seeing yourself as others see you** – you may not recognise all of your personal qualities and you may need someone else to give you a frank appraisal. This could be anyone whose opinion you value: a friend, a member of your family, a tutor, or a careers adviser.

> **Setting your own agenda** – you have the capability to widen your experience and to demonstrate relevant personal qualities through both curricular and extra-curricular activities.

Tables 8.1 and 1.1 could form a basis of this self-appraisal. Assessing your skills may be easier than critically analysing your personal characteristics. In judging your qualities, try to take a positive view and avoid being overly modest. It is important to consider your qualities in a specific context, e.g. 'I have shown that I am trustworthy by acting as treasurer for the Biological Society', as this evidence will form a vital part of your CV and job applications.

If you can identify gaps in your skills, or qualities that you would like to develop, especially in relation to the needs of your intended career, the next step is to think about ways of improving them. This will be reasonably easy in some cases, but may require some creative thinking in others. A relatively simple example would be if you decided to learn a new language or to sustain one you learned at school. There are likely to be many local college and university courses dealing with foreign languages at many different levels, so it would be a straightforward matter to join one of these. A rather more difficult case might be if you wished to demonstrate 'responsibility', because there are no courses available on this. One route to demonstrate this quality might be to put yourself up for election as an officer in a student society or club; another could be to take a leading role in a relevant activity within your community (e.g. voluntary work such as hospital radio). If you already take part in activities like these, your CV should relate them to this context.

Basic CV structures and their presentation

Box 8.1 illustrates the typical parts of a CV and explains the purpose of each part. Employers are more likely to take notice of a CV that is well organised and presented, in contrast to one that is difficult to read and assimilate. They will expect it to be concise, complete and accurate. There are many ways of presenting information in a CV, and you will be assessed partly on your choices.

- **Order.** There is some flexibility as to the order in which you can present the different parts (see Box 8.1). A chronological approach within sections helps employers gain a picture of your experience.
- **Personality and 'colour'.** Make your CV different by avoiding standard or dull phrasing. Try not to focus solely on academic aspects: you will probably have to work in a team and the social aspects of teamwork will be enhanced by your outside interests. However, make sure that the reader does not get the impression that these interests dominate your life.
- **Style.** Your CV should reflect *your* personality, but not in such a way that it indicates too idiosyncratic an approach. It is probably better to be formal in both language and presentation, as flippant or chatty expressions will not be well received.
- **Neatness.** Producing a well-presented, word-processed CV is very important. Use a laser-quality printer and good-quality paper; avoid poor-quality photocopying at all costs.
- **Layout.** Use headings for different aspects, such as personal details, education, etc. Emphasise words (e.g. with capitals, bold, italics or underlining) sparingly and with the primary aim of making the structure clearer. Remember that careful use of white space is important in design.
- **Grammar and proofreading.** Look at your CV carefully before you submit it, as sloppy errors give a very poor impression. Even if you use a

Box 8.1 **The structure and components of a typical CV and covering letter**

There is no right or wrong way to write a CV, and no single format applies. It is probably best to avoid software templates and CV 'wizards' as they can create a bland, standardised result, rather than something that demonstrates your individuality.

You should include the following with appropriate sub-headings, generally in the order given below

1. **Personal details.** This section *must* include your full name and date of birth, your address (both home and term-time, with dates, if appropriate) and a contact telephone number at each address. If you have an email account, you might also include this. You need only mention gender if your name could be either male or female.

2. **Education.** Choose either chronological order, or reverse chronological order and make sure you take the same approach in all other sections. Give educational institutions and dates (month, year) and provide more detail for your degree course than for your previous education. Remember to mention any prizes, scholarships or other academic achievements. Include your overall mark for the most recent year of your course, if it seems appropriate. Make sure you explain any gap years.

3. **Work experience.** Include all temporary, part-time, full-time or voluntary jobs. Details include dates, employer, job title and major duties involved.

4. **Skills and personal qualities.** Tables **1.1** and **8.1** give examples of the aspects you might include under this heading. Emphasise your strengths, and tailor this section to the specific requirements of the post (the 'job description'): for example, you might emphasise the practical skills you have gained during your degree studies if the post is a biological one, but concentrate on generic transferable skills and personal qualities for other jobs. Provide supporting evidence for your statements in all cases.

5. **Interests and activities.** This is an opportunity to bring out the positive aspects of your personality, and explain their relevance to the post you are applying for. Aim to keep this section short, or it may seem that your social life is more important than your education and work experience. Include up to four separate items, and provide sufficient detail to highlight the *positive* aspects of your interests (e.g. positions of responsibility, working with others, communication, etc.). Use sections 4 and 5 to demonstrate that you have the necessary attributes to fulfil the major requirements of the post.

6. **Referees.** Include the names (and titles), job descriptions, full postal addresses, contact telephone numbers and email addresses of two referees (rarely, some employers may ask for three). It is usual to include your personal tutor or course leader at university (who among other things will verify your marks), plus another person – perhaps a current or former employer, or someone who runs a club or society and who knows your personal interests and activities. Unless you have kept in touch with a particular teacher since starting university, it is probably best to choose current contacts, rather than those from your previous education.

Some other points to consider:

- Try to avoid jargon and over-complicated phrases in your CV: aim for direct, active words and phrases (see Box 16.1).

- Most employers will expect your CV to be word-processed (and spellchecked). Errors in style, grammar and presentation will count against you, so be sure to check through your final version (and ask a reliable person to second-check it for you).

- Aim for a *maximum* length of two pages, printed single-sided on A4 paper, using a 'formal' font (e.g. Times Roman or Arial) of no less than 12 point for the main text. Always print on good-quality white paper. Avoid fussy use of colour, borders or fonts.

- Do not try to cram in too much detail. Use a clear and succinct approach with short sentences and lists to improve 'readability' and create structure. Remember that your aim is to catch the eye of your potential employer, who may have many applications to work through.

- It is polite to check that people are willing to act as a referee for you and to provide them with an up-to-date copy of your CV.

Your covering letter should have four major components:

1. **Letterhead.** Include your contact details, the recipient's name and title (if known) and address, plus any job reference number.

2. **Introductory paragraph.** Explain who you are and state the post you are applying for.

3. **Main message.** This is your opportunity to sell yourself to a potential employer, highlighting particular attributes and experience. Keep it to three or four sentences at the most and relate it to the particular skills and qualities demanded in the job or person specification.

4. **Concluding paragraph.** A brief statement that you look forward to hearing the outcome of your application is sufficient.

Finally, add either 'Yours sincerely' (where the recipient's name is known) or 'Yours faithfully' (in a letter beginning 'Dear Sir or Madam') and then end with your signature.

Paying attention to the quality of **your CV** – your potential employer will regard your CV as an example of your very best work and will not be impressed if it is full of mistakes or badly presented, especially if you claim 'good written communication' as a skill!

Creating a generic CV – as you may apply for several jobs, it is useful to construct a CV in electronic format (e.g. as a *Word* file) which includes *all* information of potential relevance. This can then be modified to fit each post. Having a prepared CV on file will reduce the work each time you apply, while modifying this will help you focus on relevant skills and attributes for the particular job.

spellchecker, some errors may creep in. Ask someone whom you regard as a reliable proofreader to comment on it (many tutors will do this, if asked in advance).

- **Relevance.** If you can, slant your CV towards the job description and the qualifications required (see below). Make sure you provide evidence to back up your assertions about skills, qualities and experience.

- **Accuracy and completeness.** Check that all your dates tally; otherwise, you will seem careless. It is better to be honest about your grades and for example, a period of unemployment, than to cover this up or omit details that an employer will want to know. They may be suspicious if you leave things out.

Adjusting your CV

You should fine-tune your CV for each submission. Employers frequently use a 'person specification' to define the skills and qualities demanded in a job, often under headings such as 'essential' and 'desirable'. This will help you decide whether to apply for a position and it assists the selection panel to filter the applicants. Highlight relevant qualifications as early in your CV as possible. Be selective – do not include every detail about yourself. Emphasise relevant parts and leave out irrelevant details, according to the job. Similarly, your letter of application is not merely a formal document but is also an opportunity for persuasion (Box 8.1). You can use it to state your ambitions and highlight particular qualifications and experience. However, don't go over the top – always keep the letter to a single page.

> **KEY POINT** A well-constructed and relevant CV will not necessarily guarantee you a job, but it may well get you on to the short list for interview. A poor-quality CV is a sure route to failure.

Sources for further study

Anon. (2000) *How to Write a Curriculum Vitae.* University of London Careers Service, London.

Anon. *Graduate Prospects Ltd. CVs and cover letters How to write a CV.* Available http://www.prospects.ac.uk
Last accessed 07/03/15.
[Provides information on career opportunities and how to apply.]

Corfield, R. (2009) *Preparing the Perfect Job Application.* Kogan Page, London.

Jackson, A. and Geckeis, K. (2003) *How to Prepare your Curriculum Vitae.* McGraw-Hill, Chicago.

Purdue University. *Online Writing Lab: Writing the Curriculum vitae.* Available: https://owl.english.purdue.edu/owl/resource/641/01/
Last accessed 07/03/15.

STUDY EXERCISES

8.1 Evaluate your personal attributes. Using Table 8.1 as a source, list *five* qualities that you feel that you would best use to describe yourself, and cite the evidence you might give to a potential employer to convince them that this was the case. List *five* attributes you could develop, then indicate how you might do this.

8.2 Create a generic CV. Drawing on your school record of achievement or similar, or any CV you may already have prepared, e.g. for a summer or part-time job, create a word-processed generic CV. Save the file in an appropriate (computer) folder and make a back-up copy. Print out a copy for filing. Periodically update the word-processed version. If necessary, keep or save different versions to be used in different contexts (e.g. when you apply for a vacation job).

8.3 Think about your future career and ask for advice. Make an appointment with one of the advisers in your university's careers service. Ask what the career options might be for people with your intended degree, or determine what qualifications or module options might be appropriate for occupations that attract you.

Answers to these study exercises are available at www.pearsoned.co.uk/practicalskills

Information technology and learning resources

The ability to find scientific information is a skill required for many exercises in your degree programme. You will need to research facts and published findings as part of writing essays, literature reviews and project introductions, and when amplifying your lecture notes and revising for exams. You must also learn how to follow scientific convention in citing source material as the authority for the statements you have made.

Sources of information

For essays and revision

You are unlikely to use scientific papers (primary sources, p. 62) for these purposes – books and reviews are much more readable. If a lecturer or tutor specifies a particular book, then it should not be difficult to find out where it is shelved in your library, using the computerised index system. Library staff will generally be happy to assist with any queries. If you want to find out which books your library holds on a specified topic, use the system's subject index. You will also be able to search by author or by key words.

There are two main systems used by libraries to classify books: the Dewey Decimal system and the Library of Congress system. Libraries differ in the way they employ these systems, especially by adding further numbers and letters after the standard classification marks to signify, for example, shelving position or edition number. Enquire at your library for a full explanation of local usage.

The Web is an ever-expanding resource for gathering both general and specific information (see Chapter 11). Sites fall into analogous categories to those in the printed literature: there are sites with original information, sites that review information and bibliographic sites. One problem is that websites may be frequently updated, so information present when you first looked may be altered or even absent when the site is next consulted. Further, very little of the information on the web has been monitored or refereed. Another disadvantage is that the site information may not state the origin of the material, who wrote it or when it was written (Chapter 10, p. 62).

For literature surveys and project work

You will probably need to consult primary sources (papers). If you are starting a new research project or writing a report from scratch, you can build up a core of relevant papers by using the following methods:

- **Asking around**: supervisors or their postgraduate students will almost certainly be able to supply you with a reference or two that will start you off.
- **Searching online databases**: these cover very wide areas and are a convenient way to start a reference collection, although a charge is often made for access and sending out a listing of the papers selected (your library may or may not pass this on to you).
- **Consulting the bibliography of other papers in your collection**: an important way of finding the key papers in your field. In effect, you are taking advantage of the fact that another researcher has already done all the hard work.

Browsing in a library – this may turn up interesting material, but remember the books on the shelves are those not currently out on loan. Almost by definition, the latter may be more up to date and useful. To find out a library's full holding of books in any subject area, you need to search its catalogue (normally available as an online database). An online search will also bring up electronic versions of textbooks (e-books).

Example The book *The Selfish Gene* by Richard Dawkins (1976; Oxford University Press) is likely to be classified as follows:

Dewey Decimal system: 591.51

where 591	refers to zoology
591.5	refers to ecology of animals
591.51	refers to habits and behaviour patterns

Library of Congress system: QL751

where Q	refers to science
QL	refers to zoology
QL75	refers to animal behaviour
QL751	refers to general works and treatises

Researching a new topic – reading reviews (secondary sources, p. 62) can provide you with a useful overview at the start of a new project.

Web resources – your university library will provide you with access to a range of web-based databases and information systems. The library web pages will list these and provide links, which may be worth bookmarking on your web browser. Resources especially useful to bioscientists include:

- *ISI Web of Science, including the Science Citation Index*
- Ingentaconnect
- CSA Illumina
- Medline
- PubMed
- ScienceDirect
- Scopus
- Ovid

Most of these electronic resources operate on a subscription basis and may require an 'Athens' username and password – for details of how to obtain these, consult library staff or your library's website.

Definitions

e-book – a book published online in downloadable form.

ebrary – a commercial service offering e-books and other online resources.

e-journal – a journal published online, consisting of articles structured in the same way as a paper-based journal. A valid username and password may be required for access (arranged via your library, if it subscribes to the e-journal).

Journal/periodical/serial – any publication issued at regular intervals. In biosciences, usually containing papers (articles) describing original research findings and reviews of literature.

Monograph – a specialised book covering a single topic.

The primary literature – this comprises original research papers, published in specialist scientific periodicals. Certain prestigious general journals (e.g. *Nature*) contain important new advances from a wide subject area.

- **Referring to 'current awareness' online databases**: these are useful for keeping you up to date with current research; they usually provide a monthly listing of article details (title, authors, source, author address) arranged by subject and cross-referenced by subject and author. Current awareness databases cover a wider range of primary literature than could ever be available in any one library.

- Examples relevant to the biosciences include: *Current Contents Connect*, the *Current Advances* series (Elsevier), *Biological Abstracts* and *Zoological Record Archive* (Thompson Scientific). Some online databases also offer a service whereby they will email registered users with updates based on saved search criteria. Consult library staff or your library website to see which of these databases and services are available to you.

- **Using the *Science Citation Index* (SCI)**: this is a very valuable source for exploring the published literature in a given field, because it lets you see who has cited a given paper; in effect, SCI allows you to move forward through the literature from an existing reference. The Index is available online via ISI Web of Science.

For specialised information

You may need to consult reference works, such as encyclopedias, maps and books providing specialised information. Much of this is now available online (consult your library's information service or web pages).

Three books worth noting are:

- The *Handbook of Chemistry and Physics* (Haynes, 2014): the Chemical Rubber Company's publication (affectionately known as the 'Rubber Bible') giving all manner of physical constants, radioisotope half-lives, etc.
- The *Merck Index* (O'Neil *et al.*, 2013), which gives useful information about organic chemicals, e.g. solubility, whether poisonous, etc.
- The *Geigy Scientific Tables* (8th edition), a series of six volumes (Lentner, 1981; 1984; 1986; 1990; 1992; Lentner *et al.*, 1982) provides a wide range of information centred on biochemistry, e.g. buffer formulae, properties of constituents of living matter.

Obtaining and organising research papers

Obtaining a copy

It is usually more convenient to have personal copies of key research articles in print or electronic format for direct consultation when working in a laboratory or writing. The simplest way of obtaining these is to photocopy the originals or download and/or print off copies online (e.g. as .pdf or HTML files). For academic purposes, this is normally acceptable within copyright law. If your library does not subscribe to the journal, it may be possible for them to borrow it from a nearby institute or obtain a copy via a national borrowing centre (an 'inter-library loan'). If the latter, you will have to fill in a form giving full bibliographic details of the paper and where it was cited, as well as signing a copyright clearance statement concerning your use of the copy.

If you are unable to obtain a copy from your library, or via inter-library loan, then you might consider emailing or writing to the author to request a copy (use a Google search to find an author's address and email details).

Organising papers

Although the number of papers you accumulate in electronic and/or hard copy formats may be small to start with, it is worth putting some thought into their storage and indexing before your collection becomes disorganised and unmanageable. Few things are more frustrating than not being able to lay your hands on a vital piece of information, and this can seriously disrupt your flow when writing or revising.

Indexing your references

Whether you have obtained a printed copy, have stored downloaded files electronically, or have simply noted the bibliographic details of a reference, you will need to index each resource. This is valuable for the following reasons:

- You will probably need the bibliographic information for creating a reference list for an assignment or report.
- If the index also has database features, this can be useful, allowing you to search for key words or authors.
- If you include an 'accession number' and if you then file printed material sequentially according to this number, then it will help you to find the hard copy.
- Depending on the indexing system used, you can add comments about the reference that may be useful at a later time, e.g. when writing an introduction or conclusion.

The simplest way to create an index system is to put the details on reference cards, but database software can be more convenient and faster to sort, once the bibliographic information has been entered. If you do not feel that commercial software is appropriate for your needs, consider using a word processor or spreadsheet; their rudimentary database sorting functions (see Chapters 12 and 13) may be all that you require.

If you are likely to store lots of references and other electronic resources digitally, then you should consider carefully how this information is kept, for example by choosing file names that indicate what the file contains and that will facilitate sorting.

Making citations in text

It is particularly important to cite all sources of information used in your work – this will demonstrate to assessors that you have used appropriate source material and it will also avoid any possibility of plagiarism (using another persons work without acknowledgement, p. 60).

There are two main ways of citing articles and creating a bibliography (also referred to as 'references' or 'literature cited').

The Harvard system

For each citation, the author name(s) and the date of publication are given at the relevant point in the text. The bibliography is organised alphabetically and by date of publication for papers with the same authors. Formats normally adopted are, for example, 'Smith and Jones

Storing printed research papers – these can easily be kept in alphabetical order within filing boxes or drawers, but if your collection is likely to grow large, it will need to be refiled as it outgrows the storage space. You may therefore wish to add an 'accession number' to the record you keep in your database, and file the papers in sequence according to this as they accumulate. New filing space is only required at the 'end' and you can use the accession numbers to form the basis of a simple cross-referencing system.

Using commercial bibliographic database software to organise your references – for those with large numbers of references in their collection, and who may wish to produce lists of selected references in particular format, e.g. for inclusion in a project report or journal paper, systems like *EndNote*, *Reference Manager* or *ProCite* can reward the investment of time and money required to create a personal reference catalogue. Appropriate bibliographic data must first be entered into fields within a database (some versions assist you to search online databases and upload data from these). The database can then be searched and used to create customised lists of selected references in appropriate citation styles.

Example   Incorporating references in text – this sample shows how you might embed citations in text using the Harvard approach:
'... Brookes *et al.* (2001) proposed that protein A216 was involved in the degradation process. However, others have disputed this notion (Scott and Davis, 1997; Harley, 1998, 2000). Patel (1999a, 1999b) found that A216 is inactivated at pH values less than 5; while several authors (e.g. Hamilton, 1995; Drummond and Stewart, 2002) also report that its activity is strongly dependent on Ca^{2+} concentration ...'

(1983) stated that . . .' or 'it has been shown that . . . (Smith and Jones, 1983)'. Lists of references within parentheses are separated by semicolons, e.g. '(Smith and Jones, 1983; Jones and Smith, 1985)', normally in order of date of publication. To avoid repetition within the same paragraph, an approach such as 'the investigations of Smith and Jones indicated that' could be used following an initial citation of the paper. Where there are more than two authors it is usual to write *'et al.'*; this stands for the Latin *et alia* meaning 'and others'. If citing more than one paper with the same authors, put, for example, 'Smith and Jones (2009; 2010)' and if papers by a given set of authors appeared in the same year, letter them (e.g. Smith and Jones, 2011a; 2011b).

The numerical or Vancouver system

Papers are cited via a superscript or bracketed reference number inserted at the appropriate point. Normal format would be, for example: 'DNA sequences[4,5] have shown that ...' or 'Jones [55,82] has claimed that ...'. Repeated citations use the number from the first citation. In the true numerical method (e.g. as in *Nature*), numbers are allocated by order of citation in the text, but in the alpha-numerical method (e.g. the *Annual Review* series), the references are first ordered alphabetically in the Bibliography, then numbered, and it is this number that is used in the text. Note that with this latter method, adding or removing references is tedious, so the numbering should be done only when the text has been finalised.

KEY POINT   The main advantages of the Harvard system are that the reader might recognise the paper being referred to and that it is easily expanded if extra references are added. The main advantages of the Vancouver system are that it aids text flow and reduces length.

How to list your citations in a bibliography

Whichever citation method is used in the text, comprehensive details are required for the bibliography so that the reader has enough information to find the reference easily. Citations should be listed in alphabetical order with the priority: first author, subsequent author(s), date. Unfortunately, in terms of punctuation and layout, there are almost as many ways of citing papers as there are journals! Your department may specify an exact format for project work; if not, decide on a style and be consistent – if you do not pay attention to the details of citation you may lose marks. Take special care with the following aspects:

- **Authors and editors:** give details of *all* authors and editors in your bibliography, even if given as *'et al.'* in the text.
- **Abbreviations for journals:** while there are standard abbreviations for the titles of journals (consult library staff), it is a good idea to give the whole title, if possible.
- **Books:** the edition should always be specified as contents may change between editions. Add, for example, '(5th edition)' after the

Example Incorporating references in text – this sample shows how you might embed citations in text using the Vancouver approach:

'. . . Brookes *et al.* proposed that protein A216 was involved in the degradation process[1]. However, others have disputed this notion[2-4]. Patel[5,6] found that A216 is inactivated at pH values less than 5; while several authors[7,8] also report that its activity is strongly dependent on Ca^{2+} concentration ...'

Examples

Paper in journal:
Smith, A. B., Jones, C.D. and Professor, A. (2011). Innovative results concerning our research interest. *Journal of New Results*, **11**, 234–5.

Book:
Smith, A. B. (2009). *Summary of My Life's Work*. Megadosh Publishing Corp., Bigcity. ISBN 123-4-56-789123-4.

Chapter in edited book:
Jones, C. D. and Smith, A. B. (2010). Earth-shattering research from our laboratory. In: *Research Compendium 1998* (ed. A. Professor), pp. 123–456. Bigbucks Press, Booktown.

Thesis:
Smith, A. B. (2008). *Investigations on my Favourite Topic*. PhD thesis, University of Life, Fulchester.

Website:
Jones, A. B. (2010) *My Webpage on my Work in 2010*. Available: http://www.jonesinfo.co.uk/work2010.
Last accessed 01/04/2011.

Note: if your references are handwritten, you should indicate italics by underlining text or numerals.

Finding dates on websites quickly – when visiting a particular page, you can find occurrences of a specific date (e.g. 2011) by pressing *Control* and *F* keys together, entering 2011 in the *Find* window that appears, then carrying out a search using the *Find next* command.

title of the book. You may be asked to give the International Standard Book Number (ISBN), a unique reference number for each book published.

- **Unsigned articles,** e.g. unattributed newspaper articles and instruction manuals: refer to the author(s) in text and bibliography as 'Anon.'.
- **Websites:** there is no widely accepted format at present. You should follow departmental guidelines if these are provided, but if these are not available, we suggest providing author name(s) and date in the text when using the Harvard system (e.g. Hacker, 2010), while in the bibliography giving the URL details in the following format: Hacker, A. (2010) *University of Anytown Homepage on Aardvarks*. Available: http://www.myserver.ac.uk/homepage. Last accessed 1/4/2011. In this example, the web page was constructed in 2010, but accessed in April, 2011. If no author is identifiable, cite the sponsoring body (e.g. University of Anytown, 2010), and if there is no author or sponsoring body, write 'Anon.' for 'anonymous', e.g. Anon. (2010), and use Anon. as the 'author' in the bibliography. If the web pages are undated, *either* use the 'Last accessed' date for citation and put no date after the author name(s) in the reference list, *or* cite as 'no date' (e.g. Hacker, no date) and leave out a date after the author name(s) in the reference list – you should be consistent whichever option you choose.
- **Unread articles:** you may be forced to refer to a paper via another without having seen it. If possible, refer to another authority who has cited the paper, e.g. '... Jones (1980), cited in Smith (2009), claimed that ...'. Alternatively, you could denote such references in the bibliography by an asterisk and add a short note to explain at the start of the reference list.
- **Personal communications:** information received in a letter, seminar or conversation can be referred to in the text as, for example, '... (Smith, pers. comm.)'. These citations are not generally listed in the bibliography of papers, though in a thesis you could give a list of personal communicants and their addresses.
- **Online material:** some articles are published solely online and others online ahead of publication in printed form. The article may be given a DOI (digital object identifier), allowing it to be cited and potentially tracked before and after it is allocated to a printed issue (see http://www.doi.org/). DOIs allow for web page redirection by a central agency, and CrossRef (see http://www.crossref.org) is the official DOI registration organisation for scholarly and professional publications. DOIs can be used as 'live' hyperlinks in online articles, or cited in place of (and, when they become available, following) the volume and page numbers for the article, with the remainder of the details cited in the usual fashion: for example,

'Smith, A. and Jones, B. (2011) Our latest important research in the form of a web-published article. *Online Biosciences* 8/2011 (p. 781). Published Online: 26 March 2011. DOI: 10.1083/mabi.200680019'.

If you are using a different referencing system than Harvard, consult Pears and Shield (2010) or McMillan and Weyers (2013) for further information on how to cite websites in these systems.

Text references

Haynes, W.M. (ed.) (2014) *CRC Handbook of Chemistry and Physics,* 95th edn. CRC Press, Boca Raton.

Lentner, C. (ed.) (1981) *Geigy Scientific Tables, Vol. 1: Units of Measurement, Body Fluids, Composition of the Body. Nutrition,* 8th edn. Ciba-Geigy, Basel.

Lentner, C. (ed.) (1984) *Geigy Scientific Tables, Vol. 3: Physical Chemistry, Composition of Blood, Hematology, Somatometric Data,* 8th edn. Ciba-Geigy, Basel.

Lentner, C. (ed.) (1986) *Geigy Scientific Tables, Vol. 4: Biochemistry, Metabolism of Xenobiotics, Inborn Errors of Metabolism, Pharmacogenetics, Ecogenetics,* 8th edn. Ciba-Geigy, Basel.

Lentner, C. (ed.) (1990) *Geigy Scientific Tables, Vol. 5: Heart and Circulation,* 8th edn. Ciba-Geigy, Basel.

Lentner, C. (ed.) (1992) *Geigy Scientific Tables, Vol. 6: Bacteria, Fungi, Protozoa, Helminths,* 8th edn. Ciba-Geigy, Basel.

Lentner, C., Diem, K. and Seldrup, J. (eds) (1982) *Geigy Scientific Tables, Vol. 2: Introduction to Statistics, Statistical Tables, Mathematical Formulae,* 8th edn. Ciba-Geigy, Basel.

McMillan, K.M. and Weyers, J.D.B. (2013) *The Study Skills Book,* 3rd edn. Pearson Education, Harlow.

O'Neil, M.J., Heckelman, P.E., Koch, C.B. and Roman, K.J. (2013) *Merck Index: An Encyclopedia of Chemicals, Drugs, and Biologicals,* 15th edn. Merck & Co., Inc., Whitehouse Station, New Jersey.

Pears, R. and Shield, G. (2010) *Cite Them Right! The Essential Referencing Guide,* 8th edn. Palgrave Macmillan, Basingstoke. Also available as an online resource, see http://www.citethemright online.com/

Sources for further study

Grix, J. and Watkins, G. (2010) *Information Skills: Finding and Using the Right Resources.* Palgrave Macmillan, Basingstoke.

Neville, C. (2009) *The Complete Guide to Referencing and Avoiding Plagiarism.* Open University Press, Maidenhead.

Weselby, J.M. (2014) *Citations Made Simple: A Student's Guide to Easy Referencing: The Complete Guide.* Create Space (Amazon), Seattle.

STUDY EXERCISES

9.1 Test your library skills. This exercise relies on the fact that most university-level libraries serving biosciences departments will take the scientific journal *Nature*. To help you answer these questions, it may be beneficial to attend a library induction session, if you have not already. Alternatively, the library's help or enquiry desks may be able to assist you if you are having problems.

a) First, find out and provide the name of the classification system that your university uses for cataloguing its books and periodicals.

b) Using your library's cataloguing system (online, preferably) find out the appropriate local classification number for the journal *Nature*.

c) Where precisely is *Nature* shelved in your library? (Your answer need refer only to most recent issues if some have been archived.)

d) What is the exact title of the landmark papers in the following two volumes? (i) *Nature* **171**, 737–738 (1953); (ii) *Nature* **408**, 796–815 (2000).

9.2 Explore different methods of citing references. Go to your library and seek out the journal area for biology or life sciences. Choose *three* different journals in your subject area and from a recent edition write down how they would print a typical citation for a multi-author journal paper in the 'references' or 'literature cited' section. Where used, indicate italicised text with normal underline and bold text with wavy underline. Pay attention to punctuation. Compare these methods with each other, with the methods recommended on pp. 56–7 of this book and with the recommendation your department or your course handbook makes. Are they all the same?

9.3 Make website citations. Use a search engine (pp. 72–4) to find an informative website that covers each of the following:

a) The use of SI units.

b) Information about pollination of a specific flower by bats.

c) The different types of blood cells.

Indicate how you would cite each website at the end of an essay (follow your department's guidelines or use those in this chapter).

9.4 Compare the Harvard and Vancouver methods of citation. Pair up with a partner in your class. Each person should then pick one of the two main methods of citation and consider its pros and cons independently. Meet together and compare your lists. Given the choice, which method would you choose for (a) a hand-written essay; (b) a word-processed review; (c) an article in an academic journal, and why?

Answers to these study exercises are available at www.pearsoned.co.uk/practicalskills

10 Evaluating information

Example A Web search for the letters 'DNA' (e.g. using the search engine Google, p. 73), will reveal that this acronym appears in several hundred million websites. Not all of these deal with deoxyribonucleic acid – the listed websites include: *Doctors' Net Access*, a Web resource for American physicians; the website of the *Dermatology Nurses Association*; and that of *Delta Nu Alpha*, an international organisation for those working in the professions. These examples are easy to identify as irrelevant to research on 'DNA, the molecule', but considering the many websites that cover a given topic, which might contain the exact information you seek? How valid is the information? Is it biased towards one viewpoint or hypothesis? Does it represent current mainstream thinking on the topic? These are some of the issues that an evaluation of the information sources might deal with.

Checking the reliability of information, assessing the relative value of different ideas and thinking critically are skills essential to the scientific approach. You will need to develop your abilities to evaluate information in this way because:

- **You will be faced with many sources of information**, from which you will need to select the most appropriate material.
- **You may come across two conflicting sources of evidence** and may have to decide which is the more reliable.
- **The accuracy and validity of a specific fact may be vital** to your work.
- **You may doubt the quality of information** from a particular source.
- **You may wish to check the original source** because you are not sure whether someone else is quoting it correctly.

KEY POINT Evaluating information and thinking critically are regarded as higher-order academic skills. The ability to think deeply in this way is greatly valued in the biosciences and will consequently be assessed in coursework and exam questions (see Chapter 7).

The process of evaluating and using information can be broken down into four stages:

1. **Selecting and obtaining material.** How to find sources is covered in Chapter 9. Printed books and journals are important, but if you identify a source of this kind there may be delays in borrowing it or obtaining a photocopy. If the book or journal is available online, then downloading or printing sections or papers will be more convenient and faster. The Internet is often a first port of call if you wish to find something out quickly. For many websites, however, it can be difficult to verify the authenticity of the information given (see Box 11.5).

2. **Assessing the content.** You will need to understand fully what has been written, including any technical terms and jargon used. Establish the relevance of the information to your needs and assure yourself that the data or conclusions have been presented in an unbiased way.

3. **Modifying the information.** In order to use the information, you may need to alter it to suit your needs. This may require you to make comparisons, interpret or summarise. Some sources may require translation. Some data may require mathematical transformation before they are useful. There is a chance of error in any of these processes and also a risk of plagiarism.

4. **Analysis.** This may be your own interpretation of the information presented, or an examination of the way the original author has used the information.

Definition

Plagiarism – the unacknowledged use of another's work as if it were one's own. In this definition, the concept of 'work' includes ideas, writing, data or inventions and not simply words; and the notion of 'use' does not only mean copy 'word for word', but also 'in substance' (i.e. a copy of the ideas involved). Use of another's work is acceptable if you acknowledge the source (see Box 10.1).

KEY POINT Advances in communications and information technology mean that we can now access almost limitless knowledge. Consequently, the ability to evaluate information has become an extremely important skill.

Box 10.1 How to avoid plagiarism and copyright infringement

Plagiarism is defined on p. 60. Examples of plagiarism include:

- copying the work of a fellow student (past or present) and passing it off as your own;
- using 'essay-writing services', such as those on offer on certain websites;
- copying text or images from a source (book, journal article or website, for instance) and using this within your own work without acknowledgement;
- quoting others' words without indicating who wrote or said them;
- copying ideas and concepts from a source without acknowledgement, even if you paraphrase them.

Most students would accept that some of the above can only be described as cheating. However, many students, especially at the start of their studies, are unaware of the academic rule that they must *always* acknowledge the originators of information, ideas and concepts, and that not doing so is regarded as a form of academic dishonesty. If you adopt the appropriate conventions that avoid such accusations, you will achieve higher marks for your work as it will fulfil the markers' expectations for academic writing.

Universities have a range of mechanisms for identifying plagiarism, from employing experienced and vigilant coursework markers and external examiners to analysing students' work using sophisticated software programs. Plagiarism is always punished severely when detected. Penalties may include awarding a mark of zero to all involved – both the copier(s) and the person whose work has been copied (who is regarded as complicit in the crime). Further disciplinary measures may be taken in some instances. In severe cases, such as copying substantive parts of another's work within a thesis, a student may be dismissed from the university.

If you wish to avoid being accused of plagiarism, the remedies are relatively simple.

- **Make sure the work you present is always your own.** If you have been studying alongside a colleague, or discussing how to tackle a particular problem with your peers, make sure you write alone when working on your assignments.
- **Never be tempted to 'cut and paste'** from websites or online documents. Read these carefully, decide what the important points are, express these *in your own words* and *provide literature citations to the original sources* (see Chapters 4 and 9). In some cases, further investigations may be required to find out details of the original sources.

- **Take care when note-taking.** If you decide to copy text word for word, make sure you show this clearly in your notes with quotation marks. If you decide to make your own notes based on a source, make sure these paraphrase rather than copy phrases from the text. In both cases, write down full details of the source at the appropriate point in your notes.
- **Place appropriate citations throughout your text where required.** If you are unsure about when to do this, study reviews and articles in your subject area (see also Chapters 4 and 9).
- **Understand the differences between quotation, summarising and paraphrasing** (p. 21). Apply the relevant layout conventions: show clearly where you are quoting directly from a source by using quotation marks, and *always* cite your source(s). Try not to rely too much on quotation as this may be regarded as a lack of ability to synthesise your own ideas.

Copyright issues are often associated with plagiarism, and refer to the right to publish (and hence copy) original material, such as text, images and music. Copyright material is indicated by a symbol © and a date (see, for example, page iv of this book). Literary copyright is the aspect most relevant to students in their academic studies. UK copyright law protects authors' rights for life and gives their estates rights for a further 70 years. Publishers have 'typographical copyright' that lasts for 25 years. This means that it is illegal to photocopy, scan or print out copyright material unless you have permission, or unless your copying is limited to an extent that could be considered 'fair dealing'. For educational purposes – private study or research – in a scientific context, this generally means:

- no more than 5% in total of a work;
- one chapter of a book;
- one article per volume of an academic journal;
- 20% of a short book;
- one separate illustration or map.

You may only take one copy within the above limits, may not copy for others, and may not exceed these amounts *even if you own a copy of the original*. These rules also apply to Web-based materials, but sometimes you will find sites where the copyright is waived. Some copying may be licensed; you should consult your library's website or helpdesk to see whether it has access to licensed material. Up-to-date copyright information is generally posted close to library and departmental photocopiers.

Evaluating sources of information

One way of assessing the reliability of a piece of scientific information is to think about how it was obtained in the first place. Essentially, facts and ideas originate from someone's research or scholarship, whether they are numerical data, descriptions, concepts, or interpretations. Sources are divided into two main types:

1. **Primary sources** – those in which ideas and data are first communicated. The primary literature is generally published in the form of 'papers' (articles) in journals whether printed or online. These are usually refereed by experts in the academic peer group of the author, and they will check the accuracy and originality of the work and report their opinions back to the editors. This peer review system helps to maintain reliability, but it is not perfect. Books and, more rarely, websites and articles in magazines and newspapers, can also be primary sources but this depends on the nature of the information published rather than the medium. These sources are not formally refereed, although they may be read by editors and lawyers to check for errors and unsubstantiated or libellous allegations.

2. **Secondary sources** – those which quote, adapt, interpret, translate, develop or otherwise use information drawn from primary sources. It is the act of quoting or paraphrasing that makes the source secondary, rather than the medium. Reviews are examples of secondary scientific sources, and books and magazine articles are often of this type.

When information is modified for use in a secondary source, alterations are likely to occur, whether intentional or unintentional. Most authors do not deliberately set out to change the meaning of the primary source, but they may unwittingly do so, e.g. in changing text to avoid plagiarism or by oversimplification. Others may consciously or unconsciously exert bias in their reporting, for example, by quoting evidence that supports only one side of a debate. Therefore, the closer you can get to the primary source, the more reliable the information is likely to be. On the other hand, modification while creating a secondary source could involve correcting errors, or synthesising ideas and content from multiple sources.

Authorship and provenance

Clearly, much depends on who is writing the source and on what basis (e.g. who paid them?). Consequently, an important way of assessing sources is to investigate the ownership and provenance of the work (who and where it originated from, and why).

Can you identify who wrote the information? If it is signed or there is a 'byline' showing who wrote it, you might be able to make a judgement on the quality of what you are reading. This may be a simple decision, if you know or can assume that the writer is an authority in the area; otherwise a little research might help (for example, by putting the name into a search engine). Of course, just because Professor X thinks something does not make it true. However, if you know that this opinion is backed up by years of research and experience, then you might take it a little more seriously than the thoughts of a school pupil. If an author is not cited, effectively nobody is taking responsibility for the content. Could there be a reason for this?

Is the author's place of work cited? This might tell you whether the facts or opinions given are based on an academic study. Is there a company

---

Distinguishing between primary and secondary sources – try the 'IMRaD test'. Many primary sources contain information in the order: Introduction, Materials and Methods, Results and Discussion. If you see this format, and particularly if *data* from an experiment, study or observation are presented, then you are probably reading a primary source.

Example If a journalist wrote an article about a new 'flesh-eating bug' for the *New York Times* that was based on an article in the *British Medical Journal*, the *New York Times* article would be the secondary source, while the *British Medical Journal* article would be the primary source.

Taking account of the changing nature of websites and wikis – by their very nature, these sources may change. This means that it is important to quote accurately from them and to give a 'Last accessed' date when citing (see p. 57).

Finding out about authors and provenance – these pieces of information are easy to find in most printed sources and may even be presented just below the title, for convenience. In the case of the Web, it may not be so easy to find what you want. Relevant clues can be obtained from 'home page' links and the header, body and footer information. For example, the domain (p. 72) may be useful, while the use of the tilde symbol (,) in an address usually indicates a personal, rather than an institutional, website.

with a vested interest behind the content? If the author works for a public body, there may be publication rules to follow and they may even have to submit their work to a publications committee before it is disseminated. They are certainly more likely to get into trouble if they include controversial material.

Evaluating facts and ideas

However reliable the source of a piece of information seems to be, it is probably a good idea to retain a slight degree of scepticism about the facts or ideas involved. Even information from impeccable primary sources may not be perfect – different approaches can give different outcomes, and interpretations can change with time and with further advances in knowledge. Table 10.1 provides a checklist to use when evaluating sources.

Critically examining facts and ideas is a complex task depending on the particular issues involved, and a number of different general approaches can be applied. You will need to decide which of the following general tips are useful in your specific case:

- **Make cross-referencing checks (triangulation)** – look at more than one source and compare what is said in each. The cross-referenced sources should be as independent as possible (for example, do not compare a primary source together with a secondary review based on it). If you find that all the sources give a similar picture, then you can be more confident about the reliability of the information.

- **Look at the extent and quality of citations** – if references are quoted, these indicate that a certain amount of research has been carried out beforehand, and that the ideas or results are based on genuine scholarship. If you are doubtful about the quality of the work, these references might be worth looking at. How up to date are they? Do they cite independent work, or is the author exclusively quoting their own work, or solely the work of one person?

- **Consider the age of the source** – the fact that a source is old is not necessarily a barrier to truth, but ideas and facts may have altered since the date of publication, and methods may have improved. Can you trace changes through time in the sources available to you? What key events or publications have forced any changes in the conclusions?

- **Try to distinguish fact from opinion** – to what extent has the author supported a given viewpoint? Have relevant facts been quoted, via literature citations or the author's own researches? Are numerical data used to substantiate the points used? Are these reliable and can you verify the information, for example, by looking at the sources cited? Might the author have a reason for putting forward biased evidence to support a personal opinion?

- **Analyse the language used** – words and their use can be very revealing. Subjective wording might indicate a personal opinion rather than an objective conclusion. Propaganda and personal bias might be indicated by absolute terms, such as 'everyone knows…'; 'It can be guaranteed that …', or a seemingly one-sided consideration of the evidence. How carefully has the author considered the topic? A less studious approach might be indicated by exaggeration, ambiguity, or

---

Assessing substance over presentation – just because the information is presented well (e.g. in a glossy magazine or particularly well-constructed website), this does not necessarily tell you much about its quality. Try to look below the surface, using the methods mentioned in this chapter.

Table 10.1 Checklist for assessing information in science. How reliable is the information you have been reading? The more 'yes' answers you can give below, the more trustworthy you can assume it to be

Assessing sources

- Can you identify the author's name?
- Can you determine what relevant qualifications he/she holds?
- Can you say who employs the author?
- Do you know who paid for the work to be done?
- Is this a primary or secondary source?
- Is the content original or derived from another source?

Evaluating information

- Have you checked a range of sources?
- Is the information supported by relevant literature citation?
- Is the age of the source likely to be important regarding the accuracy of the information?
- Have you focused on the substance of the information presented rather than its packaging?
- Is the information fact or opinion?
- Have you checked for any logical fallacies in the arguments?
- Does the language used indicate anything about the status of the information?
- Have the errors associated with any numbers been taken into account?
- Have the data been analysed using appropriate statistics?
- Are any graphs constructed fairly?

the use of 'journalese' and slang. Always remember, however, that content should be judged above presentation.

- **Look closely at any numbers quoted** – if the information you are looking at is numerical in form, have statistical errors been taken into consideration, and, where appropriate, quantified? If so, does this help you arrive at a conclusion about how genuine the differences are between important values?

- **Think carefully about any hypothesis-testing statistics used** – are the methods appropriate? Are the underlying hypotheses the right ones? Have the results of any tests been interpreted correctly in arriving at the conclusion? To deal with these matters, you will need at least a basic understanding of the 'statistical approach' and of commonly used techniques (see Chapters 75 and 76).

Interpreting data

Numerical data

Information presented in public, whether as a written publication or spoken presentation, is rarely in the same form as it was when first obtained. Chapters 53 and 71 deal with processes in which data are recorded, manipulated and transformed, while Chapter 75 describes the standard descriptive statistics used to 'encapsulate' large data sets. Chapter 74 covers some relevant mathematical techniques. Sampling (essentially, obtaining representative measurements) is at the heart of many observational and experimental approaches in biology (see Chapters 27, 32 and 33), and analysis of samples is a key component of hypothesis-testing statistics (Chapter 76). Understanding these topics and carrying out the associated study exercises will help you improve your ability to interpret numerical data.

Graphs

Frequently, understanding and analysis in science depends on your ability to interpret data presented in graphical form. Sometimes, graphs may mislead. This may be unwitting, as in an unconscious effort to favour a 'pet' hypothesis of the author. Graphs may be used to 'sell' a product, e.g. in advertising, or to favour a viewpoint as, perhaps, in politics. Experience in drawing and interpreting graphs (Chapter 72) will help you spot these flawed presentations, and understanding how graphs can be erroneously presented (Box 72.3) will help you avoid the same pitfalls.

Tables

Tables, especially large ones, can appear as a mass of numbers and thus be more daunting at first sight than graphs. In essence, however, most tables are simpler than most graphs. The construction of tables is dealt with in Chapter 73.

Critical thinking

Critical thinking involves the application of logic to a problem, issue or case study. It requires a wide range of skills. Key processes involved include: acquiring and processing information; creating appropriate hypotheses and

---

Analysing a graph – this process can be split into six phases:

1. Considering the context and purpose of the graph.

2. Recognising the type of presentation and examining the axes.

3. Looking closely at the scale on each axis.

4. Examining the data presented (e.g. data points, symbols, curves).

5. Considering errors and statistics associated with the graph.

6. Reaching conclusions based on the above.

These processes are amplified in Chapter 72.

Analysing a table – as with analysing a graph, this process can be split into six phases:

1. Considering the context and purpose.

2. Examining the subheadings to see what information is contained in the rows and columns.

3. Considering the units used and checking any footnotes.

4. Comparing the data values across rows and/or down columns, looking for patterns, trends and unusual values.

5. Taking into account any statistics presented.

6. Reaching conclusions based on the above.

These processes are amplified in more detail in Chapter 73.

Learning from examples – as your lecturers introduce you to case studies, you will see how biologists have applied critical thinking to understand the nature of cells, organisms and ecosystems. Some of your laboratory sessions may mimic the processes involved – observation, hypothesis, experimental design, data gathering and analysis and formulating a conclusion (see Chapter 33). These skills and approaches can be applied in your course, e.g. when writing about a biological issue or carrying out a research project.

'You can prove anything with statistics' – leaving aside the issue that statistical methods deal with probability, not certainty (Chapter 76), it is possible to analyse and present data in such a way that they support one chosen argument or hypothesis rather than another. Detecting a bias of this kind can be difficult, but the critical thinking skills involved are essential for all scientists.

Good writing requires good logic – understanding the logic behind what you want to write is a prerequisite for creating high-quality text. Creating a plan for your writing (Chapter 16) will help you to both recognise and organise what you want to say.

formulating conclusions; and acting on the conclusions towards a specific objective.

> **KEY POINT**   Critical thinking needs reliable knowledge, but it requires you to use this appropriately to analyse a problem. It can be contrasted with rote learning – where you might memorise facts without an explicit purpose other than building your knowledge base.

Critical thinking is particularly important in biology, because the subject deals with complex and dynamic systems. These can be difficult to understand for several reasons:

- **They are often multi-faceted, involving many interactions.**
- **It can be difficult to alter one variable in an experiment without producing confounding variables** (see p. 203).
- **Many variables may be unmeasured or unmeasurable.**
- **Heterogeneity (variability) is encountered at all scales** from the molecular scale to the ecosystem.
- **Perturbation of the system can lead to unexpected ('counter-intuitive') results.**

As a result, conclusions in biological research are seldom clear-cut. Critical thinking allows you to arrive at the most probable conclusion from the results at hand; however, it also involves acknowledging that other conclusions might be possible. It allows you to weigh up these possibilities and find a working hypothesis or explanation, but also to understand that your conclusions are essentially dynamic and might alter when new facts are known. Hypothesis-testing with statistics (Chapter 76) is an important adjunct to critical thinking because it demands the formulation of simple hypotheses and provides rational reasons for making conclusions.

Recognising fallacies in arguments is an important aspect of critical thinking. Philosophers and logicians recognise different forms of argument and many different fallacies in each form. Damer (2013) provides an overview of this wide-ranging and complex topic.

Explaining your thoughts

The context for your evaluation of the literature and the associated critical thinking will normally be an essay, report or similar piece of academic writing (Chapters 17–19). The skills involved in marshalling and explaining your thoughts are regarded as highly important in research and employment.

You may have a very specific remit, as defined in the instruction for the assignment (Chapter 17), or the topic may be open. In either case, your reading around the topic should result in an overarching position or argument you wish to put forward – the 'thesis' for your writing. You may be explaining an established viewpoint or creating an original perspective

Examples

Inductive reasoning

1. The plant species growing in a mining spoil tip (Site A) are different from those in nearby land subject to similar climatic conditions (Site B).

2. The spoil tip contains a metal Z known to be toxic to a range of organisms.

3. When metal Z is added to pots containing plants from both sites, the plants common on Site B but not on Site A die, but those from Site A all survive.

4. High concentrations of metal Z in the spoil have resulted in the change in species mix due to differential sensitivity to the metal.

Possible flaw in logic: the concentration of metal Z used at step 3 might have been much higher that that found in the spoil and another potentially toxic metal that is present might not have been tested.

Deductive reasoning

1. Yellow wagtails have green backs and no throat markings.

2. This bird has a green back and no throat markings.

3. Therefore it is yellow wagtail.

Possible flaw in logic: other species of birds may also have green backs and no throat markings.

on a topic. For both situations, the same principles apply. When setting out your thoughts, you should:

- **make a clear statement of the issue being considered**, if necessary defining terms and boundaries;
- **consider the issue from different perspectives**, providing evidence for or against different propositions;
- **cite sources of evidence and ideas** that you are evaluating (Chapter 9);
- **ensure your viewpoint is logical and internally consistent;**
- **structure your writing appropriately**, for example, by first considering evidence for a particular view and then evidence against it; or by considering the development of evidence through time;
- **use an academic style of writing**, avoiding personal statements (Box 16.2);
- **arrive at a conclusion**, even if this is that the evidence is not sufficient to allow firm statements to be made.

In the sciences, the norm is to use inductive reasoning – that is, to state observed facts and assumptions at the outset, then draw a logical conclusion based on these. The alternative, deductive reasoning, starts from a general statement, premise or law which is held to be true and then reaches a conclusion by considering facts logically. You should look for these types of argument in texts and papers, and also think about possible flaws in such arguments. Chapter 33 discusses further the use of hypotheses and experiments in biological science.

Text reference

Damer, T.E. (2013) *Attacking Faulty Reasoning: A Practical Guide to Fallacy-Free Arguments*, 7th edn. Wadsworth, Belmont, California.

Sources for further study

Anon. *Critical Thinking.* Available: http://en.wikipedia.org/wiki/Critical_thinking
Last accessed 16/02/15.
[A frequently updated review of critical thinking terminology and techniques, with additional links and references.]

Gilbert, F.S., MacGregor, P.K. and Barnard, C.J., (2011) *Asking Questions in Biology: Key Skills for Practical Assessments and Project Work*, 4th edn. Prentice Hall, Harlow.

Smith, A. *Evaluation of Information Sources.* Available: http://www.vuw.ac.nz/staff/alastair_smith/evaln/evaln.htm
Last accessed 16/02/15.
[Part of the Information Quality WWW Virtual Library containing pointers to a wide range of sources giving criteria for evaluation.]

Van Gelder, T. *Critical Thinking on the Web.* Available: http://www.austhink.org/critical/
Last accessed 16/02/15.
[A useful directory of Web resources on the topic of critical thinking, associated with *Austhink,* a group based in Melbourne, Australia.]

STUDY EXERCISES

10.1 Distinguish between primary and secondary literature. Based on their titles and any research you can do in your library, determine whether the following journals are primary or secondary sources.

(a) *British Journal of Haematology*

(b) *Proceedings of the National Academy of Sciences*

(c) *Annual Review of Molecular Biology*

(d) *Essays in Biochemistry*

(e) *Microbiological Research*

(f) *Trends in Biotechnology*

(g) *The Lancet*

(h) *Oecologia*

(i) *Scientific American*

(j) *Food Safety News*

10.2 Consider a controversial issue from both sides. Select a current biological topic being discussed in the newspapers or other media. Relatively controversial issues such as 'genetically modified crops' or 'stem cell research' would be good examples. Next, write out a statement that you might use for a motion in a debate, such as 'Genetic modification of crops is a good thing' or 'Creating chimeric organisms is justified if it results in medical advances'. The exercise consists of writing at least five points in support of either side of the argument, which you should organise in tabular form. If you can find more than five points, add these to your table, but for each point that you add to one side, you should add one to the other side.

10.3 Analyse graphic presentations in the media. Many newspapers provide graphic presentations related to current issues, and graphs are frequently used in television news reports. Practise critical thinking skills by determining whether the graphs presented are a fair representation of the facts.

Answers to these study exercises are available at www.pearsoned.co.uk/practicalskills

Definitions

Browser – a program to display Web pages and other Internet resources.

FAQ – Frequently Asked Question; a file or Web page giving information on common queries, sometimes used as a file extension (.faq).

FTP – File Transfer Protocol; a mechanism for downloading files.

URL – Uniform Resource Locator; the 'address' for Web resources.

Academic use of ICT resources – a range of inappropriate activities will be identified in your university's rules for use of ICT systems. These may include: hacking, spamming, using another person's account and copyright infringement, as well as broader aspects of behaviour covered by a code of conduct or student charter.

Understanding the technology – you do not need to understand the workings of the Internet to use it – most of it is invisible to the user. To ensure you obtain the right facilities, you may need to know some jargon, such as terms for the speed of data transfer (megabits) and the nature of Internet addresses. Setting up a modem and/or local wireless network can be complex, but instructions are usually provided with the hardware. White and Downs (2014) and Gralla (2006) are useful texts if you wish to learn more about computing and the Internet.

Information and communication technology (ICT) is vital in the modern academic world and 'IT literacy' is a core skill for all bioscientists. This involves a wide range of computer-based skills, including:

- **accessing Web pages** using a Web browser such as *Internet Explorer*, *Firefox*, *Safari* or *Chrome*;
- **searching the Web** for useful information and resources using a search engine such as Google, or a meta-search engine such as Dogpile;
- **using online databases**, such as library catalogues, or complex websites, such as your university's home page;
- **downloading, storing and manipulating files;**
- **communicating via the Internet;**
- **using e-learning facilities effectively;**
- **Working with 'office'-type programs** and other software (dealt with in detail in Chapters 12 and 13).

You will probably receive an introduction to your university's networked IT systems and will be required to follow rules and regulations that are important for the operation of these systems. Whatever your level of experience with PCs and the Internet, you should also follow the basic guidelines shown in Box 11.1. Reminding yourself of these from time to time will reduce your chances of losing data.

The Internet as a global resource

The Internet is a complex network of computer networks; it is loosely organised and no one group organises it or owns it. Instead, many private organisations, universities and government organisations fund and operate discrete parts of it.

The most popular application of the Internet is the World Wide Web ('the Web'). It allows easy links to information and files which may be located on networked computers across the world. The Web enables you to access millions of 'home pages' or 'websites' – the initial point of reference with many individuals, institutions, and companies. Besides text and images, these sites may contain 'hypertext links', highlighted words or phrases that take you to another location via a single mouse click.

You can gain access to the Internet either through a local area network (LAN) at your university, at most public libraries, at a commercial 'Internet cafe', or from home via a modem connected to a broadband or dial-up Internet service provider (e.g. Virgin Media, BT or Sky).

> **KEY POINT** Most material on the Internet has not been subject to peer review or vetting. Information obtained from the Web or posted on newsgroups may be inaccurate, biased or spoof; do not assume that everything you read is true or even legal.

Box 11.1 **Important guidelines for using PCs and networks**

Hardware

- Do not drink or smoke around the computer.
- Try not to turn the computer off more than is necessary.
- Never turn off the electricity supply to the machine while in use.
- Switch off the computer and monitor when not in use (saves energy and avoids dangers of 'hijacking').
- Rest your eyes at frequent intervals if working for extended periods at a computer monitor. Consult Health and Safety Executive publications for up-to-date advice on working with display screens (http://www.hse.gov.uk/msd/dse/).
- Never try to reformat the hard disk without the help of an expert.

CDs and USB drives

- Protect CDs when not in use by keeping them in holders or boxes.
- Label USB (Universal Serial Bus) drives with your name and return details and consider adding these to a file stored on the drive.
- Try not to touch the surface of CDs, and if they need cleaning, do so carefully with a clean cloth, avoiding scratching. If floppy disks are used, keep these away from sources of magnetism (e.g. speakers).
- Keep disks and USB drives away from moisture, excess heat or cold.
- Do not use disks from others, unless you first check them for viruses.
- Do not insert or remove a disk or USB drive when it is operating (drive light on). Close all files before removing a USB drive and use the *Safely Remove Hardware* feature.
- Try not to leave a disk or USB drive in the drive when you switch the computer off.

File management

- Organise your files in an appropriate set of folders.
- Always use virus-checking programs on copied or imported files before running them.
- Make back-ups of all important files at frequent intervals (say, every ten minutes or half-hour), e.g. when using a word processor or spreadsheet.
- Periodically clear out redundant files.

Network rules

- Never attempt to 'hack' into other people's files.
- Do not give out any of your passwords to others. Change your password regularly. Make sure it is not a common word, is longer than eight characters, and includes numerical characters and punctuation symbols, as well as upper- and lower-case letters.
- Never use network computers to access or provide financial or other personal information: spyware and Trojan programs may intercept your information.
- Never open email attachments without knowing where they came from; always virus-check attachments before opening.
- Remember to log out of the network when finished; others can access your files if you forget to log out.
- Be polite when sending email messages.
- Periodically reorganise your email folder(s). These rapidly become filled with acknowledgements and redundant messages that reduce server efficiency and take up your allocated filespace.
- Do not play games without approval – they can affect the operation of the system.
- If you are setting up your own network, e.g. in your flat, always install up-to-date firewall software, antispyware and anti-virus programs.

The Golden Rule – always make back-up copies of important files and store them well away from your working copies. Ensure that the same accident cannot happen to both copies.

Spam, junk mail and phishing – these should be relatively easy to identify, and should never be responded to or forwarded. Some may look 'official' and request personal or financial details (for example, they may pretend to come from e.g. your bank, and ask for account details). Never send these details by email or your identity may be used illegally.

Online communication

You will be allocated an email account by your university and should use this routinely for communicating with staff and fellow students, rather than using a personal account. You may be asked to use email to submit work as an attachment, or you may be asked to use a 'digital drop-box' within the university's e-learning system (Box 11.2). When using email at university, follow these conventions, including etiquette, carefully:

- **Check your email account regularly** (daily). Your tutors may wish to send urgent messages to you in this way.
- **Respond promptly to emails.** Even if you are just acknowledging receipt, it is polite to indicate that you have received and understood a message.

Box 11.2 **Getting to grips with e-learning**

Some key aspects of tackling e-learning are outlined below.

1. **Develop your basic IT skills, if required.** e-learning requires only basic IT skills, such as: use of keyboard and mouse; word processing; file management; browsing and searching. If you feel weak on any of these, seek out additional courses offered by the IT administration or your department.

2. **Visit your e-learning modules regularly.** You should try to get into a routine of doing this on a daily basis at a time that suits you. Staff will present up-to-date information (e.g. lecture-room changes) via the 'announcements' section, may post information about assessments, or links to the assessments themselves, and you may wish to provide feedback or look at discussion boards and their threads.

3. **Participate.** e-learning requires an active approach.
   - At the start of each new course, spend some time getting to know what has been provided online to support your learning. As well as valuable resources, this may include crucial information such as learning outcomes (p. 30), dates of submission for coursework and weighting of marks for different assessments within the course.
   - If you are able to download lecture notes (e.g. in the form of *PowerPoint* presentations), do not think that simply reading through these will be an adequate substitute for attending lectures and making further notes (see p. 19).
   - Do not simply 'lurk' on discussion boards: take part. Ask questions; start new threads; answer points raised by others if you can.

- Try to gain as much as you can from formative online assessments (p. 30). If these include feedback on your answers, make sure you learn from this and if you do not understand it, consult your tutors.
- Learn from the critical descriptions that your lecturers provide of linked websites. These pointers may help you to evaluate such resources for yourself in future (pp. 63–5).
- Do not think that you will automatically assimilate information and concepts, just because you are viewing them online. The same principles apply as with printed media: you must apply active learning methods (p. 33).
- Help your lecturers by providing constructive feedback when they ask for it. You may find this easier to do online, rather than hurriedly filling out a feedback sheet at the end of a session.

4. **Organise files and Web links.** Take the time to create a meaningful system of folders and files for downloaded material in tandem with your own coursework files and set up folders on your browser for bookmarked websites (*Favorites* in *Internet Explorer*).

5. **Take care when submitting coursework.** Make sure you keep a back-up of any file you email or submit online and check the version you are sending carefully. Follow instructions carefully, for example regarding file type, or how to use your system's 'digital drop-box'.

- **Be polite.** Email messages can seem to be abrupt and impersonal. Take care to read your messages through before sending and if you are at all in doubt, do not send your message right away: reread at a later time and consider how others might view what you say.

- **Consider content carefully.** A useful approach is only to send what you would be happy to hear being read out loud to classmates or family.

- **Take care with language and names** when communicating with tutors and other staff. Slang phrases and text message shorthand are unlikely to be understood. Be friendly without being overfamiliar.

- **Use email for academic purposes** – this includes discussing coursework with classmates, but not forwarding questionable jokes, potentially offensive images, links to inappropriate websites, etc. In fact, doing so may break regulations and result in disciplinary action.

- **Beware of spam, junk and 'phishing'** via email.

Similar rules apply to discussion boards and the learning management system.

The Usenet Newsgroup service is an electronic discussion facility, and there are thousands of newsgroups representing different interests and

Newsgroups and online forums – these can be useful for getting answers to a specific problem: just post a query to the appropriate group and wait for someone to reply. However, always bear in mind that this is the view of an individual person.

Definition

Bookmark – a feature of browsers that allows you to save details of websites you have visited. This is termed 'add to favorites' in *Internet Explorer*. Bookmarks save you the trouble of remembering complex URL names and of typing them into the browser's address window.

Definition

m-Learning – learning delivered through mobile devices, including hand-held computers and mobile phones. m-Learning occurs independently of the location of the learner.

Examples   Useful Web portals:
Ocean Portal – http://ocean.si.edu/
British Ecological Society Portal – https://portal.britishecologicalsociety.org/portal
Bioscience education – http://www.biosciednet.org/portal/

topics. Any user can contribute to the discussion within a topic by posting their own message; it is like email, but without privacy, since your message becomes available to all other subscribers. Your university may also make use of social media, including Facebook and Twitter, or dedicated systems such as UCROO – often, the details of these systems will be explained during the initial orientation period, including how to 'follow' the university so that you keep up to date through news feeds.

Internet tools

The specific programs you will use for accessing the Internet will depend on what has been installed locally, on the network you are using, and on your Internet service provider. The best way to learn the features of the programs is to try them out, making full use of whatever help services are available.

e-learning systems

Most university departments present their courses through a mixture of face-to-face sessions (e.g. lectures, tutorials, practicals) and online resources (e.g. lecture notes, websites, discussion boards, computerised tests and assessments). This constitutes 'blended learning' on your part, with the online component also being known as e-learning. Some courses are delivered online (distance education) or by online study supplemented by block attendance on campus (intensives or residential schools).

The e-learning element is usually delivered through an online module within a virtual learning environment (e.g. Moodle, Sakai, Blackboard). It is important not to neglect the e-learning aspects of your course just because it may not be as rigidly timetabled as your face-to-face sessions. This flexibility is to your advantage, as you can work when it suits you, but it requires discipline on your part. Box 11.2 provides tips for making the most of the e-learning components of your courses. Mobile learning (m-learning), facilitated by wireless devices, is likely to grow in importance in the next few years.

Internet browsers

These are software programs that interact with remote server computers around the world to carry out the tasks of requesting, retrieving and displaying the information you require. Many different browsers exist, but the most popular are *Internet Explorer,* Mozilla *Firefox* and *Google Chrome*. These three browsers dominate the market and have plug-ins and add-on programs available that allow, for example, video sequences to be seen online. The standard functions of browsers include:

- **accessing Web documents;**
- **following links to other documents;**
- **printing the current document;**
- **maintaining a history of visited URLs** (including 'bookmarks' for key sites);
- **searching for a term in a document;**
- **viewing images and image maps.**

Browsers provide access to millions of websites. Certain sites specialise in providing catalogued links to other sites; these are known as portals

Examples Common domains and subdomains include:

.ac	academic
.com	commercial
.co	commercial
.edu	education (USA mainly)
.gov	government (USA and UK)
.mil	military (USA only)
.net	Internet-based companies
.org	organisation
.uk	United Kingdom

and can be of enormous help when searching within a particular area of interest. Your university's library website will almost certainly provide a useful portal to catalogues and search services, often arranged by subject area, and this is often the first port of call for electronic resources; get to know your way around this part of the website as early as possible during your course.

When using a Web browser program to get to a particular page of information on the Web, all you require is the location of that page, i.e. the URL (uniform resource locator). Most Web-page URLs take the form http:// or https:// (secure), followed by the various terms (domains and subdomains) that direct the system to the appropriate site. If you do not have a specific URL in mind but wish to explore appropriate sites, you will need to use a search tool with the browser.

Web search tools

With the proliferation of information on the Web, one of the main problems is finding the exact information you require. There are a variety of information services that you can use to filter the material on the network. These include:

- **search engines** (Boxes 11.3 and 11.4);
- **meta-search engines;**

Box 11.3 Useful tips for using search engines

- **Keywords should be chosen with care.** Try to make them as specific as possible, e.g. search for 'archaebacteria', rather than 'bacteria' or 'microorganisms'.

- **Most search engines are case-insensitive.** Thus 'Nobel Prize' will return the same number of hits as 'nobel prize'. If in doubt, use lower case throughout.

- **Putting keyword phrases in double quotes (e.g. "dog violet")** will result in a search for sites with the phrase as a whole rather than sites with both (all) parts of the phrase as separate words (i.e. 'dog' and 'violet' at different places within a site). This feature allows you to include common words normally excluded in the search, such as 'the'.

- **Use multiple words/phrases plus similar words to improve your search,** for example 'terrestrial crustacean "land crab" Australia'. If you can, use scientific terms, as you are likely to find more relevant sites, e.g. search for the name of a particular species such as 'Coenobita compressus'.

- **Adding words preceded with + or – will add or exclude sites with that word present** (e.g. 'blood diseases – leukaemia' will search for all blood diseases excluding leukaemia). This feature can also be used to include common words normally excluded by the search engine.

- **Check that your search terms have the correct spelling,** otherwise you may only find sites with the same misspelled word. In some cases, the search engine may prompt you with an alternative (correct) spelling. If a word

has an alternative US spelling (e.g. color, hemoglobin), then a search may only find hits from sites that use the spelling you specify.

- **Boolean operators (AND, OR, NOT) can be used with some search engines to specify combinations of keywords** to more precisely filter the sites identified (e.g. 'plant NOT engineering' will avoid sites about engineering plant and focus on botanical topics).

- **Some search engines allow 'wildcards' to be introduced with the symbol *.** For example, this will allow you to specify the root of a word and include all possible endings, as with anthropomorph*, which would find anthropomorphic, anthropomorphism, etc. If the search engine does not allow wildcards, then you will need to be especially careful with the key words used, including all possible words of relevance.

- **Numbers can be surprisingly useful in search engines.** For example, typing in EC 1.1.1.1 will find sites concerned with alcohol dehydrogenase, as this is its code number. If you know the phone number for a person, institute or company or the ISBN of the book, this can often help you find relevant pages quickly.

- **If you arrive at a large site and cannot find the point at which your searched word or phrase appears, press Control+F together** and a 'local' search window will appear, allowing you to find the point(s) where it is mentioned.

Box 11.4 Getting the most from Google searches

Google (http://www.google.com) has become the search engine of choice for millions of people, due to its simplicity and effectiveness. However, you may be able to improve your searches by understanding its default settings and how they can be changed.

- **Download the Google toolbar to your browser.** This is available from the Google homepage and will give you quick access to the Google search facility.

- **Understand how standard operators are used.** For combinations of keywords Google uses the minus operator '−' instead of NOT (exclude) and '+' instead of AND (include). Since Google usually ignores small words ('stop words' such as *in* or *the*), use '+' to include them in a search. Where no operator is specified, Google assumes that you are looking for both terms (i.e. '+' is default). If you want to search for alternative words, you can use 'OR' (e.g. *sulphur OR sulfur*). Google does not allow brackets and also ignores most punctuation marks.

- **While wildcard truncation of words using '*' is not allowed, you can use '*' to replace a whole word (or number).** For example, if you type the phrase "a *virus is approximately * nanometres*" (UK spelling) your results will give you results for Web pages where the wildcard is replaced by a number.

- **Search for exact wording.** By placing text in double inverted commas ("), you can ensure that only websites with this exact phrasing will appear at the head of your search results.

- **Search within your results to improve the outcome.** If your first search has produced a large number of results, use the *Search within results* option near the bottom of each page to type in a further word or phrase.

- **Search for words within the title of a web page.** Use the command *intitle:* to find a Web page, for example *intitle: "tissue culture"* returns Web pages with this phrase in the title (note that phrases must always be in double speech marks, not single quotes).

- **Search within a website.** Use the *site:* command to locate words/phrases on a specific website, for example *site:unicef.org dysentery* returns only those results for this disease on the UNICEF website (unicef.org). Pressing *Control+F* when visiting a web page will give you a pop-up search window.

- **Locate definitions, synonyms and spellings.** The operator *define:* enables you to find the meaning of a word. If you are unsure as to the spelling of a word, try each possibility: Google will usually return more results for the correct spelling and will often also prompt you with the correct spelling (*Did you mean…?*).

- **Find similar Web pages.** Simply click the *Similar pages* option at the end of a Google search result to list other sites (note that these sites will not necessarily include the term(s) searched for).

- **If a Web link is unavailable, try the cached (stored) page.** Clicking on *Cached* at the end of a particular result should take you to the stored page, with the additional useful feature that the search term(s) will be highlighted.

- **Use the calculator functions.** Simply enter a calculation and press *Enter* to display the result, for example '10+(2*4)' returns 18. The calculator function can also carry out simple interconversion of units, e.g. '2 *feet 6 inches in metres*' returns 0.762 (see Box 28.1 for interconversion factors between SI and non-SI units).

- **Try out the advanced search features.** In addition to the standard operators, these include the ability to specify the number of results per page (e.g. 50, to reduce the use of the *next* button), language (e.g. English), file format (e.g. for PDF files), recently updated Web pages (e.g. past three months), usage (e.g. free to use/share).

- **Find non-text material.** These include images, video and maps — always check that any material you use is not subject to copyright limitations (p. 61).

- **Use Google alerts to keep up to date.** This function (http://www.google.com/alerts) enables you to received regular updated searches by email.

- **Use Google Scholar to find articles and papers.** Go to http://scholar.google.com/ and type in either the general topic or specific details for a particular article, e.g. author names or words from the title. Results show titles/authors of articles, with links to either the full article, abstract or citation. A useful feature is the *Cited by …* link, taking you to those papers that have cited the article in their bibliography and enabling you to carry out *forward citation searching* to locate more recent papers. Also try out the advanced scholar search features to limit your search to a particular author, journal, date or subject area. However, you should note that Google Scholar provides only a basic search facility to easily accessible articles and should not be viewed as a replacement for your library's electronic journal holdings and searching software. For example, if you find the title of a paper via Google Scholar, you may be able to locate the electronic version through your own library's databases, or request it via interlibrary loan (p. 54). Another significant limitation is that older (more cited) sources are typically listed first. 'Use Advanced' Search to find sources between years *X* and *Y*.

- **Use Google Earth to explore locations.** This allows you to zoom in on satellite images to find locations.

'Dissecting' a Web address – if a URL is specified, you can often find out more about a site by progressively deleting sections of the address from the right-hand side. This will often take you to 'higher levels' of the site, or to the home page of the organisation or company involved.

Downloading files from the Internet and emails – read-only files are often available as 'pdf' files that can be viewed by Adobe reader software (available free from http://www.adobe.com), while other files may be presented as attachments to emails or as links from Web pages that can be opened by suitable software (e.g. Microsoft *Word* or 'paint' programs like *PaintShop Pro*). Take great care in the latter cases as the transfer of files can result in the transfer of associated viruses. Always check new files for viruses (especially .exe files) before running them, and make sure your virus-detecting software is kept up to date.

- **subject directories;**
- **subject gateways (portals).**

Search engines such as Google (http://www.google.com/), Bing (http://www.bing.com) and DuckDuckGo (https://duckduckgo.com) are tools designed to search, gather, index and classify Web-based information. Searching is usually by keyword(s), although specific phrases can be defined. Many search engines offer advanced searching tools such as the use of Boolean operators to specify combinations of keywords to more precisely filter the sites. Box 11.3 provides tips for refining keyword searches while Box 11.4 provides tips for enhancing searches with Google.

It is important to realise that each search engine will cover at most about 40% of the available sites; if you want to carry out an exhaustive search it is necessary to use several to cover as much of the Web as possible. Meta-search engines make this easier. These operate by combining collections of search engines. Examples include Mamma (http://www.mamma.com/) and Dogpile (http://www.dogpile.com/).

Some useful approaches to searching include the following:

- **For a comprehensive search**, use a variety of tools including search engines, meta-search engines and portals or directories.
- **For a complex, finely specified search**, employ Boolean operators and other tools to refine your keywords as fully as possible (Box 11.3). Some search engines allow you to include and exclude terms or restrict by date.
- **Use 'cascading' searching** when available – this is searching within the results of a previous search.
- **Use advanced search facilities** to limit your search, where possible, to the type of medium you are looking for (e.g. graphics, video), language, sites in a specific country (e.g. UK) or to a subject area (e.g. news only).

However well defined your search is, you will still need to evaluate the information obtained. Chapter 10 covers general aspects of this topic, while Box 11.5 provides specific advice on assessing the quality of information provided on websites.

Directories

A directory is a list of Web resources organised by subject. It can usually be browsed and may or may not have a search facility. Directories often contain better-quality information than the lists produced by search engines, as they have been evaluated, often by subject specialists or librarians. Check your library's subject resources for relevant directories.

Using the Internet as a resource

A common way of finding information on the Web is by browsing or 'surfing'. However, this can be time-consuming; try to restrict yourself to sites known to be relevant to the topic of interest. Some of the most useful

Box 11.5 How to evaluate information on the Web

It is often said that 'you can find anything on the Web'. The two main disadvantages of this are, firstly, that you may need to sift through many sources before you find what you are looking for and, secondly, that the sources you find will vary in their quality and validity. *It is important to realise that evaluating sources is a key aspect of using the Internet for academic purposes, and one that you will need to develop during the course of your studies.* The ease with which you can 'point and click' to reach various sources should not make you complacent about evaluating their information content. The following questions can help you to assess the quality of a website – the more times you can answer 'yes', the more credible the source is likely to be, and vice versa.

Authority
- Is the author identified?
- Are the author's qualifications or credentials given?
- Is the owner, publisher or sponsoring organisation identified?
- Is an address given (postal and/or email)?

It is sometimes possible to get information on authority from the site's metadata (try the 'View' 'Source' option in *Internet Explorer*, or look at the URL to see if it gives any clues as to the organisation, e.g. does the domain name end in .ac, .edu, .gov or .org, rather than .co or .com).

Content
- Is there any evidence that the information has been peer-reviewed (p. 62), edited or otherwise validated, or is it based on such sources?
- Is the information factual or based on personal opinions?
- Is the factual data original (primary) or derived from other sources (secondary)?

- Are the sources of specific factual information detailed in full (pp. 55–7)?
- Is there any indication that the information is up to date, or that the site has been recently updated?
- What is the purpose of the site and who is it aimed at?
- Is the content relevant to the question you are trying to answer?
- Is there any evidence of a potential conflict of interest, or bias? (Is the information comprehensive and balanced, or narrowly focused?)
- Did you find the information via a subject-specific website (e.g. a bioscience gateway such as BIOME), or through a more general source, such as a search engine (e.g. Google)?

The above questions are similar to those that you would use in assessing the value of a printed resource (pp. 62–3), and similar criteria should be applied to Web-based information. You should be especially wary of sites containing unattributed factual information or data whose primary source is not given.

Presentation
- What is your overall impression of how well the site has been put together?
- Are there many grammatical or spelling mistakes?
- Are there links to other websites, to support statements and factual information?

The care with which a site has been constructed can give you an indication of the credibility of the author/organisation. However, while a poorly presented site may cause you to question the credibility of the information, the reverse is not always necessarily true: do not be taken in by a slick, well-presented website – authority and content are *always* more important than presentation.

Understanding the impermanence of the Web – the temporary nature of much of the material on the Web is a disadvantage for academic purposes because it may change or even disappear after you have cited it. You may also find it difficult or impossible to find out who authored the material (p. 57). A case in point are wikis, such as Wikipedia (www.wikipedia.org). This online encyclopedia has many potential authors and the content may change rapidly as a result of new submissions or edits; nevertheless, it can be a useful resource for up-to-date general information about a wide range of topics, though it is not necessarily regarded as the best approach for researching assignments.

sites are those that provide hypertext links to other locations. Some other resources you can use on the Web are:

- **Libraries, publishers and commercial organisations.** Your university library is likely to subscribe to one or more databases providing access to scientific articles; these include BIDS (Bath Information and *Database Services* (http://www.bids.ac.uk), Web of Science (http://webofknowledge.com/), and Science Direct (http://www.sciencedirect.com/). A password is usually required, especially for off-campus use; consult your library staff for further details. Some scientific database sites give free access, without subscription or password; these include National Center for Biotechnology Information (USA) (http://www.ncbi.nlm.nih.gov/) and the Highwire Press (http://highwire.com/).

Others allow free searching, but require payment for certain articles, e.g. The Scientific World (http://www.thescientificworld.com/). The Natural History Book Service (http://www.nhbs.co.uk) provides useful information on relatively inaccessible literature such as government reports. Publishers such as Pearson and booksellers such as Amazon.com provide online catalogues and e-commerce sites that can be useful sources of information (see http://pearsoned.com/ and http://www.amazon.com).

- **Online journals and e-books.** A number of traditional journals have websites. You can keep up to date by visiting the websites of *Nature* (http://www.nature.com/), *New Scientist* (http://www.newscientist.com/) and *Scientific American* (http://www.scientificamerican.com/), or the Elsevier Science Direct website (http://www.sciencedirect.com/science/journals). Some scientific societies make their journals and other publications available via their websites, e.g. the American Society for Microbiology, at: http://www.asm.org/. Journals solely published in electronic format are also available (e.g. *Molecular Vision,* http://www.molvis.org/molvis/) but some require a subscription password for access; check whether your institute is a subscriber. The directory of open access journals (at: http://www.doaj.org/) provides a comprehensive list of all open access (free, without subscription) journals.

- **Data and images.** Archives of text material, video clips and photographs can be accessed, and much of the material is readily available. The HEA Centre for Biosciences Image Bank (http://www.bioscience.heacademy.ac.uk/imagebank) is a good example. When downloading such material, you should (i) check that you are not breaching copyright and (ii) avoid potential plagiarism by giving a full citation of the source, if you use such images in an assignment (see p. 56).

- **Biological institutions.** Many museums, botanic gardens, zoos, culture collections, scientific societies and other biological institutions around the world are now online. Use their sites to obtain specific information about collections, resources, etc. They frequently provide lists of other relevant sites or topics, e.g. the Natural History Museum (http://www.nhm.ac.uk/) or the Society for General Microbiology (http://www.sgm.ac.uk/).

- **Databases.** In addition to those covering the scientific literature, others focus on specific topics (e.g. UK Countryside and Nature Conservation, http://www.naturenet.net/index.php, or academic employment, http://jobs.ac.uk/). Box 62.2 provides further guidance on using databases in the important field of bioinformatics. Examples of nucleic acid databases include the Rutgers University database (http://ndbserver.rutgers.edu) and the FBI CODIS database, used in forensic DNA analysis (http://www.fbi.gov/about-us/lab/biometric-analysis/codis).

Using traditional sources – remember that using the Internet to find information is not a substitute for visiting your university library. Internet resources complement rather than replace CD-ROM and more traditional printed sources.

Remembering useful websites – create a 'bookmark' (= add a 'favorite') for the ones you find of value, to make revisiting easy. This can be done from the menu of your browser program. Make a copy of your bookmark file occasionally, to avoid loss of this information.

Understanding open access publication – this is a recent movement, pioneered by the Public Library of Science (PLOS). Typically authors of scholarly articles (or their institutions) pay for all costs required for online publishing, thereby providing free and unrestricted online access to all potential readers (for an example of an open access journal, see: http://www.plosbiology.org).

Note that URLs may change – make a keyword search using a search engine to find a particular site if the URL information you have does not lead you to an active page. Alternatively, use the Wayback Machine at: http://www.archive.org to locate an earlier version of a URL.

Text references

Gralla, P. (2006) *How the Internet Works,* 8th edn. Pearson, Harlow.

Sources for further study

Dussart, G. (2002) *Biosciences on the Internet: A Student's Guide*. Wiley, Chichester.

Hock, R. (2013) *The Extreme Searcher's Internet Handbook: A Guide for the Serious Searcher*, 4th edn. Cyber Age Books, New Jersey.

Grassian, E. *Thinking Critically about World Wide Web Resources*. Available: http://www.sscnet.uda.edu/library/modules/Judge/CLThinkDisc.pdf Last accessed 13/07/2015.

White, R. and Downs, T. (2014) *How Computers Work,* 10th edn. Que Publishing, Pearson, Harlow.

STUDY EXERCISES

11.1 Explore the resources of the Web using a search engine. Pick a search engine (p. 74) and use it to find the answers to the following questions.

   a) Who is Norman Borlaug, and what should he be widely famous for? What prize did he win and when?

   b) What is the common name for the species *Conolophus subcristatus*?

   c) What is the postal address of the Information Office of the Smithsonian Institution?

11.2 Compare results from a variety of search engines. First, think of an appropriate biological keyword or phrase (e.g. a species name) and enter this into several search engines. Make sure that you include metasearch engines such as Dogpile. Compare the outcomes to reveal the strengths and weaknesses of the individual search engines. Work with a colleague to compare different searches on a quantitative (i.e. number of hits) and qualitative (quality of hits) basis.

11.3 Organise your bookmarks. Enter the *'Organize favorites'* menu for your preferred browser and create folders with appropriate headings. Move existing bookmarks to these folders and save any new ones appropriately. Doing this will help you find bookmarks more easily, rather than searching through long lists.

Answers to these study exercises are available at www.pearsoned.co.uk/practicalskills

12 Using word processors, databases and other packages

Reassess your IT skills and knowledge in relation to your course – although you may be familiar with 'office' style software (e.g. Microsoft *Office* or Apple Mac *Pages*), you may be required to use features in new or different ways. For example:

- You may be asked to present documents and their components in very specific formats (such as defining fonts, line spacing, paragraph layout, margins, layout of tables and figures).
- You may benefit from using certain editing features to refine your writing (such as word counts, thesaurus, spelling and grammar checker).
- You may need to adapt your methods to longer writing exercises (such as creating an outline structure, writing parts out of sequence, creating a bibliography at the outset and contributing to it as you write).

You will probably be familiar with a range of computer programs used for academic work, and may own a laptop or PC with this software installed. To support your studies, you will also have access to networked PCs and associated programs licensed for your use. However, the specific software and versions may vary between these systems, involving different menu protocols and commands for the same or similar functions. In addition, the specific demands of academic work may involve the use of sophisticated features or even new types of program. This chapter considers some common software applications in relation to relevant tasks. Spreadsheets and presentation software are considered in detail in Chapters 13–15.

> **KEY POINT** Even when you are familiar with the basics of a software program, it may be valuable to learn more about advanced features, to ensure you are working efficiently and effectively.

Word processors

Word processors are generally available as part of an 'office' package with the advantage that they share a common interface in the different components (word processor, spreadsheet, database, etc.), allowing exchange of information (e.g. text, graphics) between applications. Examples are *WordPerfect* (Corel WordPerfect Office), *Pages* (Apple Mac) and Microsoft *Word* (Microsoft Office). Compatible open source software is also available, such as *OpenOffice*. Most word processors have similar general features but differ in operational detail. They offer specific advantages for academic work. For example, you can:

- **refine material many times before submission;**
- **insert material easily, allowing writing to take place in any sequence;**
- **use a spellchecker to check your text;**
- **use a thesaurus when composing your text;**
- **carry out ongoing checks of the word count;**
- **produce high-quality final copies;**
- **reuse part or all of the text in other documents.**

Although use of computers to compose text is now almost universal, it is worth remembering certain disadvantages of this approach:

- **the need for a reliable battery or mains power supply** (important for work in the field);
- **the need to learn specific operational details of the program;**
- **the temptation to make trivial revisions**, hence 'overworking' your text;
- **the risk of losing files or forgetting to save work appropriately.**

> **KEY POINT** It is vital to save your work frequently to a memory stick, hard drive or network drive. This should be done every 10 min or so. If you do not save regularly, you may lose hours or days of work. Most programs can be set to 'autosave' every few minutes – make sure you switch this feature on (using the appropriate commands from your program).

When using a word processor, take full advantage of the differences between word processing and handwriting (which necessarily follows a linear sequence and requires more planning):

- **Simply jot down your initial ideas for a plan,** preferably at paragraph topic level. The order can be altered easily and if a paragraph grows too much it can easily be split.
- **Start writing wherever you wish** and fill in the rest later.
- **Just put down your ideas as you think,** confident in the knowledge that it is the concepts that are important to note; their order and the way you express them can be adjusted later.
- **Do not worry about spelling and use of synonyms** – these can (and should) be checked during a separate revision of your text, using the spell-checker first to correct obvious mistakes, then the thesaurus to change words for style or meaning.
- **Use a draft printout to check (a) for pace and spacing** – difficult to correct for on-screen – and **(b) to ensure that words checked for spelling fit the required sense.**

Laying out (formatting) your document

The appearance of a document on-screen generally represents what the printout on paper will look like, but this should be checked using the 'print preview' feature before printing large documents, or where the paper size is non-standard. Menu commands can be used to adjust page margins, headers and footers, line spacing and fonts as required. Paragraph indentations should be created using the TAB or key rather than the space bar, but you can also indent or bullet whole blocks of text using special menu options. Although you can reformat your text at any time, it is good practice to enter the relevant details in the *Page Setup* when you start writing: entering them later can lead to problems due to forced reorganisation of the text layout. If you use a particular set of layout formats regularly, e.g. an A4 page with space for a letterhead, create a formatted template that can be called up whenever you start a new document. Advanced users can set up automated styles and formatting features to suit particular requirements (e.g. via *Home* tab, *Styles* group in *Word*).

Editing features

Word processors usually have many features designed to make editing documents easy. Most programs allow you to customise the toolbar to add commonly used command 'buttons' – some examples relevant to scientific writing include subscripts and superscripts, word count, A–Z sort and table column formatting. The spell-checkers and thesaurus are self-explanatory, but they generally do not include technical terms and,

---

Using online help, tutorials and textbooks – support information is usually provided in one or more of the following ways: as a help facility within the program; as a file on the installation CD; or as an online help support site. If unfamiliar with the software, it may be worthwhile investing in one of the commercial textbooks written to support users of specific programs.

Controlling widow and orphan lines – single lines or pairs of lines at the top or bottom of a page are referred to as 'widows' or 'orphans' respectively and they are generally undesirable in page design. You can control whether they appear in most word processors – for example, in *Word*, tick the *Widow/Orphan* control check box found under the *Line and Page Breaks* Tab in the *Paragraph* dialog box.

Using a spellcheck facility – do not rely on this to spot all errors. Remember that spell-check programs do not correct grammatical errors, nor do they pick up words that are incorrect but that are legitimate words e.g. 'their' for 'there', or 'form' for 'from'.

Language setting – remember to set up your document language before you start compiling text: this will ensure that the spell-check and thesaurus provide the correct spellings by default within your documents.

Quick commands for commonly used functions – most word processors provide several routes to achieve the same end result. These include: standard menus and commands; customisable toolbar commands; and key combinations. In many word processors, toolbar buttons can be added. Some useful examples of key combinations in *Word* include: *Ctrl + F* for find; *Ctrl + X* for delete, *Ctrl + P* for print, *Ctrl+C* for copy and *Ctrl + V* for paste.

Presenting your documents – it is good practice not to mix typefaces too much in a formal document; also the font size should not differ greatly for different headings, subheadings and the text.

unless set appropriately, usually suggest US spellings as default. Therefore, where a misspelling is 'incorrectly' indicated (e.g. in most versions of *Word,* by a red wavy underline) you may wish to add the word to the dictionary file.

Grammar checkers are designed to highlight text that may be incorrectly structured (e.g. in most versions of *Word,* this is shown by a green wavy underline). In general, these features are unreliable for scientific writing except when pointing out obvious errors, such as a missing verb. For example, academic style frequently demands the use of impersonal language involving passive tense and this is usually marked up as a matter for you to review, when it is perfectly acceptable. Note that the grammar-check tool is customisable in most programs (in *Word,* via the *Review* tab, *Proofing* group, *Spelling and Grammar* dialog box, *Options…* button).

The ability to move blocks of text ('cut and paste') is very valuable when reviewing and editing, and indeed can be said to have changed the nature of writing: it is now easy to write sections out of sequence, as availability of information or personal interest determines. However, there is an associated risk that the planning phase of writing (Chapter 16) is neglected, resulting in weak structure. To counteract this, you should create a plan and use temporary or permanent headings to organise your writing.

An extremely useful editing facility is the 'find' or 'search' procedure, which allows you to scan quickly through a document looking for a specified word, phrase or punctuation (in *Word,* available from the *Home* tab, *Editing* group). This is particularly valuable when combined with a replace facility so that, for example, you could replace the word 'test' with 'trial' throughout your document simply and rapidly. *Format Painter* is a convenient function found in *Word* that allows you to copy the format of one section to another, without altering the text. Most users will be familiar with the 'undo' command to rectify mistakes, but the 'redo' or 'repeat' command is equally valuable where repeated formatting or text entry is required.

Fonts and line spacing

The instructions for assignments will often define specific fonts and line spacing. The terminology involved is explained below.

- **Typeface**: the term for a family of characters of a particular design, each of which is given a particular name. The most commonly used for normal text is Times Roman (as used here for the main text) but many others are widely available, particularly for the better-quality printers. They fall into three broad groups: serif fonts with curves and flourishes at the ends of the characters (e.g. Times Roman); sans-serif fonts without such flourishes, providing a clean, modern appearance (e.g. Arial); and decorative fonts used for special purposes only, such as the production of newsletters and notices.

- **Size**: measured in units called points. There are about 28 points per cm. The standard sizes for text are 10, 11 and 12 point, but typefaces are often available up to 72 point or more.

- **Appearance**: many typefaces are available in a variety of styles and weights. Many of these are not designed for use in scientific writing, but for desktop publishing.

Preparing final documents – if no specific guidance is provided, use a 12–point proportional serif font, such as Times Roman. At this font size, 1.5 line spacing will generally be sufficient, unless otherwise specified. You can add contrast to headings by using a sans-serif typeface, such as Arial, at 11 point bold. You may wish to add extra blank lines before headings and additional half lines between paragraphs to improve readability by spacing out the text.

- **Spacing:** double or 1.5 line spacing may be specified to allow the marker to add comments or corrections. Character spacing is usually defined by the font type ('proportional' spacing, where each character takes up a different amount of space, being more readable). There are conventions for spacing after each sentence – a double space used to be the norm but single is now common. Paragraph indentation may be specified – a common format is not to indent the first paragraph in a section, but to indent thereafter (as here), but increasingly 'blocked' paragraphs are used with no indents.

- **Justification:** this is the term describing the way in which text is aligned vertically. Left justification is normal, but for formal documents, both left and right justification may be used (as here).

Table construction

Tables are widely used in the biosciences, but the conventions for their presentation are very specific (see Chapter 73). So, while you can produce basic tables via tabbing or by accessing the standard formats available on most word processor menus, this is rarely appropriate for formal submissions. Your main options here are to:

- **Use the in-built word-processor table-constructing commands.** Here, the basic table outline is constructed for you (e.g. four columns by ten rows) and can be adjusted thereafter to fit your presentational needs (e.g. by changing column widths, merging cells or deleting boundary lines)

- **Use a spreadsheet to construct the table and then copy it to the word processor file.** This requires considerably more manipulation than using the word processor directly and is best reserved for special circumstances, such as very large or complex tables of data, and especially where calculations or recalculations are involved and/or the data are already stored as a spreadsheet.

Box 73.2 provides further guidance about how to create such tables to suit your requirements using Microsoft *Office* programs.

Special characters, equations and graphics

You can draw lines and insert other small-scale graphical features such as brackets directly within most word processors, and special characters (e.g. mathematical symbols and Greek characters) should also be available. Word processors can be used to create simple charts and diagrams, but it is better to create graphs using a more flexible and fully featured application (e.g. a spreadsheet, see Box 72.2) then import the final output into the text of a document. Files must be compatible, as with 'office' suites, but if this is so, it is a relatively straightforward procedure. Microsoft *Equation Editor* may be available on your setup to assist with mathematical formulae – consult the help menu for guidance.

Inserting special characters – a wide range of symbols including Greek letters and characters from other languages are usually available within word processors.

Printing

If more than one printer is attached to your PC or network, you will need to specify which one to use from the word processor's print menu. Most printers offer choices as to text and graphics quality, so choose draft (low) quality for all but your final copy since this will save both time and materials. Use a print preview option to show the page layout if it is available. Assuming that you have entered appropriate layout and font commands, printing should then be a straightforward operation. Although document layout settings are modifiable at any time, changing the page, margin and font sizes will cause your text to be rearranged, and this can be frustrating if you have spent time carefully laying out the pages.

Databases

A database is an electronic filing system whose structure is similar to a manual record-card collection. Its collection of records is termed a file. The individual items of information on each record are termed fields. Once the database is constructed, search criteria can be used to view files through various filters according to your requirements. The computerised catalogues in your library are just such a system; you enter the filter requirements in the form of author or subject keywords.

You can use a database to catalogue, search, sort and relate collections of information. The benefits of a computerised database over a manual card-file system are:

- **the information content is easily amended or updated;**
- **printout of relevant items can be obtained;**
- **it is quick and easy to organise** through sorting and searching/selection criteria, to produce subgroups of relevant records;
- **record displays can easily be redesigned**, allowing flexible methods of presenting records according to interest;
- **relational databases can be combined**, giving the whole system immense flexibility. The older 'flat-file' databases store information in files that can be searched and sorted, but cannot be linked to other databases.

Simple database files can be constructed within spreadsheets using the columns and rows as fields and records respectively. These programs are capable of reasonably sophisticated sorting, searching and filtering operations and are probably sufficient for the types of database you are likely to require as an undergraduate. You may also make use of a bibliographic database specially constructed for that purpose (e.g. *EndNote,* see Chapter 9).

Statistical analysis packages

Statistical packages vary from small programs designed to carry out very specific statistical tasks to large sophisticated packages (e.g. *SYSTAT, SigmaStat, SPSS,* etc.) intended to provide statistical assistance, from experimental design to the analysis of results. Consider the following features when selecting a package:

- **The data entry and editing section** – this should be user-friendly, with options for transforming data.
- **Data exploration options** – these should include descriptive statistics and techniques for exploratory data analysis.
- **Hypothesis-testing techniques** – these should include ANOVA, regression analysis, multi-variate techniques, and parametric and non-parametric statistics.

Using spreadsheet statistics functions – before using a specific statistics package, check whether your spreadsheet is capable of carrying out the form of analysis you require (see Boxes 75.3 and 76.3), as this can often be the simpler option.

- **Support for research methods** – the program should provide assistance with experimental design and sampling.
- **Output facilities** – these should be suitable for graphical and tabular formats.

Some programs have very complex data entry systems, limiting the ease of using data in different tests. Ideally, the data entry and storage system should be based on a spreadsheet system, so that subsequent editing and transformation operations are straightforward.

> **KEY POINT** Make sure that you understand the theoretical basis for your statistical test and the computational techniques involved before using a particular program function.

Graphics and presentation packages

Microsoft *Office* programs or similar open source products can be used to achieve most coursework tasks requiring images, presentation slides or Web pages: *PowerPoint* is useful for creating posters (Box 14.1) and for oral presentations (Box 15.1). For more advanced tasks, additional software may be available on your network: for example:

- *SigmaPlot* can produce graphs with floating axes;
- *Adobe Illustrator* is useful for designing complex graphics;
- *Adobe Dreamweaver* enables you to produce high-quality Web pages;
- *MindGenius* can be used to produce Mind Maps.

Important points regarding the use of such packages are:

- **the learning time** required for some of the more complex operations can be considerable;
- **the quality of your printer** will limit the quality of your output;
- **not all files will readily import** into a word processor such as *Word* – you may need to save your work in a particular format. The different types of file are distinguished by the three-character filename extension, e.g. .jpg and .bmp.

Presentation using computer packages – though many computer programs enhance presentational aspects of your work, there are occasions when they can make your presentation worse. Take care to avoid the following common pitfalls:

- Default or 'chart wizard' settings for graphs may result in output that is unacceptable for the sciences (see Box 72.2).
- Fonts in labels and legends may not be consistent with other parts of your presentation.
- Some basic programs cannot produce Greek symbols (e.g. μ); do not use 'u' as a substitute. The same applies to scientific notation and superscripts: do not use 14C for ^{14}C, and replace, e.g., 1.4E + 09 with 1.4×10^9. First try cutting and pasting symbols from Word or, if this fails, leave space and draw the correct symbols by hand on the printout.

> **KEY POINT** 'Default' computer graphics are rarely satisfactory for scientific presentation. Make appropriate changes to suit scientific standards and style. Box 72.1 gives a checklist for graph drawing and Box 72.2 provides guidelines for adapting *Excel* output.

Image storage and manipulation

With the widespread use of digital images (see Chapter 50), programs that facilitate the storage and manipulation of electronic image files have become increasingly important. These programs create a library of your stored images and provide a variety of methods for organising and selecting images. Programs for image manipulation vary widely in capability, cost and associated learning time. *Adobe Photoshop* and *PaintShop Pro* are two widely available examples. Many are highly sophisticated programs intended for graphic artists. For most

Take care when copying images – you may be at risk of committing plagiarism and copyright infringement (Chapter 9).

scientific purposes, however, only relatively limited functions are required, such as image cropping and resizing. Care should be taken when manipulating images not to distort the information contained in the file (see Chapter 50).

Sources for further study

Apache. *OpenOffice.org*. Available at http://www.openoffice.org/
Last accessed 10/03/15.
[Website containing information and links for downloading freeware 'office' type programs.]

Fuller, L.U., Cook, K. and Kaufeld, J. (2010) *Access 2010 for Dummies*. Wiley, New York.

Gookin, D. (2010) *Word 2010 for Dummies*. Wiley, New York.

Rosenzweig. G. (2012) *My Pages (for Mac)*. Pearson Education Inc., USA.

Wang. W. (2010) *Office 2010 for Dummies*. Wiley, New York.

(And similar texts for other packages and release versions.)

STUDY EXERCISES

12.1 Investigate intermediate/advanced *Word* features. The tasks in the following list are likely to be useful in preparing assignments and report writing within the life sciences. Can you carry out all of the tasks? If not, use either a manual or the online *Help* feature to find out how to accomplish them. Tips are given in the answer section.

a) Sort information in a list into alphabetical order.

b) Replace a text string word or phrase with a new text string throughout your document.

c) Replace a text string in normal font with the same text string in italics throughout your document.

d) Add a 'header' and 'footer' to your document, the former showing the document's title and the latter containing page numbers on the bottom centre of the page.

e) Adjust the margins of the page to give a 5 cm margin on the left and a 2 cm margin on the right.

f) Change the type of bullets used in a list from standard (• or ■) to a different form (e.g. –, ◆ or ☑).

g) Use the 'thesaurus' option to find a different or more suitable word to express your meaning. Try, for example, to find alternatives to the word 'alternative'.

h) Carry out a spellcheck on your document.

i) Carry out a word count on your document and on a selected part of it.

j) Open two documents and switch between them.

12.2 Make precise copies of tables. Copy the following tables using a word processor such as *Word*.

Test organism	Results of analysis (units)		
	August	September	October
X			
Y			
Z			

Results of analysis (units)			
Test organism	August	September	October
X			
Y			
Z			

12.3 Investigate what programs and packages are available to you as a student. Test each program with appropriate data, images, etc.

Answers to these study exercises are available at www.pearsoned.co.uk/practicalskills

The spreadsheet is one of the most powerful and flexible computer applications. It can be described as the electronic equivalent of a paper-based long-hand calculation, where the sums are carried out automatically. Spreadsheets provide a dynamic method of storing, manipulating and analysing data sets. Advantages of spreadsheets include:

- **Ease and convenience** – especially when complex calculations are repeated on different sets of data.
- **Accuracy** – providing the entry data and cell formulae are correct, the result will be free of calculation errors.
- **Improved presentation** – data can be produced in graphical or tabular form to a high quality.
- **Integration with other programs** – graphs and tables can be exported to other compatible programs, such as a word processor in the same office suite.
- **Useful tools** – advanced features include hypothesis-testing statistics, database features and macros.

Spreadsheets can be used to:

- **manipulate raw data** by removing the drudgery of repeated calculations, allowing easy transformation of data and calculation of statistics;
- **display your data** as a graph or chart for evaluation and/or presentation – printouts can be used in practical and project reports;
- **carry out statistical analysis** using in-built procedures or by allowing construction of formulae for specific tasks;
- **model 'what if' situations** where the consequences of changes in data can be seen and evaluated;

A typical spreadsheet (Fig. 13.1) is divided into rows (identified by numbers) and columns (identified by alphabetic characters). Each individual combination of column and row forms a cell that can contain either a data item, a formula or a piece of text. Formulae can include scientific and/or statistical functions and/or a reference to other cells or groups of cells (often called a range). Complex systems of data input and analysis can be constructed. The analysis, in part or complete, can be printed out. New data can be added at any time and the sheet will recalculate automatically. The power a spreadsheet offers is related to your ability to create arrays of formulae that are accurate and templates that are easy to use.

Data entry

Spreadsheets have built-in commands that allow you to control the layout of data in the cells (see Fig. 13.2). These include number format, the number of decimal places to be shown (the spreadsheet always calculates using eight or more places), the cell width and the location of the entry within the cell (left, right or centre). An auto-entry facility assists greatly in entering large amounts of data by moving the entry cursor either vertically or horizontally as data are entered. Recalculation default is usually automatic so that when a new data value is entered the entire sheet is recalculated immediately.

Fig. 13.1 The appearance of a typical spreadsheet, showing cells, rows and columns, toolbars, etc.

Source: Screenshot of Microsoft *Excel 2007* reprinted by permission from Microsoft Corporation. Copyright © 2007 Microsoft Corporation.

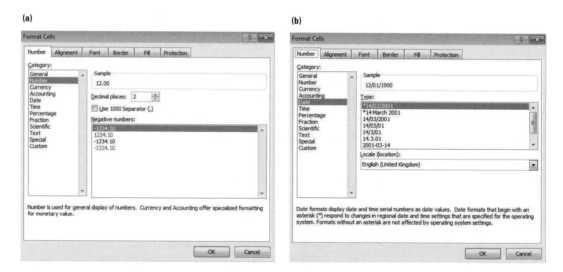

Fig. 13.2 Example of cell-formatting options within the Microsoft *Excel* spreadsheet. These menus are accessed *via* the *Format > Cell* option and would apply to all of a range of selected cells. (a) Use of the number-formatting option to specify that data will be presented to two decimal places (the underlying data will be held to greater accuracy). (b) Use of the date-formatting option to specify that dates will be presented in day/month/year format. (Spreadsheet dates are stored numerically and converted to appropriate formats. This allows a period between two dates to be calculated more easily.)

Source: Screenshot of Microsoft *Excel 2007* reprinted by permission from Microsoft Corporation. © 2007 Microsoft Corporation.

The parts of a spreadsheet

Labels

These should be used to identify parts of the spreadsheet – for example, stating what data are contained in a particular column or indicating that a cell's contents represent the end point of a calculation. It may be useful to

Using hidden columns – these are useful for storing intermediate calculations that you do not wish to be displayed on the screen or printout.

Operators and brackets in spreadsheets – the standard mathematical operators ÷ and × are usually replaced by / and * respectively, while ^ signifies 'to the power'. In complex formulae, brackets should be used to separate the elements, otherwise the results may not be what you expect. For example, *Excel* will calculate = A1*B1/C1–D1 differently from (A1*B1)/(C1–D1).

Definition

Function – a pre-programmed code for the transformation of values (mathematical or statistical functions) or selection of text characters (string functions).

Example = sin(A5) is an example of a function in *Excel*. If you write this in a cell, the spreadsheet will calculate the sine of the number in cell A5 (assuming it to be an angle in radians) and write it in the cell. Different programs may use a slightly different syntax.

Empty cells – note that these may be given the value 0 by the spreadsheet for certain functions. This may cause errors, e.g. by rendering a minimum value inappropriate. Also, an 'error return' may result for certain functions if the cell content is zero.

use the *Format* group options in the *Home* tab to delimit numerical sections of your spreadsheet. Note that spreadsheet programs have been designed to make assumptions about the nature of data entry being made. If the first character is a number, then the entry is treated as numerical data; if it is a letter, then it is treated as a text entry; and if it is a specific symbol ('=' in Microsoft *Excel*), then what follows is a formula. If you wish to enter text that starts with a number, then you must type a designated character to show this (a single quote mark in Microsoft *Excel*).

Numbers

You can also enter numbers (values) in cells for use in calculations. Many programs let you enter numbers in more than one way and you must decide which method you prefer. The way you enter the number does not affect the way it is displayed on the screen as this is controlled by the cell format at the point of entry. There are usually special ways to enter data for percentages, currency and scientific notation for very large and small numbers.

Formulae

These are the 'power tools' of the spreadsheet because they do the calculations. A cell can be referred to by its alphanumeric code, e.g. A5 (column A, row 5) and the value contained in that cell manipulated within a formula, e.g. = (A5 + 10) or (A5 + B22) in another cell. Formulae can include pre-programmed functions that can refer to a cell, so that if the value of that cell is changed, so is the result of the formula calculation. They may also include branching options through the use of logical operators (e.g. IF, TRUE, FALSE, OR, etc.).

Functions

Various functions may be offered, but only mathematical and statistical functions will be considered here.

Mathematical functions

Spreadsheets offer a range of options, including trigonometrical functions, angle functions, logarithms and random-number functions. Functions are invaluable for transforming sets of data rapidly and can be used in formulae required for more complex analyses. Spreadsheets work with an order of preference of the operators in much the same way as a standard calculator and this must always be taken into account when operators are used in formulae. They also require a very precise syntax – the program should warn you if you break this.

Statistical functions

Modern spreadsheets incorporate many sophisticated statistical functions, and if these are not appropriate, the spreadsheet can be used to facilitate the calculations required for most of the statistical tests found in textbooks. The descriptive statistics normally available include:

- **the sum of all data present in a column, row or block;**
- **the minimum and maximum of a defined range of cells;**
- **counts of cells** – a useful operation if you have an unknown or variable number of data values;

- **averages and other statistics describing location;**
- **standard deviations and other statistics describing dispersion.**

A useful function where you have large data sets allows you to create frequency distributions using predefined class intervals.

The hypothesis-testing statistical functions are usually reasonably powerful (e.g. *t*-test, ANOVA, regression) and they often return the *probability* (*P*) of obtaining the test statistic when the null hypothesis is true, so there may be no need to refer to statistical tables. Again, check on the effects of including empty cells.

Database functions

Many spreadsheets can be used as simple databases and offer a range of functions to support this, including filtering and sorting options. The rows and columns of the spreadsheet are used as the fields and records of the database (see Chapter 12). For many biological purposes, this form of database is perfectly adequate and should be seriously considered before using a full-feature database product.

Copying

All programs provide a means of copying (replicating) formulae or cell contents when required, and this is a very useful feature. This is usually accomplished by 'dragging' a cell's contents to a new range using the mouse. When copying, references to cells may be either relative, changing with the row/column as they are copied, or absolute, remaining a fixed cell reference and not changing as the formulae are copied (Fig. 13.3).

As an example, in *Excel,* copying is normally *relative* and will incrementally increase when copied: if you wish a cell reference to be *absolute* (fixed) when copied, this is done by putting a dollar ($) sign before and after the column reference letter, e.g. $C$56.

> **KEY POINT** The distinction between relative and absolute cell references is very important and must be understood; it provides one of the most common forms of error when copying formulae.

Naming blocks

When a group of cells (a block) is carrying out a particular function, it is often easier to give the block a name that can then be used in all formulae referring to that block. This powerful feature also allows the spreadsheet to be more readable.

Spreadsheet templates

A template is a preconstructed spreadsheet containing the formulae required for repeated data analysis. Data are added when they become available, and results are available as soon as the last item is entered. To create a template, the sequence of operations is:

1. **Determine what information/statistics you want to produce.**
2. **Identify the variables you will need to use,** both for original data that will be entered and for any intermediate calculations that might be required.

Statistical calculations – make sure you understand whether any functions you employ are for populations or samples (see p. 509).

Using text functions – these allow you to manipulate text within your spreadsheet and include functions such as 'search and replace' and alphabetical or numerical 'sort'.

(a)

	Cell	Formula	
Original→ cell	A1	=B1+C1	← Original formula
Copied cells	A2	=B2+C2	Copied formulae (relative)
	A3	=B3+C3	
	A4	=B4+C4	

(b)

	Cell	Formula	
Original→ cell	A1	=B1/$C$1	← Original formula
Copied cells	A2	=B2/$C$1	Copied formulae (mixed relative and absolute)
	A3	=B3/$C$1	
	A4	=B4/$C$1	

Fig. 13.3 Illustration of relative (a) and absolute (b) copying. In *Excel*, the $ sign before and after the column letter makes the cell reference absolute, as shown in (b).

Templates – these should contain:

- a data input section
- data transformation and/or calculation sections
- a results section, which can include graphics
- text in the form of headings and annotations
- a summary section.

3. **Set up areas of the spreadsheet for data entry**, calculation of intermediate values (statistical values such as sums of squares, etc.), calculation of final parameters/statistics and, if necessary, a summary area.

4. **Establish the format of the numeric data** if this is different from the default values. This can be done globally (affecting the entire spreadsheet) or locally (affecting only a specified part of the spreadsheet).

5. **Establish the column widths required** for the various activities.

6. **Add text (labels) to identify input**, intermediate formulae and output cells. This is valuable in error-tracking and when carrying out further development work. Text can be entered in designated cells, or cells can be annotated using the *New comments* feature in the *Review* tab.

7. **Enter a test set of values** to use during formula entry: use a fully worked example to check that formulae are working correctly.

8. **Enter the formulae required** to make all the calculations, both intermediate and final. Check that results are correct using the test data.

The spreadsheet is then ready for use. Delete all of the test-data values and you have created your template. Save the template to a disk and it is then available for repeated operations.

Graphics display

Most spreadsheets now offer a wide range of graph and chart options that are easy to use, and this represents an affective way to examine your data sets rapidly and comprehensively (Chapter 71). The quality of the final graphics output may be variable but is usually sufficient for data exploration and analysis. Many of the default options are business graphics styles but there are usually histogram, bar, column and pie charts, X–Y plotting, line and area graphics options available. Note that default spreadsheet graphics options often do not come up to the standards expected for the formal presentation of scientific data, unless you manipulate the option settings appropriately.

Printing spreadsheets

This is usually a straightforward menu-controlled procedure, made difficult only by the fact that your spreadsheet may be too big to fit on one piece of paper. Try to develop an area of the sheet that contains only the data that you need to print, e.g. perhaps a summary area. Remember that columns and/or rows can usually be hidden for printing purposes and you can control whether the printout is in portrait or landscape mode, and for page size and format. Use a screen-preview option, if available, to check your layout before printing. A 'print to fit' option is also available in some programs, making the output fit the page dimensions.

> Constructing a spreadsheet – start with a simple model and extend it gradually, checking for correct operation as you go.

> Preparing graphs using a spreadsheet such as Microsoft *Excel* – Box 72.2 gives step-by-step guidance for plotted curves, histograms and pie charts.

Sources for further study

Frye, C. (2010) *Microsoft Office Home and Student Step-by-Step.* Microsoft Press, Redmond.

Harvey, G. (2013) *Excel 2013 for Dummies.* Wiley, New York.

Miller, R. (2006) *Point & Click.* Openoffice.org. Prentice Hall, New Jersey.

(And similar texts for other release versions.)

STUDY EXERCISES

The instructions and tips for these problems assume that you have *Excel* (97 or later) available. If not, they should be readily modified for most advanced spreadsheet programs. If you have problems with any of the tasks, consult Box 72.2 or try using the program's *Help* facility.

13.1 Create a spreadsheet and graph (introductory).

(i) Copy the information in the table below into a spreadsheet. Name and save the spreadsheet file appropriately.

(ii) From the copied information, create a pie chart using the *Insert* tab and *Pie* (chart) function.

(iii) Adjust the colours selected so the chart will print out in black and white. Save the final version of your spreadsheet. Print the chart out directly from *Excel*.

Visits of birds to feeding area (% daily visits)

Species	% of visits
Blackbird	23
Song thrush	9
Sparrow	17
Starling	45
Other	6
Total	100

13.2 Create a spreadsheet and graph (advanced).

(i) Copy the data in the table above right into a spreadsheet. Name and save the file appropriately.

Time course of cell population growth

Time (h)	Cell count (per unit volume)
1	2.00
2	6.00
6	50.0
12	6.00×10^3
18	2.20×10^5
24	1.65×10^7

(ii) Use the spreadsheet and chart-making facilities to explore by eye which of the following transformations (see pp. 472–3) would result in the best linear fit for these data: reciprocal, square root, cube root or log.

(iii) Add a linear trend-line to the chart for the most appropriate transformation.

(iv) Copy the graph to a file in *Word* and print out.

13.3 Use a spreadsheet as a simple database.

Copy the data in the table below into cells within a spreadsheet. Modify the column widths so you can see all of the text on a single screen. Now sort the data in the following ways:

a) by subject, in alphabetical order;
b) by date and then by time of day;
c) by topic, in reverse alphabetical order.

Use the *Hide* function (right click on selected column header) so that the information in columns 4 and 6 is not displayed. Find out how to undo this operation.

My exam timetable

Subject	Date	Time	Paper	Location	Question style
Biochemistry	3 Jun	Morning	1	Great Hall	Multiple-choice
Biochemistry	17 Jun	Morning	2	Exam Hall 5	Essay paper
Biochemistry	2 Jun	Afternoon	3	Main Laboratory	Information processing
Ecology	3 Jun	Afternoon	A	Small Hall	Short-answer questions
Ecology	14 Jun	Afternoon	B	Exam Hall 5	Essay paper
Biology A	4 Jun	Morning	1	Small Hall	Short-answer questions
Biology B	1 Jul	Afternoon	1	Exam Hall 3	Short-answer questions
Biology A	13 Jun	Afternoon	2	Exam Hall 5	Essay paper
Biology B	2 Jun	Morning	2	Main Laboratory	Essay paper

Communicating information

A scientific poster is a visual display of the results of an investigation, usually mounted on a rectangular board. Posters are used in undergraduate courses, to display project results or assignment work, and at scientific meetings to communicate research findings.

In a written report you can include a reasonable amount of specific detail and the reader can go back and reread difficult passages. However, if a poster is long-winded or contains too much detail, your reader is likely to lose interest.

> **KEY POINT** A poster session is like a competition – you are competing for the attention of people in a room. Because you need to attract and hold the attention of your audience, make your poster as interesting as possible. Think of it as an advertisement for your work and you will not go far wrong.

Preliminaries

Before considering the content of your poster, you should find out:

- **the linear dimensions of your poster area**, typically up to 1.5 m wide by 1 m high;
- **the composition of the poster board** and the method of attachment, whether drawing pins, Velcro® tape, or some other form of adhesive; and whether these will be provided – in any case, it is safer to bring your own;
- **the time(s) when the poster should be set up and when you should attend;**
- **the room where the poster session will be held.**

Design

Plan your poster with your audience in mind, as this will dictate the appropriate level for your presentation. Aim to make your poster accessible to a broad audience. Since a poster is a *visual* display, you must pay particular attention to the presentation of information: work that may have taken hours to prepare can be ruined in a few minutes by the ill-considered arrangement of items (Fig. 14.1). Begin by making a draft sketch of the major elements of your poster. It is worth discussing your intended design with someone else, as constructive advice at the draft stage will save a lot of time and effort when you prepare the final version (or consult the Sources for further study on p. 97).

Layout

One approach is to divide the poster into several smaller areas, perhaps six or eight in all, and prepare each as a separate item on a piece of card. Alternatively, you can produce a single large poster on one sheet of paper or card and store it inside a protective cardboard tube. However, a single large poster may bend and crease, making it difficult to flatten out. In addition, photographs and text attached to the backing sheet may work loose; a large poster with embedded images is an alternative approach.

(a)

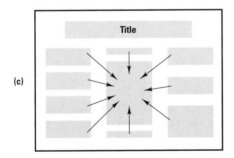

(b)

(c)

Fig. 14.1 Poster design. (a) An uninspiring design: subunits of equal area, reading left to right, are not recommended. (b) This design is more interesting and the text will be easier to read (column format). (c) Another approach, with a central focus and arrows to guide the reader.

Subdividing your poster means that each smaller area can be prepared on a separate piece of paper or card, of A4 size or slightly larger, making transport and storage easier. It also breaks the reading matter up into smaller pieces, looking less formidable to a potential reader. By using pieces of card of different colours you can provide emphasis for key aspects, or link text with figures or photographs.

You will need to guide your reader through the poster and headings/subheadings will help with this aspect. It may be appropriate to use either a numbering system, with large, clear numbers at the top of each piece of card, or a system of arrows (or thin tapes), to link sections within the poster (see Fig. 14.1). Make sure that the relationship is clear and that the arrows or tapes do not cross.

Title

Your chosen title should be concise (no more than eight words), specific and interesting, to encourage people to read the poster. Make the title large and bold – it should run across the top of your poster, in letters at least 4 cm high, so that it can be read from the other side of the room. Coloured spirit-based marker and block capitals drawn with a ruler work well, as long as your writing is readable and neat (the colour can be used to add emphasis). Alternatively, you can print out each word in large type, using a word processor. Details of authors, together with their addresses (if appropriate), should be given, usually across the top of the poster in somewhat smaller lettering than the title.

Text

Write in short sentences and avoid verbosity. Keep your poster as visual as possible and make effective use of the spaces between the blocks of text. Your final text should be double-spaced and should have a minimum capital-letter height of 8 mm (minimum type size 36 point), preferably greater, so that the poster can be read at a distance of 1 m. One method of obtaining text of the required size is to photo-enlarge standard typescript (using a good-quality photocopier), or use a high-quality (laser) printer. It is best to avoid continuous use of text in capitals, since it slows reading and makes the text less interesting to the reader. Also avoid italic, 'balloon' or decorative styles of lettering.

> **KEY POINT**   Keep text to a minimum – aim to have a *maximum* of 500 words in your poster.

Subtitles and headings

These should have a capital-letter height of 12–20 mm, and should be restricted to two or three words. They can be produced by word processor, photo-enlargement, by stencilling, or by hand, using pencilled guidelines (but make sure that no pencil marks are visible on your finished poster).

Colour

Consider the overall visual effect of your chosen display, including the relationship between your text, diagrams and the backing board. Colour can be used to highlight key aspects of your poster. However, it is very easy to ruin a poster by the inappropriate choice and application of colour. Careful use of two, or at most three, complementary colours and shades

Presenting a poster at a formal conference – it can be useful to include your photograph for identification purposes, e.g. in the top right-hand corner of the poster.

Making up your poster – text and graphics printed on good-quality paper can be glued directly onto a contrasting mounting card: use photographic spray mountant or Pritt rather than liquid glue. Trim carefully using a guillotine to give equal margins, parallel with the paper. Photographs should be placed in a window mount to avoid the tendency for their corners to curl. Another approach is to trim pages or photographs to their correct size, then encapsulate in plastic film: this gives a highly professional finish and is easy to transport.

Producing composite material for posters – *PowerPoint* is generally more useful than *Word* when you wish to include text, graphics and/or images on the same page. It is possible to use *PowerPoint* to produce a complete poster (Box 14.1), although it can be expensive to have this printed out commercially to A1 or A0 size.

will be easier on the eye and should aid comprehension. Colour can be used to link the text with the visual images (e.g. by picking out a colour in a photograph and using the same colour on the mounting board for the accompanying text). For *PowerPoint* posters, careful choice of colours for the various elements will enhance the final product (Box 14.1). Use coloured inks or water-based paints to provide colour in diagrams and figures, as felt pens rarely give satisfactory results.

Content

The typical format is that of a scientific report (see Box 18.1), i.e. with the same headings, but with a considerably reduced content. Keep references within the text to a minimum – interested parties can always ask you for further information. Also note that most posters have a summary/conclusions section at the end, rather than an abstract.

Introduction

This should give the reader background information on the broad field of study and the aims of your own work. It is vital that this section is as interesting as possible, to capture the interest of your audience. It is often worth listing your objectives as a series of numbered points.

Materials and methods

Keep this short, and describe only the principal techniques used. You might mention any special techniques, or problems of general interest.

Results

Do not present your raw data: use data reduction wherever possible, i.e. figures and simple statistical comparisons. Graphs, diagrams, histograms and pie charts give clear visual images of trends and relationships and should be used in place of data tables (see p. 478). Final copies of all figures should be produced so that the numbers can be read from a distance of 1 m. Each should have a concise title and legend, so that it is self-contained: if appropriate, a series of numbered points can be used to link a diagram with the accompanying text. Where symbols are used, provide a key on each graph (symbol size should be at least 5 mm). Avoid using graphs straight from a written version, e.g. a project report, textbook, or a paper, without considering whether they need modification to meet your requirements.

Conclusions

This is where many readers will begin, and they may go no further unless you make this section sufficiently interesting. This part needs to be the strongest part of your poster, summarising the main points. Refer to your figures here to draw the reader into the main part of your poster. A slightly larger or bolder typeface may add emphasis, though too many different typefaces can look messy. For references, smaller type can be used.

The poster session

A poster session may be organised as part of the assessment of your coursework, and this usually mirrors those held at most scientific conferences and meetings. Staff and fellow students (delegates at conferences) will mill around, looking at the posters and chatting to their authors, who are usually expected to be in attendance. If you stand at the side of your poster throughout, you are likely to discourage some readers, who may not wish to become involved in a detailed conversation about

Presenting at a scientific meeting – never be tempted to spend the minimum amount of time converting a piece of scientific writing into poster format – the least interesting posters are those where the author simply displays pages from a written communication (e.g. a journal article) on a poster board.

Designing the materials and methods section – photographs or diagrams of apparatus can help to break up the text of this section and provide visual interest. It is sometimes worth preparing this section in a smaller typeface.

Keeping graphs and diagrams simple – avoid composite graphs with different scales for the same axis, or with several trend lines (use a maximum of three trend lines per graph).

Listing your conclusions – a series of numbered points is a useful approach, if your findings fit this pattern.

Consider providing a handout – this is a useful way to summarise the main points of your poster, so that your readers have a permanent record of the information you have presented.

Box 14.1 How to create a poster using *PowerPoint*

Software such as *PowerPoint* can be used to produce a high-quality poster, providing you have access to a good colour printer. However, you should avoid the standard templates available on the Web as they encourage unnecessary uniformity and stifle creativity, leading to a less satisfying end result. The following steps give practical advice on creating a poster as a single *PowerPoint* slide.

1. **Sketch out your plans.** Decide on the main poster elements (images, graphs, tables and text sections) and their relationship with each other and draw out a one-page 'storyboard' (see Fig. 14.1). Think about colours for background, text and graphics (use two or three complementary colours) – dark text on a light background is clearer (high contrast), and uses less ink when printing. Also consider how you will link the elements in sequence, to guide readers through your 'story'.

2. **Get your material ready.** Collect together individual files for pictures, figures and tables. Make any required adjustments to images, graphs or tables before you import them into your poster.

3. **Create a new/blank slide.** Open *PowerPoint*. From the *Home* tab, select *New Slide* and then *Blank*. Then use the *Design* tab > *Page setup* group to select either *Landscape* or *Portrait* orientation and to set the correct page size (use *Width* and *Height* commands, or select a standard size like A4, A3, A2, etc.). Right-click on the slide and select *Ruler* and *Grid and Guides* (to help position elements within the slide – the horizontal and vertical guidelines can be dragged to different positions at later stages, as required). Use the templates (*Themes*) on the top line menu or select an appropriate background style and colour. In general, avoid setting a picture as your background as these tend to detract from the content of the poster. Before going further, save your work. Repeat this frequently and in more than one location (e.g. hard drive and USB memory stick).

4. **Add graphics.** For images, use the *Insert* tab, select *Picture* and browse to *Insert* the correct file. The *Insert, Object* command performs a similar function for *Excel* charts (graphs). Alternatively, use the copy-and-paste functions of complementary software. Once inserted, resize using the sizing handles in one of the corners (for photographs, take care not to alter one dimension relative to the other, or the image will be distorted). To reposition, put the mouse pointer over the image, left-click and hold, then drag to new location. While the *Insert* tab offers standard shapes and other useful features, you should avoid clipart (jaded and overused) and poor-quality images from the Web (always use the highest resolution possible) – if you do not have your final images, use blank text boxes to show their position within the poster.

5. **Add text.** Use either the *Insert* tab to select a *Text Box* and place this on your slide, then either type in your text (use the *Enter* key to provide line spacing within the box) or copy-and-paste text from a word-processed file. You will need to consider the font size for the printed poster (e.g. for an A0 poster (size 1189 × 841 mm), a printed font size of 24 point is appropriate for the main text, with larger fonts for headings and titles. If you find things difficult to read on-screen, use the *Zoom* function (either select a larger percentage in the *Zoom* box on the bottom right toolbar, or hold down the *Control* key and use the mouse wheel to scroll up (magnify) or down (reduce). Use a separate text box for each element of your poster and do not be tempted to type too much text into each box – write in phrases, using bullet points and numbered lists to keep text concise (aim for no more than 50 words per text box). Select appropriate font styles and colours using the *Home* tab, *Font* menu. For a background colour or surrounding line, right-click and use the *Format Text Effects* box and *Format Shape* box commands (line thickness and colour can then be altered). Present supplementary text elements in a smaller font – for example, details of methodology, references cited.

6. **Add boxes, lines and/or arrows** to link elements of the poster and guide the reader (e.g. Fig. 14.1). These features are available from the *Home* tab, *Drawing* group drop-down menu. Note that new inserts are overlaid on older inserts – if this proves to be a problem, select the relevant item and use the *Arrange, Order* functions to change its relative position.

7. **Review your poster.** Get feedback from another student or tutor, e.g. on a small printed version, or use a projector to view your poster without printing (adjust the distance between projector and screen to give the correct size).

8. **Revise and edit your poster.** Revisit your work and remove as much text as possible. Delete any component that is not essential to the message of the poster. Keep graphs simple and clear (p. 475 gives further advice). White space is important in providing structure.

9. **Print the final version.** Use a high-resolution colour printer (this may be costly, so you should wait until you are sure that no further changes are needed).

Note: PowerPoint commands illustrated here may vary among the various versions of this Microsoft *Office* program.

Coping with questions in assessed poster sessions – you should expect to be asked questions about your poster, and to explain details of figures, methods, etc.

the poster. Stand nearby. Find something to do – talk to someone else, or browse among the other posters, but remain aware of people reading your poster and be ready to answer any queries they may raise. Do not be too discouraged if you are not asked lots of questions: remember, the poster is meant to be a self-contained, visual story, without need for further explanation.

A poster display will never feel like an oral presentation, where the nervousness beforehand is replaced by a combination of satisfaction and relief as you unwind after the event. However, it can be a very satisfying means of communication, particularly if you follow these guidelines.

Sources for further study

Alley, M. (2007) *The Craft of Scientific Presentations: Critical Steps to Succeed and Critical Errors to Avoid*, 5th edn. Springer-Verlag, New York.

Briscoe, M.H. (1996) *Preparing Scientific Illustrations: a Guide to Better Posters, Presentations and Publications*, 2nd edn. Springer-Verlag, New York.

Davis, M.F. (2004) *Scientific Papers and Presentations*, 2nd edn. Academic Press, New York.

Gosling, P.J. (1999) *Scientist's Guide to Poster Presentations*. Kluwer, New York.

Hess, G., Tosney, K. and Liegel, L. *Creating Effective Poster Presentations*. Available: http://www.ncsu.edu/project/posters/
Last accessed 15/03/15.

Nicol, A.A.M. (2010) *Displaying Your Findings*, 6th edn. American Psychological Association, Washington.

Simmonds, D. and Reynolds, L. (1994) *Data Presentation and Visual Literacy in Medicine and Science*. Butterworth–Heineman, London.

STUDY EXERCISES

14.1 Design a poster. Working with one or more partners from your year group, decide on a suitable topic (perhaps something linked to your current teaching programme). Working individually, make an outline plan of the major elements of the poster, with appropriate subheadings and a brief indication of the content and relative size of each element (including figures, diagrams and images). Exchange draft plans with your partners and arrange a session where you can discuss their merits and disadvantages.

14.2 Prepare a checklist for assessing the quality of a poster presentation. After reading through this chapter, prepare a 10-point checklist of assessment criteria under the heading 'What makes a good poster presentation?'. Compare your list with the one that we have provided (see website below) – do you agree with our criteria, or do you prefer your own list (can you justify your preferences)?

14.3 Evaluate the posters in your university. Most universities have a wide range of academic posters on display. Some may cover general topics (e.g. course structures), while others may deal with specific research topics (e.g. poster presentations from past conferences). Consider their good and bad features (if you wish to make this a group exercise, you might compare your evaluation with that of other students in a group discussion session).

Answers to these study exercises are available at www.pearsoned.co.uk/practicalskills

Most students feel very nervous about giving talks. This is natural, since very few people are sufficiently confident and outgoing that they look forward to speaking in public. Additionally, the technical nature of the subject matter may give you cause for concern, especially if you feel that some members of the audience have a greater knowledge than you have. However, this is a fundamental method of scientific communication and an important transferable skill; therefore, it forms an important component of many courses.

The comments in this chapter apply equally to informal talks, e.g. those based on assignments and project work, and to more formal conference presentations. It is hoped that the advice and guidance given below will encourage you to make the most of your opportunities for public speaking, but there is no substitute for practice. Do not expect to find all of the answers from this, or any other, book. Rehearse, and learn from your own experience.

> **KEY POINT** The three 'Rs' of successful public speaking are: reflect – give sufficient thought to all aspects of your presentation, particularly at the planning stage; rehearse – to improve your delivery; revise – modify the content and style of your material in response to your own ideas and to the comments of others.

Preparation

Preliminary information

Begin by marshalling the details needed to plan your presentation, including:

- **the duration of the talk;**
- **whether time for questions is included;**
- **the size and location of the room;**
- **the projection/lighting facilities provided,** and whether pointers or similar aids are available.

It is especially important to find out whether the room has the necessary equipment for digital projection (e.g. PC, projector and screen, black-out curtains or blinds, appropriate lighting) or overhead projection before you prepare your audio-visual aids. If you concentrate only on the spoken part of your presentation at this stage, you are inviting trouble later on. Have a look around the room and try out the equipment at the earliest opportunity, so that you are able to use the lights, projector, etc., with confidence. For digital projection systems, check that you can load/present your material. Box 15.1 gives advice on using *PowerPoint*.

Audio-visual aids

If you plan to use overhead transparencies, find out whether your department has facilities for their preparation, whether these facilities are available for your use, and the cost of materials. Adopt the following guidelines:

- **Keep text to a minimum:** present only the key points, with up to 20 words per slide/transparency.

Opportunities for practising speaking skills – these include:

- answering lecturers' questions
- contributing in tutorials
- talking to informal groups
- giving your views at formal (committee) meetings
- demonstrating or explaining to other students, e.g. during a practical class
- asking questions in lectures/seminars
- answering an examiner's questions in an oral exam.

Learning from experience – use your own experience of good and bad lecturers to shape your performance. Some of the more common errors include:

- speaking too quickly
- reading from notes or from slides and ignoring the audience
- unexpressive, impersonal or indistinct speech
- distracting mannerisms
- poorly structured material with little emphasis on key information
- factual information too complex and detailed
- too few or too many visual aids.

Testing the room – if possible, try to rehearse your talk in the room in which it will be presented. This will help you to make allowance for layout of equipment, lighting, acoustics and sight lines that might affect the way you deliver your talk. It will also put you more at ease on the day, because of the familiarity of the surroundings.

Box 15.1 Tips on preparing and using *PowerPoint* slides in a spoken presentation

Microsoft *PowerPoint* can be used to produce high-quality visual aids, assuming a computer and digital projector are available in the room where you intend to speak. The presentation is produced as a series of electronic 'slides' on to which you can insert images, diagrams and text. When creating your slides, bear the following points in mind:

- **Plan the structure of your presentation.** Decide on the main topic areas and sketch out your ideas on paper. Think about what material you will need (e.g. pictures, graphs) and what colours to use for background and text.

- **Choose slide layouts according to purpose.** Once *PowerPoint* is running, from the *Home* tab select *Layout*. You can then add material to each new slide to suit your requirements.

- **Select your background with care.** Many of the preset background templates (*Themes*) available within the *Design* tab are best avoided, since they are overused and fussy, diverting attention from the content of the slides. Conversely, flat, dull backgrounds may seem uninteresting, while brightly coloured backgrounds can be garish and distracting. Choose whether to present your text as a light-coloured type on a dark background (more restful but perhaps less engaging if the room is dark) or a dark-coloured type on a light background (more lively).

- **Use visual images throughout.** Remember the Chinese proverb 'a picture is worth ten thousand words'. A presentation composed entirely of text-based slides will be uninteresting: adding images and diagrams will brighten up your talk considerably (use the *Insert* menu, *Picture* option). Images can be taken with a digital camera (see Chapter 50), scanned in from a printed version or copied and pasted from the Web, but you should take care not to break copyright regulations. 'Clip art' is copyright-free, but should be used sparingly, as most people will have seen the images before and they are rarely wholly relevant. Diagrams can be made from components created using the *Home*, *Drawing* group, and graphs and tables can be imported from other programs, e.g. *Excel* (Box 14.1 gives further specific practical advice on adding graphics, saving files, etc.).

- **Keep text to a minimum.** Aim for no more than 20 words on a single slide (e.g. four/five lines containing a few words per line). Use headings and subheadings to structure your talk: write only key words or phrases as 'prompts' to remind you to cover a particular point during your talk – never be tempted to type whole sentences as you will

then be reduced to reading these from the screen during your presentation, which is boring.

- **Use a large, clear font.** Use the *Slide Master* option within the *View* menu to set the default font to a non-serif style such as Arial, or Comic Sans MS. Default fonts for headings and bullet points are intentionally large, for clarity. Do not reduce these to anything less than 28-point type size (preferably larger), to cram in more words: if you have too much material, create a new slide and divide up the information.

- **Animate your material.** The *Animation* tab provides a *Custom Animation* function that enables you to introduce the various elements within a slide, e.g. text can be made to *Appear* one line at a time, to prevent the audience from reading ahead and to help maintain their attention.

- **Do not overdo the special effects.** *PowerPoint* has a wide range of features that allow complex slide transitions and animations, additional sounds, etc., but these quickly become irritating to an audience unless they have a specific purpose within your presentation.

- **Always edit your slides before use.** Check through your slides and cut out any unnecessary words, adjust the layout and animation. Remember the maxim 'less is more' – avoid too much text; too many bullet points; too many distracting visual effects or sounds.

When presenting your talk:

- **Work out the basic procedures beforehand.** Practise, to make sure that you know how to move forwards and backwards, turn the screen on and off, hide the mouse pointer, etc.

- **Do not forget to engage your audience.** Despite the technical gadgetry, *you* need to play an active role in their presentation, as explained elsewhere in this chapter.

- **Do not go too fast.** Sometimes, new users tend to deliver their material too quickly: try to speak at a normal pace, and practise beforehand.

- **Consider whether to provide a handout.** *PowerPoint* has several options, including some that provide space, for notes (e.g. Fig. 4.3). However, a handout should not be your default option, as there is a cost involved.

Note: PowerPoint commands illustrated here may vary among the various versions of this Microsoft *Office* program.

Using audio-visual aids – do not let equipment and computer gadgetry distract you from the essential rules of good speaking (pp. 98, 102). Remember that you are the presenter.

- **Make sure the text is readable: try out your material beforehand.**
- **Use several simpler figures rather than a single complex graph.**
- **Avoid too much colour** on overhead transparencies: blue and black are easier to read than red or green.
- **Do not mix slides and transparencies** as this is often distracting.
- **Use spirit-based pens for transparencies:** use alcohol for corrections.
- **Transparencies can be produced from typewritten or printed text using a photocopier**, often giving a better product than pens. Note that you must use special heat-resistant acetate sheets for photocopying.

Audience

You should consider your audience at the earliest stage, since they will determine the appropriate level for your presentation. If you are talking to fellow students you may be able to assume a common level of background knowledge. In contrast, a research lecture given to your department, or a paper at a meeting of a scientific society, will be presented to an audience from a broader range of backgrounds. An oral presentation is not the place for a complex discussion of specialised information: build up your talk from a low level. The speed at which this can be done will vary according to your audience. As long as you are not boring or patronising, you can cover basic information without losing the attention of the more knowledgeable members in your audience.

Pitching your talk at the right level – the general rule should be: 'do not overestimate the background knowledge of your audience'. This sometimes happens in student presentations, where fears about the presence of 'experts' can encourage the speaker to include too much detail, overloading the audience with facts.

Content

Although the specific details in your talk will be for you to decide, most spoken presentations share some common features of structure, as described below.

Introductory remarks

It is vital to capture the attention of your audience at the outset. Consequently, you must make sure your opening comments are strong, otherwise your audience will lose interest before you reach the main message. Remember it takes a sentence or two for an audience to establish a relationship with a new speaker. Your opening sentence should be some form of preamble and should not contain any key information. For a formal lecture, you might begin with 'Thank you for that introduction. My talk today is about …' then restate the title and acknowledge other contributors, etc. You might show a transparency or slide with the title printed on it, or an introductory photograph, if appropriate. This should provide the necessary settling-in period.

Getting the introduction right – a good idea is to have an initial slide giving your details and the title of your talk, and a second slide telling the audience how your presentation will be structured. Make eye contact with all sections of the audience during the introduction.

After these preliminaries, you should introduce your topic. Begin your story on a strong note – avoid timid or apologetic phrases.

Opening remarks are unlikely to occupy more than 10% of the talk. However, because of their significance, you might reasonably spend up to 25% of your preparation time on them.

What to cover in your introductory remarks – you should:

- explain the structure of your talk;
- set out your aims and objectives;
- explain your approach to the topic.

> **KEY POINT** Make sure you have practised your opening remarks so that you can deliver the material in a flowing style, with fewer chances of mistakes.

The main message

This section should include the bulk of your experimental results or literature findings, depending on the type of presentation. Keep details of methods to the minimum needed to explain your data. This is *not* the place for a detailed description of equipment and experimental protocol (unless it is a talk about methodology). Results should be presented in an easily digested format.

> **KEY POINT** Do not expect your audience to cope with large amounts of data; use a maximum of six numbers per slide. Remember that graphs and diagrams are usually better than tables of raw data, since the audience will be able to see the visual trends and relationships in your data.

Present summary statistics (Chapter 75) rather than individual results. Show the final results of any analyses in terms of the statistics calculated, and their significance (p. 516), rather than dwelling on details of the procedures used. Figures should not be crowded with unnecessary detail. Every diagram should have a concise title and the symbols and trend lines should be clearly labelled, with an explanatory key where necessary. When presenting graphical data (Chapter 72) always 'introduce' each graph by stating the units for each axis and describing the relationship for each trend line or data set.

> **KEY POINT** Use summary slides at regular intervals, to maintain the flow of the presentation and to emphasise the main points.

Take the audience through your story step by step at a reasonable pace. Try not to rush the delivery of your main message due to nervousness. Avoid complex, convoluted storylines – one of the most distracting things you can do is to fumble backwards through *PowerPoint* slides or overhead transparencies. If you need to use the same diagram or graph more than once then you should make two (or more) copies. In a presentation of experimental results, you should discuss each point as it is raised, in contrast to written text, where the results and discussion may be in separate sections. The main message typically occupies approximately 80% of the time allocated to an oral presentation (Fig. 15.1).

Concluding remarks

Having captured the interest of your audience in the introduction and given them the details of your story in the middle section, you must now bring your talk to a conclusion. Do not end weakly, e.g. by running out of steam on the last slide. Provide your audience with a clear 'take-home message' by returning to the key points in your presentation. It is often appropriate to prepare a slide or overhead transparency listing your main conclusions as a numbered series.

Signal the end of your talk by saying 'finally ...', 'in conclusion ...', or a similar comment and then finish speaking after that sentence. Your audience will lose interest if you extend your closing remarks beyond this point. You may add a simple end phrase (for example, 'thank you') as you put your notes into your folder, but do not say 'that's all folks!', or make any similar offhand remark. Finish as strongly and as clearly as you started. Box 15.2 gives further advice.

Allowing time for slides – as a rough guide you should allow at least two minutes per illustration, although some diagrams may need longer, depending on content. Make a note of the half-way point to help you check timing/pace.

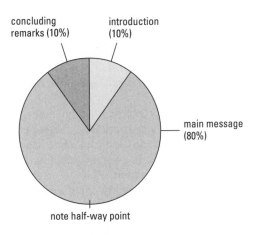

Fig. 15.1 Pie chart showing time allocation for a typical presentation.

Final remarks – make sure you give the audience sufficient time to assimilate your final slide: some of them may wish to write down the key points. Alternatively, you might provide a handout, with a brief outline of the aims of your study and the major conclusions.

Box 15.2 **Hints on spoken presentations**

In planning the delivery of your talk, bear the following aspects in mind:

- **Using notes.** Many accomplished speakers use abbreviated notes for guidance, rather than reading word for word from a prepared script. When writing your talk:

  (i) Consider preparing your first draft as a full script: write in spoken English and keep the text simple, to avoid a formal, impersonal style. Your aim should be to *talk* to your audience, not to *read* to them.

  (ii) If necessary, use note-cards with key words and phrases: it is best to avoid using a full script in the final presentation. As you rehearse and your confidence improves, a set of note-cards may be an appropriate format. Mark the position of slides/key points, etc.: each note-card should contain details of structure as well as content. Your notes should be written/printed in text large enough to be read easily during the presentation (also check that the lecture room has a lectern light or you may have problems reading your notes if the lights are dimmed). Each note-card or sheet should be clearly numbered, so that you do not lose your place.

  (iii) Decide on the layout of your talk: give each subdivision a heading in your notes, so that your audience is made aware of the structure.

  (iv) Memorise your introductory/closing remarks: you may prefer to rely on a full written version for these sections, in case your memory fails, or if you suffer 'stage fright'.

  (v) Using *PowerPoint* (Box 15.1): here, you can either use the 'notes' option (*View, Notes Page*), or you may even prefer to dispense with notes entirely, since the slides will help structure your talk, acting as an aide-mémoire for your material.

- **Work on your timing.** It is essential that your talk is the right length and the correct pace:

  (i) Rehearse your presentation: ask a friend to listen and to comment constructively on those parts that were difficult to follow, to improve your performance.

  (ii) Use 'split times' to pace yourself: following an initial run-through, add the times at which you should arrive at the key points of your talk to your notes.

These timing marks will help you keep to time during the final presentation.

  (iii) Avoid looking at your wristwatch when speaking; this sends a negative signal to the audience. Use a wall clock (where available), or take off your watch and put it beside your notes so that you can glance at it without distracting the audience.

- **Consider your image.** Make sure that the image you project is appropriate for the occasion:

  (i) Think about what to wear: aim to be respectable without 'dressing up', otherwise your message may be diminished.

  (ii) Develop a good posture: it will help your voice projection if you stand upright, rather than slouching, or leaning over a lectern.

  (iii) Deliver your material with expression: project your voice towards the audience at the back of the room and make sure you look round to make eye contact with all sections of the audience. Arm movements and subdued body language will help maintain the interest of your audience. However, you should avoid extreme gestures (it may work for some TV personalities but it is not recommended for the beginner).

  (iv) Try to identify and control any repetitive mannerisms: repeated 'empty' words/phrases, fidgeting with pens, keys, etc., will distract your audience. Note-cards held in your hand give you something to focus on, whereas laser pointers will show up any nervous hand tremors. Practising in front of a mirror may help.

- **Think about questions.** Once again, the best approach is to prepare beforehand:

  (i) Consider what questions are likely to come up, and prepare brief answers. However, do not be afraid to say 'I don't know': your audience will appreciate honesty, rather than vacillation, if you do not have an answer for a particular question.

  (ii) If no questions are asked, you might pose a question yourself and then ask for opinions from the audience: if you use this approach, you should be prepared to comment briefly if your audience has no suggestions, to avoid the presentation ending in an embarrassing silence.

Sources for further study

Alley, M. (2007) *The Craft of Scientific Presentations: Critical Steps to Succeed and Critical Errors to Avoid*, 5th edn. Springer-Verlag, New York.

Capp, C.C. and Capp, G.R. (1989) *Basic Oral Communication,* 5th edn. Prentice Hall, Harlow.

Hofmann, A.H. (2010) *Scientific Writing and Communication: Papers, Proposals and Presentations.* Oxford University Press, New York.

Jones, A.M. (2003). *The Use and Abuse of PowerPoint in Teaching and Learning in the Life Sciences: A Personal Overview.* Available at: https://www.heacademy.ac.uk/use-and-abuse-powerpoint-teaching-and-learning-life-sciences-personal-overview
Last accessed 22/03/15.

Levin, P. and Topping, G. (2006) *Perfect Presentations!* Open University Press, Maidenhead.

Radel, J. *Oral Presentations.* Available at: http://people.eku.edu/richisong/oralpres.html
Last accessed 22/03/15.

STUDY EXERCISES

15.1 Prepare a checklist for assessing the quality of an oral presentation. After reading through this chapter, prepare a 10-point checklist of assessment criteria under the heading 'What makes a good oral presentation?'. Compare your list with the one that we have provided (see website) – do you agree with our criteria, or do you prefer your checklist? Can you justify your preferences?

15.2 Evaluate the presentation styles of other speakers. There are many opportunities to assess the strengths and weaknesses of academic 'public speakers', including your university lecturers, seminar speakers, presenters of TV documentaries, etc. Decide in advance how you are going to tackle the evaluation (e.g. with a quantitative marking scheme, or a less formal procedure).

15.3 Rehearse a talk and get feedback on your performance. There are a number of approaches you might take, including: (i) recording and reviewing your presentation using a digital camera; or (ii) giving your talk to a small group of fellow students and asking them to provide constructive feedback.

Answers to these study exercises are available at www.pearsoned.co.uk/practicalskills

Monday:	morning	Lectures (University)
	afternoon	Practical (University)
	evening	Initial analysis and
		brainstorming (Home)
Tuesday:	morning	Lectures (University)
	afternoon	Locate sources (Library)
	evening	Background reading (Library)
Wednesday:	morning	Background reading (Library)
	afternoon	Squash (Sports hall)
	evening	Planning (Home)
Thursday:	morning	Lectures (University)
	afternoon	Additional reading (Library)
	evening	Prepare outline (Library)
Friday:	morning	Lab class (University)
	afternoon	Write first draft (Home)
	evening	Complete first draft, go out
		with friends (When finished)
Saturday:	morning	Shopping (Town)
	afternoon	Review first draft (Home)
	evening	Revise first draft (Home)
Sunday:	morning	Free
	afternoon	Produce final copy (Home)
	evening	Proofread then print
		essay (Home)
Monday:	morning	Final read-through and check
		Submit essay
		(deadline midday)

Fig. 16.1 Example timetable for writing a short essay.

Written communication is an essential component of all sciences. Most courses include writing exercises in which you will learn to describe ideas and results accurately, succinctly and in an appropriate style and format. The following features are common to all forms of scientific writing.

Organising your time

Making a timetable at the outset helps ensure that you give each stage adequate attention and complete the work on time (e.g. Fig. 16.1). To create and use a timetable:

1. **Break down the task into stages.**
2. **Decide on the proportion of the total time each stage should take.**
3. **Set realistic deadlines for completing each stage**, allowing some time for slippage.
4. **Refer to your timetable frequently as you work**: if you fail to meet one of your deadlines, make a serious effort to catch up as soon as possible.

> **KEY POINT** The appropriate allocation of your time to reading, planning, writing and revising will differ according to the task in hand (see Chapters 17–19).

Organising your information and ideas

Before you write, you need to gather and/or think about relevant material (Chapters 9 and 10). You must then decide:

- **what needs to be included and what does not;**
- **in what order it should appear.**

Start by jotting down headings for everything of potential relevance to the topic (this is sometimes called 'brainstorming'). A spider diagram (Fig. 16.2) or a Mind Map (Fig. 4.2) will help you organise these ideas. The next stage is to create an outline of your text (Fig. 16.3). Outlines are valuable because they:

- **force you to think about and plan the structure;**
- **provide a checklist so nothing is missed out;**
- **ensure the material is balanced in content and length;**
- **help you organise figures and tables**, by showing where they will be used.

Creating an outline – an informal outline can be made simply by indicating the order of sections on a spider diagram (as shown in Fig. 16.2).

> **KEY POINT**   A suitable structure is essential to the narrative of your writing, and should be carefully considered at the outset.

Benefits of talking about your work – discussing your topic with a friend or colleague might bring out ideas or reveal deficiencies in your knowledge.

In an essay or review, the structure of your writing should help the reader to assimilate and understand your main points. Subdivisions of the topic could simply be related to the physical nature of the subject matter (e.g. zones of an ecosystem) and should proceed logically (e.g. low-water mark to high-water mark). A chronological approach is good for evaluation of past work (e.g. the development of the concept of

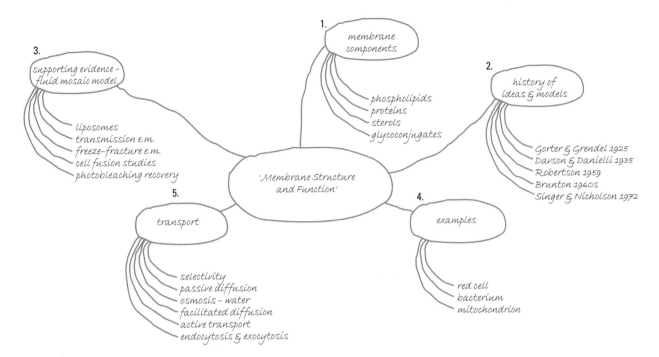

Fig. 16.2 Spider diagram showing how you might 'brainstorm' an essay with the title 'Membrane Structure and Function'. Write out the essay title in full to form the spider's body, and as you think of possible content, place headings around this to form the spider's legs. Decide which headings are relevant and which are not, and use arrows to note connections between subjects. This may influence your choice of order and may help to make your writing flow because the links between paragraphs will be natural. You can make an informal outline directly on a spider diagram by adding numbers indicating a sequence of paragraphs (as shown). This method is best when you must work quickly, as with an essay written under exam conditions.

Fig. 16.3 Formal outlines. These are useful for a long piece of work where you or the reader might otherwise lose track of the structure. The headings for sections and paragraphs are simply written in sequence with the type of lettering and level of indentation indicating their hierarchy. Two different forms of formal outline are shown, a minimal form (a) and a numbered form (b). Note that the headings used in an outline are often repeated within the essay to emphasise its structure. The content of an outline will depend on the time you have available and the nature of the work, but the most detailed hierarchy you should reasonably include is the subject of each paragraph.

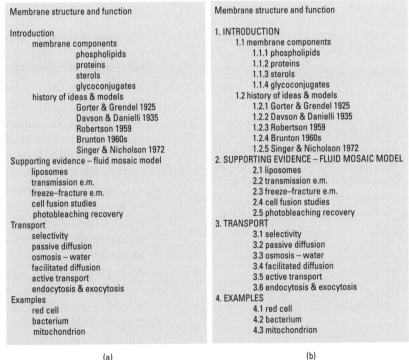

(a)

(b)

evolution), whereas a step-by-step comparison might be best for certain exam questions (e.g. 'Discuss the differences between prokaryotes and eukaryotes'). There is little choice about structure for practical and project reports (see p. 115).

Writing

Adopting a scientific style

Your main aim in developing a scientific style should be to get your message across directly and unambiguously. Although you can try to achieve this through a set of 'rules' (see Box 16.1), you may find other requirements driving your writing in a contradictory direction. For instance, the need to be accurate and complete may result in text littered with technical terms, and the flow may be continually interrupted by references to the literature. The need to be succinct also affects style and readability through the use of, for example, stacked noun-adjectives (e.g. 'restriction fragment length polymorphism') and acronyms (e.g. 'RFLP'). Finally, style is very much a matter of taste and each tutor, examiner, supervisor or editor will have pet loves and hates that you may have to accommodate. Different assignments will need different styles; Box 16.2 gives further details.

Developing technique

Writing is a skill that can be improved, but not instantly. You should analyse your deficiencies with the help of feedback from your tutors, be prepared to change work habits (e.g. start planning your work more carefully, find out more about punctuation), and willing to learn from some of the excellent texts that are available on scientific writing (p. 110).

Getting started

A common problem is 'writer's block' – inactivity or stalling brought on by a variety of causes. If blocked, ask yourself these questions:

- **Are you comfortable with your surroundings?** Make sure you are seated comfortably at a reasonably clear desk and have minimised the possibility of interruptions and distractions.
- **Are you trying to write too soon?** Have you clarified your thoughts on the subject? Have you done enough preliminary reading?
- **Are you happy with the underlying structure of your work?** If you have not made an outline, try this. If you are unhappy because you cannot think of a particular detail at the planning stage, just start writing – it is more likely to come to you while you are thinking of something else.
- **Are you trying to be too clever?** Your first sentence does not have to be earth-shattering in content or particularly smart in style. A short statement of fact or a definition is fine. If there will be time for revision, first get your ideas down on paper and then revise grammar, content and order later.
- **Do you really need to start writing at the beginning?** Try writing the opening remarks after a more straightforward part. For example, with reports of practical work, the Materials and Methods section may be the easiest place to start.
- **Are you too tired to work?** Do not try to 'sweat it out' by writing for long periods at a stretch: stop frequently for a rest.

Improving your writing skills – you need to take a long-term view if you wish to improve this aspect of your work. An essential preliminary is to invest in and *make full use of* a personal reference library (see Box 16.3).

Writing with a word processor – use the dynamic/interactive features of the word processor (Chapter 12) to help you get started: first make notes on structure and content, then expand these to form a first draft and finally revise/improve the text.

Box 16.1 How to achieve a clear, readable style

Words and phrases

- Choose short, clear words and phrases rather than long ones: e.g. use 'build' rather than 'fabricate'; 'now' rather than 'at the present time'. At certain times, technical terms must be used for precision, but do not use jargon if you do not have to.

- Do not worry too much about repeating words, especially when to introduce an alternative might subtly alter your meaning.

- Where appropriate, use the first person to describe your actions ('We decided to'; 'I conclude that'), but not if this is specifically discouraged by your supervisor.

- Favour active forms of writing ('the observer completed the survey in ten minutes') rather than a passive style ('the survey was completed by the observer in ten minutes').

- Use tenses consistently. Past tense is always used for Materials and Methods ('samples were taken from . . .') and for reviewing past work ('Smith (1990) concluded that. . .'). The present tense is used when describing data ('Fig. 1 shows. . .'), for generalisations ('Most authorities agree that. . .') and conclusions ('To conclude, . . .').

- Use statements in parentheses sparingly – they disrupt the reader's attention to your central theme.

- Avoid clichés and colloquialisms – they are usually inappropriate in a scientific context.

Punctuation

- Try to use a variety of types of punctuation, to make the text more interesting to read.

- Decide whether you wish to use 'closed' punctuation (frequent commas at the end of clauses) or 'open' punctuation (less frequent punctuation) – be consistent.

- Do not link two sentences with a comma. Use a full stop, this is an example of what not to do.

- Pay special attention to apostrophes, using the following rules:
  - To indicate possession, use an apostrophe before an 's' for a singular word (e.g. the rat's temperature was. . .') and after the s for a plural word ending in s (e.g. the rats' temperatures were = the temperatures of the rats were). If the word has a special plural (e.g. woman → women) then use the apostrophe before the s (the women's temperatures were. . .).
  - When contracting words, use an apostrophe (e.g. do not = don't; it is = it's), but remember that contractions are generally not used in formal scientific writing.
  - Do not use an apostrophe for 'its' as the possessive form of 'it' (e.g. 'the university and its surroundings') Note that 'it's' is reserved for 'it is'. This is an exception to the general rule and a very common mistake.
  - Never use an apostrophe to indicate plurals of any kind, including abbreviations.

Sentences

- Do not make them overlong or complicated.

- Introduce variety in structure and length.

- If unhappy with the structure of a sentence, try chopping it into a series of shorter sentences.

Paragraphs

- Get the paragraph length right – five sentences or so. Do not submit an essay that consists of a single paragraph, nor one that contains many single sentence paragraphs.

- Make sure each paragraph is logical, dealing with a single topic or theme.

- Take care with the first sentence in a paragraph (the 'topic' sentence); this introduces the theme of the paragraph. Further sentences should then develop this theme, e.g. by providing supporting information, examples, or contrasting cases.

- Use 'linking' words or phrases to maintain the flow of the text within a paragraph (e.g. 'for example'; 'in contrast'; 'however'; 'on the other hand').

- Make your text more readable by adopting modern layout style. The first paragraph in any section of text is usually not indented, but following paragraphs may be (by the equivalent of three character spaces). In addition, the space between paragraphs should be slightly larger than the space between lines. Follow departmental guidelines if these specify a format.

- Group paragraphs in sections under appropriate headings and subheadings to reinforce the structure underlying your writing.

- Think carefully about the first and last paragraphs in any piece of writing: these are often the most important as they respectively set the aims and report the conclusions.

Note: If you are not sure what is meant by any of the terms used here, consult a guide on writing (see p. 110).

Box 16.2 Using appropriate writing styles for different purposes (with examples)

Note that courses tend to move from assignments that are predominantly descriptive in the early years to a more analytical approach towards the final year (see Chapter 5). Also, different styles may be required in different sections of a write-up, e.g. descriptive for introductory historical aspects, becoming more analytical in later sections.

Descriptive writing

This is the most straightforward style, providing factual information on a particular subject, and is most appropriate:

- in essays where you are asked to 'describe' or 'explain' (p. 112);

- when describing the results of a practical exercise, e.g.: 'The experiment shown in Figure 1 confirmed that enzyme activity was strongly influenced by temperature, as the rate observed at 37 °C was more than double that seen at 20 °C.'

However, in literature reviews and essays where you are asked to 'discuss' (p. 112) a particular topic, the descriptive approach is mostly inappropriate, as in the following example, where a large amount of specific information from a single scientific paper has been used without any attempt to highlight the most important points:

In a study carried out between July and October 2005, a total of 225 sputum samples from patients attending 25 different clinics in England and Wales were screened. Bacteria were isolated from 67.6% of these samples, with 47.42% of the samples giving *Pseudomonas aeruginosa*, 11.76% *Burkholderia cepacia* and 8.59% *Stentrophomonas maltophilia* (Grey and Gray, 2009).

In the most extreme examples, whole paragraphs or pages of essays may be based on descriptive factual detail from a single source, often with a single citation at the end of the material, as above. Such essays often score low marks in essays where evidence of deeper thinking is required (Chapter 5).

Comparative writing

This technique is an important component of academic writing, and it will be important to develop your comparative writing skills as you progress through your course. Its applications include:

- answering essay questions and assignments of the 'compare and contrast' type (p. 112);

- comparing your results with previously published work in the Discussion section of a practical report.

To use this style, first decide on those aspects you wish to compare and then consider the material (e.g. different literature sources) from these aspects – in what ways do they agree or disagree with each other? One approach is to compare/contrast a different aspect in each paragraph. At a practical level, you can use 'linking' words and phrases to help orientate your reader as you move between aspects where there is agreement and disagreement. These include, for agreement: 'in both cases'; 'in agreement with'; 'is also shown by the study of'; 'similarly'; 'in the same way'; and for disagreement: 'however'; 'although'; 'in contrast to'; 'on the other hand'; 'which differs from'. The comparative style is fairly straightforward, once you have decided on the aspects to be compared. The following brief example compares two different studies using this style:

While Grey and Gray (2009) reported that *Pseudomonas aeruginosa* was present in 47.4% of 225 UK sputum samples, Black and White (2010) showed that 89.1% of sputum samples from 2592 patients were positive for this bacterium.

Comparative text typically makes use of two or more references per paragraph.

Analytical writing

Typically, this is the most appropriate form of writing for:

- a review of scientific literature on a particular topic;

- an essay where you are asked to 'discuss' (p. 112) different aspects of a particular topic;

- evaluating a number of different published sources within the Discussion section of a final-year project dissertation.

By considering the significance of the information provided in the various sources you have read, you will be able to take a more critical approach. Your writing should evaluate the importance of the material in the context of your topic (see also Chapter 10). In analytical writing, you need to demonstrate critical thinking (p. 64) and personal input about the topic in a well-structured text that provides clear messages, presented in a logical order and demonstrating synthesis from a number of sources by appropriate use of citations (p. 55). Detailed information and relevant examples are used only to explain or develop a particular aspect, and not simply as 'padding' to bulk up the essay. The following example shows how detail can be used for explanation.

Pseudomonas aeruginosa is often isolated from sputum samples of cystic fibrosis patients: a short-term UK study with a relatively small size sample (225 patients) isolated this bacterium from around half of all samples (Grey and Gray, 2009), while a longer-term study with a far larger sample size (2592 patients) gave an isolation rate of almost 90% (Black and White, 2010).

Analytical writing is based on a broad range of sources, typically with several citations per paragraph.

Box 16.3 **Improve your writing ability by consulting a personal reference library**

Using dictionaries

We all know that a dictionary helps with spelling and definitions, but how many of us use one effectively? You should:

- Keep a dictionary beside you when writing and always use it if in any doubt about spelling or definitions.
- Use it to prepare a list of words that you have difficulty in spelling: apart from speeding up the checking process, the act of writing out the words helps commit them to memory.
- Use it to write out a personal glossary of terms. This can help you memorise definitions. From time to time, test yourself.

Not all dictionaries are the same! Ask your tutor or supervisor whether he/she has a preference and why. Try out the *Oxford Advanced Learner's Dictionary*, which is particularly useful because it gives examples of use of all words and helps with grammar, e.g. by indicating which prepositions to use with verbs. Dictionaries of biology tend to be variable in quality, possibly because the subject is so wide and new terms are continually being coined. *Henderson's Dictionary of Biological Terms* (Benjamin Cummings) is a useful example.

Using a thesaurus

A thesaurus contains lists of words of similar meaning grouped thematically; words of opposite meaning always appear nearby.

- Use a thesaurus to find a more precise and appropriate word to fit your meaning, but check definitions of unfamiliar words with a dictionary.
- Use it to find a word or phrase 'on the tip of your tongue' by looking up a word of similar meaning.
- Use it to increase your vocabulary.

Roget's Thesaurus is the standard. Several publishers produce a combined dictionary and thesaurus.

Using guides for written English

Guides provide help with the use of words.

- Use them to solve grammatical problems such as when to use 'shall' or 'will', 'which' or 'that', 'effect' or 'affect', 'can' or 'may', etc.
- Use them for help with the paragraph concept and the correct use of punctuation.
- Use them to learn how to structure writing for different tasks.

Learning from others – ask a colleague to read through your draft and comment on its content and overall structure.

Reviewing your text

Detailed review of each draft is strongly advised for all writing, apart from in exams. Using a word processor, this can be a simple process. Where possible, schedule your writing so you can leave each draft to 'settle' for at least a couple of days. When you return to it fresh, you will see more easily where improvements can be made. Try the following structured review process, each stage being covered in a separate scan of your text:

1. **Examine content.** Have you included everything you need to? Is all the material relevant?
2. **Check the grammar and spelling.** Can you spot any 'howlers'?
3. **Focus on clarity.** Is the text clear and unambiguous? Does each sentence really say what you want it to say?
4. **Be succinct.** What could be missed out without spoiling the essence of your work? It might help to imagine an editor has set you the target of reducing the text by 15%.
5. **Improve style.** Could the text read better? Consider the sentence and paragraph structure and the way your text develops to its conclusion.

Reviewing your text – to improve clarity and shorten your text, 'distil' each sentence by taking away unnecessary words and 'condense' words or phrases by choosing a shorter alternative.

Common errors

These include (with examples):

- **Problems over singular and plural words** ('a bacteria is'; 'the results shows').

- **Verbose text** ('One definition that can be employed in this situation is given in the following sentence').
- **Misconstructed sentences** ('Health and safety regulations should be made aware of').
- **Misuse of punctuation,** especially commas and apostrophes (for examples, see Box 16.1).
- **Poorly constructed paragraphs** (for advice/examples, see Box 16.1).

Sources for further study

Butterfield, J. (ed.) (2015) *Fowler's Dictionary of Modern English Usage*, 4th edn. Oxford University Press, Oxford.

Clark, R. *The English Style Book. A Guide to the Writing of Scholarly English*. Available: http://www.litencyc.com/stylebook/stylebook.php Last accessed 16/02/15.

Kane, T.S. (1994) *The New Oxford Guide to Writing*. Oxford University Press, New York. [This is excellent for the basics of English – it covers grammar, usage and the construction of sentences and paragraphs.]

Matthews, J.R. and Matthews, R.W. (2014) *Successful Scientific Writing. A Step-by-Step Guide for the Biological and Medical Sciences*, 4th edn. Cambridge University Press, Cambridge.

McMillan, K.M. and Weyers, J.D.B. (2011) *How to Write Essays and Assignments,* 2nd edn. Pearson Education, London.

McMillan, K.M. and Weyers, J.D.B. (2013) *The Study Skills Book,* 3rd edn. Pearson, Harlow.

Partridge, E. (1978) *You Have a Point There*. Routledge, London. [This covers punctuation in a very readable manner.]

Pechnick, J.A. and Lamb, B.C. (1996) *How to Write about Biology*. HarperCollins, London.

STUDY EXERCISES

16.1 'Brainstorm' an essay title. Pair up with a partner in your class. Together, pick a suitable essay title from a past exam paper. Using the spider diagram or another technique, individually 'brainstorm' the title. Meet afterwards, compare your ideas, and discuss their relative merits and disadvantages.

16.2 Improve your writing technique. From the following checklist, identify the *three* weakest aspects of your writing, either in your own opinion or from essay/assignment feedback:

- grammar;
- paragraph organisation;
- presentation of work;
- punctuation;
- scientific style;
- sentence structure/variety;
- spelling;
- structure and flow;
- vocabulary.

Now either borrow a book from a local library or buy a book that deals with your weakest aspects of writing. Read the relevant chapters or sections and for each aspect write down some tips that should help you in future.

16.3 Improve your spelling and vocabulary with two lists. Create a pair of lists and pin these up beside your desk. One should be entitled *Spelling Mistakes* and the other *New Words*. Now, whenever you make a mistake in spelling or have to look up how to spell a word in a dictionary, add the problem word to your spelling list, showing where you made the mistake. Also, whenever you come across a word whose meaning is unclear to you, look it up in a dictionary and write the word and its meaning in the 'new words' vocabulary list.

Answers to these study exercises are available at www.pearsoned.co.uk/practicalskills

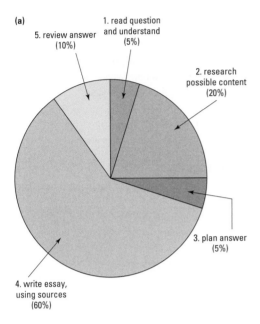

(a)
1. read question and understand (5%)
5. review answer (10%)
2. research possible content (20%)
3. plan answer (5%)
4. write essay, using sources (60%)

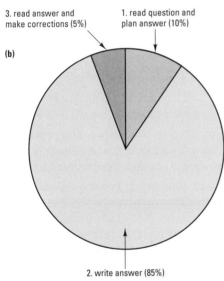

(b)
3. read answer and make corrections (5%)
1. read question and plan answer (10%)
2. write answer (85%)

Fig. 17.1 Typical division of time for an essay written (a) as part of in-course assessment or (b) under exam conditions.

> **Considering essay content** – it is rarely enough simply to lay down facts for the reader – you must analyse them and comment on their significance (see p. 24).

The main functions of an essay are to show how much you understand about a topic and how well you can organise and express your knowledge.

Organising your time

The way you should divide your time when producing an essay depends on whether you are writing it for in-course assessment or under exam conditions (Fig. 17.1). Essays written over a long period with access to books and other resources will probably involve a research element, not only before the planning phase but also when writing (Fig. 17.1(a)). For exams, it is assumed that you have revised appropriately (Chapter 6) and essentially have all the information at your fingertips. To keep things uncomplicated, the time allocated for each essay should be divided into three components – planning, writing and reviewing (Fig. 17.1(b)), and you should adopt time-saving techniques whenever possible (Box 7.2).

Making a plan for your essay

Dissect the meaning of the essay question or title

Read the title very carefully and think about the topic before starting to write. Consider the definitions of each of the important nouns (this can help in approaching the introductory section). Also think about the meaning of the verb(s) used and try to follow each instruction precisely (see Table 17.1). Do not get sidetracked because you know something about one word or phrase in the title: consider the whole title and all its ramifications. If there are two or more parts to the question, make sure you give adequate attention to each part.

Consider possible content and examples

Research content using the methods described in Chapters 9 and 10. If you have time to read several sources, consider their content in relation to the essay title. Can you spot different approaches to the same subject? Which do you prefer as a means of treating the topic in relation to your title? Which examples are most relevant to your case, and why?

> **KEY POINT** Most marks for essays are lost because the written material is badly organised or is irrelevant. An essay plan, by definition, creates order and, if thought about carefully, should ensure relevance.

Construct an outline

Every essay should have a structure related to its title. Your plan should be written down (but scored through later if written in an exam book). Think about an essay's content in three parts:

1. **The introductory section,** in which you should include definitions and some background information and context for the topic being considered. You should also tell your reader how you plan to approach the subject.

Ten Golden Rules for essay writing – these are framed for in-course assessments (p. 35), though many are also relevant to exams (see also Box 7.2).

1. Read the question carefully, and decide exactly what the assessor wants you to achieve in your answer.

2. Make sure you understand the question by considering all aspects – discuss your approach with colleagues or a tutor.

3. Carry out the necessary research (using books, journals, the web), taking appropriate notes. Gain an overview of the topic before getting involved with the details.

4. Always plan your work in outline before you start writing. Check that your plan covers the main points and that it flows logically.

5. Introduce your essay by showing that you understand the topic and stating how you intend to approach it.

6. As you write the main content, ensure it is relevant by continually looking back at the question.

7. Use headings and subheadings to organise and structure your essay.

8. Support your statements with relevant examples, diagrams and references where appropriate.

9. Conclude by summarising the key points of the topic, indicating the present state of knowledge, what we still need to find out and how this might be achieved.

10. Always reread your essay before submitting it. Check grammar and spelling and confirm that you have answered all aspects of the question.

Using diagrams – give a title and legend to each diagram so that it makes sense in isolation and point out in the text when the reader should consult it (e.g. 'as shown in Fig. 1 . . .' or 'as can be seen in the accompanying diagram, . . .').

Table 17.1 Instructions often used in essay questions and their meanings. When more than one instruction is given (e.g. compare and contrast; describe and explain), make sure you carry out *both* or you may lose a large proportion of the available marks (see also Table 5.1)

Account for:	give the reasons for
Analyse:	examine in depth and describe the main characteristics of
Assess:	weigh up the elements of and arrive at a conclusion about
Comment:	give an opinion on and provide evidence for your views
Compare:	bring out the similarities between
Contrast:	bring out dissimilarities between
Criticise:	judge the worth of (give both positive and negative aspects)
Define:	explain the exact meaning of
Describe:	use words and diagrams to illustrate
Discuss:	provide evidence or opinions about, arriving at a balanced conclusion
Enumerate:	list in outline form
Evaluate:	weigh up or appraise; find a numerical value for
Explain:	make the meaning of something clear
Illustrate:	use diagrams or examples to make clear
Interpret:	express in simple terms, providing a judgement
Justify:	show that an idea or statement is correct
List:	provide an itemised series of statements about
Outline:	describe the essential parts only, stressing the classification
Prove:	establish the truth of
Relate:	show the connection between
Review:	examine critically, perhaps concentrating on the stages in the development of an idea or method
State:	express clearly
Summarise:	without illustrations, provide a brief account of
Trace:	describe a sequence of events from a defined point of origin

2. **The middle of the essay**, where you develop your answer and provide relevant examples. Decide whether a broad analytical approach is appropriate or whether the essay should contain more factual detail.

3. **The conclusion,** which you can make quite short. You should use this part to summarise and draw together the components of the essay, without merely repeating previous phrases. You might mention such things as: the broader significance of the topic; its future; its relevance to other important areas of biology. Always try to mention both sides of any debate you have touched on, but beware of 'sitting on the fence'.

KEY POINT Use paragraphs to make the essay's structure obvious. Emphasise them with headings and subheadings unless the material beneath the headings would be too short or trivial.

Start writing

- **Never lose track of the importance of content and its relevance.** Repeatedly ask yourself: 'Am I really answering this question?' Never waffle just to increase the length of an essay. Quality, rather than quantity, is important.

- **Illustrate your answer appropriately.** Use examples to make your points clear, but remember that too many similar examples can stifle the flow of an essay. Use diagrams where a written description would be difficult or take too long. Use tables to condense information.

- **Where a submission is hand-written, take care with your handwriting.** You cannot get marks if your writing is illegible. Try to cultivate an open form of handwriting, making the individual letters large and distinct. If there is time, make out a rough draft from which a tidy version can be copied.

Reviewing your answer

Make sure that you leave enough time to:

- **reread the question** to check that you have answered all points;
- **reread your essay** to check for errors in punctuation, spelling and content. Make any corrections obvious. In an exam do not panic if you suddenly realise you have missed a large chunk out as the reader can be redirected to a supplementary paragraph if necessary.

> **Learning from lecturers' and tutors' comments** – ask for further explanations if you do not understand a comment or why an essay was less successful than you thought it should have been.

Sources for further study

Good, S. and Jensen, B. (1996) *The Student's Only Survival Guide to Essay Writing*. Orca Book Publishers, Victoria.

McMillan, K.M. and Weyers, J.D.B. (2011) *How to Write Essays and Assignments*, 2nd edn. Pearson Education, London.

McMillan, K.M. and Weyers, J.D.B. (2013) *The Study Skills Book,* 3rd edn. Prentice Hall, Harlow.

Royal Literary Fund. *Writing Essays*. Available: http://www.rlf.org.uk/resource/writing-essays Last accessed 17/02/15.
[An accessible, informal guide.]

STUDY EXERCISES

17.1 Practise dissecting essay titles. Using past exam papers, take each essay title and carefully 'dissect' the wording, working out exactly what you think the assessor expects you to do (see e.g. Table 17.1).

17.2 Write essay plans under self-imposed time limits. Continuing from Table 17.1, outline plans for essays from a past exam paper. Allow yourself a maximum of 5 minutes per outline. Within this time your main goal is to create an essay plan. To do this, you may need to 'brainstorm' the topic. Alternatively, if you allocate 10 minutes per essay, you may be able to provide more details, e.g. list the examples you could describe.

17.3 Practise reviewing your work carefully. For the next assignment you write, review it fully as part of the writing process. This will require you to finish the first draft about one week before the hand-in date, e.g. by setting yourself an earlier deadline than the submission date. This exercise is best done with a word-processed essay. Do not worry if it is a little over the word limit at this stage.

(a) Print out a copy of the essay. Do not look at it for at least two days after finishing this version.

(b) Review 1: spelling, grammar and sense. Read through the draft critically (try to imagine it had been written by someone else) and correct any obvious errors that strike you. Does the text make sense? Do sentences/paragraphs flow smoothly?

(c) Review 2: structure and relevance. Consider again the structure of the essay, asking yourself whether you have really answered the question that was set. Are all the parts in the right order? Is anything missed out? Have you followed precisely the instruction(s) in the title? Are the different parts of the essay linked together well?

(d) Review 3: shorten and improve style. Check the word count. Shorten the essay if required. Look critically at phrasing and, even if the essay is within the word limit, ask yourself whether any of the words are unnecessary or whether the text could be made more concise, more precise or more apt.

Answers to these study exercises are available at www.pearsoned.co.uk/practicalskills

Practical reports, project reports, theses and scientific papers differ greatly in depth, scope and size, but they all have the same basic structure. Some variation is permitted, however (see Box 18.1), and you should always follow the advice or rules provided by your department.

Additional parts may be specified: for project reports, dissertations and theses, a Title Page is often required and a List of Figures and Tables as part of the Contents section. When work is submitted for certain degrees, you may need to include declarations and statements made by the student and supervisor. In scientific papers, a list of Keywords is often added following the Abstract: this information may be combined with words in the title for computer cross-referencing systems.

> **KEY POINT** Department, school or faculty regulations may specify a precise format for your report or thesis. Consult these rules at an early stage and follow them closely to avoid losing marks.

Practical and project reports

These are exercises designed to make you think more deeply about your experiments and to practise and test the skills necessary for writing up research work. Special features are:

- **Introductory material is generally short** and unless otherwise specified should outline the aims of the experiment(s) with a minimum of background material.
- **Materials and methods may be provided by your supervisor for practical reports.** If you make changes to this, you should state clearly what you did. With project work, your lab notebook (see p. 209) should provide the basis for writing this section.
- **Great attention in assessment will be paid to presentation and analysis of data.** Take special care over graphs (see p. 475 for further advice). Make sure your conclusions are justified by the evidence you present.

Theses and dissertations

These are submitted as part of the examination for a degree following an extended period of research. They act to place on record full details about your experimental work and will normally only be read by those with a direct interest in it – your examiners or colleagues. Note the following:

- **The Introduction and Materials and Methods Sections are lengthy and detailed** providing the full context for your research.
- **You are allowed scope to expand on your findings** and to include detail that might otherwise be omitted in a scientific paper.
- **You may have problems with the volume of information** that has to be organised. One method of coping with this is to divide your thesis into chapters, each having the standard format (as in Box 18.1). A General Introduction can be given at the start and a General Discussion at the end. Discuss this with your supervisor.

Basic structure of scientific reports – this usually follows the 'IMRAD' acronym: Introduction, Materials and Methods, Results and Discussion.

Options for discussing data – the main optional variants of the general structure include combining *Results* and *Discussion* into a single section and adding a separate *Conclusions* section.

- The main advantage of a joint *Results and Discussion* section is that you can link together different experiments, perhaps explaining why a particular result led to a new hypothesis and the next experiment. However, a combined *Results and Discussion* section may contravene your department's regulations, so you should check before using this approach.
- The main advantage of having a separate *Conclusions* section is to draw together and emphasise the chief points arising from your work, when these may have been 'buried' in an extensive *Discussion* section.

Box 18.1 The structure of reports of experimental work

Undergraduate practical and project reports are generally modelled on this arrangement or a close variant of it, because this is the structure used for nearly all research papers and theses. The more common variations include Results and Discussion combined into a single section and Conclusions appearing separately as a series of points arising from the work. In scientific papers, a list of Keywords (for computer cross-referencing systems) may be included following the Abstract. Acknowledgements may appear after the Contents section, rather than near the end. Department or faculty regulations for producing theses and reports often require a Title Page to be inserted at the start and a List of Figures and Tables as part of the Contents section, and may specify declarations and statements to be made by the student and supervisor.

Part (in order)	Contents/purpose	Checklist for reviewing content
Title	Explains what the project was about	❑ Does it explain what the text is about succinctly? ❑ Are all the details correct?
Authors plus their institutions	Explains who did the work and where; also where they can be contacted now	
Abstract/ Summary	Synopsis of methods, results and conclusion of work described. Allows the reader to grasp quickly the essence of the work	❑ Does it explain why the work was done? ❑ Does it outline the whole of your work and your findings?
Contents	Shows the organisation of the text (not required for short papers)	❑ Are all the sections covered? ❑ Are the page numbers correct?
Abbreviations	Lists all the abbreviations used (but not those of SI, chemical elements, or standard biochemical terms)	❑ Have they all been explained? ❑ Are they all in the accepted form? ❑ Are they in alphabetical order?
Introduction	Orientates the reader, explains why the work has been done and its context in the literature, why the methods used were chosen, why the experimental organisms were chosen. Indicates the central hypothesis behind the experiments	❑ Does it provide enough background information and cite all the relevant references? ❑ Is it of the correct depth for the readership? ❑ Have all the technical terms been defined? ❑ Have you explained why you investigated the problem? ❑ Have you outlined your aims and objectives? ❑ Have you explained your methodological approach? ❑ Have you stated your hypothesis?
Materials and Methods	Explains how the work was done. Should contain sufficient detail to allow another competent worker to repeat the work	❑ Is each experiment covered and have you avoided unnecessary duplication? ❑ Is there sufficient detail to allow repetition of the work? ❑ Are proper scientific names and authorities given for all organisms? ❑ Have you explained where you got them from? ❑ Are the correct names, sources and grades given for all chemicals?
Results	Displays and describes the data obtained. Should be presented in a form that is easily assimilated (graphs rather than tables, small tables rather than large ones)	❑ Is the sequence of experiments logical? Are the parts adequately linked? ❑ Are the data presented in the clearest possible way? ❑ Have SI units been used properly throughout? ❑ Has adequate statistical analysis been carried out? ❑ Is all the material relevant? ❑ Are the figures and tables all numbered in the order of their appearance? ❑ Are their titles appropriate? ❑ Do the figure and table legends provide all the information necessary to interpret the data without reference to the text? ❑ Have you presented the same data more than once?
Discussion/ Conclusions	Discusses the results: their meaning, their importance; compares the results with those of others; suggests what to do next	❑ Have you explained the significance of the results? ❑ Have you compared your data with other published work? ❑ Are your conclusions justified by the data presented?
Acknowledge- ments	Gives credit to those who helped carry out the work	❑ Have you listed everyone that helped, including any grant-awarding bodies?
Literature cited (Bibliography)	Lists all references cited in appropriate format: provides enough information to allow the reader to find the reference in a library	❑ Do all the references in the text appear on the list? ❑ Do all the listed references appear in the text? ❑ Do the years of publications and authors match? ❑ Are the journal details complete and in the correct format? ❑ Is the list in alphabetical order, or correct numerical order?

Steps in the production of a practical report or thesis

Choose the observations or experiments you wish to describe and decide how best to present them

Try to start this process before your lab work ends, because at the stage of reviewing your experiments, a gap may become apparent (e.g. a missing control) and you might still have time to rectify the deficiency. Irrelevant material should be ruthlessly eliminated, at the same time bearing in mind that negative results can be extremely important (see p. 201). Use as many different forms of data presentation as are appropriate, but avoid presenting the same data in more than one form. Relegate large tables of primary data to an appendix and summarise the important points within the main text (with a cross-reference to the appendix). Make sure that the experiments you describe are representative: always state the number of times they were repeated and how consistent your findings were.

Make up plans or outlines for the component parts

The overall structure of practical and project reports is well defined (see Box 18.1), but individual parts will need to be well organised, as with any other form of writing (see Chapter 16).

Write

The Materials and Methods section is often the easiest to write once you have decided what to report. Remember to use the past tense and do not allow results or discussion to creep in. The Results section is the next easiest as it should only involve description. At this stage, you may benefit from jotting down ideas for the Discussion – this may be the hardest part to compose as you need an overview both of your own work and of the relevant literature. It is also liable to become wordy, so try hard to make it succinct. The Introduction should not be too difficult if you have fully understood the aims of the experiments. Write the Abstract and complete the list of references at the end. To assist with the latter, it is a good idea to collate appropriately formatted details of the references (Chapter 9) as you progress.

Revise the text

Once your first draft is complete, try to answer all the questions given in Box 18.1. Show your work to your supervisors and learn from their comments. Let a friend or colleague who is unfamiliar with your subject read your text; they may be able to pinpoint obscure wording and show where information or explanation is missing. If writing a thesis, doublecheck that you are adhering to your institution's thesis regulations.

Prepare the final version

Markers appreciate neatly produced work but a well-presented document will not disguise poor science. If using a word processor, print the final version with the best printer available. Make sure figures are clear and in the correct size and format.

Submit your work

Your department will specify when to submit a thesis or project report, so plan your work carefully to meet this deadline or you may lose marks. Tell your supervisor early of any circumstances that may cause delay and check to see whether any forms must be completed for late submission, or evidence of extenuating circumstances.

Choosing between graphs and tables – graphs are generally easier for the reader to assimilate, whereas tables can be used to condense a lot of data into a small space.

Repeating your experiments – remember, if you do an experiment twice, you have repeated it only once.

Presenting your results – remember that the order of results presented in a report need not correspond with the order in which you carried out the experiments: you are expected to rearrange them to provide a logical sequence of findings.

Using the correct tense – always use the past tense to describe the methodology used in your work, since it is now complete. Use the present tense only for generalisations and conclusions (p. 107).

Oral assessments – there may be an oral exam (*'viva voce'*) associated with the submission of a thesis or dissertation. The primary aim of the examiners will be to ensure that you understand what you did and why you did it.

Producing a scientific paper

Scientific papers are the means by which research findings are communicated to others. Peer-reviewed papers are published in journals; each covers a well-defined subject area and publishes details of the format they expect.

> Peer review – the process of evaluation and review of a colleague's work. In scientific communication, a paper is reviewed by two or more expert reviewers ('referees') for comments on quality and significance as a key component of the validation procedure.

> **KEY POINT** Peer review is an important component of the process of scientific publication; only those papers whose worth is confirmed by the peer-review process will be published.

It would be very unusual for an undergraduate to submit a paper on his or her own – this would normally be done in collaboration with your project supervisor and only then if your research has satisfied appropriate criteria. However, it is important to understand the process whereby a paper comes into being (Box 18.2), as this can help you understand and interpret the primary literature.

Box 18.2 Steps in producing a scientific paper

Scientific papers are the lifeblood of any science and it is a major landmark in your scientific career to publish your first paper. The main steps in doing this should include the following:

Assessing potential content

The work must be of an appropriate standard to be published and should be 'new, true and meaningful'. Therefore, before starting, the authors need to review their work critically under these headings. The material included in a scientific paper will generally be a subset of the total work done during a project, so it must be carefully selected for relevance to a clear central hypothesis – if the authors will not prune, the referees and editors of the journal certainly will!

Choosing a journal

There are thousands of journals covering biology and each covers a specific area (which may change through time). The main factors in deciding on an appropriate journal are the range of subjects it covers, the quality of its content and the number and geographical distribution of its readers. The choice of journal always dictates the format of a paper since authors must follow to the letter the journal's 'Instructions to Authors'.

Deciding on authorship

In multi-author papers, a contentious issue is often who should appear as an author and in what order they should be cited. Where authors make an equal contribution, an alphabetical order of names may be used. Otherwise, each author should have made a substantial contribution to the paper and should be prepared to defend it in public. Ideally, the order of appearance will reflect the amount of work done rather than seniority. This may not happen in practice!

Writing

The paper's format will be similar to that shown in Box 18.1, and the process of writing will include outlining, reviewing, etc., as discussed elsewhere in this chapter. Figures must be finished to an appropriate standard and this may involve preparing photographs or digital images of them.

Submitting

When completed, copies of the paper are submitted to the editor of the chosen journal with a simple covering letter. A delay of one to two months usually follows while the manuscript is sent to two or more anonymous referees who will be asked by the editor to check that the paper is novel, scientifically correct and that its length is justified.

Responding to referees' comments

The editor will send on the referees' comments to the authors who will then have a chance to respond. The editor will decide on the basis of the comments and replies to them whether the paper should be published. Sometimes quite heated correspondence can result if the authors and referees disagree.

Checking proofs and waiting for publication

If a paper is accepted, it will be sent off to the publishers. The next the authors see of it is the proofs (first printed version in style of journal), which have to be checked carefully for errors and returned. Eventually, the paper will appear in print, but a delay of six months following acceptance is not unusual. Nowadays, papers are often available electronically, via the Web, in pdf format – see p. 57 for advice on how to cite such online papers.

Sources for further study

Berry, R. (2004) *The Research Project: How to Write It,* 5th edn. Routledge, London.

Davis, M. and Davis, K.J. (2012) *Scientific Papers and Presentations. Navigating Scientific Communication in Today's World,* 3rd edn. Academic Press.

Day, R.A. and Gastel, B. (2012) *How to Write and Publish a Scientific Paper,* 7th edn. Cambridge University Press, Cambridge.

Hofmann, A.H. (2014) *Scientific Writing and Communication: Papers, Proposals, and Presentations,* 2nd edn. Oxford University Press, New York

Lobban, C.S. and Schefter, M. (1992) *Successful Lab Reports: A Manual for Science Students.* Cambridge University Press, Cambridge.

Luey, B. (2009) *Handbook for Academic Authors,* 5th edn. Cambridge University Press, Cambridge.

Matthews, J.R., Bowen, J.M. and Matthews, R.W. (2007) *Successful Science Writing: A Step-by-step Guide for the Biological and Medical Sciences,* 3rd edn. Cambridge University Press, Cambridge.

McMillan, K.M. and Weyers, J.D.B. (2014) *How to Complete a Successful Research Project.* Pearson Education, London.

Valiela, I. (2009) *Doing Science: Design, Analysis and Communication of Scientific Research,* 2nd edn. Oxford University Press, Oxford.
[Covers scientific communication, graphical presentations and aspects of statistics.]

STUDY EXERCISES

18.1 **Write a formal 'Materials and Methods' section.** Adopting the style of a research paper (e.g. past tense, all relevant detail reported such that a competent colleague could repeat your work), write out the Materials and Methods for a practical you have recently carried out. Ask a colleague or tutor to comment on what you have written.

18.2 **Describe a set of results in words.** Again adopting the style of a research paper, write a paragraph describing the results contained in a particular table or graph. Ask a colleague or tutor to comment on your description, to identify what is missing or unclear.

18.3 **Write an abstract for a paper.** Pair up with a colleague. Each of you should independently choose a different research paper in a current journal. Copy the paper, but mask over the abstract section, having first counted the words used. Swap papers. Now, working to the same amount of words as in the original, read the paper and provide an abstract of its contents. Then, compare this with the real abstract.

Answers to these study exercises are available at www.pearsoned.co.uk/practicalskills

19 Writing literature surveys and reviews

The literature survey or review (sometimes 'dissertation') is a specialised form of essay that summarises and reviews the evidence and concepts concerning a particular area of research.

> **KEY POINT** A literature review should **not** be a simple recitation of facts. The best reviews are those that analyse information rather than simply describe it.

Organising your time

Figure 19.1 illustrates how you might divide up your time for writing a literature survey. There are many subdivisions in this chart because of the size of the task: in general, for lengthy tasks, it is best to divide up the work into manageable chunks. Note also that proportionately less time is allocated to writing itself than with an essay. In a literature survey, make sure that you spend adequate time on research and revision.

Selecting a topic

You may have no choice in the topic to be covered, but if you do, carry out your selection as a three-stage process:

1. **Identify a broad subject area that interests you.**
2. **Find and read relevant literature in that area.** Try to gain a broad impression of the field from books and general review articles. Discuss your ideas with your supervisor.
3. **Select a relevant and concise title.** The wording should be considered very carefully as it will define the content expected by the reader. A narrow subject area will cut down on the amount of literature you will be expected to review, but will also restrict the scope of the conclusions you can make (and vice versa for a wide subject area).

Scanning the literature and organising your references

You will need to carry out a thorough investigation of the literature before you start to write. The key issues are as follows:

- **Getting an initial toe-hold in the literature.** Seek help from your supervisor, who may be willing to supply a few key papers to get you started. Hints on expanding your collection of references are given on p. 53.

- **Assessing the relevance and value of each article.** This is the essence of writing a review, but it is difficult unless you already have a good understanding of the field. Try reading earlier reviews in your area and discussing the topic with your supervisor or other academic staff.

- **Clarifying your thoughts.** One approach is to use the SPSER method (Box 19.1). Another way to gain a better overview of a complex topic literature is to subdivide the literature into a smaller subject areas.

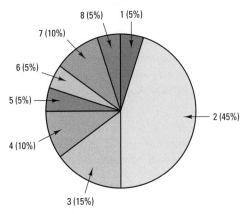

Fig. 19.1 Pie chart showing how you might allocate time for a literature survey:

1. select a topic
2. scan the literature
3. plan the review
4. write first draft
5. leave to settle
6. prepare a structured review of text
7. write final draft
8. produce top copy.

Creating a glossary – one barrier to developing an understanding of a new topic is the jargon used. To overcome this, create your own glossary. You may wish to cross-reference a range of sources to ensure the definitions are reliable and context-specific. Remember to note your sources in case you wish to use the definition within your review.

Using index cards – these can help you organise large numbers of references. Write key points and author information on each card – this helps when considering where the reference fits into the literature. Arrange the cards in subject piles, eliminating irrelevant ones. Order the cards in the sequence in which you wish to write.

Box 19.1 How to analyse a topic using the SPSER approach

This method is useful when trying to get to grips with a new and complex subject. The approach helps you to 'deconstruct' or 'unpack' the topic and involves considering the task in five discrete stages given the acronym SPSER. Your thoughts could be noted using a table or the 'Pattern' or 'Mind Map' approaches (p. 18), ready for converting into text.

1. **Situation.** Briefly outline the context of the topic and the history of its development. For some complex topics you may also need to refer to an introductory text or website to build a foundation for your understanding.

2. **Problem.** State the essential issues or problems that researchers or others are tackling. In a research paper, this might be stated succinctly in the Abstract, but you might also find it at the end of the Introduction. In some cases this will be inherent in the title of the exercise you have been given.

3. **Solution.** Outline possible solutions or ways of tacking the problem. These will probably arise from your reading of papers, reviews and other sources.

4. **Evaluation.** Note positive or negative features of each solution and provide evidence to support your viewpoint. A tabular approach might be useful, listing positives on one side of a vertical line and negatives on the other.

5. **Recommendation.** Arrive at a conclusion by deciding what might/should happen next, outlining your reasoning.

Deciding on structure and content

The general structure and content of a literature survey are described below. The *Annual Review* series (available in most university libraries) provides good examples of the style expected in reviews of biosciences.

Introduction

The Introduction should give the general background to the research area, concentrating on its development and importance. You should also make a statement about the scope of your survey; as well as defining the subject matter to be discussed, you may wish to restrict the period being considered.

Main body of text

The review itself should discuss the published work in the selected field and may be subdivided into appropriate sections. Within each portion of a review, the approach is usually chronological, with appropriate linking phrases (e.g. 'Following on from this, …'; 'Meanwhile, Bloggs (1980) tackled the problem from a different angle …'). However, a good review is much more than a chronological list of work done. It should:

- **allow the reader to obtain an overall view of the current state of the research area,** identifying the key areas where knowledge is advancing;

- **show how techniques are developing** and discuss the benefits and disadvantages of using particular organisms or experimental systems;

- **assess the relative worth of different types of evidence** – this is the most important aspect (see Chapter 10). Do not be intimidated from taking a critical approach as the conclusions you may read in the primary literature are not always correct;

- **indicate where there is conflict in findings or theories,** suggesting if possible which side has the stronger case;

- **indicate gaps in current knowledge.**

You do not need to wait until you have read all the sources available to you before starting to write the main body. Word processors allow you

Defining terms – the Introduction is a good place to explain the meaning of the key terms used in your survey or review.

Balancing opposing views – even if you favour one side of a disagreement in the literature, your review should provide a balanced and fair description of all the published views of the topic. Having done this, if you do wish to state a preference, give reasons for your opinion.

to modify and move pieces of text at any point and it will be useful to write paragraphs about key sources, or groups of related papers, as you read them. Try to create a general plan for your review as soon as possible. Place your draft sections of text under an appropriate set of subheadings that reflects your plan, but be prepared to rearrange these and retitle or reorder sections as you proceed. Not only will working in this way help to clarify your thoughts, but it may help you avoid a last-minute rush of writing near to the submission date.

Conclusions

The conclusions should draw together the threads of the preceding parts and point the way forward, perhaps listing areas of ignorance or where the application of new techniques may lead to advances.

References, etc.

The References or Literature Cited section should provide full details of all papers referred to in the text (see p. 56). The regulations for your department may also specify a format and position for the title page, contents page, acknowledgements, etc.

> **Making citations** – a review of literature poses stylistic problems because of the need to cite large numbers of papers; in the *Annual Review* series this is overcome by using numbered references (see p. 56).

Sources for further study

McMillan, K.M. and Weyers, J.D.B. (2013) *The Study Skills Book,* 3rd edn. Pearson Education, Harlow.

McMillan, K.M. and Weyers, J.D.B. (2011) *How to Write Dissertations and Project Reports,* 2nd edn. Prentice Hall, Harlow.

Rudner, L.M. and Schafer, W.D. (1999) How to write a scholarly research report, *Practical Assessment, Research & Evaluation,* 6(13). Available: http://pareonline.net/getvn.asp?v=6&n=13. Last accessed 17/02/15.

STUDY EXERCISES

19.1 Summarise the main differences between a review and a scientific paper. From the many subject areas in the *Annual Review* series (find via your library's periodical indexing system), pick one that matches your subject interests, and within this find a review that seems relevant or interesting. Read the review and write down *five* ways in which the writing style and content differs from that seen in scientific papers.

19.2 Gather a collection of primary sources for a topic. From the journal section of the library, select an interesting scientific paper published about 5–10 years ago. First, work *back* from the references cited by that paper: can you identify from the text or the article titles which are the most important and relevant to the topic? List five of these, using the proper conventions for citing articles in a reference list (see Chapter 9). Note that each of these papers will also cite other articles, always going back in time. Now, using the Institute of Scientific Information (ISI) Science Citation Index or a similar system (see pp. 53–4), work *forward* and find out who has cited your selected article in the time since its publication. Again list the five most important articles found.

19.3 Write a synopsis of a review. Using one of the *Annual Review* series as a source, allow yourself just *five* single-sentence bullet points to summarise the key points reported in a particular review.

Answers to these study exercises are available at www.pearsoned.co.uk/practicalskills

Fundamental laboratory techniques

All knowledge and theory in science have originated from practical observation and experimentation: this is equally true for disciplines as diverse as microscopy and molecular genetics. Practical work is an important part of most courses and often accounts for a significant proportion of the assessment marks. The abilities developed in practical classes will continue to be useful throughout your course and beyond, some within science and others in any career you choose (see Chapter 1).

Being prepared

The main points to remember are:

- **Read any handouts in advance:** make sure you understand the purpose of the practical and the particular skills involved. Does the practical relate to, or expand upon, a current topic in your lectures? Is there any additional preparatory reading that will help?
- **Take along appropriate textbooks,** to explain aspects in the practical.
- **Consider what safety hazards might be involved,** and any precautions you might need to take, before you begin (p. 137).
- **Listen carefully to any introductory guidance and note any important points:** adjust your schedule/handout as necessary.
- **Organise your bench space during the practical session** – make sure your lab book is adjacent to, but not within, your working area. You will often find it easiest to keep clean items of glassware etc. on one side of your working space, with used equipment on the other side.
- **Write up your work as soon as possible,** and submit it on time, or you may lose marks.
- **Catch up on any work you have missed as soon as possible** – preferably, before the next practical session.

> **KEY POINT** You will get the most out of practicals if you prepare well in advance. Do not go into a practical session assuming that everything will be provided, without any input on your part.

Ethical, safety and legal aspects

You will need to consider the ethical, safety and legal implications of biological work at several points during your studies:

- The ethical aspects of laboratory work that involves working with humans or animals must be carefully considered; Chapter 21 considers this in more detail.
- Safe working means following a code of safe practice, supported by legislation, alongside a moral obligation to avoid harm to yourself and others, as discussed in Chapter 22.
- Fieldwork can have a legal aspect, e.g. in relation to specimen collection (Chapter 36), since legislation may protect particular habitats and/or species. Ethical and human aspects might include:

Developing practical skills – these will include:
- designing experiments
- observing and measuring
- recording data
- analysing and interpreting data
- reporting/presenting.

Using textbooks in the lab – take this book (or photocopies of relevant chapters) along to the relevant classes, so that you can make full use of the information during the practical sessions.

SAFETY NOTE Mobile phones should never be used in a lab class, as there is a risk of contamination from hazardous substances. Always switch off your mobile phone before entering a laboratory. Conversely, they are an extremely useful accessory for fieldwork.

Getting to grips with bioethics – in addition to any moral implications of your lab practicals and fieldwork, you may have the opportunity to address broader issues within your course (see Chapter 21). Professional scientists must always consider the consequences of their work, and it is therefore important that you develop your understanding of these issues alongside your academic studies.

possible damage to fauna and flora; safety of others within your group; implications of your work to others, e.g. landowners or other users (Chapter 31). These must all be considered before any work is carried out.

Basic requirements

Recording practical results

When you start your degree studies in biology, you may be provided with a printed notebook to guide you through lab and field procedures. This may include an introduction to the practical or field work; instructions for handling specimens, making observations or using instruments; details of any experiments to be carried out; and space for drawing diagrams, recording your results and making notes. The notebook may also include 'prompts' for analysis or conclusions, sometimes in the form of questions.

In later years, you will be expected to organise your own laboratory notes. An A4 loose-leaf ring binder offers flexibility, since you can insert laboratory handouts or lined and graph paper at appropriate points. The danger of losing one or more pages from a loose-leaf system is the main drawback. Bound books avoid this problem, although those containing alternating lined/graph or lined/blank pages tend to be wasteful – it is often better to paste sheets of graph paper into a bound book, as required.

A good-quality HB pencil or propelling pencil is recommended for recording your raw data, making diagrams, etc., as mistakes are easily corrected. Buy a black, spirit-based (permanent) marker for labelling experimental glassware, Petri plates, etc. Fibre-tipped fine-line drawing/lettering pens are useful for preparing final versions of graphs and diagrams for assessment purposes. Use a see-through ruler with an undamaged edge for graph drawing, so that you can see data points and information below the ruler as you draw. Alternatively, use graphdrawing software (see Chapter 72).

Calculators

In the early stages, it is usually better to avoid complex calculators with impressive-looking but often unused pre-programmed functions. Opt for more memory, parentheses or statistical functions rather than engineering or mathematical constants. Programmable calculators can be useful for advanced studies but note that they are usually unacceptable for exam use. The following may also be helpful when choosing a calculator:

- **Power sources.** Choose a battery-powered machine, rather than a mains-operated or solar-powered type. You will need one with basic mathematical/scientific operations, including powers, logarithms (p. 500), roots and parentheses (brackets), together with statistical functions such as sample means and standard deviations (Chapter 75).
- **Mode of operation.** Spend some time finding out how your calculator operates. Does it have true algebraic logic (e.g. $\sqrt{}$ then number, rather than number then $\sqrt{}$)? How does it deal with (and display) scientific notation and logarithms (pp. 499–500)?
- **Display.** Some calculators will display an entire mathematical operation (e.g. '2 + 4 = 6'), while others simply display the last number/operation. The former type may offer advantages in tracing errors.

Presenting results – though you do not need to be a graphic designer to produce work of a satisfactory standard, presentation and layout are important and you will lose marks for poorly presented work. Chapter 72 gives further practical advice.

Using inexpensive calculators – many unsophisticated calculators have a restricted display for exponential numbers and do not show the 'power of 10', e.g. displaying 2.4×10^{-5} as 2.4^{-05}, or even $2.4E-05$, or $2.4-05$.

Using calculators for numerical problems – Chapter 74 gives further advice.

Buying and using a calculator – many university departments specify a particular make and model of calculator for students to use from first year onwards. The advantage of this approach is that you can seek help from staff during tutorials and lab classes, if you are unsure of specific functions. Conversely, staff will not be fully familiar with other types of calculator, so may not be able to answer such questions if you are using an alternative make or model.

Presenting graphs and diagrams – ensure these are large enough to be easily read: a common error is to present graphs or diagrams that are too small, with poorly chosen scales (see p. 190).

Presenting more advanced practical work

In some practical reports and in project work, you may need to use more sophisticated presentation equipment. Computer-based graphics packages can be useful – choose easily read fonts such as Arial for posters and consider the layout and content carefully (p. 80). Alternatively, you could use fine-line drawing pens and dry-transfer lettering/symbols, although this can be more time-consuming than computer-based systems, e.g. using Microsoft *Excel* (pp. 480–3).

For a spoken presentation, Microsoft *PowerPoint* is widely used – Box 15.1 gives practical advice on preparing *PowerPoint* slides. Alternative approaches include using transparencies and an overhead projector (use permanent markers and acetate sheets, or photocopy onto special transparencies), or paper-based images/text presented using a document camera (check that the camera is centred and in focus before you begin). One advantage of these approaches is that you can write on the projected material during the presentation, for example, to add emphasis. Further advice on oral presentations is given in Chapter 15.

Sources for further study

Barnard, C.J., Gilbert, F.S. and MacGregor, P.K. (2011) *Asking Questions in Biology,* 4th edn. Prentice Hall, Harlow.

Bonner, P. and Hargreaves A. (2011) *Basic Bioscience Laboratory Techniques*: *A Pocket Guide.* Wiley, New York.

Dickey, J.L. (2002) *Laboratory Investigations in Biology,* 2nd edn. Benjamin Cummings, San Francisco.

STUDY EXERCISES

20.1 **Consider the value of practical work.** Spend a few minutes thinking about the purpose of practical work within a specific part of your course (e.g. a particular first-year module) and then write a list of the six most important points. Compare your list with the generic list that we have provided on the website, which is based on our experience as lecturers – does it differ much from your list, which is drawn up from a student perspective?

20.2 **Make a list of items required for a particular practical exercise.** This exercise is likely to be most useful if you can relate it to an appropriate practical session on your course. However, we have given a model list for a snail dissection, to help you get started. (Note - see also Chapter 31, for a similar study exercise related to fieldwork.)

20.3 **Check your calculator skills.** Carry out the following mathematical operations, using either a hand-held calculator or a PC with appropriate 'calculator' software.

(a) $5 \times (2 + 6)$

(b) $(8.3 \div [6.4 - 1.9]) \times 24$ (to four significant figures)

(c) $(1 \div 32) \times (5 \div 8)$ (to three significant figures)

(d) $1.2 \times 10^5 + 4.0 \times 10^4$ in scientific notation (see p. 408)

(e) $3.4 \times 10^{-2} - 2.7 \times 10^{-3}$ in 'normal' notation (i.e. conventional notation, not scientific format) and to three decimal places.

(See also the numerical exercises in Chapter 67.)

Answers to these study exercises are available at www.pearsoned.co.uk/practicalskills

Bioethics is the philosophical study of ethical issues arising from recent advances in biology and medicine, though its scope also extends to cover environmental and global issues. It aims to provide a framework for making decisions, based on specific moral principles that take into account the reasons underlying different choices and the consequences of specific decisions. Particular groups of individuals will often adopt a set of ethical rules and standards as part of their professional code of conduct, e.g. medical ethics.

This chapter aims to provide you with a basic framework that will help you understand the various ethical theories and principles, so that you can consider some of the issues in contemporary bioethics. Without an ethical framework, individuals can only make subjective 'value judgements', based solely on personal opinions and viewpoints, rather than objective analysis.

One starting point is to understand the following fundamental terms:

- **Values** – the various beliefs and views held by individuals and groups, either for or against something; for example, religious, cultural or political values.
- **Morals** – individual or collective beliefs about whether something is 'right' or 'wrong'; for example, is it right to use animals in testing of pharmaceutical products?
- **Ethics** (moral philosophy) – a system of moral values and decision-making, described in terms of principles, rules and standards that are adopted by a specific group of people; for example, research ethics, legal ethics.

Note that sometimes the practical distinction between morals and ethics is less clear-cut than described above, and the terms are occasionally used interchangeably.

Bioscience degree programmes place increasing emphasis on the ethical and social impacts of contemporary scientific advances and on the need for scientists engaged in potentially controversial work to communicate their ideas and decision-making to the general public. Bioscience research raises many moral, ethical and legal dilemmas, requiring difficult choices to be made, e.g. in relation to animal testing of medical products, and students are likely to be asked to reflect on bioethical questions and moral dilemmas, e.g. in group discussions and debates. In discussing such issues, you will also gain experience that will have broader value, in terms of understanding some of the issues linked to the public understanding of science, and of how these issues can be addressed through communication with the public. You will find links and resources addressing this broader topic on the websites of a number of professional scientific organisations such as the Royal Society of London, and the Wellcome Trust in the UK, or the Coalition on the Public Understanding of Science (COPUS) in the USA.

Recognising the difference between morals and ethics – consider a biomedical scientist who believes that murder is fundamentally wrong (immoral), but the scientist's research into a specific disease requires experiments that will lead to the death of some animals, to be carried out under a research ethics code that specifies that the animals should not endure unnecessary pain or suffering. What should the scientist do? You will find that many aspects of bioethics are best considered through specific examples of moral/ethical dilemmas such as this.

Understanding the terminology of issues and dilemmas – these topics are sometimes referred to as ethical, legal and social issues (ELSI).

> **KEY POINT** In contrast to numerical calculations (Chapter 74) there is rarely a single 'right' or 'wrong' answer to bioethical issues. Consequently, it is important to use ethical principles and systems to consider each issue in a logical manner. Your aim should be to develop your skills in critical thinking and reasoning (pp. 64–5), and in explaining how you have arrived at a particular decision on a bioethical issue, with sound evidence and a reasoned argument (pp. 65–6) to support the position that you have taken.

Bioethics can be subdivided into three main areas:

1. **Environmental ethics** – for example, the use of genetically modified organisms (GMOs), the hunting and killing of endangered species.
2. **Animal ethics** – examples include factory farming, the use of transgenic animals and xenotransplantation.
3. **Human ethics** – dealing with medical and social issues, such as human cloning, embryo and stem cell research, genetic testing and the storage of DNA profiles on databases.

Ethical theories

A basic understanding of the various theories of ethics may help you formulate your ideas in relation to specific questions and issues. The major theories include:

- **Utilitarianism** – the notion that it is ethical to choose the action that produces the greatest good (happiness) for the greatest number. For example, the use of atomic bombs against civilians in the Japanese cities of Hiroshima and Nagasaki in World War II was justified in terms of its utilitarian value, in terms of ending the war and preventing Japan's development and use of similar weapons, with potentially greater loss of life on both sides.
- **Deontology** – the theory that states that a particular action is either intrinsically good (right) or bad (wrong), and that such principles are absolute. According to deontological theory, decisions must be based on the actions themselves, rather than on their overall consequences. Such an approach is enshrined in the Judaeo-Christian religious value system (for example, in the ten commandments of the Old Testament). While this approach appears simple to apply, for example, in upholding the principle that 'it is always wrong to lie', it does not allow for particular circumstances, e.g. is it wrong to lie if, in doing so, you could prevent a murder? Perhaps in this example of an ethical dilemma, most people would agree that the greater 'good' of preventing a murder would outweigh the lesser 'good' of not lying, though this then moves towards utilitarianism.

- **Consequentialism** – where decisions are made and choices are selected by considering their likely outcomes and consequences, rather than from any intrinsic viewpoint. From a consequentialist viewpoint, a lie would be the right course of action choice to take if it prevented a murder.
- **Virtue theory** – arising from the philosophy of Plato and Aristotle, this enshrines the notion that making decisions according to established virtues, such as honesty, wisdom, justice and altruism will lead to ethically valid choices. As with consequentialism and utilitarianism,

Examples of current bioethical issues and dilemmas:

- Whether to locate wind farms in the Scottish highlands. (Weigh up the benefits of 'green energy' against those of damage to environmentally sensitive areas.)
- Whether vivisection (using dissection procedures on living animals) can be justified if it may lead to medical advances that reduce human suffering.
- Whether the potential advantages of using GMOs outweigh the possible disadvantages (e.g. the use of 'golden rice', which has been genetically modified to contain genes for vitamin A synthesis, as a means of tackling vitamin A deficiency in some developing countries).

virtue theory is distinct from deontology, since it is not based on rigid absolutes.

- **Objectivism** – the theory that what is right and wrong is intrinsic and applies equally to all people, places and times. The alternative is that morality is subjective, being entirely dependent upon the views of the individual (**subjectivism**).

Ethical principles

The following provides a framework of fundamental principles for considering particular topics and questions from different perspectives, in no particular order:

- **Beneficence** – the obligation to do good. For example, if it is possible to prevent suffering by a particular course of action, then it should be carried out. However, dilemmas arise in particular circumstances when conflicts arise, e.g. is it acceptable to cause suffering to animals used in medical research that aims to prevent suffering to humans in the future?
- **Non-maleficence** – the duty to do no harm. This is enshrined within the Hippocratic oath of medical practitioners, but is not without difficulty in its practical application in specific circumstances. For example, should a medical practitioner continue to maintain someone who is on a life-support machine and who has no reasonable expectation of recovery, if this is causing anguish to the patient's family, and if the practitioner knows that another person is awaiting life support? Here, the practitioner has to weigh up the different aspects of the case, and make a balanced judgement.
- **Justice** – the obligation to treat all people fairly and impartially; for example, lack of discrimination between people on the grounds of race, colour, sex, language, religion, age or other status. This principle is also enshrined in Article 2 of the Universal Declaration of Human Rights (United Nations, 1948).
- **Autonomy** – the duty to allow an individual to make their own choices, without constraints. This principle underlies the notion of 'informed consent' in medical and scientific research.
- **Respect** – the need to show due regard for others; for example, the need to take into account the rights and beliefs of all people equally. This is also an intrinsic component of the UN Universal Declaration of Human Rights. It is also an essential component of the principle of confidentiality of personal information about participants in medical and scientific research (see p. 133).
- **Rationality** – the notion that a particular action of choice should be based on reason and logic. Many scientists would argue that the scientific method itself (pp. 201–2) is a good example of the principle of rationality in action.
- **Honesty** – the obligation to be truthful. This principle also underpins 'informed consent', since individuals cannot make informed choices if they have not been provided with information that is true and complete. It is also a fundamental aspect of scientific research ethics, with scientists having a duty to present their results as they truly are, without falsification or bias. When this principle is breached, the misconduct is often termed 'scientific fraud'.

Examples of scientific fraud – these include the case of Andrew Wakefield, a former doctor who was struck off the UK medical register for 'misleading, dishonest and irresponsible' claims of a link between measles-mumps-rubella (MMR) vaccination and autism, based on seriously flawed research. For more examples, see journalist Ben Goldacre's 'Bad Science' website at: http://www.badscience.net/

- **Precautionary principle** – the notion that it is better not to carry out an action if there is any risk of harm. For example, many people would apply the precautionary principle in deciding that the risks of building nuclear power stations outweigh their potential benefits. It is enshrined in principle 15 of the Rio Declaration (United Nations, 1992) as follows:

> to protect the environment, the precautionary approach shall be widely applied by States according to their capabilities. Where there are threats of serious or irreversible damage, lack of full scientific certainty shall not be used as a reason for postponing cost-effective measures to prevent environmental degradation.

It is also an aspect of the sustainable approach to global issues, such as global warming. While the precautionary principle is widely applied to environmental ethics, it is also relevant to other areas. For example, in medical ethics, what should a doctor recommend where only limited information is available about the possible relationship between exposure to a particular factor and a specific disease?

It is clear from the above listing, that several different principles may apply to a particular bioethical issue, and it will be up to individuals to weigh up the relative merits and disadvantages of particular choices and options in terms of the different outcomes against these principles when faced with an 'ethical dilemma'. Box 21.1 outlines a practical and sequential approach to making ethical decisions.

Box 21.1 **A step-wise approach to making ethical decisions**

The following text is based on the framework of Shamoo and Resnik (2009):

1. **Define the issue or dilemma.** Typically this will be in the form of a question that requires a choice to be made, for example: 'Should the UK allow euthanasia?' The problem/question will often be given to you as part of a coursework assignment.

2. **Research the literature and gather together relevant information.** This may include (i) finding scholarly papers that consider the ethics of the topic, e.g. medical ethicists who have considered euthanasia from a particular standpoint, (ii) seeking other perspectives, e.g. those of organisations in favour of and opposed to euthanasia.

3. **Work out what options are available.** For example, you might consider under what circumstances euthanasia might be permitted, and how this would be regulated, rather than restricting yourself to a 'yes' or 'no' answer.

4. **Consider the options in relation to ethical theories and principles.** For example, what outcomes would fit with deontological theory, or utilitarianism? How do the principles of non-maleficence, autonomy and respect for the individual apply in relation to euthanasia?

5. **Consider whether anything else should be taken into account.** This step provides an opportunity for a 'second check' on whether you have considered all aspects of the topic, e.g. in relation to euthanasia, have you considered whether the views of relatives would carry any weight? Or whether your decision would apply to an individual suffering from a psychological disorder?

6. **Make your decision** – choose a particular response to the question posed at the outset. In relation to euthanasia, this might be an unqualified 'yes' or 'no', or might take the form of a conditional response, e.g. 'only in the following circumstances...'

7. **Explain your choice** – use the evidence gained in steps 2–5 to justify the decision made at step 6. In an assessed coursework assignment, this component of your written text will probably represent the greatest proportion of the marks available (always remember that it is not about reaching the 'right' or 'wrong' answer, but about reaching a logical and reasoned decision, based on interpretation of the issue in relation to sound ethical principles and reasoned discussion, supported by published sources).

Seeking ethical approval – note that many universities have separate committees for human ethics and animal ethics – you should consult your supervisor for specific advice well in advance of your planned start date, since your work cannot proceed until it has ethical approval, and this will need to fit in with the timelines of the relevant ethics committee.

Research involving human subjects

As explained on pp. 130–1, a number of ethical principles apply to the conduct of any research involving human subjects. Such principles apply in addition to those of health and safety (Chapter 22). If you carry out any research involving human participants as part of your degree programme, you will need to obtain ethical approval before you begin any practical work. Usually, this will involve submission of a request for ethical approval of your proposed project to your institution's ethics committee.

> **KEY POINT** For the purposes of most undergraduate research proposals, while a student may be the investigator, the responsibility for following appropriate ethical approval procedures will remain with the supervisor.

Outlined below are the three main aspects that you should consider in all research involving human participants, irrespective of whether the research is 'invasive' (e.g. collecting mouth swabs from participants to study a particular DNA sequence) or 'non-invasive' (e.g. questioning participants about their drinking habits as part of a survey on the effects of alcohol consumption on study patterns and exam performance).

Providing information

All participants should be informed in writing and through oral explanation about the main aspects of the research study, typically through the use of a Participant Information Sheet that should include the following aspects:

- **An invitation to participate in the study,** with an explanation that participation is entirely voluntary and that the person is free to withdraw from the study at any stage.
- **A succinct outline of the purpose of the study,** and its likely outcomes.
- **A summary of what the participant will be expected to do,** and the timelines involved.
- **A statement of any possible benefits or risks involved in participation in the study.**
- **An assurance of confidentiality and anonymity in use of data from the study,** including details of how long the data will be retained within the institution.
- **Information about the funding source(s).**
- **Names and contact details of researchers/supervisors.**

Many institutions provide templates for such information sheets, which helps students to address all of the key aspects of the proposed study. The Participant Information Sheet should be written in plain English so that it can be readily understood. Researchers should also answer any additional questions that participants might have, based on their reading of the information sheet.

Definition

Plain English – a clear, succinct style that aims to be direct and natural, avoiding jargon, difficult laguage or complex structure. This will allow the meaning of the text to be understood by people from all walks of life.

Seeking informed consent

After reading the Participant Information Sheet, subjects are then requested to complete and sign an Informed Consent Form. As with the Participant Information Sheet, most institutions will provide a suitable template for student use – for broader examples, see the World Health Organization templates at: http://www.who.int/rpc/research_ethics/informed_consent/en/contents.

Assuring confidentiality

All participants must be assured that their personal details will be protected through processes that ensure anonymity. This means that all data recording and storage must be de-identified, e.g. through the use of randomly assigned codes to individuals, rather than using names. In addition to ethical considerations relating to storage and use of personal information, you must also comply with any legislation relating to data storage and protection, in terms of how it is held and used, the time-limit for maintaining the information and how it will be destroyed at the end of this period.

Box 21.2 provides practical advice that should help ensure that your research study is conducted in accordance with ethical principles.

> **UK data protection legislation** – the Data Protection Act 1998 covers the procedures that must be followed in relation to storage of data, particularly sensitive personal data, such as racial origins, religious beliefs and sexuality (see: http://www.legislation.gov.ukukpga/1998/29).

Box 21.2 **A step-wise approach to conducting ethical research**

The following text is based on McMillan and Weyers (2013):

1. **Research your institution's guidelines and procedures for research ethics.** Typically, such material, including general guidelines, templates and forms, will be available through the university website. For discipline-specific information, you may need to consult your departmental website, or seek advice from your supervisor. Investigating these procedures and guidelines at the outset will help you understand (i) the aspects of your own study that require ethical approval and (ii) how these might be addressed.

2. **Consider the ethical dimensions of your project.** Are there any examples of unethical behaviour in the scientific literature or popular media that might impact on the design or implementation of your study? By taking a 'devil's advocate' stance and thinking about your study from different perspectives, you should be able to consider any potential ethical flaws in your proposed experimental design (see also Box 21.1).

3. **Discuss your proposed research with other students and staff.** Such debate will help raise your awareness of any issues that might impact on your proposed study, and will help you 'defend' your proposal in your application for ethical approval and also in your final report.

4. **Modify the proposed study to avoid ethical problems.** If steps 2 or 3 raise any ethical issues, then it is best to consider how best to modify your experimental design to address these. For example, if self-analysis or peer-group discussion has identified that the research might pose particular risks to specific groups of participants, then you might wish to consider excluding such groups. However, you would also need to consider the implication of such a move to your study outcomes.

5. **Complete all necessary paperwork in advance of your study.** Remember that no experimental work can begin until you have formal written approval from the relevant ethics committee.

6. **Carry out all practical procedures and data collection/storage as required by guidelines and legislation.** Remember that this is an important aspect of your personal development and a key aspect of professional practice that will apply when you graduate. All research scientists are bound by the same requirements to conduct their research within an ethical framework.

Animal welfare legislation in the UK – the Animal Welfare Act 2006 is designed to prevent animal cruelty and directive 2010/63/EU is designed to protect animals used for scientific purposes. Before any research activities are carried out, they must be approved through a licence issued by the Home Office, under the Animals (Scientific Procedures) Act 1976.

Using animals in teaching and research

In additional to complying with any legislative requirements, such as laws relating to animal welfare, all teaching activities and research studies involving vertebrates and cephalopods (which have been recognised as being capable of feeling pain, suffering distress and suffering lasting harm) must follow the ethical guidelines of the institution, which will include seeking ethical approval in advance of any procedure involving live animals. Such approval must also be sought in relation to killing of animals for teaching or research purposes: Chapter 42 considers the ethics of dissection in the context of zoology teaching.

If you are involved in animal experimentation as part of a research project (Chapter 35), you must take care to follow all conditions of the animal licence obtained by your institution – consult your supervisor if you are unsure of any aspect of the regulations.

Text references

McMillan, K. and Weyers, J.D.B. (2013) *The Study Skills Book,* 3rd edn. Pearson Education, Harlow.

Shamoo, A.E. and Resnik, D.B. (2009) *Responsible Conduct of Research,* 2nd edn. Oxford University Press, Oxford.

United Nations (1948) *Universal Declaration of Human Rights: Homepage.* Available: http://www.un.org/en/documents/udhr/index.shtml Last accessed 23/03/15.

United Nations (1992) *Rio Declaration: Homepage.* Available: http://www.un.org/documents/ga/conf151/aconf15126-1annex1.htm Last accessed 23/03/15.

Sources for further study

Barry, V. (2011) *Bioethics in a Cultural Context.* Wadsworth Cengage Publishing, Boston.

Bioethics Education Project, BEEP, Bristol University (2011) *Bioethics Homepage.* Available: http://www.beep.ac.uk/ Last accessed 23/03/15.

Bryant, J., la Velle, L.B. and Searle, J. (2005) *Introduction to Bioethics.* Wiley, Chichester.

DeGrazia, D., Mappes, T. and Ballard, J. (2010) *Biomedical Ethics.* McGraw-Hill, New York.

National Health Museum, Atlanta (2009) *Bioethics.* Available: http://www.accessexcellence.org/RC/AB/IE/bioethics.php Last accessed 23/03/15.

National Institutes of Health (2010) *Bioethics Resources on the Web.* Available: http://bioethics.od.nih.gov/ Last accessed 23/03/15.

Vaughn, T. (2009) *Bioethics, Principles, Issues and Cases.* Oxford University Press, Oxford.

STUDY EXERCISES

21.1 Test your understanding of the terminology and theories of ethics. Distinguish between the following pairs of terms:

a) morals and ethics;

b) value judgements and ethical decisions;

c) utilitarianism and deontology;

d) consequentialism and virtue theory;

e) objectivism and subjectivism.

21.2 Test your understanding of ethical principles. Explain each of the following terms, in the context of ethics:

a) beneficence;

b) non-maleficence;

c) justice;

d) autonomy;

e) precautionary principle.

21.3 Locate web-based information on bioethics posted by relevant organisations and learned societies. This could be important if you are researching a topic with an ethical dimension, e.g. an essay on the pros and cons of genetic modification of farm animals, or if you are considering a research project involving human or animal subjects.

21.4 'Brainstorm' the ethical issues within a research project or topic. Make a 'spider diagram' (p. 105) or 'mind map' (p. 19) with your title or topic at the centre. Then, use the ethical framework described in this chapter to consider the various aspects of the title/topic, and the choices, decisions and actions required. You could carry out this exercise on an individual basis, or as a group exercise (share your spider diagrams or mind maps with other students, and discuss your views and those of others within the group).

Answers to these study exercises are available at www.pearsoned.co.uk/practicalskills

Definitions

Hazard – the intrinsic ability of a substance or biological agent to cause harm.

Risk – the likelihood that a substance or biological agent might be harmful under specific circumstances.

Distinguishing between hazard and risk – one of the hazards associated with water is drowning. However, the risk of drowning in a few drops of water is negligible.

inhalation

ingestion

inoculation or absorption

absorption from spillage

Fig. 22.1 Major routes of entry of harmful substances into the body.

Health and safety law requires institutions to provide a working environment that is safe and without risk to health. Where appropriate, training and information on safe working practices must be provided. Students and staff must take reasonable care to ensure the health and safety of themselves and of others, and must not misuse any safety equipment.

> **KEY POINT** All practical work must be carried out with safety in mind, to minimise the risk of harm to yourself and to others – safety is everyone's responsibility.

Risk assessment

The most widespread approach to safe working practice involves the use of risk assessment, which aims to establish:

- **The intrinsic chemical, biological and physical hazards,** together with any maximum exposure limits (MELs) or occupational exposure standards (OESs), where appropriate. Chemical manufacturers provide data sheets listing the hazards associated with particular chemical compounds, while pathogenic (disease-causing) microbes are categorised according to their ability to cause illness (pp. 286–9).
- **The risks involved,** by taking into account the amount of substance to be used, the way in which it will be used and the possible routes of entry into the body (Fig. 22.1).

> **KEY POINT** It is important to distinguish between the intrinsic *hazards* of a particular substance and the *risks* involved in its use in a particular exercise.

- **The persons at risk,** and the ways in which they might be exposed to hazardous substances, including accidental exposure (spillage).
- **The steps required to prevent or control exposure.** Ideally, a non-hazardous or less hazardous alternative should be used. If this is not feasible, adequate control measures must be used, e.g. a fume cupboard or other containment system. Personal protective equipment (e.g. lab coats, safety glasses) must be used in addition to such containment measures. A safe means of disposal is also required.

The outcome of the risk-assessment process must be recorded and appropriate safety information must be passed on to those at risk. For most practical classes, risk assessments will have been carried out in advance by the person in charge: the information necessary to minimise the risks to students may be given in the practical schedule. Make sure you know how your department provides such information and that you have read the appropriate material before you begin your practical work. You should also pay close attention to the person in charge at the beginning of the practical session, as they may emphasise the major hazards and risks. In project work, you will need to be involved in the risk assessment process along with your supervisor, before you carry out any practical work.

In addition to specific risk assessments, most institutions will have a safety handbook, or equivalent, giving general details of safe working practices, together with the names and telephone numbers of safety

Fig. 22.2 Warning labels for specific chemical hazards. These appear on suppliers' containers and on tape used to label working vessels.

personnel, first aiders, hospitals, etc. Make sure you read the relevant sections of this document and follow any instructions given.

Basic rules for laboratory work

- Make sure you know what to do in case of fire, including exit routes, how to raise the alarm, and where to gather on leaving the building. Remember that the most important consideration at all times is human safety: do not attempt to fight a fire unless it is safe to do so.
- All laboratories should display notices telling you where to find the first-aid kit and who to contact in case of accident/emergency. Report all accidents, even those appearing insignificant – your department will have a formal recording procedure to comply with safety legislation.
- Wear appropriate protective clothing at all times – a clean lab coat (buttoned up), plus safety glasses if there is any risk to the eyes.
- Never smoke, eat or drink in any laboratory, because of the risks of contamination by inhalation or ingestion (Fig. 22.1).
- Never mouth-pipette any liquid. Use a pipette filler (see p. 139) or, if appropriate, a pipettor (p. 140).
- Take care when handling glassware – see p. 143 for details.
- Know the chemical hazards warning symbols (Fig. 22.2).
- Use a fume hood for hazardous chemicals. Make sure that it is working and then open the front only as far as is necessary: many fume hoods are marked with a maximum opening.
- Always use the minimum quantity of any hazardous materials.
- Work in a logical, tidy manner and minimise risks by thinking ahead.
- Always clear up at the end of each session. This is an important aspect of safety, encouraging responsible laboratory work.
- Dispose of waste in appropriate containers. Most labs will have bins for sharps, glassware, hazardous solutions and radioactive waste.

Basic rules for fieldwork safety

- Make sure you understand your objectives, the potential hazards and appropriate responses to them, before you set out.
- Your work must be designed carefully, to allow for the experience of the participants and the locations visited. Do not overestimate what can be achieved – fieldwork is often physically demanding.
- Any physical disabilities must be brought to the attention of the organiser, so that appropriate precautions can be taken.
- A comprehensive first-aid kit must be carried: at least two participants should have training in first aid.
- Never work alone without permission from your organiser.
- Make sure you can read a map and use a compass: carry both.
- Your clothing and equipment must be suitable for all of the weather conditions likely to be encountered during the work.
- Check the weather forecast before departure: look out for changes in the weather at all times and do not hesitate to turn back if necessary.
- Leave full details of your intended working locations, routes and times. Never change these arrangements without informing someone.
- Make sure that you know the international distress signals (Table 22.1). Chapter 31 covers general aspects of fieldwork.
- Check mobile phone coverage for the field location – some departments provide satellite mobile phones for fieldwork in remote areas.

Table 22.1   International distress signals

On land

1. *Six* long whistle blasts, torch flashes, arm waves or shouts for help in succession. *Pause* for about 1 min, then repeat, as necessary

On water

1. Using a whistle or torch

   The Morse code signal 'SOS' should be signalled, as follows:

	3 *short* blasts/flashes
then	3 *long* blasts/flashes
then	3 *short* blasts/flashes.

   Pause, then repeat, as necessary

2. Raise and lower arms slowly and repeatedly

3. Wave an oar with a brightly coloured cloth tied to it, slowly from side to side

4. Fire *red* flares or *orange* smoke

Sources for further study

Anon. *Biological Hazards at Work*. Available: http://www.hse.gov.uk/biosafety/information.htm Last accessed 21/03/15.
[Provides links to relevant resources.]

Anon. *COSHH basics.* Available: http://www.hse.gov.uk/coshh/basics.htm Last accessed: 09/07/2015.
[UK Health and Safety Executive website giving an overview of the principles of risk assessment and management.]

Anon. (2004) *Laboratory Biosafety Manual*, 3rd edn. World Health Organisation, Geneva.

Fleming, D.O. and Hunt, D.L. (2006) *Biological Safety: Principles and Practice.* ASM Press, Washington.

Health and Safety Executive (2005) *Control of Substances Hazardous to Health Regulations.* HSE, London.

STUDY EXERCISES

22.1 Test your knowledge of safe working procedures. After reading the appropriate sections of this book, can you remember the following:

(a) The four main steps involved in the process of risk assessment?

(b) The major routes of entry of harmful substances into the body?

(c) The warning labels for the major chemical hazard symbols (either describe them or draw them from memory)?

(d) The international symbol for a biohazard?

(e) The international symbol for radioactivity?

22.2 Locate relevant health and safety features in a laboratory. Find each of the following in one of the laboratories used as part of your course (draw a simple location map, if this seems appropriate):

(a) fire exit(s);

(b) fire-fighting equipment;

(c) first-aid kit;

(d) 'sharps' container;

(e) container for disposal of broken glassware;

(f) eye wash station (where appropriate).

22.3 Investigate the health and safety procedures in operation at your university. Can you find out the following?

(a) Your university's procedure in case of fire.

(b) The colour coding for fire extinguishers available in your department and their recommendations for use.

(c) The accident-reporting procedure used in your department.

(d) Your department's Code of Safe Practice relating to a specific aspect of bioscience, e.g. working with micro-organisms.

22.4 Carry out risk assessments for specific chemical hazards. Look up the hazards associated with the use of the following chemicals and list the appropriate protective measures required to minimise risk during use in a lab class:

(a) formaldehyde solution, used as a preservative for animals collected during a marine biology field trip (Chapter 31);

(b) acetone, used as a solvent for the quantitative analysis of plant pigments (Chapter 68);

(c) sodium hydroxide, used in solid form to prepare a dilute solution to be used for pH adjustment (Chapter 26).

Answers to these study exercises are available at www.pearsoned.co.uk/practicalskills

Reading any volumetric scale – make sure your eye is level with the bottom of the liquid's meniscus and take the reading from this point.

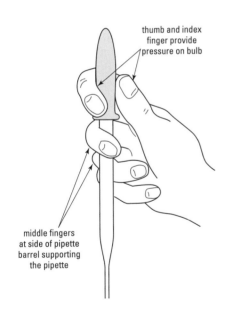

thumb and index finger provide pressure on bulb

middle fingers at side of pipette barrel supporting the pipette

Fig. 23.1 How to hold a Pasteur pipette.

Measuring and dispensing liquids

The equipment you should choose to measure out liquids depends on the volumes being dispensed, the accuracy required and the number of times the job must be done (Table 23.1).

Table 23.1 Criteria for choosing a method for measuring out a liquid

Method	Best volume range	Accuracy	Usefulness for repetitive measurement
Pasteur pipette	0.03–2 mL	Low	Convenient
Conical flask/beaker	25–5000 mL	Very low	Convenient
Measuring cylinder	5–2000 mL	Medium	Convenient
Volumetric flask	5–2000 mL	High	Convenient
Burette	1–100 mL	High	Convenient
Glass pipette	1–100 mL	High	Convenient
Mechanical pipettor	5–1000 μL	High*	Convenient
Syringe	0.5–20 μL	Medium**	Convenient
Microsyringe	0.5–50 μL	High	Convenient
Weighing	Any (depends on accuracy of balance)	Very high	Inconvenient

* If correctly calibrated and used properly (see p. 140).
** Accuracy depends on width of barrel: large volumes are less accurate.

Certain liquids may cause problems:

- **High-viscosity liquids** are difficult to dispense: allow time for all the liquid to transfer.
- **Organic solvents** may evaporate rapidly, making measurements inaccurate: work quickly; seal containers without delay.
- **Solutions prone to frothing** (e.g. protein and detergent solutions) are difficult to measure and dispense: avoid forming bubbles due to overagitation; do not transfer quickly.
- **Suspensions** (e.g. cell cultures) may sediment: thoroughly mix them before dispensing.

Pasteur pipettes

Hold correctly during use (Fig. 23.1) – keep the pipette vertical, with the middle fingers gripping the barrel while the thumb and index finger provide controlled pressure on the bulb. Squeeze gently to dispense individual drops. To avoid the risk of cross-contamination, take care not to draw up solution into the bulb or to lie the pipette on its side. Alternatively, use a plastic disposable 'Pastette'.

Measuring cylinders and volumetric flasks

These must be used on a level surface so that the scale is horizontal; you should first fill with solution until just below the desired mark; then fill slowly (e.g. using a Pasteur pipette) until the meniscus is level with the mark. Allow time for the solution to run down the walls of the vessel.

Fig. 23.2 Glass pipettes: (a) graduated pipette, reading from zero to shoulder; (b) graduated pipette, reading from maximum to tip, by gravity; (c) bulb (volumetric pipette, showing volume (calibration mark to tip, by gravity) above the bulb.

Fig. 23.3 A pipettor (autopipettor) – the Gilson Pipetman. The pipettor is designed to fit comfortably into your 'dominant' hand, with all four fingers wrapped around the upper (coloured) section and your thumb on the push-button.

Burettes

These should be mounted vertically on clamp stands – do not overtighten. First ensure the tap is closed and fill the body with solution using a funnel. Open the tap and allow some liquid to fill the tubing below the tap before first use. Take a meniscus reading, noting the value in your notebook. Dispense the solution via the tap and measure the new meniscus reading. The volume dispensed is the difference between the two readings. Titrations using a burette are usually performed on a magnetic stirrer (pp. 146, 150).

Pipettes

These come in various designs, including graduated and bulb (volumetric) pipettes (Fig. 23.2). Take care to check the volume scale before use: some empty from full volume to zero, others from zero to full volume; some scales refer to the shoulder of the tip, others to the tip either by gravity or after blowing out.

> **KEY POINT** For safety reasons, never mouth-pipette – various aids are available such as the Pi-pump.

Pipettors (autopipettors or micropipettes)

These come in two basic types:

- **Air displacement pipettors.** For routine work with dilute aqueous solutions. One of the most widely used examples is the Gilson Pipetman (Fig. 23.3). Box 23.1 gives practical guidance on its use.
- **Positive displacement pipettors.** For non-standard applications, including dispensing viscous, dense or volatile liquids, or certain procedures in molecular genetics, e.g. the polymerase chain reaction (PCR) (p. 403), where an air displacement pipettor might create aerosols, leading to errors.

Air displacement and positive displacement pipettors may be:

- **Fixed volume:** capable of delivering a single factory-set volume.
- **Adjustable:** where the volume is determined by the operator across a particular range of values.
- **Preset:** movable between a limited number of values.
- **Multi-channel:** able to deliver several replicate volumes at the same time.

Whichever type you use, you must ensure that you understand the operating principles of the volume scale and the method for changing the volume delivered – some pipettors are easily misread.

A pipettor must be fitted with the correct disposable tip before use: each manufacturer produces different tips to fit particular models. Specialised tips are available for particular applications, e.g. PCR (p. 403).

Syringes

Syringes should be used by placing the tip of the needle in the solution and drawing the plunger up slowly to the required point on the scale. Check the barrel to make sure no air bubbles have been drawn up. Expel slowly and touch the syringe on the edge of the vessel to remove any liquid adhering to the end of the needle. Microsyringes should always be cleaned before and after use by repeatedly drawing up and

Box 23.1 Using a pipettor (autopipettor) to deliver accurate, reproducible volumes of liquid

A pipettor can be used to dispense volumes with accuracy and precision, by following this stepwise procedure:

1. **Select a pipettor that operates over the appropriate range.** Most adjustable pipettors are accurate only over a particular working range and should not be used to deliver volumes below the manufacturer's specifications (minimum volume is usually 10–20% of maximum value). Do not attempt to set the volume above the maximum limit, or the pipettor may be damaged.

2. **Set the volume to be delivered.** In some pipettors, you 'dial up' the required volume. Types like the Gilson Pipetman have a system where the scale (or 'volumeter') consists of three numbers, read from top to bottom of the barrel, and adjusted using the black knurled adjustment ring (Fig. 23.3). This number gives the first three digits of the volume scale and thus can only be understood by establishing the maximum volume of the Pipetman, as shown on the push-button on the end of the plunger (Fig. 23.3). The following examples illustrate the principle for two common sizes of Pipetman:

P1000 Pipetman
(maximum volume 1000 µL)
if you dial up

the volume is set at 1000 µL

P20 Pipetman
(maximum volume 20 µL)
if you dial up

the volume is set at 10.0 µL

3. **Fit a new disposable tip to the end of the barrel.** Make sure that it is the appropriate type for your pipettor and that it is correctly fitted. Press the tip on firmly using a slight twisting motion – if not, you will take up less than the set volume and liquid will drip from the tip during use. Tips are often supplied in boxes, for ease of use: if sterility is important, make sure you use appropriate sterile technique at all times (p. 285). *Never, ever, try to use a pipettor without its disposable tip.*

4. **Check your delivery.** Confirm that the pipettor delivers the correct volume by dispensing volumes of distilled water and weighing on a balance, assuming 1 mg = 1 µL = 1 mm³. The value should be within 1% of the selected volume. For small volumes, measure several 'squirts' together, e.g. 20 'squirts' of 5 µL = 100 mg. If

the pipettor is inaccurate (p. 175) giving a biased result (e.g. delivering significantly more or less than the volume set), you can make a temporary correction by adjusting the volumeter scale down or up accordingly (the volume *delivered* is more important than the value *displayed* on the volumeter), or have the pipettor recalibrated. If the pipettor is imprecise (p. 175), delivering a variable amount of liquid each time, it may need to be serviced. After calibration, fit a clean (sterile) tip if necessary.

5. **Draw up the appropriate volume.** Holding the pipettor *vertically*, press down on the plunger/push-button until a resistance (spring-loaded stop) is met. Then place the end of the tip in the liquid. Keeping your thumb on the plunger/push-button, release the pressure slowly and evenly: watch the liquid being drawn up into the tip, to confirm that no air bubbles are present. Wait a second or so, to confirm that the liquid has been taken up, then withdraw the end of the tip from the liquid. Inexperienced users often have problems caused by drawing up the liquid too quickly/carelessly. If you accidentally draw liquid into the barrel, seek assistance from your demonstrator or supervisor as the barrel will need to be cleaned before further use.

6. **Make a quick visual check on the liquid in the tip.** Does the volume seem reasonable? (For example, a 100 µL volume should occupy approximately half the volume of a P200 tip.) The liquid will remain in the tip, without dripping, as long as the tip is fitted correctly and the pipettor is not tilted too far from a vertical position.

7. **Deliver the liquid.** Place the end of the tip against the wall of the vessel at a slight angle (10–15° from vertical) and press the plunger/push-button slowly and smoothly to the first (spring-loaded) stop. Wait a second or two, to allow any residual liquid to run down the inside of the tip, then press again to the final stop, dispensing any remaining liquid. Remove from the vessel with the plunger/push-button still depressed.

8. **Eject the tip.** Press the tip ejector button if present (Fig. 23.3). If the tip is contaminated, eject directly into an appropriate container, e.g. a beaker of disinfectant, for microbiological work, or a labelled container for hazardous solutions (p. 137). For repeat delivery, fit a new tip if necessary and begin again at step 5 above. Always make sure that the tip is ejected before putting a pipettor on the bench.

Example Using eqn [23.1], 9 g of a liquid with a density of 1.2 g mL^{-1} = 7.5 mL.

Fig. 23.4 Stabilising pipettor (autopipettor) with the thumb of your second hand (note: thumb is positioned well away from the disposable tip).

expelling pure solvent. The dead space in the syringe needle can occupy up to 4% of the nominal syringe volume. A way of avoiding such problems is to fill the dead space with an inert substance (e.g. silicone oil) after sampling. Alternatively, use a syringe where the plunger occupies the needle space (small volumes only).

Balances

These can be used to weigh accurately (p. 150) how much liquid you have dispensed. Convert mass to volume using the equation:

$$\text{mass/density} = \text{volume} \qquad [23.1]$$

Densities of common solvents can be found in Haynes (2014). You will also need to know the liquid's temperature, as density is temperature-dependent.

Holding and storing liquids

Test tubes

These are used for colour tests, small-scale reactions, holding cultures, etc. The tube can be sterilised by heating (pp. 283, 285) and maintained in this state with a cap or cotton-wool plug.

Beakers

Beakers are used for general purposes, e.g. heating a solvent while the solute dissolves, carrying out a titration.

Conical (Erlenmeyer) flasks

These are used for storage of solutions: their wide base makes them stable, while their small mouth reduces evaporation and is easily sealed.

Bottles and vials

These are used when the solution needs to be sealed for safety, sterility or to prevent evaporation or oxidation. They usually have a screw top or ground-glass stopper to prevent evaporation and contamination. Many types are available, including 'bijou', 'McCartney', 'universal' and 'Winkler'.

You should clearly label all stored solutions (see p. 150), including relevant hazard information, preferably marking with hazard warning tape (p. 137). Seal vessels in an appropriate way, e.g. using a stopper or a sealing film such as Parafilm or Nescofilm to prevent evaporation. To avoid degradation store your solution in a fridge, but allow it to reach room temperature before use.

Creating specialised apparatus

Glassware systems incorporating ground-glass connections such as Quickfit are useful for setting up combinations of standard glass components, e.g. for chemical reactions. In project work, you may need to adapt standard forms of glassware for a special need. A glassblowing service (often available in chemistry departments) can make special items to order.

Choosing between glass and plastic

Bear in mind the following points:

- **Reactivity.** Plastic vessels often distort at relatively low temperatures; they may be flammable, may dissolve in certain organic solvents and may be affected by prolonged exposure to ultraviolet (UV) light. Some plasticisers may leach from vessels and have been shown to have

Table 23.2 Typical spectral cut-off values for glass and plastics (λ_{50} = wavelength at which transmission of EMR is reduced to 50%)

Material	λ_{50} (nm)
Routine glassware	340
Pyrex glass	292
Polycarbonate	396
Acrylic	342
Polyester	318
Quartz	220

SAFETY NOTE Special cleaning of glass – for an acid wash use dilute acid, e.g. 100 mmolL^{-1} (100 mol m^{-3}) HCl. Rinse thoroughly at least three times with distilled or deionised water. Glassware that must be exceptionally clean (e.g. for a micro-nutrient study) should be washed in a chromic acid bath, but this involves toxic and corrosive chemicals and should only be used under supervision.

biological activity. Glass may adsorb ions and other molecules and then leach them into solutions, especially in alkaline conditions. Pyrex glass is stronger than ordinary soda glass and can withstand temperatures up to 500 °C.

- **Rigidity and resilience.** Plastic vessels are not recommended where volume is critical as they may distort through time: use class A volumetric glassware for accurate work, e.g. preparing solutions (Chapter 25). Glass vessels are more easily broken than plastic (Box 23.2), which is particularly important for centrifugation (see p. 445).
- **Opacity.** Both glass and plastic absorb light in the UV range of the EMR spectrum (Table 23.2). Quartz should be used where this is important, e.g. in cuvettes for UV spectrophotometry (see p. 449).
- **Disposability.** Plastic items may be cheap enough to make them disposable, an advantage where there is a risk of chemical or microbial contamination.

Cleaning glass and plastic

Take care to avoid the possibility of contamination arising from prior use of chemicals or inadequate rinsing following washing. A thorough rinse with distilled or deionized water immediately before use will remove dust and other deposits and is good practice in quantitative work, but ensure that the rinsing solution is not left in the vessel. 'Strong' basic detergents (e.g. Pyroneg) are good for solubilising acidic deposits. If there is a risk of basic deposits

---

Box 23.2 Safe working with glass

Many minor accidents in the laboratory are due to lack of care with glassware. You should follow these general precautions:

- Always wear safety glasses when there is any risk of glass breakage – e.g. when using low pressures, or heating solutions.
- Take care when attaching tubing to glass tubes and when putting glass tubes into bungs – always hold the tubing and glassware close together, as shown in Fig. 23.5, and wear thick gloves when appropriate.
- Use a 'soft' Bunsen flame when heating glassware – this avoids creating a hot spot, where cracks may start: always use tongs or special heat-resistant gloves when handling hot glassware (never use a rolled-up paper towel).
- Do not use chipped or cracked glassware – it may break under very slight strain and should be disposed of in the broken glassware bin.
- Never carry large glass bottles/flasks by their necks – support them with a hand underneath or, better still, carry them in a basket.
- Do not force bungs too firmly into bottles – these can be extremely difficult to remove. If you need a tight seal, use a screw-top bottle with a rubber or plastic seal.

Fig. 23.5 Handling glass pipettes and tubing.

- Dispose of broken glass thoroughly and carefully – use disposable paper towels and wear thick gloves. Always put pieces of broken glass in the correct bin.

remaining, use an acid wash. If there is a risk of contamination from organic deposits, a rinse with Analar grade ethanol is recommended. Glassware can be disinfected by washing with a sodium hypochlorite bleach such as Chloros, with sodium metabisulphite or a blended commercial product such as Virkon – dilute as recommended before use and rinse thoroughly with sterile water after use. Alternatively, to sterilise, heat glassware to at least 121 °C for 15 min, in an autoclave or 160 °C for 3 h in an oven.

Text reference

Haynes, W.M. (ed.) (2014) *CRC Handbook of Chemistry and Physics,* 5th edn. CRC Press, Boca Raton.

Sources for further study

Boyer, R.F. (2011) *Biochemistry Laboratory: Modern Theory and Techniques,* 2nd edn. Prentice Hall, New Jersey.

Hendrickson, C., Byrd, L.C., Hunter, N.W. and Henrickson, C. (2005) *A Laboratory Manual for General, Organic and Biochemistry,* 5th edn. McGraw Hill, New York.

Ninfa, A.J., Ballou, D.P. and Benore, M. (2010) *Fundamental Laboratory Approaches for Biochemistry and Biotechnology*, 2nd edn. Wiley, New York.

Seidman, L.A. and Moore, C.J. (2008) *Basic Laboratory Methods for Biotechnology,* 2nd edn. Benjamin Cummings, San Francisco.

STUDY EXERCISES

23.1 Decide on the appropriate methods and equipment for the following procedures:

(a) Preparing one litre of ethanol at approximately 70% v/v in water, for use as a general preservative (Chapter 37).

(b) Adding 10μL of a sample to the well of an agarose gel during a molecular biology procedure (Chapter 62).

(c) Preparing a calibration standard of 100 mL of DNA, to contain 200.0 μg mL^{-1}, for spectrophotometry (Chapter 68).

(d) Carrying out a titration curve for a buffer solution (Chapter 26).

23.2 Write a protocol for calibrating and using a pipettor. After reading this chapter, prepare a detailed stepwise protocol explaining how to use a pipettor to deliver a specific volume, say of 500μL (e.g. using a Gilson Pipetman, or an alternative if your department does not use this type). Ask another student to evaluate your protocol and provide you with written feedback – either simply by reading through your protocol, or by trying it out with a pipettor as part of a class exercise (check with a member of staff before you attempt this in a laboratory).

23.3 Determine the accuracy and precision of a pipette. Using the following data for three different models of pipettor, determine which pipettor is most *accurate* and which is most *precise* (check p. 175 if you are unsure of the definitions of these two terms). All three pipettors were set to deliver 1000 μL (1.000 mL), and ten repetitive measurements of the weight of the volume of water in grammes delivered were made using a three-place balance:

Model A pipettor:	0.986; 0.971; 0.993; 0.964; 0.983; 0.996; 0.977; 0.969; 0.982; 0.974
Model B pipettor:	1.013; 1.011; 1.010; 1.009; 1.011; 1.010; 1.011; 1.009; 1.011; 1.012
Model C pipettor:	0.985; 1.022; 1.051; 1.067; 0.973; 0.982; 0.894; 1.045; 1.062; 0.928

In your answer, you should support your conclusions with appropriate numerical (statistical) evidence (see Chapter 75 for appropriate measures of location and dispersion).

Answers to these study exercises are available at www.pearsoned.co.uk/practicalskills

Using chemicals

Safety aspects

In practical classes, the person in charge has a responsibility to inform you of any hazards associated with the use of chemicals. For routine practical procedures, a risk assessment (p. 136) will have been carried out by a member of staff and relevant safety information will be included in the practical schedule: an example is shown in Table 24.1.

In project work, your first duty when using an unfamiliar chemical is to find out about its properties, especially those relating to safety. Your department must provide the relevant information to allow you to do this. If your supervisor has filled out the form, read it carefully before signing. Box 24.1 gives further advice.

> **KEY POINT** Before you use any chemical you must find out whether safety precautions need to be taken and complete the appropriate forms confirming that you appreciate the risks involved.

Selection

Chemicals are supplied in various degrees of purity and this is always stated on the manufacturer's containers. Suppliers differ in the names given to the grades and there is no conformity in purity standards. Very pure chemicals cost more, sometimes a lot more, and should only be used if the situation demands. If you need to order a chemical, your department will have a defined procedure for doing this.

Preparing solutions

Solutions are usually prepared with respect to their molar concentrations (e.g. mmol L^{-1}, or mol m^{-3}), or mass concentrations (e.g. g L^{-1}, or kg m^{-3}):

Using chemicals responsibly – be considerate to others: always return storeroom chemicals promptly to the correct place. Report when supplies are getting low to the person who looks after storage/ordering. If you empty an aspirator or wash bottle, fill it up from the appropriate source.

Finding out about chemicals – *The Merck Index* (O'Neil *et al.*, 2013) and the *CRC Handbook of Chemistry and Physics* (Haynes, 2014) are useful sources of information on the physical and biological properties of chemicals, including melting and boiling points, solubility, toxicity, etc. (see Fig. 24.1).

8599. Sodium Chloride. [7647-14-5] Salt; common salt. ClNa; mol wt 58.44. Cl 60.67%, Na 39.34%. NaCl. The article of commerce is also known as *table salt, rock salt* or *sea salt.* Occurs in nature as the mineral *halite.* Produced by mining (rock salt), by evaporation of brine from underground salt deposits and from sea water by solar evaporation: *Faith, Keyes & Clark's Industrial Chemicals,* F. A. Lowenheim, M. K. Moran, Eds. (Wiley-Interscience, New York, 4th ed., 1975) pp. 722-730. Toxicity studies: E. M. Boyd, M. N. Shanas, *Arch. Int. Pharmacodyn.* **144,** 86 (1963). Comprehensive monograph: D. W. Kaufmann, *Sodium Chloride,* ACS Monograph Series no. **145** (Reinhold, New York, 1960) 743 pp.

Cubic, white crystals, granules, or powder; colorless and transparent or translucent when in large crystals. d 2.17. The salt of commerce usually contains some calcium and magnesium chlorides which absorb moisture and make it cake. mp 804° and begins to volatilize at a little above this temp. One gram dissolves in 2.8 ml water at 25°, in 2.6 ml boiling water, in 10 ml glycerol; very slightly sol in alcohol. Its soly in water is decreased by HCl. Almost insol in concd HCl. Its aq soln is neutral. pH: 6.7-7.3. d of satd aq soln at 25° is 1.202. A 23% aq soln of sodium chloride freezes at −20.5°C (5°F). LD$_{50}$ orally in rats: 3.75 ±0.43 g/kg (Boyd, Shanas).

Note: Blusalt, a brand of sodium chloride contg trace amounts of cobalt, iodine, iron, copper, manganese, zinc is used in farm animals.

USE: Natural salt is the source of chlorine and of sodium as well as of all, or practically all, their compds, e.g., hydrochloric acid, chlorates, sodium carbonate, hydroxide, etc.; for preserving foods; manuf soap, to salt out dyes in freezing mixtures; for dyeing and printing fabrics, glazing pottery, curing hides; metallurgy of tin and other metals.

THERAP CAT: Electrolyte replenisher; emetic; topical anti-inflammatory.

THERAP CAT (VET): Essential nutrient factor. May be given orally as emetic, stomachic, laxative or to stimulate thirst (prevention of calculi). Intravenously as isotonic solution to raise blood volume, to combat dehydration. Locally as wound irrigant, rectal douche.

Fig. 24.1 Example of a typical *Merck Index* entry showing the type of information given for each chemical.

Table 24.1 Representative risk-assessment information for a practical exercise in molecular biology, involving the isolation of DNA

Substance	Hazards	Comments
Sodium dodecyl sulphate (SDS)	Irritant Toxic	Wear gloves
Sodium hydroxide (NaOH)	Highly corrosive Severe irritant	Wear gloves
Isopropanol	Highly flammable Irritant/corrosive Potential carcinogen	No naked flames Wear gloves
Phenol	Highly toxic Causes skin burns Potential carcinogen	Use in fume hood Wear gloves
Chloroform	Volatile and toxic Irritant/corrosive Potential carcinogen	Use in fume hood Wear gloves

Box 24.1 Safe working with chemicals

You should always treat chemicals as potentially dangerous, following these general precautions:

- Do not use any chemical until you have considered the risks involved – for lab classes, you should carefully read all hazard and risk information provided *before* you start work. In project work, you may need to be involved in the risk-assessment process with your supervisor.

- Wear a laboratory coat at all times – the coat should be fully fastened and cleaned appropriately, should any chemical compound be spilled on it. Closed-toe footwear will protect your feet should spillages occur.

- Make sure you know where the safety apparatus is kept before you begin working – this includes eye bath, fire extinguisher, blanket, first-aid kit.

- Wear safety glasses and gloves when working with toxic, irritant or corrosive chemicals, and for any substances where the hazards are not yet fully characterised – make sure you understand the hazard warning signs (p. 137) along with any specific hazard-coding system used in your department. Carry out procedures with solid material in a fume hood.

- Use aids such as pipette fillers to minimise the risk of contact with hazardous solutions – these aids are further detailed on pp. 139–40.

- Never smoke, eat, drink or chew gum in a lab where chemicals are handled – this will minimise the risk of ingestion.

- Label all solutions appropriately – use the appropriate hazard warning information (see pp. 137, 150).

- Report all spillages of chemicals/solutions – make sure that spillages are cleaned up properly.

- Store hazardous chemicals only in the appropriate locations – for example, a spark-proof fridge is required for flammable liquids; acids and solvents should not be stored together.

- Dispose of chemicals in the correct manner – if unsure, ask a member of staff (do not assume that it is safe to use the lab waste bin or the sink for disposal).

- Wash hands after any direct contact with chemicals or biochemical material – always wash your hands at the end of a lab session.

Examples Using eqn [24.1], 25 g of a substance dissolved in 400 mL of water would have a mass concentration (p. 155) of $25 \div 400 = 0.0625$ g mL^{-1} ($\equiv 62.5$ mg mL^{-1} $\equiv$ 62.5 g L^{-1})

Using eqn [24.1], 0.4 mol of a substance dissolved in 0.5 L of water would have a molar concentration of $0.4 \div 0.5 = 0.8$ mol L^{-1} ($\equiv 800$ mmol L^{-1}).

Solving solubility problems – if your chemical does not dissolve after a reasonable time:

- check the limits of solubility for your compound (see *Merck Index*, O'Neil *et al.*, 2013),
- check the pH of the solution – solubility often changes with pH, e.g. you may be able to dissolve the compound in an acidic or basic solution.

---

both can be regarded as an amount of *substance* per unit volume of *solution*, in accordance with the relationship:

$$\text{concentration} = \frac{\text{amount}}{\text{volume}} \qquad [24.1]$$

The most important aspect of eqn [24.1] is to recognise clearly the units involved, and to prepare the solution accordingly: for molar concentrations, you will need the relative molecular mass of the compound, so that you can determine the mass of substance required. Further advice on concentrations and interconversion of units is given on p. 156.

Box 24.2 shows the steps involved in making up a solution. The concentration you require is likely to be defined by a protocol you are following and the grade of chemical and supplier may also be specified. Success may depend on using the same source and quality, e.g. with enzyme work. To avoid waste, think carefully about the volume of solution you require, though it is always a good idea to err on the high side because you may spill some or make a mistake when dispensing it. Try to choose one of the standard volumes for vessels, as this will make measuring-out easier.

Use distilled or deionised water to make up aqueous solutions and stir to make sure all the chemical is dissolved. Magnetic stirrers are the most convenient means of doing this: carefully drop a clean magnetic stirrer bar ('flea') in the beaker, avoiding splashing; place the beaker centrally on the stirrer plate, switch on the stirrer and gradually increase the speed of stirring. When the crystals or powder have completely dissolved, switch off and retrieve the flea with a magnet or another flea. Take care not to contaminate your solution when you do this and rinse the flea with distilled water.

Box 24.2 How to make up an aqueous solution of known concentration from solid material

1. Find out or decide the concentration of chemical required and the degree of purity necessary.

2. Decide on the volume of solution required.

3. Find out the relative molecular mass of the chemical (M_r). This is the sum of the atomic (elemental) masses of the component elements and can be found on the container. If the chemical is hydrated, i.e. has water molecules associated with it, these must be included when calculating the mass required.

4. Work out the mass of chemical that will give the concentration desired in the volume required.

    Suppose your procedure requires you to prepare 250 mL of 0.1 mol L^{-1} NaCl.

    (a) Begin by expressing all volumes in the same units, either millilitres or litres (e.g. 250 mL as 0.25 L).

    (b) Calculate the number of moles required from eqn [24.1]: 0.1 = amount (mol) ÷ 0.25.
    By rearrangement, the required number of moles is thus 0.1 × 0.25 = 0.025 mol.

    (c) Convert from mol to g by multiplying by the relative molecular mass (M_r for NaCl = 58.44).

    (d) Therefore, you need to make up 0.025 × 58.44 = 1.461 g to a final volume of 250 mL of solution, using distilled water.

    In some instances, it may be easier to work in SI units, though you must be careful when using exponential numbers (p. 499).

    Suppose your protocol states that you need 100 mL of 10 mmol L^{-1} KCl.

    (a) Start by converting this to 100 × 10^{-6} m^3 of 10 mol m^{-3} KCl.

    (b) The required number of mol is thus (100 × 10^{-6}) × (10) = 10^3.

    (c) Each mol of KCl weighs 72.56 g (M_r, the relative molecular mass).

    (d) Therefore you need to make up 72.56 × 10^{-3} g = 72.56 mg KCl to 100 × 10^{-6} m^3 (100 mL) with distilled water.

    See Box 25.1 for additional information.

5. Weigh out the required mass of chemical to an appropriate accuracy. If the mass is too small to weigh to the desired degree of accuracy, consider the following options:

    (a) Make up a greater volume of solution.

    (b) Make up a stock solution that can be diluted at a later stage (see below).

    (c) Weigh the mass first, and calculate what volume to make the solution up to afterwards using eqn [24.1].

6. Add the chemical to a beaker or conical flask then add a little less water than the final amount required. If some of the chemical sticks to the paper, foil or weighing boat, use some of the water to wash it off.

7. Stir and, if necessary, heat the solution to ensure all the chemical dissolves. You can visually determine when this has happened by observing the disappearance of the crystals or powder.

8. If required, check and adjust the pH of the solution when cool (see Chapter 26).

9. Make up the solution to the desired volume. If the concentration needs to be accurate, use a class A volumetric flask; if a high degree of accuracy is not required, use a measuring cylinder (class B).

    (a) Pour the solution from the beaker into the measuring vessel using a funnel to avoid spillage.

    (b) Make up the volume so that the meniscus comes up to the appropriate measurement line. For accurate work, rinse out the original vessel and use this liquid to make up the volume.

10. Transfer the solution to a reagent bottle or a conical flask and label the vessel clearly.

'Obstinate' solutions may require heating, but do this only if you know that the chemical will not be damaged at the temperature used. Use a stirrer-heater to keep the solution mixed as you heat it. Allow the solution to cool before you measure volume or pH as these are affected by temperature.

Stock solutions

Stock solutions are valuable when making up a range of solutions containing different concentrations of a reagent or if the solutions have some common ingredients. They also save work if the same solution is used over a prolonged period (e.g. a nutrient solution). The stock solution is more

Working with chemicals and stock solutions – it is poor laboratory practice to return unused chemicals or stock solutions to their original containers after use. They should be discarded, following the appropriate safety procedures.

Table 24.2 Use of stock solutions. Suppose you need a set of solutions 10 mL in volume containing differing concentrations of KCl, with and without reagent Q. You decide to make up a stock of KCl at twice the maximum required concentration (50 mmol L^{-1} = 50 mol m^{-3}) and a stock of reagent Q at twice its required concentration. The table shows how you might use these stocks to make up the media you require. Note that the total volumes of stock you require can be calculated from the table (end column)

Stock solutions	Volume of stock required to make required solutions (mL)						Total volume of stock required (mL)
	No KCl plus Q	No KCl minus Q	15 mmol L^{-1} KCl plus Q	15 mmol L^{-1} KCl minus Q	25 mmol L^{-1} KCl plus Q	25 mmol L^{-1} KCl minus Q	
50 mmol L^{-1} KCl	0	0	3	3	5	5	16
[reagent Q] × 2	5	0	5	0	5	0	15
Water	5	10	2	7	0	5	29
Total	10	10	10	10	10	10	60

Storing stock solutions – these are best kept in darkness in a refrigerator to minimise the growth of photoautotrophic microbes.

Making a dilution – use the relationship $[C_1]V_1 = [C_2]V_2$ to determine volume or concentration (see p. 156).

Removing a magnetic flea from a volumetric flask – use a strong magnet to bring the flea to the top of the flask, to avoid contamination during removal.

Using the correct volumes for dilutions – it is important to distinguish between the volumes of the various liquids: a one-in-ten dilution is obtained using 1 volume of stock solution plus 9 volumes of diluent $(1 + 9 = 10)$. Note that when this is shown as a ratio, it may represent the initial and final volumes (e.g. 1:10) or, sometimes, the volumes of stock solution and diluent (e.g. 1:9).

Using diluents – various liquids are used, including distilled or deionised water, salt solutions, buffers, Ringer's solution (p. 298), etc., according to the specific requirements of the procedure.

concentrated than the final requirement and is diluted as appropriate when the final solutions are made up. The principle is best illustrated with an example (Table 24.2).

Preparing dilutions

Making a single dilution

You may need to dilute a stock solution to give a particular mass concentration, or molar concentration. Use the following procedure:

1. **Transfer an accurate volume of stock solution to a volumetric flask,** using appropriate equipment (Table 23.1).
2. **Make up to the calibration mark with solvent** – add the last few drops from a pipette or solvent bottle, until the meniscus is level with the calibration mark.
3. **Mix thoroughly,** either by repeated inversion (holding the stopper firmly) or by prolonged stirring, using a magnetic stirrer. Make sure you add the magnetic flea *after* the volume adjustment step.

For routine work using dilute aqueous solutions where the highest degree of accuracy is not required, it may be acceptable to substitute test tubes or conical flasks for volumetric flasks. In such cases, you would calculate the volumes of stock solution and diluent required, with the assumption that the final volume is determined by the sum of the individual volumes of stock and diluent used (e.g. Table 24.2). Thus, a two-fold dilution would be prepared using 1 volume of stock solution and 1 volume of diluent. The dilution factor is obtained from the ratio of the initial concentration of the stock solution and the final concentration of the diluted solution. The dilution factor can be used to determine the volumes of stock and diluent required in a particular instance. For example, suppose you wanted to prepare 100 mL of a solution of NaCl at 0.2 mol L^{-1}. Using a stock solution containing 4.0 mol L^{-1} NaCl, the dilution factor is $0.2 \div 4.0 = 0.05 = 1/20$ (a twenty-fold dilution). Therefore, the amount of stock solution required is 1/20th of 100 mL = 5 mL and the amount of diluent needed is 19/20th of 100 mL = 95 mL.

Preparing a dilution series

Dilution series are used in a wide range of procedures, including the preparation of standard curves for calibration of analytical instruments (p. 327), and in microbiology and immunoassay, where a range of dilutions of

Using a linear dilution series – this is useful when preparing a calibration curve (p. 327) for spectrophotometric analysis of a substance, based on the Beer–Lambert relationship (p. 447).

Using a logarithmic dilution series – this is useful when a broad range of concentrations is required, e.g. for titration of biologically active substances (p. 166), making a plate count of a suspension of microbes (p. 295), or when a process is logarithmically related to concentration (Chapter 53).

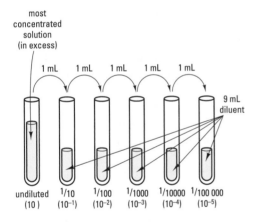

Fig. 24.2 Preparation of a dilution series. The example shown is a decimal dilution series, down to 1/100 000 (10^{-5}) of the solution in the first (left-hand) tube. Note that all solutions must be mixed thoroughly before transferring the volume to the next in the series. In microbiology and cell culture, sterile solutions and appropriate aseptic technique will be required (Chapter 47).

a particular sample is often required (pp. 295, 380). A variety of different approaches can be used:

Linear dilution series Here, the concentrations are separated by an equal amount, e.g. a series containing protein at 0, 0.2, 0.4, 0.6, 0.8, 1.0 µg mL^{-1}. Such a dilution series might be used to prepare a calibration curve for spectrophotometric assay of protein concentration (Box 60.1), or an enzyme assay (p. 387). Use $[C_1]V_1 = [C_2]V_2$ (p. 156) to determine the amount of stock solution required for each member of the series, with the volume of diluent being determined by subtraction.

Logarithmic dilution series Here, the concentrations are separated by a constant proportion, often referred to as the step interval.

The most common examples are:

- **Doubling dilutions** – where each concentration is half that of the previous one (two-fold step interval, $\log_2$ dilution series). First, make up the most concentrated solution at twice the volume required. Measure out half of this volume into a vessel containing the same volume of diluent, mix thoroughly and repeat, for as many doubling dilutions as are required. The concentrations obtained will be 1/2, 1/4, 1/8, 1/16, etc., times the original (i.e. the dilutions will be two-, four-, eight- and sixteen-fold, etc.).

- **Decimal dilutions** – where each concentration is one-tenth that of the previous one (ten-fold step interval, $\log_{10}$ dilution series). First, make up the most concentrated solution required, with at least a 10% excess. Measure out one-tenth of the volume required into a vessel containing nine times as much diluent, mix thoroughly and repeat. The concentrations obtained will be 1/10, 1/100, 1/1000, etc., times the original (i.e. dilutions of 10^{-1}, 10^{-2}, 10^{-3}, etc.). To calculate the actual concentration of solute, multiply by the appropriate dilution factor.

When preparing serial doubling or decimal dilutions, it is often easiest to add the appropriate amount of diluent to several vessels beforehand, as shown in the worked example in Fig. 24.2. When preparing a dilution series, it is essential that all volumes are dispensed accurately, e.g. using calibrated pipettors (p. 140), otherwise any inaccuracies will be compounded, leading to gross errors in the most dilute solutions.

Harmonic dilution series Here, the concentrations in the series take the values of the reciprocals of successive whole numbers, e.g. 1, 1/2, 1/3, 1/4, 1/5, etc. The individual dilutions are simply achieved by a stepwise increase in the volume of diluent in successive vessels, e.g. by adding 0, 1, 2, 3, 4 and 5 times the volume of diluent to a set of test tubes, then adding a constant unit volume of stock solution to each vessel. Although there is no dilution transfer error between individual dilutions, the main disadvantage is that the series is non-linear, with a step interval that becomes progressively smaller as the series is extended.

Solutions must be thoroughly mixed before measuring out volumes for the next dilution. Use a fresh measuring vessel for each dilution to avoid contamination, or wash your vessel thoroughly between dilutions. Clearly label the vessel containing each dilution when it is made: it is easy to get

Preparing a dilution series using pipettes or pipettors – use a fresh pipette or disposable tip for each dilution, to prevent carry-over of solutions.

SAFETY NOTE When using a vortex mixer with open and capped test tubes – do not vortex too vigorously or liquid will spill from the top of the tube, creating a contamination risk.

SAFETY NOTE You must always clean up any spillages of chemicals, as you are the only person who knows the risks from the spilled material.

Deciding on which balance to use – select a balance that weighs to an appropriate number of decimal places. For example, you should use a top-loading balance weighing to one decimal place for less accurate work. Note that a weight of 6.4 g on such a balance may represent a true value of between 6.350 g and 6.449 g (to three decimal places).

confused. When deciding on the volumes required, allow for the aliquot removed when making up the next member in the series. Remember to discard any excess from the last in the series if volumes are critical.

Mixing solutions and suspensions

Various devices may be used, including:

- **Magnetic stirrers and fleas.** Magnetic fleas come in a range of shapes and sizes, and some stirrers have integral heaters. During use, stirrer speed may increase as the instrument warms up.
- **Orbital shakers and shaking water baths.** Use these to provide controlled mixing at a particular temperature, e.g. for long-term incubation and cell-growth studies (p. 291).
- **Vortex mixers.** For vigorous mixing of small volumes of solution, e.g. when preparing a dilution series in test tubes. Take care when adjusting the mixing speed – if the setting is too low, the test tube will vibrate rather than creating a vortex, giving inadequate mixing. If the setting is too high, the test tube may slip from your hand.
- **Bottle rollers.** For cell-culture work, ensuring gentle, continuous mixing.

Storing chemicals and solutions

Labile chemicals may be stored in a fridge or freezer. Take special care when using chemicals that have been stored at low temperature: the container and its contents must be warmed up to room temperature before use, otherwise water vapour will condense on the chemical. This may render any weighing you do meaningless and it could ruin the chemical. Other chemicals may need to be kept in a desiccator, especially if they are deliquescent.

KEY POINT Label all stored chemicals clearly with the following information: the chemical name (if a solution, state solute(s), concentration(s) and pH if measured), plus any relevant hazard-warning information, the date made up and your name.

Using balances

Electronic balances with digital readouts are now favoured over mechanical types: they are easy to read and their self-taring feature means the mass of the weighing boat or container can be subtracted automatically before weighing an object. The most common type offers accuracy down to 1 mg over the range 1 mg to 160 g, which is suitable for most biological applications.

To operate a standard self-taring balance:

1. **Check that the balance is level,** using the adjustable feet to centre the bubble in the spirit level (usually at the back of the machine). For accurate work, make sure a draught shield is on the balance.
2. **Place an empty vessel in the middle of the balance pan** and allow the reading to stabilise. *If the object is larger than the pan, take care that no part rests on the body of the balance or the draught shield as this will invalidate the reading.* Press the tare bar to bring the reading to zero.

Box 24.3 How to use Vernier calipers

Note that numbers on the scale refer to centimetres. Vernier scales consist of two numerical scales running side by side, the moving one being shorter with ten divisions compressed into the length for nine on the longer, static one. Use Vernier calipers to measure objects to the nearest 0.1 mm:

1. Clamp the stops lightly over the object as in Fig. 24.3(a), taking care not to deform it.

2. **Read off the number of whole millimetres** by taking the value on the fixed scale lying to the left of the first line on the moving (short) scale, i.e. 8 mm in Fig. 24.3(b).

3. **Read off 0.1 mm value** by finding which line in the moving scale corresponds most closely with a line on the fixed scale, i.e. 0.5 mm in Fig. 24.3(b). If the zero of the short scale corresponded to a whole number on the static scale, then record 0.0 mm as this shows fully the precision of the measurement.

4. Add these numbers to give the final reading, i.e. 8.5 mm in Fig. 24.3(b).

Fig. 24.3 (a) Vernier caliper (b) Vernier measurement scale.

Weighing – never weigh anything directly on a balance's pan: you may contaminate it for other users. Use a weighing boat or a small piece of aluminium foil. Otherwise, choose a suitable vessel such as a beaker, conical flask or aluminium tray.

3. **Place the chemical or object carefully in the vessel** (powdered chemicals should be dispensed with a suitably sized clean spatula). Take care to avoid spillages.

4. **Allow the reading to stabilise and make a note of the value.**

5. **If you add excess chemical, take great care when removing it.** Do not return any excess chemical to the original container – dispose of it following appropriate safety measures. Switch off if you need to clean any deposit accidentally left on or around the balance.

Larger masses should be weighed on a top-loading balance to an appropriate degree of accuracy. Take care to note the limits for the balance: while most have devices to protect against overloading, you may damage the mechanism. In the field, spring or battery-operated balances may be preferred. Try to find a place out of the wind to use them.

For extremely small masses, there are electrical balances that can weigh down to 1 μg, but these are very delicate and must be used under supervision.

Measuring length and area

When measuring linear dimensions, the device you need depends on the size of object you are measuring and the precision demanded (Table 24.3).

For many regularly shaped objects, area can be estimated from linear dimensions (see p. 498). The areas of irregular shapes can be measured with an optical measuring device or a planimeter. These have the benefits of speed and ease of use; instructions are machine-specific. A simple 'low tech' method is to trace objects on to good-quality paper or to photocopy

Table 24.3 Suitability of devices for measuring linear dimensions

Measurement device	Suitable lengths	Degree of precision
Eyepiece graticule (light microscopy)	1 μm to 10mm	0.5 μm
Vernier calipers	1–100 mm	0.1 mm
Ruler	10 mm to 1 m	1.0 mm
Tape measure	10 mm to 30 m	1.0 mm
Optical surveying devices	1 m to 100 m	0.1 m

See Box 24.3 for method of using Vernier calipers.

them. If the outline is then cut round, the area can be estimated by weighing the cutout and comparing to the mass of a piece of the same paper of known area. Avoid getting moisture from the specimen on to the paper as this will affect the reading.

Measuring and controlling temperature

Heating specimens

Care is required when heating specimens. Safety glasses should always be worn. Use a thermostatically controlled electric stirrer-heater if possible. If using a Bunsen burner, keep the flame well away from yourself and your clothing (tie back long hair). Use a non-flammable mat beneath a Bunsen to protect the bench. Switch off when no longer required. To light a Bunsen, close the air hole first, then apply a lit match or lighter. Open the air hole if you need a hotter, more concentrated flame: the hottest part of the flame is just above the apex of the blue cone in its centre.

Ovens and drying cabinets may be used to dry specimens or glassware. They are normally thermostatically controlled. If drying organic material for dry weight measurement, do so at about 80 °C to avoid caramelising the specimen. Always state the actual temperature used as this affects results. Check that all water has been driven off by weighing until a constant mass is reached.

Cooling specimens

Fridges and freezers are used for storing stock solutions and chemicals that would either break down or become contaminated at room temperature. Normal fridge and freezer temperatures are about 4°C and –15°C respectively. Ice baths can be used when reactants must be kept close to 0 °C. Most bioscience departments will have a machine that provides flaked ice for use in these baths. If common salt is mixed with ice, temperatures below 0 °C can be achieved. A mixture of ethanol and solid CO_2 will provide a temperature of –72 °C if required. To freeze a specimen quickly, immerse in liquid N_2 (–196 °C) using tongs and wearing an apron and thick gloves, as splashes will damage your skin. Always work in a well-ventilated room.

Maintaining cultures or specimens at constant temperature

Thermostatically controlled temperature rooms and incubators can be used to maintain temperature at a desired level. Always check with a thermometer or thermograph that the thermostat is accurate enough for your study. To achieve a controlled temperature on a smaller scale, e.g. for an oxygen electrode (p. 425), use a water bath. These usually incorporate heating elements, a circulating mechanism and a thermostat. Baths for subambient temperatures have a cooling element.

Controlling atmospheric conditions

Gas composition

The atmosphere may be 'scrubbed' of certain gases by passing through a U-tube or Dreschel bottle containing an appropriate chemical or solution.

For accurate control of gas concentrations, use cylinders of pure gas; the contents can be mixed to give specified concentrations by controlling individual flow rates. The cylinder-head regulator (Fig. 24.4) allows you to

---

SAFETY NOTE There is a danger of fire whenever organic material is heated and a danger of scalding from heated liquids.

SAFETY NOTE Heating/cooling glass vessels – take care if heating or cooling glass vessels rapidly as they may break when heat-stressed. Freezing aqueous solutions in thin-walled glass vessels is risky because ice expansion may break the glass.

Understanding thermometer calibration – some are calibrated for use in air, others require partial immersion in liquid and others total immersion – check before use.

SAFETY NOTE If a mercury thermometer is broken, report the spillage, as mercury is a poison.

Example Water vapour can be removed by passing gas over dehydrated $CaCO_3$ and CO_2 may be removed by bubbling through KOH solution.

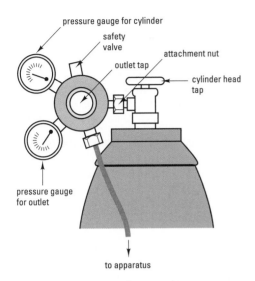

Fig. 24.4 Parts of a cylinder-head regulator. The regulator is normally attached by tightening the attachment nut clockwise; the exception is with cylinders of hydrogen, where the special regulator is tightened *anticlockwise* to avoid the chance of this potentially explosive gas being incorrectly used.

> **Using a timer** – always set the alarm before the critical time, so that you have adequate time to react.

control the pressure (and hence flow rate) of gas; adjust using the controls on the regulator or with spanners of appropriate size. Before use, ensure the regulator outlet tap is off (turn anticlockwise), then switch on at the cylinder (turn clockwise) – the cylinder dial will give you the pressure reading for the cylinder contents. Now switch on at the regulator outlet (turn clockwise) and adjust to desired pressure/flow setting. To switch off, carry out the above directions in reverse order.

Pressure

Many forms of pump are used to pressurise or provide a partial vacuum, usually to force gas or liquid movement. Each has specific instructions for use. Many laboratories are supplied with 'vacuum' (suction) and pressurised air lines that are useful for procedures such as vacuum-assisted filtration. Make sure you switch off the taps after use. Take special care with glass items kept at very low or high pressures. These should be contained within a metal cage to minimise the risk of injury.

Measuring time

Many experiments and observations need to be carefully timed. Large-faced stopclocks allow you to set and follow 'experimental time' and remove the potential difficulties in calculating this from 'real time' on a watch or clock. Some timers incorporate an alarm that you can set to warn when readings or operations must be carried out; 24-h timers are available for controlling light and temperature regimes.

Miscellaneous methods for treating samples

Homogenising

This involves breaking up and mixing specimens to give a uniform preparation. Blenders are used to homogenise animal and plant material and work best when an adequate volume of liquid is present: buffer solution may be added to specimens for this purpose. Use in short bursts to avoid overheating the motor and the sample. A mortar and pestle Fig. 24.5 are used for grinding up specimens. Acid-washed sand grains can be added to help break up the tissues. For quantitative work with brittle samples, care must be taken not to lose material when the sample breaks into fragments.

Separation of components of mixtures and solutions

Particulate solids (e.g. soils) can be separated on the basis of size using sieves. These are available in stacking forms that fit on automatic shakers. Sieves with the largest pores are placed at the top and the assembly is shaken for a fixed time until the sample separates. Suspensions of solids in liquids may be separated out by centrifugation (see p. 442) or filtration. Various forms of filter paper are available having different porosities and purities. Vacuum-assisted filtration speeds up the process and is best carried out with a filter funnel attached to a filter flask. Filtration through pre-sterilised membranes with very small pores (e.g. the Millipore type) is an excellent method of sterilising small volumes of solution. Solvents can be removed from solutes by heating, using rotary film evaporation under low pressure and, for water, by freeze drying. The last two are especially useful for heat-labile solutes – follow the manufacturers' specific instructions for use.

Fig. 24.5 Mortar (bowl) and pestle (grinding tool).

Text references

Haynes, W.M. (ed.) (2014) *CRC Handbook of Chemistry and Physics,* 95th edn. CRC Press, Boca Raton.

O'Neil, M.J., Smith, A. and Henckelman, P.E. (eds.) (2013) *The Merck Index: An Encyclopedia of Chemicals, Drugs and Biologicals,* 15th edn. Merck & Co., Inc., New Jersey.

Sources for further study

Boyer, R.F. (2011) *Biochemistry Laboratory: Modern Theory and Techniques*, 2nd edn. Prentice Hall, New Jersey.

Ninfa, A.J., Ballou, D.P. and Benore, M. (2009) *Fundamental Laboratory Approaches for Biochemistry and Biotechnology*. Wiley, New York.

Seidman, L.A. and Moore, C.J. (2008) *Basic Laboratory Methods for Biotechnology: Textbook and Laboratory Reference*, 2nd edn. Prentice-Hall, New Jersey.

Turgeon, M.L. (2010) *Clinical Laboratory Science: The Basics and Routine Techniques*, 6th edn. Mosby, St. Louis.

STUDY EXERCISES

24.1 Practise the calculations involved in preparing specific volumes of aqueous solutions (see also Study exercise 25.1). What mass of substance would be required to prepare each of the following (answer in each case to three significant figures):

(a) 100 mL of NaCl at 50 mmol L^{-1}? (M_r of NaCl = 58.44.)

(b) 250 mL of mannitol at 0.10 mol L^{-1}? (M_r of mannitol = 182.17.)

(c) 200 mL of a bovine serum albumin solution at 800 μg mL^{-1}?

(d) 0.5 L of $MgCl_2$ prepared from the hexahydrate salt at 22.5 mmol L^{-1}? (M_r of $MgCl_2.6H_2O$ = 203.30.)

(e) 400 mL of DNA at 20 ng μL^{-1}?

24.2 Practise the calculations involved in preparing dilutions (answer in each case to three significant figures).

(a) If you added 1.0 mL of an aqueous solution of NaCl at 0.4 mol L^{-1} to 9.0 mL of water, what would be the final concentration of NaCl in mmol L^{-1}?

(b) If you added 25 mL of an aqueous solution of DNA at 10 μg mL^{-1} to a 500 mL volumetric flask and made it up to the specified volume with water, what would be the final concentration of DNA, in ng mL^{-1}?

(c) If you added 10 μL of an aqueous solution of sucrose at 200 mmol L^{-1} to a 250 mL volumetric flask and made it up to the specified volume with

water, what would be the final concentration of sucrose, in nmol mL^{-1}?

(d) How would you prepare 250 mL of KCl at a final concentration of 20.0 mmol L^{-1} from a solution containing KCl at 0.2 mol L^{-1}?

(e) How would you prepare 1×10^{-3} m^3 of glucose at a final concentration of 50 μmol m^{-3} from a stock solution containing glucose at 20.0 gm^{-3} (M_r of glucose = 180.16)?

24.3 Practise the calculations involved in using stock solutions. Suppose you had the following stock solutions: NaCl 100.0 mmol L^{-1}; KCl 200.0 mmol L^{-1}; $CaCl_2$ 160.0 mmol L^{-1}; glucose 5.0 mmol L^{-1}. Calculate the volumes of each stock solution and the volume of water required to prepare each of the following (answer in each case to three significant figures):

(a) 1.0 mL of a solution containing only KCl at 10.0 mmol L^{-1}

(b) 50 mL of a solution containing NaCl at 2.5 mmol L^{-1} and glucose at 0.5 mmol L^{-1}

(c) 100 mL of a solution containing NaCl at 5.0 mmol L^{-1}, KCl at 2.5 mmol L^{-1}, $CaCl_2$ at 40.0 mmol L^{-1} and glucose at 0.25 mmol L^{-1}

(d) 10 mL of a solution containing NaCl, $CaCl_2$ and KCl, all at 20.0 mmol mL^{-1}

(e) 25 mL of a solution containing CaCl2 at 8.0 mmol L^{-1}, KCl at 50.0 mmol L^{-1} and glucose at 20.0 nmol mL^{-1}.

Answers to these study exercises are available at www.pearsoned.co.uk/practicalskills

Definitions

Electrolyte – a substance that dissociates, either fully or partially, in water to give two or more ions.

Mole (of a substance) – the equivalent in mass to relative molecular mass, in grams.

Relative atomic mass (A_r) – the mass of an atom relative to $^{12}C = 12$.

Relative molecular mass (M_r) – the mass of a compound's formula unit relative to $^{12}C = 12$.

Expressing solute concentrations – you should use SI units wherever possible. However, you are likely to meet non-SI concentrations and you must be able to deal with these units too.

Example   A 1.0 molar solution of NaCl would contain 58.44g NaCl (the relative molecular mass) per litre of solution.

A solution is a homogeneous liquid, formed by the addition of solutes to a solvent (usually water in biological systems). The behaviour of solutions is determined by the types of solutes involved and by their proportions, relative to the solvent. Many laboratory exercises involve calculation of concentrations, e.g. when preparing an experimental solution at a particular concentration, or when expressing data in terms of solute concentration (see pp. 156–8; 162). Make sure that you understand the basic principles set out in this chapter before you tackle such exercises.

Solutes can affect the properties of solutions in several ways, including:

Electrolytic dissociation

This occurs where individual molecules of an electrolyte dissociate to give charged particles (ions). For a strong electrolyte, e.g. NaCl, dissociation is essentially complete. In contrast, a weak electrolyte, e.g. acetic acid, will be only partly dissociated, depending upon the pH and temperature of the solution (p. 163).

Osmotic effects

These are the result of solute particles lowering the effective concentration of the solvent (water). These effects are particularly relevant to biological systems since membranes are far more permeable to water than to most solutes. Water moves across biological membranes from the solution with the higher effective water concentration to that with the lower effective water concentration (osmosis).

Ideal/non-ideal behaviour

This occurs because solutions of real substances do not necessarily conform to the theoretical relationships predicted for dilute solutions of so-called ideal solutes. It is often necessary to take account of the non-ideal behaviour of real solutions, especially at high solute concentrations (see Haynes, 2014, and Robinson and Stokes, 2002, for appropriate data).

Concentration

In SI units (p. 180), the concentration of a solute in a solution is expressed in mol m^{-3}, which is convenient for most biological purposes. The concentration of a solute is usually symbolised by square brackets, e.g. [NaCl]. Details of how to prepare a solution using SI and non-SI units are given on p. 147.

A number of alternative ways of expressing the relative amounts of solute and solvent are in general use, and you may come across these terms in your practical work, or in the literature:

Molarity

This is the term used to denote molar concentration, [C], expressed as moles of solute per litre volume of solution (mol L^{-1}). This non-SI term continues to find widespread usage, in part because of the familiarity of working scientists with the term, but also because laboratory glassware is calibrated in millilitres and litres, making the preparation of molar and millimolar

---

> ### Box 25.1 Useful procedures for calculations involving molar concentrations
>
> 1. **Preparing a solution of defined molarity.** For a solute of known relative molecular mass, M_r, the following relationship can be applied:
>
> $$[C] = \frac{\text{mass of solute/relative molecular mass}}{\text{volume of solution}} \quad [25.1]$$
>
> So, if you wanted to make up 200 mL (0.2 l) of an aqueous solution of NaCl (M_r 58.44) at a concentration of 500 mmol L⁻¹ (0.5 mol L⁻¹), you could calculate the amount of NaCl required by inserting these values into eqn [25.1]:
>
> $$0.5 = \frac{\text{mass of solute/58.44}}{0.2}$$
>
> which can be rearranged to
>
> $$\text{mass of solute} = 0.5 \times 0.2 \times 58.44 = 5.844 \text{ g}$$
>
> The same relationship can be used to calculate the concentration of a solution containing a known amount of a solute, e.g. if 21.1 g of NaCl were made up to a volume of 100 mL (0.1L), this would give
>
> $$[\text{NaCl}] = \frac{21.1/58.44}{0.1} = 3.61 \text{ mol L}^{-1}$$
>
> 2. **Dilutions and concentrations.** The following relationship is very useful if you are diluting (or concentrating) a solution:
>
> $$[C_1]V_1 = [C_2]V_2 \quad [25.2]$$
>
> where $[C_1]$ and $[C_2]$ are the initial and final concentrations, while V_1 and V_2 are their respective volumes: each pair must be expressed in the same units. Thus, if you wanted to dilute 200mL of 0.5 mol L⁻¹ NaCl to give a final molarity of 0.1 mol L⁻¹, then, by substitution into eqn [25.2]:
>
> $$0.5 \times 200 = 0.1 \times V_2$$
>
> Thus $V_2 = 1000$ mL (in other words, you would have to add water to 200 mL of 0.5 mol L⁻¹ NaCl to give a final volume of 1000 mL to obtain a 0.1 mol L⁻¹ solution).
>
> 3. **Interconversion.** A simple way of interconverting amounts and volumes of any particular solution is to divide the amount and volume by a factor of 10³: thus a molar solution of a substance contains 1 mol L⁻¹, which is equivalent to 1 mmol mL⁻¹, or 1 μmol μL⁻¹, or 1 nmol nL⁻¹, etc. You may find this technique useful when calculating the amount of substance present in a small volume of solution of known concentration, e.g. to calculate the amount of NaCl present in 50 μL of a solution with a concentration (molarity) of 0.5 mol L⁻¹ NaCl:
>
> (a) this is equivalent to 0.5 μmol μL⁻¹;
>
> (b) therefore 50 μL will contain 50 × 0.5 μmol L = 25 μmol L.
>
> Alternatively, you may prefer to convert to primary SI units, for ease of calculation (see Box 28.1).
> The 'unitary method' (p. 501) is an alternative approach to these calculations.

> **Example** A 1.0 molar solution can also be expressed as a 1.0 mol L⁻¹ solution.

solutions relatively straightforward. However, the symbols in common use for molar (M) and millimolar (mM) solutions are at odds with the SI system and many people now prefer to use mol L⁻¹ and mmol L⁻¹ respectively, to avoid confusion (you may also see mol L⁻¹ and mmol L⁻¹). Box 25.1 gives details of some useful approaches to calculations involving molarities.

Molality

> **Example** A 0.5 molal solution of NaCl would contain 58.44 × 0.5 = 29.22 g NaCl per kg of water.

This is used to express the concentration of solute relative to the *mass* of solvent, i.e. mol kg⁻¹. Molality is a temperature-independent means of expressing solute concentration, rarely used except when the osmotic properties of a solution are of interest (pp. 158–61).

Per cent composition (% w/w)

> **Example** A 5% w/w sucrose solution contains 5 g sucrose and 95 g water (= 95 mL water, assuming a density of 1 g mL⁻¹) to give 100 g of solution.

This is the solute mass (in g) per 100 g solution. The advantage of this expression is the ease with which a solution can be prepared, since it simply requires each component to be pre-weighed (for water, a volumetric measurement may be used, e.g. using a measuring cylinder) and then mixed together. Similar terms are parts per thousand (‰), i.e. mg g⁻¹, and parts per million (ppm), i.e. μgg⁻¹.

Example A 5% w/v sucrose solution contains 5 g sucrose in 100 mL of solution. A 5% v/v glycerol solution would contain 5 mL glycerol in 100 mL of solution.

Note that when water is the solvent this is often not specified in the expression, e.g. a 20% v/v ethanol solution contains 20% ethanol made up to 100 mL of solution using water.

Per cent concentration (% w/v and % v/v)

For solutes added in solid form, this is the number of grams of solute per 100 mL solution. This is more commonly used than per cent composition, since solutions can be accurately prepared by weighing out the required amount of solute and then making this up to a known volume using a volumetric flask. The equivalent expression for liquid solutes is % v/v.

The principal use of mass/mass or mass/volume terms (including gL^{-1}) is for solutes whose relative molecular mass is unknown (e.g. cellular proteins), or for mixtures of certain classes of substance (e.g. total salt in sea water). You should *never* use the per cent term without specifying how the solution was prepared, i.e. by using the qualifier w/w, w/v or v/v. For mass concentrations, it is simpler to use mass per unit volume, e.g. $mg\,L^{-1}$, $\mu g\,\mu L^{-1}$, etc.

Parts per million concentration (ppm)

This is a non-SI weight per volume (w/v) concentration term commonly used in quantitative analysis such as flame photometry, atomic absorption spectroscopy and gas chromatography, where low concentrations of solutes are analysed. The term ppm is equivalent to the expression of concentration as $\mu g\,mL^{-1}$ ($10^{-6}\,\mu g\,mL^{-1}$) and a 1.0 ppm solution of a substance will have a concentration of $1.0\,\mu g\,mL^{-1}$ ($1.0 \times 10^{-6}\,\mu g\,mL^{-1}$).

Parts per billion (ppb) is an extension of this concentration term as $ng\,mL^{-1}$ ($10^{-9}g\,mL^{-1}$) and is commonly used to express concentrations of very dilute solutions. For example, the allowable concentration of arsenic in water is 0.05 ppm, but it is more conveniently expressed as 50 ppb.

Example The concentration of a NaCl solution is stated as 3 ppm. This is equivalent to 3 $\mu g\,mL^{-1}$ (3 mg L^{-1}). The relative molecular mass of NaCl is 58.44 g mol^{-1}, so the solution has a concentration of $3 \times 10^{-6} \div 58.44$ mol $mL^{-1} = 5.13 \times 10^{-8}$ mol $mL^{-1} = 0.0513\,\mu mol$ $mL^{-1} = 51.3\,\mu mol\,L^{-1}$.

Activity (*a*)

This is a term used to describe the *effective* concentration of a solute. In dilute solutions, solutes can be considered to behave according to ideal (thermodynamic) principles, i.e. they will have an effective concentration equivalent to the actual concentration. However, in concentrated solutions (≥ 500 mol m^{-3}), the behaviour of solutes is often non-ideal, and their effective concentration (activity) will be less than the actual concentration [*C*]. The ratio between the effective concentration and the actual concentration is called the activity coefficient (γ) where

Table 25.1 Activity coefficient of NaCl solutions as a function of molality. Data from Robinson and Stokes (2002)

Molality	Activity coefficient at 25 °C
0.1	0.778
0.5	0.681
1.0	0.657
2.0	0.668
4.0	0.783
6.0	0.986

$$\gamma = \frac{a}{[C]} \qquad [25.3]$$

Equation [25.3] can be used for SI units (mol m^{-3}), molarity (mol L^{-1}) or molality (mol kg^{-1}). In all cases, γ is a dimensionless term, since *a* and [*C*] are expressed in the same units. The activity coefficient of a solute is effectively unity in dilute solution, decreasing as the solute concentration increases (Table 25.1). At high concentrations of certain ionic solutes, γ may increase to become greater than unity.

Example A solution of NaCl with a molality of 0.5 mol kg^{-1} has an activity coefficient of 0.681 at 25 °C and a molal activity of $0.5 \times 0.681 = 0.340$ mol kg^{-1}.

KEY POINT Activity is often the correct expression for theoretical relationships involving solute concentration (e.g. where a property of the solution is dependent on concentration). However, for most practical purposes, it is possible to use the actual concentration of a solute rather than the activity, since the difference between the two terms can be ignored for dilute solutions.

The particular use of the term 'water activity' is considered below (p. 159), since it is based on the mole fraction of solvent, rather than the effective concentration of solute.

Equivalent mass (equivalent weight)

Equivalence and normality are outdated terms, although you may come across them in older texts. They apply to certain solutes whose reactions involve the transfer of charged ions, e.g. acids and alkalis (which may be involved in H^+ or OH^- transfer), and electrolytes (which form cations and anions that may take part in further reactions). These two terms take into account the valency of the charged solutes. Thus the equivalent mass of an ion is its relative molecular mass divided by its valency (ignoring the sign), expressed in grams per equivalent (eq) according to the relationship:

$$\text{equivalent mass} = \frac{\text{relative molecular mass}}{\text{valency}} \qquad [25.4]$$

For acids and alkalis, the equivalent mass is the mass of substance that will provide 1 mol of either H^+ or OH^- ions in a reaction, obtained by dividing the molecular mass by the number of available ions (n), using n instead of valency as the denominator in eqn [25.4].

Normality

A 1 normal solution (1 N) is one that contains one equivalent mass of a substance per litre of solution. The general formula is:

$$\text{normality} = \frac{\text{mass of substance per litre}}{\text{equivalent mass}} \qquad [25.5]$$

Osmolarity

This non-SI expression is used to describe the number of moles of osmotically active solute particles per litre of solution (osmol L^{-1}). The need for such a term arises because some molecules dissociate to give more than one osmotically active particle in aqueous solution.

Osmolality

This term describes the number of moles of osmotically active solute particles per unit mass of solvent (osmol kg^{-1}). For an ideal solute, the osmolality can be determined by multiplying the molality by n, the number of solute particles produced in solution (e.g. for NaCl, $n = 2$). However, for real (i.e. non-ideal) solutes, a correction factor (the osmotic coefficient, ϕ) is used:

$$\text{osmolality} = \text{molality} \times n \times \phi \qquad [25.6]$$

If necessary, the osmotic coefficients of a particular solute can be obtained from tables (e.g. Table 25.2): non-ideal behaviour means that ϕ may have values >1 at high concentrations. Alternatively, the osmolality of a solution can be measured using an osmometer.

Colligative properties and their use in osmometry

Several properties vary in direct proportion to the effective number of osmotically active solute particles per unit mass of solvent and can be used to determine the osmolality of a solution. These colligative properties include freezing point, boiling point and vapour pressure.

Examples For carbonate ions (CO_3^{2-}), with a relative molecular mass of 60.00 and a valency of 2, the equivalent mass is 60.00/2 = 30.00 g eq^{-1}.

For sulphuric acid (H_2SO_4, relative molecular mass 98.08), where 2 hydrogen ions are available, the equivalent mass is 98.08/2 = 49.04 g eq^{-1}.

Example A 0.5 N solution of sulphuric acid would contain $0.5 \times 49.04 = 24.52$ g L^{-1}.

Example Under ideal conditions, 1 mol of NaCl dissolved in water would give 1 mol of Na^+ ions and 1 mol of Cl^- ions, equivalent to a theoretical osmolarity of 2 osmol L^{-1}.

Example A 1.0 mol kg^{-1} solution of NaCl has an osmotic coefficient of 0.936 at 25°C and an osmolality of $1.0 \times 2 \times 0.936 = 1.872$ osmol kg^{-1}.

Table 25.2 Osmotic coefficients of NaCl solutions as a function of molality. Data from Robinson and Stokes (2002)

Molality	Osmotic coefficient at 25 °C
0.1	0.932
0.5	0.921
1.0	0.936
2.0	0.983
4.0	1.116
6.0	1.271

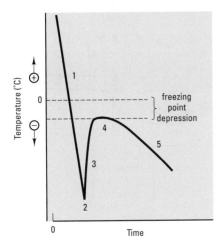

Fig. 25.1 Temperature responses of a cryoscopic osmometer. The response can be subdivided into:

1. initial supercooling
2. initiation of crystallisation
3. crystallisation/freezing
4. plateau, at the freezing point
5. slow temperature decrease.

Example A 1.0 mol kg^{-1} solution of NaCl at 25 °C has an osmolalilty of 1.872 osmol kg^{-1} and an osmotic pressure of 1.872 × 2.479 = 4.641 MPa.

Table 25.3 Water activity (a_w) of NaCl solutions as a function of molality. Data from Robinson and Stokes (2002)

Molality	a_w
0.1	0.997
0.5	0.984
1.0	0.967
2.0	0.932
4.0	0.852
6.0	0.760

An osmometer is an instrument that measures the osmolality of a solution, usually by determining the freezing point depression of the solution in relation to pure water, a technique known as cryoscopic osmometry. A small amount of sample is cooled rapidly and then brought to the freezing point (Fig. 25.1), which is measured by a temperature-sensitive thermistor probe calibrated in mosmol kg^{-1}. An alternative method is used in vapour pressure osmometry, which measures the relative decrease in the vapour pressure produced in the gas phase when a small sample of the solution is equilibrated within a chamber.

Osmotic properties of solutions

Several interrelated terms can be used to describe the osmotic status of a solution. In addition to osmolality, you may come across the following:

Osmotic pressure

This is based on the concept of a membrane permeable to water, but not to solute molecules. For example, if a sucrose solution is placed on one side and pure water on the other, then a passive driving force will be created and water will diffuse across the membrane into the sucrose solution, since the effective water concentration in the sucrose solution will be lower (see Fig. 25.2). The tendency for water to diffuse into the sucrose solution could be counteracted by applying a hydrostatic pressure equivalent to the passive driving force. Thus, the osmotic pressure of a solution is the excess hydrostatic pressure required to prevent the net flow of water into a vessel containing the solution. The SI unit of osmotic pressure is the pascal, Pa (= kg m^{-1} s^{-2}). Older sources may use atmospheres, or bars, and conversion factors are given in Box 28.1. Osmotic pressure and osmolality can be interconverted using the expression 1 osmol kg^{-1} = 2.479 MPa at 25 °C.

The use of osmotic pressure has been criticised as misleading, since a solution does not exhibit an 'osmotic pressure' unless it is placed on the other side of a selectively permeable membrane from pure water.

Water activity (a_w)

This is a term often used to describe the osmotic behaviour of microbial cells. It is a measure of the relative proportion of water in a solution, expressed in terms of its mole fraction, i.e. the ratio of the number of moles of water (n_w) to the total number of moles of all substances (i.e. water and solutes) in solution (n_t), taking into account the molal activity coefficient of the solvent, water (i.e. γ_w):

$$a_w = \gamma_w \frac{n_w}{n_t} \qquad [25.7]$$

The water activity of pure water is unity, decreasing as solutes are added. One disadvantage of a_w is the limited change that occurs in response to a change in solute concentration: a 1.0 mol kg^{-1} solution of NaCl has a water activity of 0.967 (Table 25.3).

Osmolality, osmotic pressure and water activity are measurements based solely on the osmotic properties of a solution, with no regard for any other driving forces, e.g. hydrostatic and gravitational forces. In circumstances where such other forces are important, you will need to measure a variable that takes into account these aspects of water status, namely water potential.

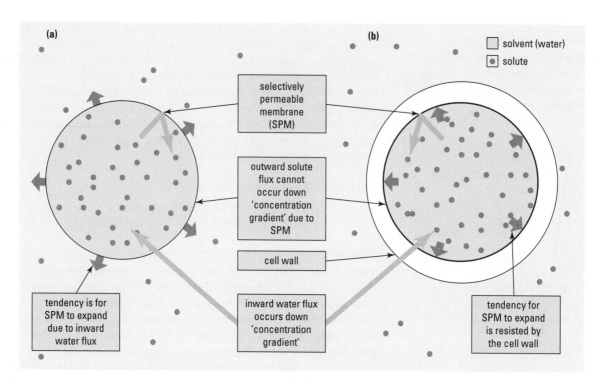

Fig. 25.2 Illustration of forces driving solvent (water) and solute movement across a selectively permeable membrane (SPM). Energetically, both solutes and solvents tend to move down their respective 'concentration gradient' (strictly, down their chemical potential gradient). However, solute molecules cannot leave the model cells illustrated because they cannot pass through the SPM. In the situation illustrated in (a), water will tend to move from outside the cell to within because the solute molecules have effectively 'diluted' the water within the cell (illustrated by the density of point shading), creating a gradient in 'concentration' and because this molecule is able to pass through the SPM. The result will be an expansion of this model cell (short arrows). The osmotic pressure is the (theoretical) pressure that would need to be applied to prevent this. If the model cell were surrounded by a cell wall, as in (b), this would resist expansion, leading to internal pressurisation (turgor pressure, p. 161).

Water potential (hydraulic potential) and its applications

Water potential, Ψ_w, is the most appropriate measure of osmotic status in many areas of bioscience. It is a term derived from the chemical potential of water. It expresses the difference between the chemical potential of water in the test system and that of pure water under standard conditions and has units of pressure (i.e. Pa). It is a more appropriate term than osmotic pressure because it is based on sound theoretical principles and because it can be used to predict the direction of passive movement of water, since water will flow down a gradient of chemical potential (i.e. osmosis occurs from a solution with a higher water potential to one with a lower water potential). A solution of pure water at 20 °C and at 0.1 MPa pressure (i.e. ≈ atmospheric) has a water potential of zero. The addition of solutes will lower the water potential (i.e. make it negative), while the application of pressure, e.g. from hydrostatic or gravitational forces, will raise it (i.e. make it positive).

Often, the two principal components of water potential are referred to as the solute potential, or osmotic potential (Ψ_s, sometimes symbolised as Ψ_π or π) and the hydrostatic pressure potential (Ψ_p) respectively. For a solution at atmospheric pressure, the water potential is due solely to the presence of osmotically active solute molecules (osmotic potential) and

Examples A 1.0 mol kg^{-1} solution of NaCl has a (negative) water potential of −4.641 MPa.
Pure water at 0.2 MPa pressure (about 0.1 MPa above atmospheric pressure) has a (positive) water potential of 0.1 MPa.

Measuring water potential – eqn [25.9] ignores the effects of gravitational forces – for systems where gravitational effects are important an additional term is required (Nobel, 2005).

Fig. 25.3 Vapour pressure osmometer. Courtesy of KNAUER, Germany, www.knauer.net

may be calculated from the measured osmolality (osmol kg^{-1}) at 25 °C, using the relationship:

$$\Psi_w(MPa) = \Psi_s(MPa) = -2.479 \times \text{osmolality} \qquad [25.8]$$

For aquatic microbial cells, e.g. algae, fungi and bacteria, equilibrated in their growth medium at atmospheric pressure, the water potential of the external medium will be equal to the cellular water potential ('isotonic') and the latter can be derived from the measured osmolality of the medium (eqn [25.8]) by osmometry (pp. 158–9). The water potential of such cells can be subdivided into two major parts, the cell solute potential (Ψ_s) and the cell turgor pressure (Ψ_p) as follows:

$$\Psi_w = \Psi_s + \Psi_p \qquad [25.9]$$

To calculate the relative contribution of the osmotic and pressure terms in eqn [25.9], an estimate of the internal osmolality is required, e.g. by measuring the freezing point depression of expressed intracellular fluid. Once you have values for Ψ_w and Ψ_s, the turgor pressure can be calculated by substitution into eqn [25.9].

For terrestrial plant cells, the water potential may be determined directly using a vapour pressure osmometer, by placing a sample of the material within the osmometer chamber and allowing it to equilibrate. If Ψ_s of expressed sap is then measured, Ψ_p can be determined from eqn [25.9].

The van't Hoff relationship can be used to estimate Ψ_s, by summation of the osmotic potentials due to the major solutes, determined from their concentrations, as:

$$\Psi_s = -RTn\phi[C] \qquad [25.10]$$

where RT is the product of the universal gas constant and absolute temperature (2479 J mol^{-1} at 25 °C), n and ϕ are as previously defined and $[C]$ is expressed in SI terms as mol m^{-3}.

Text references

Haynes, W.M. (ed.) (2014) *CRC Handbook of Chemistry and Physics*, 95th edn. CRC Press, Boca Raton.

Nobel, P.S. (2005) *Physicochemical and Environmental Plant Physiology*, 3rd edn. Academic Press, New York.

Robinson, R.A. and Stokes, R.H. (2002) *Electrolyte Solutions*, 2nd edn. Dover Publications, New York.

Sources for further study

Burtis, C.A. and Ashwood, E.R. (2014) *Fundamentals of Clinical Chemistry*, 7th edn. Saunders, Philadelphia.

Chapman, C. (1998) *Basic Chemistry for Biology*. McGraw-Hill, New York.

Postma, J.M., Roberts, J.L. and Hollenberg, J.L. (2009) *Chemistry in the Laboratory*, 7th edn. Freeman, New York.

Seidman, L.A. and Moore, C.J. (2008) *Basic Laboratory Methods for Biotechnology: Textbook and Laboratory Reference*, 2nd edn. Prentice Hall, New Jersey.

Timberlake, K.C. (2013) *Laboratory Manual for General, Organic and Biological Chemistry*, 3rd edn. Prentice Hall, New Jersey.

STUDY EXERCISES

25.1 Practise calculations involving molar concentrations (see also Study exercises 24.1 and 24.2). What mass of substance would be required to prepare each of the following aqueous solutions (answer in grams, to three decimal places in each case):

(a) 1 L of NaCl at a concentration of 1 molar? (M_r of NaCl = 58.44.)

(b) 250 mL of $CaCl_2$ at 100 mmol L^{-1}? (M_r of $CaCl_2$ = 110.99.)

(c) 2.5 L of mannitol at 10 nmol μL^{-1}? (M_r of mannitol = 182.17.)

(d) 400 mL of KCl at 5% w/v?

(e) 250 mL of glucose at 2.50 mol m^{-3}? (M_r of glucose = 180.16.)

25.2 Practise expressing concentrations in different ways. Express all answers to three significant figures:

(a) What is 5 g L^{-1} sucrose, expressed in terms of molarity? (M_r of sucrose = 342.3.)

(b) What is 1.0 mol m^{-3} NaCl, expressed in g L^{-1}? (M_r of NaCl = 58.44.)

(c) What is 5% v/v ethanol, expressed in terms of molarity? (M_r of ethanol = 46.06 and density of ethanol at 25 °C = 0.789 gm L^{-1}.)

(d) What is 150 mmol L^{-1} glucose, expressed in terms of per cent concentration (% w/v)? (M_r of glucose = 180.16.)

(e) What is a 1.0 molal solution of KCl, expressed as per cent composition (% w/w)? (M_r of KCl = 74.55.)

25.3 Calculate osmolality and osmotic potentials.

(a) Assuming NaCl, KCl and $CaCl_2$ behave according to ideal thermodynamic principles, what would be the predicted osmolality of a solution containing:

(i) NaCl alone, at 50 mmol kg^{-1}?

(ii) KCl at 200 mmol kg^{-1} and $CaCl_2$ at 40 mmol kg^{-1}?

(iii) NaCl at 100 mmol kg^{-1}, KCl at 60 mmol kg^{-1} and $CaCl_2$ at 75 mmol kg^{-1}?

(b) What is the predicted osmotic pressure and osmotic potential of each of the solutions in (a) at 25 °C? (answer to three significant figures in all cases).

Answers to these study exercises are available at www.pearsoned.co.uk/practicalskills

pH is a measure of the amount of hydrogen ions (H$^+$) in a solution: this affects the solubility of many substances and the activity of most biological systems, from individual molecules to whole organisms. It is usual to think of aqueous solutions as containing H$^+$ ions (protons), though protons actually exist in their hydrated form, as hydronium ions (H$_3$O$^+$). The proton concentration of an aqueous solution [H$^+$] is affected by several factors:

- **Ionisation (dissociation) of water,** which liberates protons and hydroxyl ions in equal quantities, according to the reversible relationship:

$$H_2O \rightleftharpoons H^+ + OH^- \qquad [26.1]$$

- **Dissociation of acids,** according to the equation:

$$H{-}A \rightleftharpoons H^+ + A^- \qquad [26.2]$$

where H$-$A represents the acid and A$^-$ is the corresponding conjugate base. The dissociation of an acid in water will increase the amount of protons, reducing the amount of hydroxyl ions as water molecules are formed (eqn [26.1]). The addition of a base (usually, as its salt) to water will decrease the amount of H$^+$, due to the formation of the conjugate acid (eqn [26.2]).

- **Dissociation of alkalis,** according to the relationship:

$$X{-}OH \rightleftharpoons X^+ + OH^- \qquad [26.3]$$

where X$-$OH represents the undissociated alkali. Since the dissociation of water is reversible (eqn [26.1]), in an aqueous solution the production of hydroxyl ions will effectively act to 'mop up' protons, lowering the proton concentration.

Many compounds act as acids, bases or alkalis: those that are almost completely ionised in solution are usually called strong acids or bases, while weak acids or bases are only slightly ionised in solution (p. 167).

In an aqueous solution, most of the water molecules are not ionised. In fact, the extent of ionisation of pure water is constant at any given temperature and is usually expressed in terms of the ion product (or ionisation constant) of water, K_w:

$$K_w = [H^+][OH^-] \qquad [26.4]$$

where [H$^+$] and [OH$^-$] represent the molar concentration (strictly, the activity) of protons and hydroxyl ions in solution, expressed as mol L^{-1}. At 25 °C, the ion product of pure water (Table 26.1) is 10^{-14} mol^2 L^{-2} (i.e. 10^{-8} mol^2 m^{-6}). This means that the concentration of protons in solution will be 10^{-7} mol L^{-1} (10^{-4} mol m^{-3}), with an equivalent concentration of hydroxyl ions (eqn [26.1]). Since these values are very low and involve negative powers of 10, it is customary to use the pH scale, where:

$$pH = -\log_{10}[H^+] \qquad [26.5]$$

and [H$^+$] is the proton activity in mol L^{-1} (see p. 157).

Definitions

Acid – a compound that acts as a proton donor in aqueous solution.

Alkali – a compound that liberates hydroxyl ions when it dissociates. Since hydroxyl ions are strongly basic, this will reduce the proton concentration.

Ampholyte – a compound that can act as both an acid and a base. Water is an ampholyte since it may dissociate to give a proton and a hydroxyl ion (amphoteric behaviour).

Base – a compound that acts as a proton acceptor in aqueous solution.

Conjugate pair – an acid together with its corresponding base.

Table 26.1 Effects of temperature on the ion product of water (K_w), H$^+$ ion concentration and pH at neutrality. Values calculated from Haynes (2014)

Temp. (° C)	K_w (mol^2 L^{-2})	[H$^+$] at neutrality (nmol L^{-1})	pH at neutrality
0	0.11×10^{-14}	33.9	7.47
4	0.17×10^{-14}	40.7	7.39
10	0.29×10^{-14}	53.7	7.27
20	0.68×10^{-14}	83.2	7.08
25	1.01×10^{-14}	100.4	7.00
30	1.47×10^{-14}	120.2	6.92
37	2.39×10^{-14}	154.9	6.81
45	4.02×10^{-14}	199.5	6.70

Example   Human plasma has a typical H$^+$ concentration of approximately 0.4×10^{-7} mol L^{-1} (=$10^{-7.4}$ mol L^{-1}), giving a pH of 7.4.

Measuring pH – the pH of a neutral solution changes with temperature (Table 26.1), due to the enhanced dissociation of water with increasing temperature. This must be taken into account when measuring the pH of any solution and when interpreting your results.

SAFETY NOTE Preparing a dilute acid solution using concentrated acid – always slowly add the concentrated acid to water, not the reverse, since the strongly exothermic process can trigger a violent reaction with water.

SAFETY NOTE Preparing an alkali solution – typically, the alkali will be in solid form (e.g. NaOH) and addition to water will rapidly raise the temperature of the solution: use only heat-resistant glassware, cooled with water if necessary.

Table 26.2 Properties of some pH indicator dyes

Dye	Acid–base colour change	Useful pH range
Thymol blue (acid)	red–yellow	1.2–6.8
Bromophenol blue	yellow–blue	1.2–6.8
Congo red	blue–red	3.0–5.2
Bromocresol green	yellow–blue	3.8–5.4
Resazurin	orange–violet	3.8–6.5
Methyl red	red–yellow	4.3–6.1
Litmus	red–blue	4.5–8.3
Bromocresol purple	yellow–purple	5.8–6.8
Bromothymol blue	yellow–blue	6.0–7.6
Neutral red	red–yellow	6.8–8.0
Phenol red	yellow–red	6.8–8.2
Thymol blue (alkaline)	yellow–blue	8.0–9.6
Phenolphthalein	none–red	8.3–10.0

KEY POINT While pH is strictly the negative logarithm (to the base 10) of H^+ activity, in practice H^+ concentration in $mol\ L^{-1}$ (equivalent to $kmol\ m^{-3}$ in SI terminology) is most often used in place of activity, since the two are virtually the same, given the limited dissociation of H_2O. The pH scale is not SI: nevertheless, it continues to be used widely in biological science.

The value where an equal amount of H^+ and OH^- ions are present is termed neutrality: at 25 °C the pH of pure water at neutrality is 7.0. At this temperature, pH values below 7.0 are acidic while values above 7.0 are alkaline.

Always remember that the pH scale is a *logarithmic* one, not a *linear* one: a solution with a pH of 3.0 is not twice as acidic as a solution of pH 6.0, but one thousand times as acidic (i.e. contains 1000 times the amount of H^+ ions). Therefore, you may need to convert pH values into proton concentrations before you carry out mathematical manipulations (see Box 75.2). For similar reasons, it is important that pH change is expressed in terms of the original and final pH values, rather than simply quoting the difference between the values: a pH change of 0.1 has little meaning unless the initial or final pH is known.

Measuring pH

pH electrodes

Accurate pH measurements can be made using a pH electrode, coupled to a pH meter. The pH electrode is usually a combination electrode, comprising two separate systems: an H^+-sensitive glass electrode and a reference electrode which is unaffected by H^+ ion concentration (Fig. 26.2). When this is immersed in a solution, a pH-dependent voltage between the two electrodes can be measured using a potentiometer. In most cases, the pH electrode assembly (containing the glass and reference electrodes) is connected to a separate pH meter by a cable, although some hand-held instruments (pH probes) have the electrodes and meter within the same assembly, often using an H^+-sensitive field effect transistor in place of a glass electrode, to improve durability and portability.

Box 26.1 gives details of the steps involved in making a pH measurement with a glass pH electrode and meter.

pH indicator dyes

These compounds (usually weak acids) change colour in a pH-dependent manner. They may be added in small amounts to a solution, or they can be used in paper strip form. Each indicator dye usually changes colour over a restricted pH range, typically 1–2 pH units (Table 26.2): universal indicator dyes/papers make use of a combination of individual dyes to measure a wider pH range. Dyes are not suitable for accurate pH measurement as they are affected by other components of the solution including oxidising and reducing agents and salts. However, they are useful for:

- **estimating the approximate pH of a solution;**
- **determining a change in pH,** for example at the end point of a titration or the production of acids during bacterial metabolism;
- **establishing the approximate pH of intracellular compartments,** for example the use of neutral red as a 'vital' stain (p. 263).

Buffers

Rather than simply measuring the pH of a solution, you may wish to *control* the pH, e.g. in metabolic experiments, or in a growth medium for cell culture (p. 291). In fact, you should consider whether you need to control pH in any experiment involving a biological system, whether whole organisms, isolated cells, subcellular components or biomolecules. One of the most effective ways to control pH is to use a buffer solution.

A buffer solution is usually a mixture of a weak acid and its conjugate base. Added protons will be neutralised by the anionic base while a reduction in protons, e.g. due to the addition of hydroxyl ions, will be counterbalanced by dissociation of the acid (eqn [26.2]); thus the conjugate pair acts as a 'buffer' to pH change. The innate resistance of most biological fluids to pH change is due to the presence of cellular constituents that act as buffers, e.g. proteins, which have a large number of weakly acidic and basic groups in their amino acid side chains.

Buffer capacity and the effects of pH

The extent of resistance to pH change is called the buffer capacity of a solution. The buffer capacity is measured experimentally at a particular pH by titration against a strong acid or alkali: the resultant curve will be strongly sigmoidal, with a plateau where the buffer capacity is greatest (Fig. 26.1). The mid-point of the plateau represents the pH where equal quantities of acid and conjugate base are present, and is given the symbol pK_a, which refers to the negative logarithm (to the base 10) of the acid dissociation constant, K_a, where

$$K_a = \frac{[H^+][A^-]}{[HA]}$$ [26.6]

By rearranging eqn [26.6] and taking negative logarithms, we obtain:

$$pH = pK_a + \log_{10}\frac{[A^-]}{[HA]}$$ [26.7]

This relationship is known as the Henderson–Hasselbalch equation and it shows that the pH will be equal to the pK_a when the ratio of conjugate base to acid is unity, since the final term in eqn [26.7] will be zero. Consequently, pK_a is an important factor in determining buffer capacity at a particular pH. In practical terms, a buffer solution will work most effectively at pH values about one unit either side of the pK_a.

Selecting an appropriate buffer

When selecting a buffer, you should be aware of certain limitations to their use. Citric acid and phosphate buffers readily form insoluble complexes with divalent cations, while phosphate can also act as a substrate, activator or inhibitor of certain enzymes. Both of these buffers contain biologically significant quantities of cations, e.g. Na⁺ or K⁺. TRIS (Table 26.3) is often toxic to biological systems: due to its high lipid solubility it can penetrate membranes, uncoupling electron transport reactions in whole cells and isolated organelles. In addition, it is markedly affected by temperature, with a tenfold increase in H⁺ concentration from 4 °C to 37 °C. A number of zwitterionic molecules (having both positive and negative groups) have

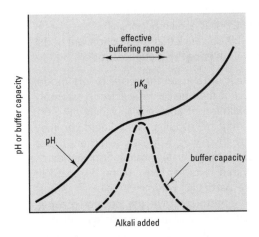

Fig. 26.1 Theoretical pH titration curve for a buffer solution. pH change is lowest and buffer capacity is greatest at the pK_a of the buffer solution.

Features of an ideal buffer – for biological purposes a buffer would possess the following characteristics:

- impermeability to biological membranes;
- biological stability and lack of interference with metabolic and biological processes;
- lack of significant absorption of ultraviolet or visible light;
- lack of formation of insoluble complexes with cations;
- minimal effect of ionic composition or salt concentration;
- limited pH change in response to temperature.

Box 26.1 Using a glass pH electrode and meter to measure the pH of a solution

The following procedure should be used whenever you make a pH measurement: consult the manufacturer's handbook for specific information, where necessary. Do not be tempted to miss out any of the steps detailed below, particularly those relating to the effects of temperature, or your measurements are likely to be inaccurate.

1. Stir the test solution thoroughly before you make any measurement: it is often best to use a magnetic stirrer. Leave the solution for sufficient time to allow equilibration at lab temperature.

2. Record the temperature of every solution you use, including all calibration standards and samples, since this will affect K_w, neutrality and pH.

3. Set the temperature compensator on the meter to the appropriate value. This control makes an allowance for the effect of temperature on the electrical potential difference recorded by the

meter: it does *not* allow for the other temperature-dependent effects mentioned elsewhere. Basic instruments have no temperature compensator, and should only be used at a specified temperature, either 20 °C or 25 °C, otherwise they will not give an accurate measurement. More sophisticated systems have automatic temperature compensation.

4. Rinse the electrode assembly (Fig. 26.2) with distilled water and gently dab off the excess water on to a clean tissue: check for visible damage or contamination of the glass electrode (consult a member of staff if the glass is broken or dirty). Also check that the solution within the glass assembly is covering the metal electrode.

5. Calibrate the instrument: set the meter to 'pH' mode, if appropriate, and then place the electrode assembly in a standard solution of known pH, usually pH 7.00. This solution may be supplied as a liquid, or may be prepared by dissolving a measured amount of a calibration standard in water: calibration standards are often provided in tablet form, to be dissolved in water to give a particular volume of solution. Adjust the calibration control to give the correct reading. Remember that your calibration standards will only give the specified pH at a particular temperature, usually either 20 °C or 25 °C. If you are working at a different temperature, you must establish the actual pH of your calibration standards, either from the supplier, or from literature information.

6. Remove the electrode assembly from the calibration solution and rinse again with distilled water: dab off the excess water. Basic instruments have no further calibration steps (single-point calibration), while the more refined pH meters have additional calibration procedures.

   If you are using a basic instrument, you should check that your apparatus is accurate over the appropriate pH range by measuring the pH of another standard whose pH is close to that expected for the test solution. If the standard does not give the expected reading, the instrument is not functioning correctly: consult a member of staff.

   If you are using an instrument with a slope control function, this will allow you to correct for any deviation in electrical potential from that predicted by the theoretical relationship (at 25 °C, a change in pH of 1.00 unit should result in a change in electrical potential of 59.16mV) by performing a two-point calibration. Having calibrated the instrument at pH 7.00, immerse in a second standard at the same temperature as that of the first standard,

Fig 26.2 Measurement of pH using a combination pH electrode and meter. The electrical potential difference recorded by the potentiometer is directly proportional to the pH of the test solution.

(continued)

Box 26.1 (continued)

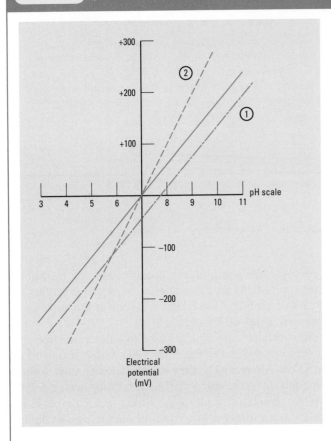

Fig. 26.3 The relationship between electrical potential and pH. The solid line shows the response of a calibrated electrode while the other plots are for instruments requiring calibration: 1 has the correct slope but incorrect isopotential point (calibration control adjustment is needed); 2 has the correct isopotential point but incorrect slope (slope control adjustment is needed).

usually buffered to either pH 4.00 or pH 9.00, depending upon the expected pH of your samples. Adjust the slope control until the exact value of the second standard is achieved (Fig. 26.3). A pH electrode and meter calibrated using the two-point method will give accurate readings over the pH range from 3 to 11: laboratory pH electrodes are not accurate outside this range, since the theoretical relationship between electrical potential and pH is valid.

7. Once the instrument is calibrated, measure the pH of your solution(s), making sure that the electrode assembly is rinsed thoroughly between measurements. You should be particularly aware of this requirement if your solutions contain organic biological material, e.g. soil, tissue fluids, protein solutions, etc., since these may adhere to the glass electrode and affect the calibration of your instrument. If your electrode becomes contaminated during use, check with a member of staff before cleaning: avoid touching the surface of the glass electrode with abrasive material. Allow sufficient time for the pH reading to stabilise in each solution before taking a measurement: for unbuffered solutions, this may take several minutes, so do not take inaccurate pH readings due to impatience!

8. After use, the electrode assembly must not be allowed to dry out. Most pH electrodes should be stored in a neutral solution of KCl, either by suspending the assembly in a small beaker, or by using an electrode cap filled with the appropriate solution (typically $1.0 \ mol \ L^{-1}$ KCl buffered at pH 7.0). However, many labs simply use distilled water as a storage solution, leading to loss of ions from the interior of the electrode assembly. In practice, this means that pH electrodes stored in distilled water will take far longer to give a stable reading than those stored in KCl.

9. Switch the meter to zero (where appropriate), but do not turn off the power: pH meters give more stable readings if they are left on during normal working hours.

Problems (and solutions) include: inaccurate and/or unstable pH readings caused by cross-contamination (rinse electrode assembly with distilled water and blot dry between measurements); development of a protein film on the surface of the electrode (soak in 1% w/v pepsin in $0.1 \ mol \ L^{-1}$ HCl for at least an hour); deposition of organic or inorganic contaminants on the glass bulb (use an organic solvent, such as acetone, or a solution of $0.1 \ mol \ L^{-1}$ disodium ethylenediamine tetraacetic acid, respectively); drying out of the internal reference solutions (drain, flush and refill with fresh solution, then allow to equilibrate in $0.1 \ mol \ L^{-1}$ HCl for at least an hour); cracks or chips to the surface of the glass bulb (use a replacement electrode).

Table 26.3 pK_a values at 25 °C and M_r of some acids and bases (upper section) and some large organic zwitterions (lower section) commonly used in buffer solutions. For polyprotic acids, where more than one proton may dissociate, the pK_a values are given for each ionisation step. Only the trivial acronyms of the larger molecules are provided: their full names can be obtained from the catalogues of most chemical suppliers

Acid or base	pK_a value(s)	M_r
Acetic acid	4.8	60.1
Boric acid	9.2	61.8
Citric acid	3.1, 4.8, 5.4	191.2
Glycylglycine	3.1, 8.2	132.1
Phosphoric acid	2.1, 7.1, 12.3	98.0
Phthalic acid	2.9, 5.5	166.1
Succinic acid	4.2, 5.6	118.1
TRIS (base)*	8.1	121.1
CAPS (free acid)	10.4	221.3
CHES (free acid)	9.3	207.3
HEPES (free acid)	7.5	238.3
MES (free acid)	6.1	213.2
MOPS (free acid)	7.2	209.3
PIPES (free acid)	6.8	302.4
TAPS (free acid)	8.4	243.3
TRICINE (free acid)	8.1	179.2

*Note that this compound is hygroscopic and should be stored in a desiccator; also see text regarding its potential toxicity (p. 167).

Table 26.4 Preparation of sodium phosphate buffer solutions for use at 25 °C. Prepare separate stock solutions of (a) disodium hydrogen phosphate and (b) sodium dihydrogen phosphate, both at 200 mol m^{-3}. Buffer solutions (at 100 mol m^{-3}) are then prepared at the required pH by mixing together the volume of each stock solution shown in the table, then diluting to a final volume of 100 mL using distilled or deionised water

Required pH (at 25 °C)	Volume of stock (a) Na$_2$HPO$_4$ (mL)	Volume of stock (b) NaH$_2$PO$_4$ (mL)
6.0	6.2	43.8
6.2	9.3	40.7
6.4	13.3	36.7
6.6	18.8	31.2
6.8	24.5	25.5
7.0	30.5	19.5
7.2	36.0	14.0
7.4	40.5	9.5
7.6	43.5	6.5
7.8	45.8	4.2
8.0	47.4	2.6

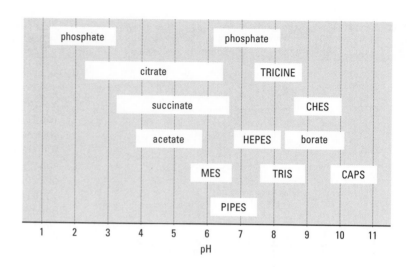

Fig. 26.4 Useful pH ranges of some commonly used buffers.

been introduced to overcome some of the disadvantages of traditional buffers. These newer compounds are often referred to as 'Good buffers', to acknowledge the work of Dr N.E. Good: HEPES is one of the most useful zwitterionic buffers, with a pK_a of 7.5 at 25 °C.

These zwitterionic substances are usually added to water as the free acid: the solution must then be adjusted to the correct pH with a strong alkali, usually NaOH or KOH. Alternatively, they may be used as their sodium or potassium salts, adjusted to the correct pH with a strong acid, e.g. HCl. Consequently, you may need to consider what effects such changes in ion concentration may have in a solution where zwitterions are used as buffers. In addition, zwitterionic buffers can interfere with protein determinations (e.g. Lowry method, p. 388).

Figure 26.4 shows a number of traditional and zwitterionic buffers and their effective pH ranges. When selecting one of these buffers, aim for a pK_a that is in the direction of the expected pH change (Table 26.3). For example, HEPES buffer would be a better choice of buffer than PIPES for use at pH 7.2 for experimental systems where a pH increase is anticipated, while PIPES would be a better choice where acidification is expected.

Preparation of buffer solutions

Having selected an appropriate buffer, you will need to make up your solution to give the desired pH. You will need to consider two factors:

1. **The ratio of acid and conjugate base** required to give the correct pH.

2. **The amount of buffering required;** buffer capacity depends upon the absolute quantities of acid and base, as well as their relative proportions.

In most instances, buffer solutions are prepared to contain between 10 mmol L^{-1} and 200 mmol L^{-1} of the conjugate pair. Although it is possible to calculate the quantities required from first principles using the Henderson–Hasselbalch equation, there are sources that tabulate the amount of substance required to give a particular volume of solution with a specific pH value for a range of buffers (e.g. Anon., 2006, 2011). For traditional buffers, it is customary to mix stock solutions of acidic and basic components in the correct proportions to give the required pH (Table 26.4). For zwitterionic acids, the usual procedure is to add the compound to water, then bring the solution to the required pH by adding a specific amount of strong alkali or acid (obtained from tables). Alternatively, the required pH can be obtained by dropwise addition of alkali or acid, using a meter to check the pH, until the correct value is reached. When preparing solutions of zwitterionic buffers, the acid may be relatively insoluble. Do not wait for it to dissolve fully before adding alkali to change the pH – the addition of alkali will help bring the acid into solution, but make sure it has all dissolved before the desired pH is reached.

Finally, when preparing a buffer solution based on tabulated information, always confirm the pH with a pH meter before use.

Text references

Anon. (2006) *pH Theory and Practice: A Radiometer Analytical Guide.* Available: http://www.radiometer-analytical.com/pdf/ph_theory.pdf Last accessed 10/07/2015.

Anon. (2011) *Buffer Reference Center.* Available: http://www.sigmaaldrich.com/life-science/core-bioreagents/biological-buffers/learning-center/buffer-reference-center.html Last accessed: 10/07/2015.

Haynes, W.M. (ed.) (2014) *CRC Handbook of Chemistry and Physics,* 95th edn. CRC Press, Boca Raton.

Source for further study

Anon. (2003) *Buffers. A Guide for the Preparation and use of Buffers in Biological Systems.* Available: http://wolfson.huji.ac.il/purification/PDF/Buffers/Calbiochem_Buffers_Booklet.pdf Last accessed 20/07/2015.

STUDY EXERCISES

26.1 Practise interconverting pH values and proton concentrations. Express all answers to three significant figures.

(a) What is pH 7.4 expressed as [H⁺] in mol L⁻¹?

(b) What is pH 4.1, expressed as [H⁺] in mol m⁻³?

(c) What is the pH of a solution containing H⁺ at 2×10^{-5} mol L⁻¹?

(d) What is the pH of a solution containing H⁺ at $10^{-12.5}$ mol L⁻¹?

(e) What is the pH of a solution containing H⁺ at 2.8×10^{-5} mol m⁻³?

26.2 Decide on a suitable buffer to use. In the following instances, choose a buffer that would be suitable:

(a) Maintaining the pH at 8.5 during an enzyme assay of a cell-free extract at 25 °C.

(b) Keeping a stable pH of 6.5 in an experiment to measure the uptake of radiolabelled glucose by a dense suspension of *E. coli*.

(c) Carrying out an assay of photosynthetic activity at pH 7.2 at temperatures of 10 °C, 20 °C and 30 °C.

(d) Stabilising pH at 5.5 during enzyme extraction, in a solution where you intend to measure total protein concentration at a later stage.

26.3 Practise using the Henderson–Hasselbalch equation. What are the relative proportions of deprotonated (A⁻) and protonated (HA) forms of each substance at the following pH values:

(a) acetic acid ($pK_a = 4.8$) for use in an experiment at pH 3.8?

(b) boric acid ($pK_a = 9.2$) for use in an experiment at pH 9.5?

(c) HEPES ($pK_a = 7.5$) for use in an experiment at pH 8.1?

Answers to these study exercises are available at www.pearsoned.co.uk/practicalskills

The investigative approach

27 The principles of measurement

The term data (singular = datum, or data value) refers to items of information, and you will use different types of data from a wide range of sources during your practical work. Consequently, it is important to appreciate the underlying features of data collection and measurement.

Variables

Biological variables (Fig. 27.1) can be classified as follows:

Quantitative variables

These are characteristics whose differing states can be described by means of a number. They are of two basic types:

1. **Continuous variables,** such as length; these are usually measured against a numerical scale. Theoretically, they can take any value on the measurement scale. In practice, the number of significant figures of a measurement is directly related to the precision of your measuring system; for example, dimensions measured with Vernier calipers will provide readings of greater precision than a millimetre ruler (p. 151).
2. **Discontinuous (discrete) variables,** such as the number of eggs in a nest; these are always obtained by counting and therefore the data values must be whole numbers (integers). There are no intermediate values – for example, you never find 1.25 eggs in a nest.

Ranked variables

These provide data that can be listed in order of magnitude (i.e. ranked). A familiar example is the abundance of an organism in a sample, which is often expressed as a series of ranks, e.g. rare = 1, occasional = 2, frequent = 3, common = 4 and abundant = 5. When such data are given numerical ranks, rather than descriptive terms, they are sometimes called 'semi-quantitative data'. Note that the difference in magnitude between ranks need not be consistent. For example, regardless of whether there was a one-year or a five-year gap between offspring in a family, their rank in order of birth would be the same.

Qualitative variables (attributes)

These are non-numerical and descriptive; they have no order of preference, and therefore are not measured on a numerical scale nor ranked in order of magnitude, but are described in terms of categories. Examples include viability (i.e. dead or alive) and shape (e.g. round, flat, elongated, etc.).

Variables may be independent or dependent. Usually, the variable under the control of the experimenter (e.g. time) is the independent variable, while the variable being measured is the dependent variable (p. 202). Sometimes it is not appropriate to describe variables in this way, and they are then referred to as interdependent variables (e.g. the length and breadth of an organism).

The majority of data values are recorded as direct measurements, readings or counts, but there is an important group, called derived (or computed), that results from calculations based on two or more data values, e.g. ratios, percentages, indices and rates.

Definition

Variable – any characteristic or property that can take one of a range of values (contrast this definition with that for a **parameter**, which is a numerical constant in any particular instance).

Working with discontinuous variables – note that while the original data values must be integers, derived data and statistical values do not have to be whole numbers. Thus, it is perfectly acceptable to express the *mean* number of children per family as 2.4.

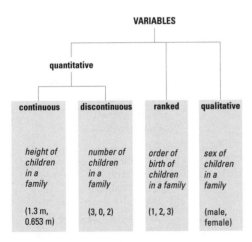

Fig. 27.1 Examples of the different types of variables as used to describe some characteristics of families.

Measurement scales

Variables may be measured on different types of scale:

- **Nominal scale:** this classifies objects into categories based on a descriptive characteristic. It is the only scale suitable for qualitative data.
- **Ordinal scale:** this classifies by rank. There is a logical order in any number scale used.
- **Interval scale:** this is used for quantitative variables. Numbers on an equal-unit scale are related to an arbitrary zero point.
- **Ratio scale:** this is similar to the interval scale, except that the zero point now represents an absence of that character (i.e. it is an absolute zero). In contrast to the interval scale, the ratio of two values is meaningful (e.g. a temperature of 200 K is twice that of 100 K).

The measurement scale is important in determining the mathematical and statistical methods used to analyse your data. Table 27.1 presents a summary of the important properties of these scales. Note that you may be able to measure a characteristic in more than one way, or you may be able to convert data collected in one form to a different form. For instance, you might measure light in terms of the photon flux density (p. 372) between particular wavelengths of the EMR spectrum (ratio scale), or simply as 'blue' or 'red' (nominal scale); you could find out the dates of birth of individuals (interval scale) but then use this information to rank them in order of birth (ordinal scale). Where there are no other constraints, you should use a ratio scale to measure a quantitative variable, since this will allow you to use the broadest range of mathematical and statistical procedures (Table 27.1).

Table 27.1 Some important features of scales of measurement

	Measurement scale			
	Nominal	Ordinal	Interval	Ratio
Type of variable	Qualitative (Ranked)* (Quantitative)*	Ranked (Quantitative)*	Quantitative	Quantitative
Examples	Species Sex Colour	Abundance scales Reproductive condition Optical assessment of colour development	Fahrenheit temperature scale Date (BC/AD)	Kelvin temperature scale Weight Length Response time Most physical measurements
Mathematical properties	Identity	Identity Magnitude	Identity Magnitude Equal intervals	Identity Magnitude Equal intervals True zero point
Mathematical operations possible on data	None	Rank	Rank Addition Subtraction	Rank Addition Subtraction Multiplication Division
Typical statistics used	Only those based on frequency of counts made: contingency tables, frequency distributions, etc. Chi-square test	Non-parametric methods, sign tests Mann-Whitney U-test	Almost all types of test, t-test, analysis of variance (ANOVA), etc. (check distribution before using, Chapter 76)	Almost all types of test, t-test, ANOVA, etc. (check distribution before using, Chapter 76)

*In some instances (see text for examples).

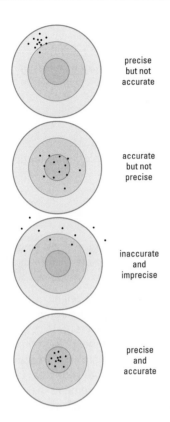

precise
but not
accurate

accurate
but not
precise

inaccurate
and
imprecise

precise
and
accurate

Fig. 27.2 'Target' diagrams illustrating precision and accuracy.

Minimising errors – determine early in your study what the dominant errors are likely to be and concentrate your time and effort on reducing these.

Working with derived data – special effort should be made to reduce measurement errors because their effects can be magnified when differences, ratios, indices or rates are calculated.

Accuracy and precision

Accuracy is the closeness of a measured or derived data value to its true value, while precision is the closeness of repeated measurements to each other (Fig. 27.2). A balance with a fault in it (i.e. a bias, see below) could give precise (i.e. very repeatable) but inaccurate (i.e. untrue) results. Unless there is bias in a measuring system, precision will lead to accuracy and it is precision that is generally the most important practical consideration, if there is no reason to suspect bias. You can investigate the precision of any measuring system by repeated measurements of individual samples.

Absolute accuracy and precision are impossible to achieve, due to both the limitations of measuring systems for continuous quantitative data and the fact that you are usually working with incomplete data sets (samples, p. 196). It is particularly important to avoid spurious accuracy in the presentation of results; include only those digits that the accuracy of the measuring system implies (p. 497). This type of error is common when changing units (e.g. inches to metres) and in derived data, especially when calculators give results to a large number of decimal places.

Bias (systematic error) and consistency

Bias is a systematic or non-random distortion and is one of the most troublesome difficulties in using numerical data. Biases may be associated with incorrectly calibrated instruments, e.g. a faulty pipettor, or with experimental manipulations, e.g. shrinkage during the preservation of a specimen. Bias in measurement can also be subjective, or personal, e.g. an experimenter's preconceived ideas about an 'expected' result.

Bias can be minimised by using a carefully standardised procedure, with fully calibrated instruments. You can investigate bias in 'trial runs' by measuring a single variable in several different ways, to see whether the same result is obtained.

If a personal bias is possible, 'blind' measurements should be made where the identity of individual samples is not known to the operator, e.g. using a coding system.

Measurement error

All measurements are subject to error, but the dangers of misinterpretation are reduced by recognising and understanding the likely sources of error and by adopting appropriate protocols and calculation procedures.

A common source of measurement error is carelessness, e.g. reading a scale in the wrong direction or parallax errors. This can be reduced greatly by careful recording and may be detected by repeating the measurement. Other errors arise from faulty or inaccurate equipment, but even a perfectly functioning machine has distinct limits to the accuracy and precision of its measurements. These limits are often quoted in manufacturers' specifications and are applicable when an instrument is new; however, you should allow for some deterioration with age. Further errors are introduced when the subject being studied is open to influences outside your control. Resolving such problems requires appropriate experimental design (Chapter 33) and sampling procedures (Chapter 32).

One major influence virtually impossible to eliminate is the effect of the investigation itself: even putting a thermometer in a liquid may change the temperature of the liquid. The very act of measurement may give rise to a confounding variable (p. 203) as discussed in Chapter 31.

Sources for further study

Anon. *Measurement.* Available: http://wikipedia.org/wiki/Measurement
Last accessed 26/03/15.

Erikson, B.H. and Nosanchuk, T.A. (1992) *Understanding Data,* 2nd edn. Open University Press, Milton Keynes. [A text aimed at social science students but with clear explanations of issues that are generic, including information on analysis of data.]

Rowlett, R., *How Many? A Dictionary of Units of Measurement.* Available: http://www.unc.edu/~rowlett/units
Last accessed 26/03/15.

Shuttleworth, M. *Choosing Scientific Measurements.* Available https://explorable.com/scientific-measurements
Last accessed 26/03/15.

STUDY EXERCISES

27.1 Classify variables. Decide on the type of variables used for the following measures, indicating whether they are quantitative or qualitative, continuous or discontinuous, and the type of scale that would be used.

(a) Number of organisms in a population.

(b) Length of individuals in a population.

(c) Colour of flowers.

(d) Species present in a sample.

(e) Date of a sample.

(f) Reproductive condition of an animal.

27.2 Assess errors and accuracy of a set of measurements. Assume that you are asked to measure the length, breadth and height of each of a sample of 20 limpet shells (see figure) using (a) a ruler and (b) a pair of Vernier calipers (p. 151). Identify the sources of error likely to be present in your measurements using each tool and the precision to which you would be able to quote your data. Devise a protocol for taking these measurements that would help you to minimise the errors and maximise the accuracy and precision.

27.3 Investigate types of errors. A student weighed a set of standard masses on two electronic balances and obtained the readings shown in the table below. Explain these results in terms of the type of error involved in each case.

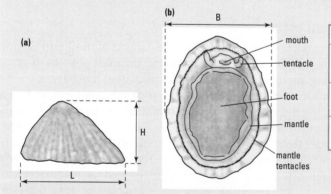

Lateral (a) and ventral (b) views of a limpet to show measurement variables length (L), breadth (B) and height (H). Note also the uneven edge of the shell.

Comparison of weights of masses on two balances					
	Standard mass (g)				
	10	25	50	100	250
Reading (balance A)	10.050	25.049	50.051	100.048	250.052
Reading (balance B)	10.004	25.011	50.021	100.039	250.102

Answers to these study exercises are available at www.pearsoned.co.uk/practicalskills

28   SI units and their use

Dimensionless measurements –
some quantities can be expressed as
dimensionless ratios or logarithms
(e.g. pH), and in these cases you do not
need to use a qualifying unit.

When describing a measurement, you normally state both a number and a unit (e.g. 'the length is 1.85 metres'). The number expresses the ratio of the measured quantity to a fixed standard, while the unit identifies that standard measure or dimension. Clearly, a single unified system of units is essential for efficient communication of such data within the scientific community. The Système International d'Unités (SI) is the internationally ratified form of the metre-kilogram-second system of measurement and represents the accepted scientific convention for measurements of physical quantities.

Another important reason for adopting consistent units is to simplify complex calculations where you may be dealing with several measured quantities (see pp. 494 and 496). Although the rules of the SI are complex and the scale of the base units is sometimes inconvenient, to gain the full benefits of the system you should observe its conventions strictly.

The description of measurements in SI involves:

- **seven base units and two supplementary units,** each having a specified abbreviation or symbol (Table 28.1);
- **derived units,** obtained from combinations of base and supplementary units, which may also be given special symbols (Table 28.2);
- **a set of prefixes** to denote multiplication factors of 10^3, used for convenience to express multiples or fractions of units (Table 28.3).

Table 28.1 The base and supplementary SI units

Measured quantity	Name of SI unit	Symbol
Base units		
Length	metre	m
Mass	kilogram	kg
Amount of substance	mole	mol
Time	second	s
Electric current	ampere	A
Temperature	kelvin	K
Luminous intensity	candela	cd
Supplementary units		
Plane angle	radian	rad
Solid angle	steradian	sr

Table 28.2 Some important derived SI units

Measured quantity	Name of unit	Symbol	Definition in base units	Alternative in derived units
Energy	joule	J	$m^2\,kg\,s^{-2}$	N m
Force	newton	N	$m\,kg\,s^{-2}$	$J\,m^{-1}$
Pressure	pascal	Pa	$kg\,m^{-1}\,s^{-2}$	$N\,m^{-2}$
Power	watt	W	$m^2\,kg\,s^{-3}$	$J\,s^{-1}$
Electric charge	coulomb	C	As	$J\,V^{-1}$
Electric potential difference	volt	V	$m^2\,kg\,A^{-1}\,s^{-3}$	$J\,C^{-1}$
Electric resistance	ohm	Ω	$m^2\,kg\,A^{-2}\,s^{-3}$	VA^{-1}
Electric conductance	siemens	S	$s^3\,A^2\,kg^{-1}\,m^{-2}$	$A\,V^{-1}$ or Ω^{-1}
Electric capacitance	farad	F	$s^4\,A^2\,kg^{-1}\,m^{-2}$	$C\,V^{-1}$
Luminous flux	lumen	lm	cd sr	
Illumination	lux	lx	$cd\,sr\,m^{-2}$	$lm\,m^{-2}$
Frequency	hertz	Hz	s^{-1}	
Radioactivity	becquerel	Bq	s^{-1}	
Enzyme activity	katal	kat	$mol\ substrate\ s^{-1}$	

Table 28.3 Prefixes used in the SI

Multiple	Prefix	Symbol	Multiple	Prefix	Symbol
10^{-3}	milli	m	10^3	kilo	k
10^{-6}	micro	μ	10^6	mega	M
10^{-9}	nano	n	10^9	giga	G
10^{-12}	pico	p	10^{12}	tera	T
10^{-15}	femto	f	10^{15}	peta	P
10^{-18}	atto	a	10^{18}	exa	E
10^{-21}	zepto	z	10^{21}	zetta	Z
10^{-24}	yocto	y	10^{24}	yotta	Y

Recommendations for describing measurements in SI units

Basic format

- **Express each measurement as a number separated from its units by a space.** If a prefix is required, no space is left between the prefix and the unit it refers to. Symbols for units are only written in their singular

Example   10 μg is correct, while 10μg, 10 μg. and 10μ g are incorrect; 2.6 mol is right, but 2.6 mols is wrong.

> **Example** n stands for nano and N for newtons.

> **Example** 1982 963.192 309 kg (perhaps better expressed as 1.982 963 192 309 Gg).

> Units combined by multiplication – an alternative convention is to use a raised dot · between multiplied units to avoid ambiguity. Thus, m · s indicates metre-second, rather than millisecond (ms).

> **Examples**
> 10 μm is preferred to 0.000 01 m or 0.010 mm.
> 1 mm^2 = 10^{-6} m^2 (not one-thousandth of a square metre).
> 1 dm^3 (1 L) is more properly expressed as 1×10^{-3} m^3.
> The mass of a neutrino is 10^{-36} kg.
> State as MW m^{-2} rather than W mm^{-2}.

form and do not require full stops to show that they are abbreviated or that they are being multiplied together.

- **Give symbols and prefixes appropriate upper- or lower-case initial letters** as this may define their meaning. Upper-case symbols are named after persons but when written out in full they are not given initial capital letters.

- **Show the decimal sign as a full point on the line.** Some metric countries continue to use the comma for this purpose and you may come across this in the literature: commas should not therefore be used to separate groups of thousands. In numbers that contain many significant figures, you should separate multiples of 10^3 by spaces rather than commas.

Compound expressions for derived units

- **Take care to separate symbols in compound expressions by a space** to avoid the potential for confusion with prefixes. Note, for example, that 200 m s (metre-seconds) is different from 200 ms (milliseconds).

- **Express compound units by using negative powers** rather than a solidus (/): for example, write mol m^{-3} rather than mol/m^3. The solidus is reserved for separating a descriptive label from its units (see p. 475).

- **Use parentheses to enclose expressions being raised to a power** if this avoids confusion: for example, a photosynthetic rate might be given in mol CO_2 (mol photons)$^{-1}$ s^{-1}.

- **Where there is a choice, select relevant (natural) combinations of derived and base units:** for example you might choose units of Pa m^{-1} to describe a hydrostatic pressure gradient rather than kg m^{-2} s^{-2}, even though these units are equivalent and the measurements are numerically the same.

Use of prefixes

- **Use prefixes to denote multiples of 10^3** (Table 28.3) so that numbers are kept between 0.1 and 1000.

- **Treat a combination of a prefix and a symbol as a single symbol.** Thus, when a modified unit is raised to a power, this refers to the whole unit including the prefix.

- **Avoid the prefixes deci (d) for 10^{-1}, centi (c) for 10^{-2}, deca (da) for 10 and hecto (h) for 100** as they are not strictly SI.

- **Express very large or small numbers as a number between 1 and 10 multiplied by a power of 10** if they are outside the range of prefixes shown in Table 28.3.

- **Do not use prefixes in the middle of derived units:** they should be attached only to a unit in the numerator (the exception is in the unit for mass, kg).

> **KEY POINT** For the foreseeable future, you will need to make conversions from other units to SI units, as much of the literature quotes data using imperial, c.g.s. or other systems. You will need to recognise these units and find the conversion factors required. Examples relevant to biology are given in Box 28.1. Table 28.4 provides values of some important physical constants in SI units.

Box 28.1	How to interconvert values between some redundant units and the SI

Quantity	SI unit/symbol	Old unit/symbol	Multiply number in old unit by this factor for equivalent in SI unit*	Multiply number in SI unit by this factor for equivalent in old unit*
Area	square metre/m^2	acre	4.04686×10^3	0.247105×10^{-3}
		hectare/ha	10×10^3	0.1×10^{-3}
		square foot/ft^2	0.092903	10.7639
		square inch/in^2	645.16×10^{-9}	1.55000×10^6
		square yard/yd^2	0.836127	1.19599
Angle	radian/rad	degree/°	17.4532×10^{-3}	57.2958
Energy	joule/J	erg	0.1×10^{-6}	10×10^6
		kilowatt hour/kWh	3.6×10^6	0.277778×10^{-6}
		calorie/cal	4.1868	0.2388
Length	metre/m	Ångstrom/Å	0.1×10^{-9}	10×10^9
		foot/ft	0.3048	3.28084
		inch/in	25.4×10^{-3}	39.3701
		mile	1.60934×10^3	0.621373×10^{-3}
		yard/yd	0.9144	1.09361
Mass	kilogram/kg	ounce/oz	28.3495×10^{-3}	35.2740
		pound/lb	0.453592	2.20462
		stone	6.35029	0.157473
		hundredweight/cwt	50.8024	19.6841×10^{-3}
		ton (UK)	1.01605×10^3	0.984203×10^{-3}
Pressure	pascal/Pa	atmosphere/atm	101325	9.86923×10^{-6}
		bar/b	100000	10×10^{-6}
		millimetre of mercury/mmHg	133.322	7.50064×10^{-3}
		torr/Torr	133.322	7.50064×10^{-3}
Radioactivity	becquerel/Bq	curie/Ci	37×10^9	27.0270×10^{-12}
Temperature	kelvin/K	centigrade (Celsius) degree/°C	°C + 273.15	K − 273.15
		Fahrenheit degree/°F	(°F + 459.67) × 5/9	(K × 9/5) − 459.67
Volume	cubic metre/m^3	cubic foot/ft^3	0.0283168	35.3147
		cubic inch/in^3	16.3871×10^{-6}	61.0236×10^3
		cubic yard/yd^3	0.764555	1.30795
		UK pint/pt	0.568261×10^{-3}	1759.75
		US pint/liq pt	0.473176×10^{-3}	2113.38
		UK gallon/gal	4.54609×10^{-3}	219.969
		US gallon/gal	3.78541×10^{-3}	264.172

*In the case of temperature measurements, use formulae shown.

Table 28.4 Some physical constants in SI terms

Physical constant	Symbol	Value and units
Avogadro's constant	N_A	6.022174×10^{23} mol^{-1}
Boltzmann's constant	k	1.380626×10^{-23}
Charge of electron	e	1.602192×10^{-19} C
Gas constant	R	8.31443 J K^{-1} mol^{-1}
Faraday's constant	F	9.648675×10^4 C mol^{-1}
Molar volume of ideal gas at STP	V_0	0.022414 m^3 mol^{-1}
Speed of light *in vacuo*	c	2.997924×10^8 ms^{-1}
Planck's constant	h	6.626205×10^{-34} J s

Expressing units of volume – in this book we use the symbol 'L' rather than 'l' for litre to avoid potential confusion between 'i' and l, but use SI units where this simplifies calculations. In formal scientific writing, constructions such as 1×10^{-6} m^3 (= 1 mL) and 1 mm^3 (= 1 µL) may be used.

Expressing enzyme activity – the derived SI unit is the katal (kat), which is the amount of enzyme that will transform 1 mol of substrate in 1 s (see Chapter 60).

Converting between concentration units – being able to express concentrations in different units is important as this skill is frequently used when following instructions and interpreting data (Chapter 25).

Definition

STP – Standard Temperature and Pressure = 293.15K and 0.101 325MPa.

Unit oddities – examples of nomenclature that do not fit easily into the SI:
- **Dates.** The UK convention is 10/03/11 for 10 March 2011, but in the US this would mean 3 October 2011. Avoid confusion by stating 3 Mar 2011 or making it clear which convention you are adopting in your Materials and Methods or table/figure footnotes/legend.
- **Ohm.** If the symbol Ω is not available in your word processor, then write 'ohm' (not capital O).
- **Computing terminology.** b is used for bit, and B for Byte. Mega and giga can be ambiguous as they may refer loosely to 2^{20} or 2^{30} respectively for binary data. The International Electrotechnical Commission has attempted to rationalise this with a new set of prefixes (Rowlett, 2005).

Some implications of SI in biology

Volume

The SI unit of volume is the cubic metre, m^3, which is rather large for practical purposes. The litre (l) and the millilitre (ml) are technically obsolete, but are widely used and glassware is still calibrated using them. Note also that the US spelling is liter. You may find litre given the symbol L, as in this text.

Mass

The SI unit for mass is the kilogram (kg) rather than the gram (g): this is unusual because the base unit has a prefix applied.

Amount of substance

You should use the mole (mol, i.e. Avogadro's constant, see Table 28.4) to express very large numbers. The mole gives the number of atoms in the atomic mass, a convenient constant. Always specify the elementary unit referred to in other situations (e.g. mol photons m^{-2} s^{-1}).

Concentration

The SI unit of concentration, mol m^{-3}, is quite convenient for biological systems. It is equivalent to the non-SI term 'millimolar' (mM = mmol L^{-1}), while 'molar' (M = mol L^{-1}) becomes kmol m^{-3}. Note that the symbol M in the SI is reserved for mega and hence should not be used for concentrations. If the solvent is not specified, then it is assumed to be water (see Chapter 25).

Time

In general, use the second (s) when reporting physical quantities having a time element (e.g. give photosynthetic rates in mol CO$_2$ m^{-2} s^{-1}). Hours (h), days (d) and years should be used if seconds are clearly absurd (e.g., samples were taken over a 5-year period). Note, however, that you may have to convert these units to seconds when doing calculations.

Temperature

The SI unit is the kelvin, K. The degree Celsius scale has units of the same magnitude, °C, but starts at 273.15K, the melting point of ice at STP. Temperature is similar to time in that the Celsius scale is in widespread use, but note that conversions to K may be required for calculations. Note also that you must not use the degree sign (°) with K and that this symbol must be in upper case to avoid confusion with k for kilo; however, you *should* retain the degree sign with °C to avoid confusion with the coulomb, C.

Light

While the first six base units in Table 28.1 have standards of high precision, the SI base unit for luminous intensity, the candela (cd) and the derived units lm and lx (Table 28.2), are defined in 'human' terms. They are, in fact, based on the spectral responses of the eyes of 52 American infantrymen measured in 1923. Clearly, few organisms 'see' light in the same way as this sample of humans. Also, light sources differ in their spectral quality. For these reasons, it is better to use expressions based on energy or photon content (e.g. W m^{-2} or mol photons m^{-2} s^{-1}) in studies other than those on human vision. Ideally you should specify the photon wavelength spectrum involved (see Chapter 58).

Text reference

Rowlett, R. (2005) *How Many? A Dictionary of Units of Measurement.* Available: http://www.unc.edu/~rowlett/units/ Last accessed 17/02/15.

Sources for further study

Meatyard, B. (ed.) (2009) *Biological Nomenclature: Standard Terms and Expressions Used in the Teaching of Biology*, 4th edn. Institute of Biology, London.

The NIST Reference on Constants, Units and Uncertainty: International System of Units (SI). Available: http://physics.nist.gov/cuu/Units/index .html Last accessed 17/02/15.

Pennycuick, C.J. (1988) *Conversion Factors: SI Units and Many Others: Over 2100 Conversion Factors for Biologists and Mechanical Engineers Arranged in 21 Quick-Reference Tables.* University of Chicago Press, Chicago.

Tapson, F. *A Dictionary of Units.* Available: http:// www.cleavebooks.co.uk/dictunit/dictunit1.htm Last accessed 17/02/15.

STUDY EXERCISES

28.1 Practise converting between units. Using Box 28.1 as a source, convert the following amounts into the units shown. Give your answers to three significant figures.

(a) 101 000 Pa into atmospheres.

(b) One square yard into square millimetres.

(c) One UK pint into millilitres.

(d) 37 °C into kelvin.

(e) 11 stone 6 pounds into kilograms.

28.2 Practise using prefixes appropriately. Simplify the following number/unit combinations using an appropriate prefix so that the number component lies between 0.1 and 1000.

(a) 10 000 mm

(b) 0.015 mL

(c) 5×10^9 J

(d) 65 000 ms^{-1}

(e) 0.000 000 0001 g

28.3 Check units in an equation. The Hagan-Poiseuille equation describes water flux in smooth cylindrical pipes assuming laminar flow. This equation can be expressed as:

$$J_v = \frac{r^2 + \delta P}{8\eta \times \delta x}$$

where:

r = the radius of the cylindrical pipe (m);

η ('eta') = the viscosity of the liquid (Pa s);

δP = the pressure difference across the ends of the pipe (Pa);

δx = the length of the pipe (m).

Verify, by putting relevant units in place of the variables in this equation and simplifying the resulting relationship, that appropriate units for J_v, the mean flow rate, are ms^{-1}.

Answers to these study exercises are available at www.pearsoned.co.uk/practicalskills

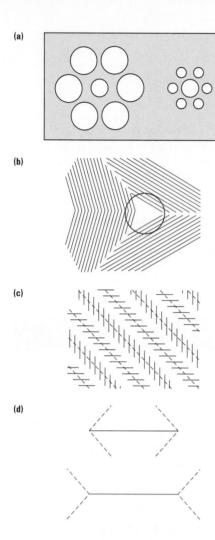

Fig. 29.1 Examples of 'optical illusions' caused by problems of perception. Image (a) shows how the sizes of adjacent objects can distort a simple comparison of size: the central circles in either hexagonal pattern are the same size. Image (b) reveals how shapes can be distorted by adjacent linear objects: the inner shape is a perfect circle. Image (c) illustrates how directional cues can lead to confusion: the dashed lines are parallel. Image (d) shows how adjacent shapes can make comparison of simple linear dimensions difficult: the two solid lines are the same length. In all cases, the correct perceptions can be confirmed only with a measurement aid such as a ruler or compass.

Observations provide the basic information leading to the formulation of hypotheses, an essential step in the scientific method (see Fig. 33.1). Observations are obtained either directly by our senses or indirectly through the use of instruments that extend our senses and may be either:

- **Qualitative:** described by words or terms rather than by numbers and including subjective descriptions in terms of variables such as colour, shape and smell; often recorded using photographs and drawings.
- **Quantitative:** numerical values derived from counts or measurements of a variable (see Chapter 27), frequently requiring use of some kind of instrument.

> **KEY POINT** Although qualitative and quantitative observations are useful in biology, you should try to make numerical counts or measurements wherever possible, as this allows you to define your observations more rigorously and make objective comparisons using statistical tools.

Factors influencing the quality of observations

Perception

Observation is highly dependent upon the perception of the observer (Fig. 29.1). Perception involves both visual and intuitive processes, so your interpretation of what you see is very dependent upon what you already know or have seen before. Thus, two persons observing the same event or object may 'see' it differently, a good example of bias. This is frequently true in microscopy where experience is an important factor in interpretation.

When you start biology, your knowledge base will be limited and your experience restricted. Practical training in observation provides the opportunity to develop both aspects of your skills in a process that is effectively a positive feedback loop – the more you know/see as a result of practice, the better will your observations become.

Precision and error

Obviously very important for interpretive accuracy, with both human and non-human components. These are dealt with in Chapter 27.

Artefacts

These are artificial features introduced usually during some treatment process such as chemical fixation prior to microscopic examination. They may be included in the interpretive process if their presence is not recognised – again, prior experience and knowledge are important factors in spotting artefacts (see Chapter 46 and especially Fig. 46.1 in relation to microscopy).

Developing observational skills

You must develop your knowledge and observational skills to benefit properly from your practical work. The only way to acquire these skills is through extensive practice.

Make sure your observations are:

- **relevant,** i.e. directed towards a clearly defined objective;
- **accurate,** i.e. related to a scale whenever possible;
- **repeatable,** i.e. as error-free (precise) as possible.

Much biological work attempts to relate structure to function, often through careful analysis of structure at different levels. One of the best ways to develop observational skills is by making accurate drawings or diagrams, forcing you to look more carefully than is usual (see Chapter 30). An important observational skill to develop is the interpretation of two-dimensional images, such as sections through plant/animal material and photographs, in terms of the three-dimensional forms from which they are derived. This requires a clear understanding of the nature of the image in terms of both scale and orientation (see Chapter 46).

Counting

Counting is an observational skill that requires practice to become both accurate and efficient. It is easy to make errors or lose count when working with large numbers of objects. Use a counting aid whenever human error might be significant. There are many such aids such as tally counters, tally charts and specialised counting devices like colony counters. It is important to avoid counting items twice. For example, when counting microbial colonies on Petri plates, each colony can be marked off as it is counted on the base of the plate using a spirit-based marker pen.

Another valuable technique is to use a grid system to organise the counting procedure. This has become formalised in equipment such as the haemocytometer used for counting blood cells (see Box 48.1). Remember that you must decide on a protocol for sampling, particularly with regard to the direction of counting within the grid and for dealing with boundary overlaps to prevent double counting at the edges of the grid squares (see Fig. 32.4).

Observation during examinations

Making appropriate observations during practical examinations often causes difficulty, particularly when qualitative observations are needed, e.g. when asked to classify a specimen, giving reasons. Answering such questions clearly requires biological knowledge but also requires a strategy to provide the relevant observations. Thus for the above example, observations relevant to determining each taxonomic level should be made and recorded: set out your observations in a logical sequence so that you show the examiner how you arrived at your conclusion.

A similar approach is needed when asked to make observations related to structure or function. For example, if comments on a method of locomotion are required, do not make observations on irrelevant structures. You will obtain maximum marks only if your answers are concise and relevant (Chapter 7).

Observer effects – remember that your presence, or the act of observation itself, may influence the event you are observing. Take appropriate precautions, such as using a hide when observing animal behaviour.

Preparation – thorough theoretical groundwork before a practical class or examination is vitally important for improving the quality of your observations.

Perceptual illusions – beware deceptions such as holes appearing as bumps in photographs, misinterpretations due to lack of information on scale and other simple but well-known tricks of vision (see Fig. 29.1).

Making counts – acetate sheets can be used as overlays for photographs, drawings and other images, then as each object is counted it can be marked off using a water-based marker pen. Remember to mark identification points in case the sheet slips!

Text reference

Edwards, B. (2000). *The New Drawing on the Right Side of your Brain.* Souvenir Press, London.

Sources for further study

Akins, K. (ed.) (1996) *Perception.* Vancouver Studies in Cognitive Science, Vol. 5. Oxford University Press, Oxford.
[A detailed examination of the problems of perception from a psychological perspective.]

Raulin, M.L. *Naturalistic Observation and Case-Study Research.* Available: http://www.abacon.com/graziano/ch06/
Last accessed 26/03/15.

STUDY EXERCISES

29.1 Assess the difficulties of making counts of large numbers. The dots in the area shown in the figure below represent bacterial colonies on a Petri dish. To appreciate the difficulties of counting numbers accurately, make three rapid independent attempts at counting the dots by eye (without using any aids). Now subdivide the area, such that the number of colonies per subdivision is more easily counted (i.e. is a number between 1 and 20). Again carry out three further estimates using these subdivisions, summing the counts obtained each time. Compare the results using both methods – which is the more precise and which the more accurate?

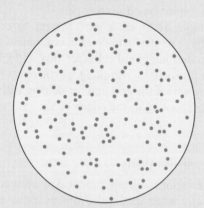

Bacterial colonies on a Petri dish

29.2 Practise observing and hypothesising. Find a grassy area nearby, such as a lawn, play area or a kerb-side verge. Observe this area from 30 m distance.

What can you see? How would you describe it? Now walk on the grass, look down and observe it from 1–2 m distance. What do you see now? Is this a monoculture? Almost certainly not. What other plant species are present? Would you need to get even closer to observe them? Even if you have a rough idea of their generic (and species) names, how could you be certain? How would you describe the distribution of individual plants of each type? In the case of the rarer yet easily identified species, is it random, regular, or patchy? How would you gather evidence about this? How would you test this hypothesis? Finally, what biological reasons might you conclude were responsible for the distribution?

29.3 Trick your brain into letting you do better drawings. One of the problems with making drawings is that the brain thinks it 'knows' how things should look. This may cause distortion in diagrams. One way to produce realistic drawings of relatively complex objects is to turn the object upside down, then try to draw it. Because the brain no longer recognises the object, it focuses on the relationships between spaces and lines and allows you to make a reasonable copy (Edwards, 2000). Try this out by taking a photograph of an object that you would normally find difficult to draw, e.g. a person's face or an animal, and using this technique to copy it.

Answers to these study exercises are available at www.pearsoned.co.uk/practicalskills

30 Drawings and diagrams

Drawing has an important place in biological teaching because of its role in developing observational skills. You need to look at a specimen very carefully to be able to draw it accurately, while labelling a diagram forces you to think about the component structures and their positions. If your observation of a specimen is poor, so too will be your diagram.

Strictly, a *drawing* is a detailed and accurate representation of a specimen, requiring no previous biological knowledge. This level of artwork is never required for normal practical work. A *diagram*, on the other hand, needs to be accurate in its general proportions, but is otherwise very stylised, showing only the most important features. Biological knowledge is required to select items for inclusion and to decide what detail to ignore. Diagrams are often called figures in formal scientific writing, but may sometimes be referred to loosely as drawings, since the above distinction is frequently ignored.

> **KEY POINT** You may not feel confident about being able to produce quality artwork in practicals, especially when the time allowed is limited. However, the requirements of biological drawing are not as demanding as you might think and the skills required can be learned. By following the guidelines and techniques explained below, most students should be able to produce good diagrams.

The main types of figure

Cell diagrams

In a cell diagram, your aim is to show accurately the details of the individual cells in a tissue. You would normally draw a cell diagram from a specimen viewed by light microscopy at a magnification of 200× or more. Your diagram should be detailed, but need not comprise more than a few cells, especially if they are all similar (Fig. 30.1(a)). You may be asked to draw only one complete cell, plus *part* of any neighbouring cells, so that the interrelationship between cells can be seen. Use labels to note any structures that are stained, and what this might mean in terms of cell chemistry. For cells with thin walls, you may need to exaggerate the wall thickness to show it on your diagram. If this is necessary, add an explanatory note to your diagram (see Fig. 30.1(a)).

Tissue diagrams or maps

The purpose of a tissue diagram is to show how the tissues are placed in a section through a whole organism or one of its organs (compare Figs 30.1(b) and 30.1(c)). You should not include any cellular detail unless specifically instructed, and shading or hatching should be kept to a minimum. The main difficulty you will encounter is deciding where to draw the boundary line between tissues: cell differentiation is rarely discrete, so cells at the boundary may show characteristics of both tissues. A certain amount of background knowledge and interpretive skill is required for this.

Figure numbering in reports – if producing several diagrams, graphs, etc., number them consecutively as Figure 1, Figure 2, etc., and use this system to refer to them in the text.

Definitions

When drawing cell diagrams, tissue maps and body plans, it might help to bear in mind the following definitions:

Cells – the 'building blocks' of life. Always separated from their environment and other cells by a membrane ± a cell wall. Typical dimensions: 10 to 100 μm. Examples: leaf mesophyll cell; kidney nephron cell; yeast cell.

Organs – structures consisting of several tissues working as a group to perform a specific function. Examples: leaf; kidney.

Tissues – collections of cells having similar structure and function. From one cell thick to many cell layers. Examples: palisade mesophyll in a leaf; renal cortex in the kidney.

Vicia faba spongy mesophyll cell TS

Transverse section of a dicotyledon leaf

Fig. 30.1 Cell and tissue diagrams compared with a photomicrograph: (a) cell diagram showing representative spongy mesophyll cells, based on an electron microscope image of a leaf section; (b) photomicrograph of a transverse section through a leaf (light microscopy). Different scales are used, but in both cases they are indicated by a bar (see p. 278). Note how the cell wall structure is clarified in (a) by slightly exaggerating its thickness. In a light microscope section, less detail of organelles would be visible. (c) Tissue diagram of part of a transverse leaf section. Note that this diagram is uncluttered by detail and the individual tissues are not shaded – the scale is indicated by a magnification factor relevant to the actual size of the diagram (reduced here) (see p. 277).

Choosing what to include in a morphological diagram – since the essence of this type of drawing is to show only the important features, it is worth making a list of the items to be included, or to highlight these where mentioned in your practical schedule. Do not forget, however, to add enough anatomical detail, e.g. of a body plan, to place outline features in the appropriate context. Always add a scale and effective labelling. Note that if you wish to show fine detail, consider doing this on a separate higher-scale diagram.

Morphological diagrams and body plans

In a morphological diagram, the objective is to provide a lifelike representation indicating the main surface features (Fig. 30.2). For a body plan, your aim is to show the relationships between segments, organs or other body parts, often following a dissection (Fig. 30.3). In both cases, shading should be avoided, unless it serves to highlight a particular feature. You should fully label both types of diagram and add notes where relevant. The main problem you are likely to encounter is keeping the different parts in proportion to each other – this can be solved by using construction lines and frames (see below).

KEY POINT Remember that in drawing some 'artistic licence' is possible, allowing you to merge features seen on different specimens, or show, for example, the inside or underside of one of the parts.

Continuous tone drawings are more refined versions of the above where shading is used to provide a pictorial representation of the item of interest, providing fine detail and realism (Fig. 30.4). These drawings are generally only used for presentation work and are not normally expected from students. It can be argued that a properly taken photograph (Chapter 50) is often more appropriate and less time-consuming.

Apparatus diagrams

Here, your aim is to portray the components of some experimental set-up as a diagram (Fig. 30.5). Note that these figures are normally drawn as a section rather than a perspective drawing. Also, you may be more concerned with the relationship between parts than with showing them

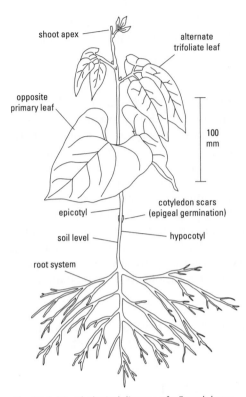

Fig. 30.2 Morphological diagram of a French bean seedling, *Phaseolus vulgaris* L. Note the lack of shading compared with Fig. 30.4.

Fig. 30.4 Continuous tone drawing of a French bean seedling, *Phaseolus vulgaris* L. Compare with Fig. 30.2.

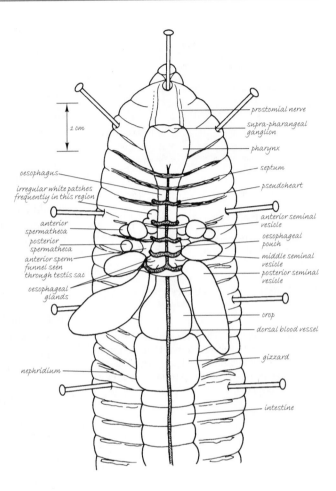

Fig. 30.3 Example of a body plan. This is a diagram of a general dissection of an earthworm, *Lumbricus terrestris* L., dorsal view. Note the lack of shading and the use of labels and annotations to identify clearly what has been drawn.

Fig. 30.5 Two-dimensional apparatus diagram of a diffusion porometer (not drawn to uniform scale). Note the use of letters to simplify labelling – these should be explained in the figure legend (e.g. L = leaf, P = pump).

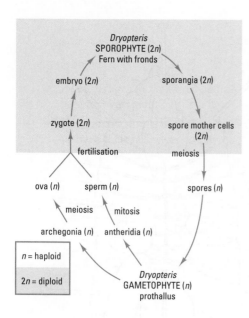

Fig. 30.6 Example of a chart: diagrammatic life cycle of the fern *Dryopteris filix-mas*. Note the use of shading and chromosome number to denote haploid and diploid phases of the life cycle, and how a key is used to explain this.

to a uniform scale. For example, in Fig. 30.5, the leaf clamp (LC) is exaggerated in size compared with other components so that its internal detail can be shown.

Charts

The main purpose of a chart is to organise information (e.g. Fig. 30.6). You can use charts to communicate complex ideas, procedures or lists of facts by simplifying, grouping and appropriate layout. In biology, charts are particularly useful for illustrating life cycles, metabolic pathways and organisational hierarchies. Flowcharts (e.g. Fig. 35.1) are a specialised form. To be effective, charts must be logically organised. A good chart should clarify the parts and their relationships and its presentation should be simple, clear and visually pleasing. Make several rough sketches with different arrangements before deciding on the final version. You should use appropriate words or symbols to denote the components and link them with lines or arrows to show sequences or interrelationships. Computer software packages can be used to enhance the quality of presentation.

Graphs and histograms

These are used to display numerical information (data) in a form that is easily assimilated. Chapter 72 covers the main types of graphs and how they should be constructed.

Steps towards drawing a good diagram

To produce good figures, both planning and careful execution are needed (Box 30.1).

Planning

The first stage of any drawing is to decide exactly what to draw. To focus your thoughts, you will need to decide on the answers to the following questions:

- What is the purpose of the drawing?
- What type of drawing is required?
- What should go into it?
- What magnification or reduction is required?

Box 30.1 Checklist for making a good diagram

1. Decide exactly what you are going to draw and why.

2. Decide how large the diagram should be.

3. Decide where you are going to place the diagram on the page.

4. Start drawing:

   (a) Draw what you see, not what you expect to see.

   (b) Use carefully measured construction lines to provide the correct proportions.

   (c) Avoid shading.

   (d) Avoid excess detail, especially in tissue diagrams.

   (e) Use conventions where appropriate, e.g. cut edges represented by a double line.

5. Label the drawing/diagram carefully and comprehensively.

6. Give the diagram a title, scale and legend. Include organism, classification, part drawn, orientation, stain(s), magnification, etc.

Once these decisions have been made, you can determine the position and size of your diagram. Your diagram should be as large as possible, but remember to leave space for legends and labels.

Producing labels for black and white diagrams in formal reports – create the text using a word processor and high-quality printer. This can then be stuck on your diagram using e.g. Pritt Stick, taking care to keep all the text parallel. If the diagram is now photocopied or photographed with high contrast, the use of different pieces of paper cannot be detected. Alternatively, scan the diagram and add labels using a suitable program such as *PowerPoint* or *Word*.

Materials

Most diagrams for practicals are drawn in pencil, to allow corrections to be made. Propelling pencils are valuable for ensuring constant line thickness but they do not allow you the flexibility to vary line thickness as you can by changing the angle of an ordinary pencil. If you prefer to use an ordinary pencil, sharpen it frequently. Invest in a good-quality eraser – those of poor quality tend to smudge badly – and frequently clean its working surface on a spare piece of paper. Always use plain paper for drawing, and if you are asked to supply your own, make sure it is of good quality. Use pen and ink to create line drawings for illustration purposes in posters, project reports, etc. Such diagrams should be in black and white only. Computer drawing programs can also provide good-quality output suitable for these tasks.

Constructing a diagram

1. **Draw a faint rectangle in pencil** to show the figure boundaries.

2. **Draw very faint 'construction lines'** using a 2H pencil with a sharp point to get the basic proportions and outlines correct before progressing. These should be erased once the basic drawing is complete. To lay in construction lines, use a ruler or pair of dividers to determine the actual proportions of the object to be drawn and then, using these dimensions, construct a scaled frame to allow further important reference points to be located (Fig. 30.7).

Using construction lines – these are vital for producing well-proportioned drawings.

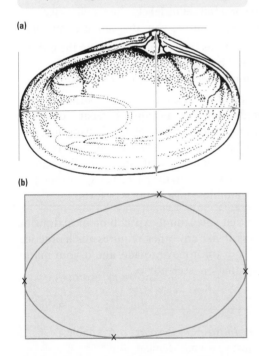

Fig. 30.7 How to draw an object in proportion: (a) determine linear dimensions; (b) construct a frame and outline, using reference points (x) determined from scaled measurements of the original specimen.

3. **Draw the main outlines faintly with your 2H pencil. When satisfied, go over the lines with a sharp HB or 2B pencil.** Draw firm, continuous lines, not hesitant, scratchy ones and make sure that junctions between lines are properly drawn. If you need to distinguish between different regions within your drawing, use hatching or stippling (but not shading), and avoid drawing unnecessary detail such as large numbers of cells (see Fig. 30.8). Always draw what you see, rather than what you think you should see, and seek advice at an early stage if you cannot identify a particular structure, or if you think your specimen is atypical. When drawing a specimen that is symmetrical or that contains repeated forms, it will save time if you draw its outline, and only provide detail in one of the replicated elements.

4. **Complete your drawing by adding labels.** This requires you to interpret your observations and helps you to remember what the structures look like. Careful and accurate labelling is as important as the drawing itself. It should be done clearly and neatly using either radiating or horizontal lines ending in arrowheads or large dots to indicate exact label references. The lines should not cross. Labels should be written clearly in one orientation, so that they can be read without needing to turn the paper. Annotations (short explanatory notes in brackets below the labels) are strongly recommended – regard them as notes to yourself about what you have seen, or what has been pointed out by tutors. For practical work, use a pencil for labelling in case your demonstrator or tutor corrects your work.

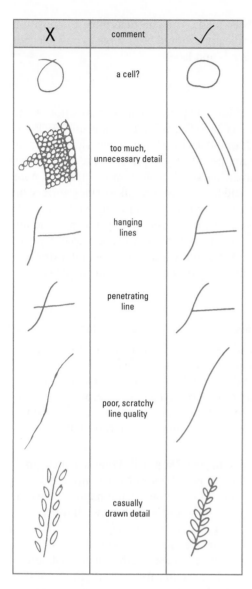

X	comment	✓
	a cell?	
	too much, unnecessary detail	
	hanging lines	
	penetrating line	
	poor, scratchy line quality	
	casually drawn detail	

Fig. 30.8 Examples of common errors in biological drawings. Most of these mistakes are due to lack of care or attention to detail – easily solved!

5. **Add a title,** a scale or magnification factor (see Chapter 46), and a legend. The legend should provide all relevant information, including:

   (a) the binomial Latin name for the organism and a taxonomic classification, if appropriate;

   (b) details of the preparation of the subject, e.g. TS, whole mount, ventral dissection, etc., and any stains used.

Drawing from the microscope

Begin by positioning the paper beside your drawing hand and use the 'opposite' eye for examining the specimen; thus for a right-handed person, the paper is placed on the right of the microscope and you use your left eye. With a binocular microscope, use only one of the eyepieces; if you keep both eyes open, it is possible with practice to learn to draw and see the page with one eye while observing the specimen with the other. For specimens that need to be drawn very accurately in project work, projection devices such as the *camera lucida* may be required.

Avoiding mistakes

There are four main categories of error in student diagrams:

- **Incorrect positioning and proportions** – solve these problems by following steps 1–3 above.
- **Forgetting to add a title, scale or a full set of labels** – use a checklist such as that provided in Box 30.1 to ensure your diagram is complete.
- **Untidiness in presentation** – avoid this type of error by using the correct materials as discussed above and by taking care – untidiness is frequently due to lack of attention to detail, as illustrated in Fig. 30.8.
- **Biological inaccuracies** – this kind of mistake is the most important, and will lose you most marks. Avoiding these errors requires preparation before the practical, so that you know more about what you are drawing and, for instance, have a good idea which parts are which *before* you start. Try to focus clearly on the objectives of the practical, and listen carefully to any tips given by your tutors, which may relate to the particular specimen(s) available on that day rather than to those in your notes or texts.

Finally, it is important to realise that you cannot expect biological drawing skills to develop overnight. This skill, like any other, requires much practice. Try to learn from any feedback your tutor may provide, and if your marks are consistently low without explanation, seek advice.

Sources for further study

Jepson, M. (1942) *Biological Drawings with Notes, Parts 1 and 2,* 5th edn. John Murray, London. [Although this book is old, you may find it in your library. Provides examples of labelled biological drawings in an exemplary style.]

Leslie, C.W. (1995) *The Art of Field Sketching: A Naturalist's Sketchbook.* Kendal/Hunt, Dubuque.

Sodt, J. *Botanical Illustration Resources.* Available: http://www.library.wwu/botanical_illus
Last accessed 18/02/15.
[A wide-ranging source of information about drawing plant life. Although the focus of this site is on botanical illustration, it contains many useful references.]

Zweifel, F.W. (1988) *A Handbook of Biological Illustration,* 2nd edn. University of Chicago Press, Chicago.

STUDY EXERCISES

30.1 Practise making well-proportioned drawings. Choose a picture of an animal or plant from a biology textbook or a printout from the Web. Using the techniques described on pp. 188–90 and Fig 30.7, make your own drawing based on this image.

30.2 Devise a flowchart. Present the instructions given in one of the Boxes in this text (for example, Boxes 41.1, 48.1 and 60.1) in the form of a flowchart.

30.3 Compare and contrast the main types of biological diagrams. Complete the following table.

Comparison of the main types of biological diagrams.

Feature	Type of diagram		
	Cell diagram	Tissue diagram	Morphological diagram
Should the diagram carry a scale?			
What might be an appropriate scale in m?			
Approximate number of cells visible			
Are organelles visible?			
Should shading be used?			
Should labelling be used?			

Answers to these study exercises are available at www.pearsoned.co.uk/practicalskills

Fieldwork is often one of the most enjoyable and rewarding aspects of undergraduate courses in biology, although it can also be arduous, especially in bad weather. The aims of fieldwork include:

- **introducing a broad range of living organisms,** with consideration of their roles in natural communities;
- **developing skills in field techniques,** with particular reference to sampling strategies and methods, and to the measurement of environmental variables;
- **providing experience in the use of identification keys** and fostering diagnostic skills;
- **developing skills in data handling, analysis, interpretation and presentation,** together with the writing of scientific reports;
- **experiencing teamwork over an extended period;**
- **broadening environmental awareness.**

To make the most of your time and effort while carrying out field investigations, you will need clear objectives and adequate preparation. You can minimise problems by carefully thinking your activities through beforehand.

> **KEY POINT** Field studies are often time-restricted, and access to equipment and apparatus may be limited, so anticipation is the key to successful and safe fieldwork.

Field excursions

You are likely to make short visits to field sites during the early part of your course. These excursions provide an opportunity to gain an overview of a particular environment in a single visit, e.g. an afternoon excursion to a marine rocky shore. Note the following points:

- **Carry out the recommended background reading** – you will get much more out of the visit if you read any specific handouts or notes in advance. If you plan to take these into the field, you can put them in separate polythene pockets – if necessary, you can make these fully waterproof by sealing the top opening with adhesive tape. It might also be useful to carry out some general background reading on the area to be visited. You should pay particular attention to any risk assessment information and safety guidance notes.
- **Make sure you are well prepared for the weather and the terrain** – listen to the weather forecast and take appropriate clothing to keep you warm and dry. Just because your visit is short, you should not dismiss this aspect of the trip. It is better to take extra clothing that you do not use than to leave it at home. Match your footwear with the environment, e.g. wear waterproof boots, not trainers, on a visit to a rocky shore. Brightly coloured clothing is a good idea, since this will be easier to spot if you become lost, or separated from the main group. Take account of the exposure of the site – in bright sunshine,

Typical coursework assessments for field excursions – these include:

- For specimens you have seen, recalling the names, diagnostic features, habitat, growth form, behaviour and physiology, etc.
- Making a diagrammatic representation of the site visited, e.g. a labelled map or transect.
- Interpreting key features of a photograph of the area visited.
- Interpreting data representing aspects of the location, based on the knowledge you have gained about the organisms and their environment.

Preventing rain damage to printed information in the field – use laminated sheets for keys, maps and any other items you wish to protect.

you may need a hat and sun protection cream, but in chilly and windy conditions a hat, gloves and a warm coat would be required. A large pair of domestic rubber gloves can be worn over woollen gloves for work in cold and wet conditions, e.g. sampling a lake in winter. Students without appropriate clothing may be unable to participate in all or some of the field activities.

- **During the excursion, keep up with the field leader** – you will need to be close enough to hear any commentary. Take notes as you go, to get maximum value from the visit: your course assessment may test recall and understanding and this may be your only opportunity to visit this location with a knowledgeable 'guide'. Remember to bring a suitable notebook and pencil for making notes and recording data – for further details, see Chapter 34. Rewrite your field notes soon after returning from the excursion, adding additional detail, as required.

- **Act responsibly at all times** – consider your own safety and that of others in the group, and minimise the impact of your visit on the environment, e.g. keep to paths, where appropriate. Follow at all times any specific safety instructions you are given during the visit.

> **SAFETY NOTE** Anticipate possible hazards – for example, there may be a significant risk of slipping on seaweed-covered rocks, on wet grass or on loose scree slopes. Also consider the risks involved in handling plants and animals (poisons, allergies, bites, stings, etc.) – if in doubt, ask your instructor before picking up any unknown specimen.

Planning and preparation for field-based project work

In an individual or group project involving a fieldwork component, you are likely to carry out a more involved and detailed study, in comparison to a field excursion: Box 31.1 gives details of the key questions to consider.

> **KEY POINT** Always construct a checklist (e.g. Table 31.1) of the required fieldwork equipment, paying attention to the smallest detail.

Table 31.1 A checklist of generic equipment for fieldwork

- Notebook and pencil
- Datasheets
- Indelible marker pens
- Camera
- Food and drink (plus emergency rations)
- First-aid equipment
- Safety (protective) gear
- Whistle
- Two-way radio or mobile phone
- Watch or clock
- Torch
- Multi-purpose knife (e.g. Swiss Army knife)
- Maps and compass or GPS (position locating device)
- Hand lens
- Specialised measuring equipment
- Sampling equipment (including spares)
- Specimen storage materials and labels
- Rucksack or other carrying devices
- Laminated guides and plans
- Specific items such as field guides, biological keys, etc.
- Survival bag (e.g. for longer trips, or in remote locations)
- Buoyancy aid and line (e.g. for trips near deep water)

Making a preliminary inspection of a field site

It is unlikely that you will be able to answer the questions in Box 31.1 if you or your instructor have not visited the field area in advance. The location, for example, may be remote, rugged or dangerous; it may be intertidal or subject to problems of access; the area may present social hazards (e.g. in some urban areas) or be culturally sensitive (e.g. nature reserves or archaeologically important sites). All fieldwork programmes, therefore, should be preceded by a 'scoping' or orientation study involving a preliminary inspection of the area. Make sure that the organisms that you wish to study are present and accessible. Any problems relating to access problems should be resolved at this stage and permissions obtained, where appropriate. Always make sure that landowners know who you are, where you are from, what you intend to do, and when you propose to be on their land. Be open and honest with them and involve them in your project.

> **KEY POINT** Never enter private land without permission: this is potentially trespassing. After your fieldwork, always thank the landowner for their help and support.

Box 33.1 Questions you should consider before carrying out a field project

- **What are the aims and objectives of the study?** Your fieldwork must be realistic in its scope and purpose; most people overestimate what can be done in a given time period, especially if the weather is poor. Recognise your own limitations and those of others who might be involved and do not overstretch yourself.

- **How long and how many periods of fieldwork are required?** If your work requires more than one visit, what time interval will be needed between visits? This may affect the logistics and cost of the work.

- **What are the safety implications of the work?** It is always essential to take full account of safety issues, and you should read and take note of the basic rules in Chapter 22. Make sure that a risk assessment has been carried out before any field-based project work is carried out: the major hazards should be considered, and any steps required to minimise risk must be identified in advance (see p. 136).

- **Am I likely to encounter any difficult conditions or environments?** Plans should always be based upon a preliminary site inspection and you should always discuss your intentions with more experienced fieldworkers.

- **What samples are needed and how will they be collected?** The number, frequency, nature and spatial or temporal distribution of samples must be consistent with the purpose of the fieldwork and may require considerable planning in advance. Too many samples can be as much of a problem as too few, so determine the minimum required sample size with statistical evaluation in mind. Chapter 32 considers aspects of sampling strategy. The sampling or environmental measurement and recording protocol that you choose will influence your equipment requirements.

- **How should samples be stored for transport to the laboratory?** If the work involves collection of samples or specimens, make sure you include appropriate storage vessels and a means of labelling them (e.g. a spirit-based marker). You may also need to identify temporary storage locations in the field – note that samples may be ruined by inappropriate field storage (e.g. inactivation of bacteria by brief exposure to bright sunlight) and this damage may not be obvious on return to the laboratory. Water samples may require refrigeration to minimise changes in biological composition. Remember too that large numbers of water, sediment or soil samples can be very heavy; make sure that you have considered the best way to transport them. Chapters 36, 37 and 38 give further advice on collecting and preserving biological specimens.

- **What equipment will be required?** You should compile a detailed list well in advance, and then check that everything is available and in working order.

- **What transport will be needed for personnel, equipment and samples?** You may need to arrange for a vehicle or a boat of appropriate size. On land, choose a vehicle appropriate for the terrain. In shallow water, use only an inflatable dinghy or shallow-draught boat. In deep water, make sure that the boat can withstand any possible sea conditions. Take account of the capacity of your means of transport – to overload a vehicle or boat is a risk to safety.

SAFETY NOTE Remember your responsibilities during fieldwork – although your activities will normally be organised in consultation with a member of staff, this does not free you from a personal responsibility for your own and the group's safety.

No sampling or survey work is carried out at the inspection stage. However, the field area should be walked over and examined. Appropriate notes should be taken, and, where applicable, the site should be photographed, so that the strategy for subsequent sampling can be determined.

Working safely and responsibly

Always wear appropriate clothing and safety equipment and let someone else know where you will be working and when you expect to return. Do not then alter this plan without informing that person. Remember, too, that what may be a comfortable environment for one person may not

SAFETY NOTE Safe and responsible fieldwork also means:

- Do not drink alcohol or take drugs (other than prescribed medication) in the field.
- Do not annoy or antagonise the local population.
- Leave gates as you find them.
- Do not drop litter.
- Avoid fire risks, especially when working in wooded areas, sand dunes or heathland.
- Do not frighten or disturb livestock or domestic animals.
- Do not attempt to carry too much. Keep rucksack loads below 14 kg. Make several trips rather than risk exhaustion or injury by trying to carry too great a load.
- You should also observe the general safety points discussed in Chapter 22.

be so for another, depending on their outdoor experience and interests. Do not attempt to work from a boat or climb a rock face without prior training. If your work requires specialist support, you must seek assistance from an appropriately qualified individual (e.g. a qualified rock climber or a scuba diver).

Sources for further study

Anon. (1995) *Field Work Code of Practice*. CVCP, London.

Field Studies Council. *Biology or Science Fieldwork.* Available: http//www.field-studies.council.org/ outdoor-classroom/biology-fieldwork.aspx Last accessed 26/03/15

Jones, A.M., Duck, R., Reed, R. and Weyers, J.D.B. (2000). *Practical Skills in Environmental Science.* Prentice Hall, Harlow.

Nichols, D. (1999) *Safety in Biological Fieldwork. Guidance Notes for Codes of Practice,* 4th edn. Institute of Biology, London.

University of York. *Risk Assessment of Fieldwork Activities.* Available: http//www.york.ac.uk/biology/intranet/ health-safety/fieldwork/risks/ Last accessed 26/03/15

Watts, S. and Halliwell, L. (eds) (1996) *Essential Environmental Science: Methods and Techniques.* Routledge, London.

STUDY EXERCISES

31.1 **Check your knowledge of generic fieldwork requirements.** What items are missing from the following list of generic fieldwork items? Notebook and pencil; datasheets; camera; food and drink (plus emergency rations); safety (protective) gear; whistle, watch or clock; torch; multi-purpose knife; map(s) and compass/GPS device; hand lens; specialised measuring equipment; specimen storage materials and labels; rucksack or other carrying devices; laminated guides and plans; field guides.

31.2 **List specific items that would be needed to carry out an ecological investigation.** This could be, for example, of the fauna and flora of a marine rocky shore or of plants in a meadow. Make a note of why each item is required and any problems that might be associated with its use.

31.3 **Consider the hazards of a particular field study.** List the potential hazards of carrying out a survey of woodlice in an urban estate environment.

Answers to these study exercises are available at www.pearsoned.co.uk/practicalskills

When carrying out research, it is unlikely that you can observe or measure every individual in the population of organisms in which you are interested. Hence, statistics obtained from a subset (or sample) are normally used to estimate relevant parameters for the biological population. Samples consist of data values for a particular variable (e.g. length), each recorded from an individual sampling unit (e.g. a limpet) in a sample of n units (e.g. $n = 50$ limpets) taken from the population under investigation (e.g. those limpets on a particular rocky shore). The term 'replicate' can be applied either to the measurement (i.e. repeated measurement or observation taken at the same time or location) or the actual sampling unit.

Symbols are used to represent each type of sample statistic: these are given Roman-character symbols, e.g. $\overline{Y}$ for the sample mean, while the equivalent population parameter is given a Greek symbol, e.g. μ for the population mean. When estimating population parameters from sample statistics, the sample size is important, larger sample sizes allowing greater statistical confidence. However, the optimum sample size is normally a balance between statistical and practical considerations.

This chapter deals mainly with sampling in fieldwork where natural populations are to be observed under undisturbed conditions; however, the same principles may also apply in a laboratory context (see Chapter 33).

Definitions

Biological population – all those individuals within a specified time or space about which inferences are to be made, specified according to some biological definition (perhaps related to life history, growth stage, or sex), and normally investigated at a particular location and time.

Parameter – a numerical constant or mathematical function used to describe a particular population (e.g. the mean height of 18-year-old females).

Statistic – an estimate of a parameter obtained from a sample (e.g. the height of 18-year-old females based on those in your class).

Specifying the population being sampled

At the outset, it is important to provide a complete description of the biological population being sampled. Failure to do this will make your results difficult to interpret or to compare with other observations, including your own. When populations are to be compared, ideally only the variable under consideration should differ if maximum significance is to be placed upon the analysis.

Choosing relevant population factors to specify – these should include:

- geographical location;
- type of habitat;
- date and time of sampling;
- age, sex, physiological condition and health of sampled organisms;
- other details relevant to your work, e.g. an index of pollution.

This information might apply to all members of the population or to a matrix of data associated with each replicate.

Deciding on a sampling strategy

Selecting a sample involves the formulation of rules and methods (the sampling protocol) by which some members of the population are included in the sample. The chosen sample is then measured using defined procedures to obtain relevant data. Finally, the information so obtained is processed to calculate appropriate statistics. Good sampling design takes into account all of these procedures, and it should relate to the objectives of your investigation by providing valid estimates of features of the population(s) of interest.

Truly representative samples should be:

- **taken at random,** or in a manner that ensures that every member of the population has an equal chance of being selected;
- **large enough to provide sufficient precision in estimation of population characteristics;**
- **unbiased by the sampling procedure or equipment.**

Selecting specimens – do not include data for which the appropriate population specification is unavailable.

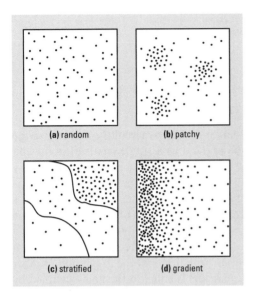

Fig. 32.1 Types of distribution.

(a) random (b) patchy
(c) stratified (d) gradient

Definitions

Gradient – a distribution that varies smoothly over the sampling area.

Homogeneous – evenly distributed.

Patchy – showing clustered (contagious) distribution (e.g. numbers of parasites within hosts).

Stratified – showing a distribution with discrete levels or strata (e.g. algae on a rocky shore)

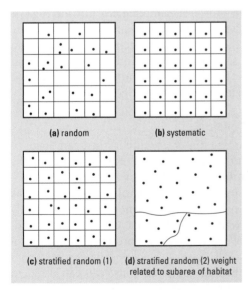

(a) random (b) systematic
(c) stratified random (1) (d) stratified random (2) weight related to subarea of habitat

Fig. 32.2 Basic methods of sampling.

These requirements may well conflict and a degree of compromise may be necessary. You should, however, take great care to minimise bias in sampling, or population parameters inferred from your observations will be unrealistic and this may invalidate your work and its conclusions.

> **KEY POINT** Sampling involves choices on the part of the investigator. A 'sampling strategy' should allow you to obtain reliable and useful information about your particular population(s), while using your resources efficiently.

The sampling protocol

You should decide on a sampling protocol before any investigation proceeds. The main aspects to be determined are: the position of samples; the size and shape of the sampling area; and the number of sampling units in each sample. Before this can be done, however, information is required about the likely distribution of organisms. This can be even (homogeneous), patchy (contagious), stratified (homogeneous within subareas) or present as a gradient (Fig. 32.1). You might decide which type applies from a pilot study, published research or by analogy with other systems.

Locating your samples

The positions where sampling takes place can be determined either randomly, systematically or in some stratified manner (Fig. 32.2).

In simple random sampling (Fig. 32.2(a)), a 2-D coordinate grid is superimposed on the area to be investigated. The required number of grid reference data pairs is then obtained using random numbers (p. 205) and samples taken at these points. Every organism in the population thus has an equal chance of selection, but the area may not be covered evenly. This method is best if the distribution of organisms is homogeneous.

Systematic sampling (Fig. 32.2(b)) involves selecting the location of the first sampling position at random and then taking samples at fixed distances from this. This method has the advantage of simplicity and it is often used where the intention is to map data. The disadvantages are, firstly, that the results can be biased if the interval between sampling positions coincides with some periodic distribution of the population, and, secondly, that there is no reliable method of estimating the standard error of the sample mean.

Stratified random sampling may be preferable if you wish to avoid these disadvantages yet still ensure that each part of the area is represented (e.g. where you suspect there to be stratified micro-habitats, as in a tidal seashore). The area is divided into subareas within each of which random sampling is carried out. These areas can either be constant in size (Fig. 32.2(c)) or related to known features in the sampling area (Fig. 32.2(d)). If the latter, strata are normally sampled in proportion to their area. 'Weighting' is the general term applied to sampling procedures that allow the calculated statistics to represent the population better by accounting for differences in the distribution of the chosen character. You can analyse data from different strata by a one-way analysis of variance (see Chapter 76).

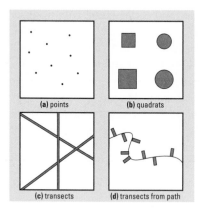

Fig. 32.3 Methods of positioning samples.

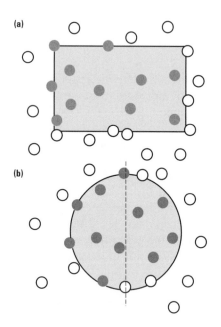

Fig. 32.4 Examples of sampling protocols for reducing edge effects. Filled circles represent objects to be counted or measured, open circles those to be ignored. (a) Rectangular area: all objects touching the top and left-hand sides including the top left-hand corner are included, as well as those clearly within the perimeter. (b) Circular area: all objects clearly within the perimeter are included as well as those touching the perimeter on the left-hand side. In the bottom half, those that would touch both the perimeter and the imaginary plane of symmetry are ignored.

The dimensions of the sampling area

The chief options are:

- **point sampling** (Fig. 32.3(a));
- **quadrat sampling** (Fig. 32.3(b));
- **transect (traverse) sampling** (Fig. 32.3(c)).

Quadrats are usually either circular or square. A circular quadrat has the advantage that its position can be marked as a single (central) point and the area defined by use of a tape measure, whereas a square quadrat may require marking at each corner. Transects are generally used when it is difficult to move through the site to position quadrats. If a defined path is present, transects taken at right angles to the path (Fig. 32.3(d)) will save time in reaching the sites.

The problems involved are best illustrated by reference to the selection of a quadrat for fieldwork sampling. Note first that the maximum number of independent sampling units (quadrats) is equal to the area occupied by the population under study (total theoretical sampling area) divided by the area of the sampling units (quadrat area).

When the distribution is truly random, then all quadrat sizes are equally effective for estimating population parameters (assuming the total number of individuals sampled is equal). If the distribution is patchy, a smaller quadrat size may be more effective than a larger one: too large an area might obscure the true nature of the clumped distribution. Alternatively, you may wish to exclude a patchy distribution from your investigation and should thus choose a relatively large sampling area. If the distribution is stratified or graded, then sampling area is generally less important than sampling position.

The distribution and size of the organisms must be considered: it is obvious that you would require different-sized quadrats for trees in a forest than for daisies on a lawn. Otherwise, small sample areas have the advantage that more small samples can usually be taken for the same amount of labour. This may result in increased precision and many small areas will cover a wider range of the habitat than few large ones, so the catch can be more representative. However, sampling error at the edge of quadrats is proportionately greater as sample area diminishes, increasing as the scale of the quadrat and the sampled item become closer. To avoid such effects, you need to establish a protocol for dealing with items that overlap the edge of the quadrats (Fig. 32.4). These protocols are also valid when sampling objects in e.g. microscope fields.

Number of sampling units per sample

When small numbers of sampling units are used, this can lead to imprecise estimates of population parameters because the values of the sample statistics will be susceptible to the effects of random variation – this is especially true if the underlying spatial distribution is patchy, as is often the case. You may then be unable to demonstrate statistically that there are differences between the populations. On the other hand, measuring very large numbers of replicates may represent an impractical workload.

To estimate appropriate numbers of sampling units, you can use data from a pilot study to work out the probability of detecting a specified difference in the measured variable at a specified confidence level (see Sokal

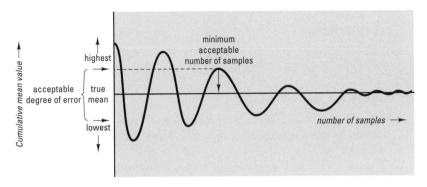

Fig. 32.5 Determination of the number of sampling units required.

and Rohlf, 2012). However, the formulae involved are complex. A simpler method, usually performed as a pilot study, is as follows:

1. **Take five sampling units at random and calculate the arithmetic mean of the measured variable for this sample.**

2. **Take five more units and calculate the mean for the ten units you have now collected.**

3. **Continue sampling in five-unit steps and plot the cumulative mean value against the number of samples.** When the mean fluctuates within acceptable limits, say ±5%, a suitable number of sampling units has been reached (Fig. 32.5).

The sequence of events in creating a sampling strategy is shown in Fig. 32.6.

Sampling in time

Sampling in time presents a different set of problems. If examining a phenomenon that fluctuates regularly (e.g. with a period governed by day and night, or high and low tide), then the frequency of sampling has to be determined with that periodicity in mind. In fact, you should consider the possible existence of periodic phenomena, even if they are not immediately obvious. If the phenomenon you are investigating is likely to change through time by some complex function such as the logarithmic decay in pollutant concentration, then your sampling intervals should be spaced according to an appropriate logarithmic series.

Subsampling

If you do not wish to sample the whole of a quadrat, perhaps because the density of sampling units is too high and it would take too long, then you can employ subsampling by studying a defined part of the quadrat – also known as two-stage sampling. If studying the density of organisms, this can simplify counting: assuming the organisms are randomly distributed, and only a small proportion of the population is sampled in each sub-unit, then the counts should follow a Poisson distribution (see Chapter 76). If this distribution is confirmed, only items present in single subsamples need to be counted to estimate the total numbers in each quadrat. The precision of the estimate then depends on the size of the count.

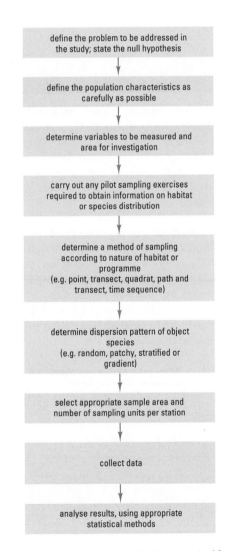

Fig. 32.6 Flowchart outlining decisions required for field sampling studies.

Text reference

Sokal, R.R. and Rohlf, F.J. (2012) *Biometry,* 4th edn. W.H. Freeman, San Francisco.

Sources for further study

Baker, J.M. and Wolff, W.J. (1987) *Biological Surveys of Estuaries and Coasts.* Cambridge University Press, Cambridge.
[Includes chapters on planning biological surveys and safety as well as numerous sections on sampling specific habitats.]

Keith, L.H. (1991) *Environmental Sampling and Analysis: A Practical Guide.* Lewis Publishers Inc., Michigan.
[Advanced text that is particularly good on how to devise protocols.]

New, T.R. (1998) *Invertebrate Surveys for Conservation.* Oxford University Press, Oxford.
[Excellent source for most aspects of sampling.]

Slingsby, D. and Cook, C. (1986) *Practical Ecology.* Macmillan Education, London.
[Contains a general introduction to sampling and other aspects of fieldwork.]

Zhang, C. (2007) *Fundamentals of Environmental Sampling and Analysis.* Wiley, Hoboken.

STUDY EXERCISES

32.1 Calculate the density of subsampled copepods. A sample of plankton was collected using a 5 L water bottle sampler. It was filtered upon collection through a 200 μm mesh net and the contents retained and preserved. Because the resulting volume of material was too large to count all the individuals present, the material was subsampled by taking ten replicate 1% aliquots. The numbers of copepods collected for each subsample determined using a counting chamber are given in the table below. From these data, calculate the mean density of copepods as (a) density per litre and (b) density per cubic metre of water.

Subsamples of copepods

1	2	3	4	5	6	7	8	9	10
234	255	342	276	288	295	324	301	299	273

32.2 Devise a protocol for counting limpets on a rocky platform using some form of quadrat. Take into account the size of the organism (up to 50 mm long), their normal density (up to 50 per square metre) and the significance of edge effects at the periphery of your sampling device. Assume that the limpets are randomly distributed.

32.3 Calculate the number of samples required for a given level of accuracy. The table below presents data on the effect of number of samples on the estimate of the mean for a population of limpets. Calculate the minimum number of samples that would be necessary to provide a measure of the mean (sample mean) that is within 10% of the true (population) mean (which is 50).

Estimated mean after taking different numbers of samples

Number of samples	1	3	5	10	15	20	25	30	35	40
Estimated mean (limpets m^{-2})	90	62	42.8	56.1	43.3	54.8	52.7	48.9	50.6	50

Answers to these study exercises are available at www.pearsoned.co.uk/practicalskills

33 Scientific method and design of experiments

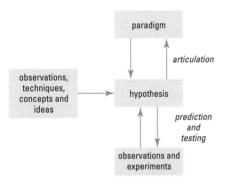

Fig. 33.1 A model of scientific method as used when testing hypotheses on a small scale. Hypotheses can arise as a result of various thought processes on the part of the scientist, and are consistent with the overlying paradigm. Each hypothesis is testable by experiment or observation, leading to its confirmation or rejection. Confirmed hypotheses act to strengthen the status of the paradigm, but rejected ones do not immediately result in the paradigm's replacement.

Biology is a body of knowledge based on observation and experiment. Biologists attempt to explain life in terms of theories and hypotheses. They make predictions from these hypotheses and test them by experiment or further observations. The philosophy and sociology that underlie this process are complex topics (see e.g. Chalmers, 2013). Any brief description must involve simplifications.

Figure 33.1 models the scientific process you are most likely to be involved in – testing 'small-scale' hypotheses. These represent the sorts of explanations that can give rise to predictions that can be tested by an experiment or a series of observations. For example, you might put forward the hypothesis that the rate of K^+ efflux from a particular cell type is dependent on the intracellular concentration of calcium ions. This might then lead to a prediction that the application of a substance known to decrease the intracellular concentration of calcium ions would reduce K^+ efflux from the cells. An experiment could be set up to test this hypothesis and the results would either confirm or falsify the hypothesis.

If confirmed, a hypothesis is retained with greater confidence. If falsified, it is either rejected outright as false, or modified and retested. Alternatively, it might be decided that the experiment was not a valid test of the hypothesis (perhaps because it was later found that the applied substance could not penetrate the cell membrane to a presumed site of action).

Nearly all scientific research deals with the testing of small-scale hypotheses. These hypotheses operate within a theoretical framework that has proven to be successful (i.e. is confirmed by many experiments and is consistently predictive). This operating model or 'paradigm' is not changed readily, and, even if a result appears that seems to challenge the conventional view, would not be overturned immediately. The conflicting result would be 'shelved' until an explanation was found after further investigation. In the example used above, a relevant paradigm could be the notion that life processes are ultimately chemical in nature.

Although changes in paradigms are rare, they are important, and the scientists who recognise them become famous. For example, a 'paradigm shift' can be said to have occurred when Darwin's ideas about Natural Selection replaced Special Creation as an explanation for the origin of species. Generally, however, results from hypothesis-testing tend to support and develop ('articulate') the paradigm, enhancing its relevance and strengthening its status. Thus, research in the area of population genetics has developed and refined Darwinism.

Where do ideas for small-scale hypotheses come from? They arise from one or more thought processes on the part of a scientist:

- **analogy with other systems;**
- **recognition of a pattern;**
- **recognition of departure from a pattern;**
- **invention of new analytical methods;**
- **development of a mathematical model;**
- **intuition;**
- **imagination.**

Recently, it has been recognised that the process of science is not an entirely objective one. For instance, the choice of analogy that led to a new hypothesis might well be subjective, depending on past knowledge or understanding. Also, science is a social activity, where researchers put forward and defend viewpoints against those who hold an opposing view; where groups may work together towards a common goal; and where effort may depend on externally dictated financial opportunities and constraints. As with any other human activity, science is bound to involve an element of subjectivity.

No hypothesis can ever be rejected with certainty. Statistics might allow us to quantify as vanishingly small the probability of an erroneous conclusion, but we are nevertheless left in the position of never being 100% certain that we have rejected all relevant alternative hypotheses, nor 100% certain that our decision to reject some alternative hypotheses was correct. However, despite these problems, experimental science has yielded and continues to yield many important findings.

> **KEY POINT** The fallibility of scientific 'facts' is essential to grasp. No explanation can ever be 100% certain as it is always possible for a new alternative hypothesis to be generated. Our understanding of biology changes all the time as new observations and methods force old hypotheses to be retested.

Quantitative hypotheses, those involving a mathematical description of a system, are important in biology. They can be formulated concisely by mathematical models. Formulating models is often useful because it forces deeper thought about mechanisms and encourages simplification of the system. A mathematical model:

- **is inherently testable through experiment;**
- **identifies areas where information is lacking or uncertain;**
- **encapsulates many observations;**
- **allows you to predict the behaviour of the system.**

Remember, however, that assumptions and simplifications required to create a model may result in it being unrealistic. Further, the results obtained from any model are only as good as the information put into it.

The terminology of experimentation

In many experiments, the aim is to provide evidence for causality. If x causes y, we expect, repeatedly, to find that a change in x results in a change in y. Hence, the ideal experiment of this kind involves measurement of y, the dependent (measured) variable, at one or more values of x, the independent variable, and subsequent demonstration of some relationship between them. Experiments therefore involve comparisons of the results of treatments – changes in the independent variable as applied to an experimental subject. The change is engineered by the experimenter under controlled conditions.

Subjects given the same treatment are known as replicates (they may be called plots). A block is a grouping of replicates or plots. The blocks are contained in a field, i.e. the whole area (or time) available for the experiment (Fig. 33.2). These terms originated from the statistical analysis of agricultural experiments, but they are now used for all areas of biology.

Deciding whether to accept or reject a hypothesis – this is sometimes clear-cut, as in some areas of genetics, where experiments can be set up to result in a binary outcome (Chapters 61 and 62). In many other cases, the existence of 'biological variation' means that statistical techniques need to be employed, with implicit acceptance of uncertainty in conclusions (Chapters 75 and 76; Box 61.2).

Definition

Mathematical model – an algebraic summary of the relationship between the variables in a system.

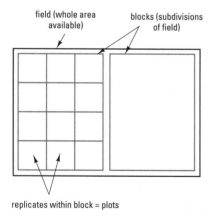

field (whole area available)   blocks (subdivisions of field)

replicates within block = plots

Fig. 33.2 Terminology and physical arrangements of elements in an experiment. Each block should contain the complete range of treatments (treatments may be replicated more than once in each block).

Why you need to control variables in experiments

Interpretation of experiments is seldom clear-cut because uncontrolled variables always change when treatments are given.

Confounding variables

These increase or decrease systematically as the independent variable increases or decreases. Their effects are known as systematic variation. This form of variation can be disentangled from that caused directly by treatments by incorporating appropriate controls in the experiment. A control is really just another treatment where a potentially confounding variable is adjusted so that its effects, if any, can be taken into account. The results from a control may therefore allow an alternative hypothesis to be rejected. There are often many potential controls for any experiment.

The consequence of systematic variation is that you can never be certain that the treatment, and the treatment alone, has caused an observed result. By careful design, you can, however, 'minimise the uncertainty' involved in your conclusion. Methods available include:

- **ensuring, through experimental design, that the independent variable is the only major factor that changes in any treatment;**
- **incorporating appropriate controls** to show that potential confounding variables have little or no effect;
- **selecting experimental subjects randomly** to cancel out systematic variation arising from biased selection;
- **matching or pairing individuals among treatments** so that differences in response due to their initial status are eliminated;
- **arranging subjects and treatments randomly** so that responses to systematic differences in conditions do not influence the results;
- **ensuring that experimental conditions are uniform** so that responses to systematic differences in conditions are minimised (when attempting this, beware 'edge effects' where subjects on the periphery of the layout receive substantially different conditions from those in the centre).

Nuisance variables

These are uncontrolled variables that cause differences in the value of y independently of the value of x, resulting in random variation. Experimental biology is characterised by the high number of nuisance variables that are present and their relatively great influence on results: biological data tend to have large errors. To reduce and assess the consequences of nuisance variables:

- **incorporate replicates to allow random variation to be quantified;**
- **choose subjects that are as similar as possible;**
- **control random fluctuations in environmental conditions.**

Constraints on experimental design

Box 33.1 outlines the important stages in designing an experiment. At an early stage, you should find out how resources may constrain the design. For example, limits may be set by availability of subjects, cost of treatment, availability of a chemical or bench space. Logistics may be a factor (e.g. time taken to record or analyse data).

Example Suppose you wish to investigate the effect of a metal ion on the growth of a culture. If you add the metal as a salt to the culture and then measure the growth, you will immediately introduce at least two confounding variables, compared with a control that has no salt added.

Firstly, you will introduce an anion that may also affect growth in its own right, or in combination with the metal ion; secondly, you will alter the osmotic potential of the medium (see pp. 160–1). Both of these effects could be tested using appropriate controls.

Reducing edge effects – one way to do this is to incorporate a 'buffer zone' of untreated subjects around the experiment proper.

Evaluating design constraints – a good way to do this is by processing an individual subject through the experimental procedures – this type of 'preliminary run' can help to identify potential difficulties.

Box 33.1 **Checklist for designing and performing an experiment**

1. **Preliminaries**
   (a) Read background material and decide on a subject area to investigate.
   (b) Formulate a simple hypothesis to test. It is preferable to have a clear answer to one question than to be uncertain about several questions.
   (c) Decide which dependent variable you are going to measure and how. Is it relevant to the problem? Can you measure it accurately, precisely and without bias?
   (d) Think about and plan the statistical analysis of your results. Will this affect your design?

2. **Designing**
   (a) Find out the limitations on your resources.
   (b) Choose treatments and conditions that alter the minimum of confounding variables.
   (c) Incorporate as many effective controls as possible.
   (d) Keep the number of replicates as high as is feasible.
   (e) Ensure that the same number of replicates is present in each treatment.
   (f) Use effective randomisation and blocking arrangements.

3. **Planning**
   (a) List all the materials you will need. Order any chemicals and make up solutions; grow, collect or breed the experimental subjects you require; check equipment is available.
   (b) Organise space and/or time in which to do the experiment.
   (c) Account for the time taken to apply treatments and record results. Make out a timesheet if things will be hectic.

4. **Carrying out the experiment**
   (a) Record the results and make careful notes of everything you do (see Chapter 34). Make additional observations to those planned if interesting things happen.
   (b) Repeat experiment if time and resources allow.

5. **Analysing**
   (a) Graph data as soon as possible (during the experiment if you can). This will allow you to visualise what has happened and make adjustments to the design (e.g. timing of measurements).
   (b) Carry out the planned statistical analysis.
   (c) Jot down conclusions and new hypotheses arising from the experiment.

Your equipment or facilities may affect design because you cannot regulate conditions as well as you might desire. For example, you may be unable to ensure that temperature and lighting are equal over an experiment laid out in a glasshouse or you may have to accept a great deal of initial variability in your subjects. These problems are especially acute for experiments carried out in the field or subjects collected from the wild.

Use of replicates

> Deciding the number of replicates in each treatment – try to:
> - maximise the number of replicates in each treatment;
> - make the number of replicates the same for each treatment.

Replicate results show how variable the response is within treatments. They allow you to compare the differences among treatments in the context of the variability within treatments – you can do this via statistical tests such as analysis of variance (Chapter 76). Larger sample sizes tend to increase the precision of estimates of parameters and increase the chances of showing a significant difference between treatments if one exists. For statistical reasons (weighting, ease of calculation, fitting data to certain tests), it is often best to keep the number of replicates even. Remember that the degree of independence of replicates is important: subsamples cannot act as replicate samples – they tell you about variability in the measurement method but not in the quantity being measured.

Pseudo-replication – this term applies to a situation where replicates are used, but they are not fully independent of each other. An example might occur when testing the effect of a chemical on liver function, if several samples from one animal's organ were treated as independent replicates, rather than a sample from the livers of each of several animals. In practice, an approach which examined both intra- and inter-animal variability might be the best way to proceed, using several samples from each of several animals, then analysing the results using ANOVA (Chapter 76).

If the total number of replicates available for an experiment is limited by resources, you may need to compromise between the number of treatments and the number of replicates per treatment. Statistics can help here, for it is possible to work out the minimum number of replicates you would need to show a certain difference between pairs of means (say 10%) at a specified level of significance (say $P = 0.05$). For this, you need to obtain a prior estimate of variability within treatments perhaps from an initial trial run (see Sokal and Rohlf, 2012).

Randomisation of treatments

The two aspects of randomisation you must consider are:

- **positioning of treatments within experimental blocks;**
- **allocation of treatments to the experimental subjects.**

For relatively simple experiments, you can adopt a completely randomised design; here, the position and treatment assigned to any subject is defined randomly. You can draw lots, use a random number generator on a calculator, or use the random number tables that can be found in most books of statistical tables (see Box 33.2).

Box 33.2 **How to use random number tables to assign subjects to positions and treatments**

This is one method of many that could be used. It requires two sets of n random numbers – where n is the total number of subjects used.

1. Number the subjects in any arbitrary order but in such a way that you know which is which (i.e. mark or tag them).

2. Decide how treatments will be assigned, e.g. first five subjects selected, treatment A; second five – treatment B, etc.

3. Use the first set of random numbers in the sequence obtained to identify subjects and allocate them to treatment groups in order of selection as decided in (2).

4. Map the positions for subjects in the block or field. Assign numbers to these positions using the second set of random numbers, working through the positions in some arbitrary order, e.g. top left to bottom right.

5. Match the original numbers given to subjects with the position numbers.

To obtain a sequence of random numbers:

1. Decide on the range of random numbers you need.

2. Decide how you wish to sample the random number tables (e.g. row by row and top to bottom) and your starting point.

3. Moving in the selected manner, read the sequence of numbers until you come to a group that fits your needs (e.g. in the sequence 978186, 18 represents a number between 1 and 20). Write this down and continue sampling until you get a new number. If a number is repeated, ignore it. Small numbers need to have the appropriate number of zeros preceding (e.g. 5 = 05 for a range in the tens, 21 = 021 for a range in the hundreds).

4. When you come to the last number required, you do not need to sample any more: simply write it down.

Example: You find the following random number sequence in a table and wish to select numbers between 1 and 10 from it.

9059146823	4862925166	1063260345
1277423810	9948040676	6430247598
8357945137	2490145183	5946242208
6588812379	2325701558	3260726568

Working left to right and top to bottom, the order of numbers found is 5, 10, 3, 9, 4, 6, 2, 1, 8, 7 as indicated by blue type. If the table is sampled by working row by row right to left from bottom to top, the order is 6, 10, 7, 2, 9, 3, 4, 8, 1, 5.

Example If you knew that soil type varied in a graded fashion across a field, you might arrange blocks to be long thin rectangles at right angles to the gradient to ensure conditions within the block were as even as possible.

A completely randomised layout has the advantage of simplicity but cannot show how confounding variables alter in space or time. This information can be obtained if you use a blocked design in which the degree of randomisation is restricted. Here, the experimental space or time is divided into blocks, each of which accommodates the complete set of treatments (Fig. 33.2). When analysed appropriately, the results for the blocks can be compared to test for differences in the confounding variables across the field and these effects can be separated out from the effects of the treatments. The size and shape (or timing) of the block you choose is important: besides being able to accommodate the number of replicates desired, the suspected confounding variable should be relatively uniform within the block.

A Latin square is a method of placing treatments so that they appear in a balanced fashion within a square block or field. Treatments appear once in each column and row (see Fig. 33.3), so the effects of confounding variables can be 'cancelled out' in two directions at right angles to each other. This is effective if there is a smooth gradient in some confounding variable over the field. It is less useful if the variable has a patchy distribution, where a randomised block design might be better.

Latin square designs are useful in serial experiments where different treatments are given to the same subjects in a sequence (e.g. Fig. 33.4). A disadvantage of Latin squares is the fact that the number of plots is equal to the number of replicates, so increases in the number of replicates can only be made by the use of further Latin squares.

3 × 3   4 × 4

Fig. 33.3 Examples of Latin square arrangements for 3 and 4 treatments. Letters indicate treatments; the number of possible arrangements for each size of square increases greatly as the size increases.

Pairing and matching subjects

The paired comparison is used to reduce systematic variation when there are two treatments. Examples of its use are:

- **'Before and after' comparison.** Here, the pairing removes variability arising from the initial state of the subjects, e.g. weight gain of mice on a diet, where the weight gain may depend on the initial weight.

- **Application of a treatment and control to parts of the same subject or to closely related subjects.** This allows comparison without complications arising from different origin of subjects, e.g. drug or placebo given to sibling rats, virus-containing or control solution swabbed on left or right halves of a leaf.

- **Application of treatment and control under shared conditions.** This allows comparison without complications arising from different environments of subjects, e.g. rats in a cage, plants in a pot.

Matched samples represent a restriction on randomisation where you make a balanced selection of subjects for treatments on the basis of some attribute or attributes that may influence results, e.g. age, sex, prior history. The effect of matching should be to 'cancel out' the unwanted source(s) of variation. Disadvantages include the subjective element in choice of character(s) to be balanced, inexact matching of quantitative characteristics, the time matching takes and possible wastage of unmatched subjects.

When analysed statistically, both paired comparisons and matched samples can show up differences between treatments that might otherwise be rejected on the basis of a fully randomised design, but note that the statistical analysis may be different.

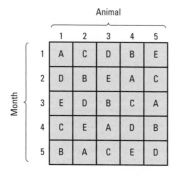

Animal

Month

Fig. 33.4 Example of how to use a Latin square design to arrange sequential treatments. The experimenter wishes to test the effect of drugs A–E on weight gain, but has only five animals available. Each animal is fed on the control diet for the first 3 weeks of each month, then on the control diet plus drug for the last week. Weights are taken at the start and finish of each treatment. Each animal receives all treatments.

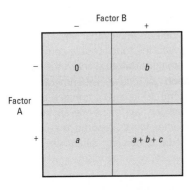

Fig. 33.5 Design of a simple multifactorial experiment. Factors A and B have effects a and b when applied alone. When both are applied together, the effect is denoted by $a + b + c$.
- If $c = 0$, there is no interaction (e.g. $2 + 2 + c = 4$).
- If c is positive, there is a positive interaction (synergism) between A and B (e.g. $2 + 2 + c = 5$).
- If c is negative, there is a negative interaction (antagonism) between A and B (e.g. $2 + 2 + c = 3$).

Reporting correctly – it is good practice to report how many times your experiments were repeated (in Materials and Methods); in the Results section, you should add a statement saying that the illustrated experiment is representative.

Multi-factorial experiments

The simplest experiments are those in which one treatment (factor) is applied at a time to the subjects. This approach is likely to give clear-cut answers, but it could be criticised for lacking realism. In particular, it cannot take account of interactions among two or more conditions that are likely to occur in real life. A multi-factorial experiment (Fig. 33.5) is an attempt to do this; the interactions among treatments can be analysed by specialised statistics.

Multi-factorial experiments are economical on resources because of 'hidden replication'. This arises when two or more treatments are given to a subject because the result acts statistically as a replicate for each treatment. Choice of relevant treatments to combine is important in multi-factorial experiments; for instance, an interaction may be present at certain concentrations of a chemical but not at others (perhaps because the response is saturated). It is also important that the measurement scale for the response is consistent, otherwise spurious interactions may occur. Beware when planning a multi-factorial experiment that the numbers of replicates do not get out of hand: you may have to restrict the treatments to 'plus' or 'minus' the factor of interest (as in Fig. 33.5).

Repetition of experiments

Even if your experiment is well designed and analysed, you should recognise that only limited conclusions can be made. Firstly, what you can say is valid for a particular place and time, with a particular investigator, experimental subject and method of applying treatments. Secondly, if your results were significant at the 5% level of probability (p. 516), there is still an approximately 1-in-20 chance that the results did arise by chance. To guard against these possibilities, it is important that experiments are repeated. Ideally, this would be done by an independent scientist with independent materials. However, it makes sense to repeat work yourself so that you can have full confidence in your conclusions. Many scientists recommend that experiments are done three times in total, but this may not be possible in undergraduate work.

Text references

Chalmers, A.F. (2013) *What is this Thing called Science?*, 4th edn. Open University Press, Buckingham.

Sokal, R.R. and Rohlf, F.J. (2012) *Biometry*, 4th edn. W.H. Freeman, San Francisco.

Sources for further study

Heath, D. (1995) *An Introduction to Experimental Design and Statistics for Biology*. UCL Press, London.

Quinn, G.P. and Keough, M.J. (2002) *Experimental Design and Data Analysis for Biologists*. Cambridge University Press, Cambridge.

STUDY EXERCISES

33.1 Create a Latin square design. Treatments W, X, Y and Z are to be applied to potted plants in a glasshouse where the researcher suspects there may be a slight gradation in temperature and light over an oblong bench that can hold a total of 48 pots. You decide to use an experimental design consisting of three 4 × 4 blocks of plants, each arranged in a different Latin square design (see below). Assign treatments to the locations in the diagram below. Explain why this design will help to eliminate the effects of the confounding variables.

suspected temperature gradient

suspected light gradient

33.2 Generate random numbers. Produce a list of 20 random whole numbers between 1 and 5. This can be carried out using a spreadsheet. If using Microsoft *Excel*, investigate the RAND() and INT functions. The RAND() function produces a random number between 0 and 1, so you will need to multiply by a constant factor to scale your final output appropriately. Copy your test formula(e) to several cells to test empirically whether it works.

33.3 Investigate possible interactions. Treatments A, B and C involve tests of three different nutrients on proliferation of cells in a culture of carrot cells. Analyse the results in the table below to determine whether any interactions have occurred between the four possible combinations of treatments. No complex statistical analysis is required, simply a calculation of mean experimental effect (treatment minus controls) in each combined treatment and a comparison with the data for the relevant treatments on their own (in other words, assume that observed differences reflect true underlying differences). Classify the results as 'no interaction', 'antagonism' or 'synergism'

Results of cell proliferation experiment

| Treatment | Replicate (growth of callus in g) | | | | |
	1	2	3	4	5
Control	4.8	5.2	5.2	4.8	5.0
A	6.5	7.3	7.0	7.1	7.1
B	7.7	8.3	8.5	7.5	8.0
C	10.7	10.0	9.9	9.8	9.6
A + B	6.6	7.2	6.8	7.2	7.2
A + C	11.9	11.7	12.0	12.0	12.4
B + C	17.3	16.8	17.1	16.6	17.2
A + B + C	12.8	12.9	13.1	13.3	12.9

Answers to these study exercises are available at www.pearsoned.co.uk/practicalskills

34 Making notes of practical work

When carrying out advanced lab work or research projects, you will need to master the important skill of managing data and observations and learn how to keep a reliable record of your studies. This is important for the following reasons:

- **An accurate record helps when using information later,** perhaps for exam purposes or when writing a report.
- **It allows you to practise important skills** such as scientific writing, drawing diagrams, preparing graphs and tables and interpreting results.
- **Analysing and writing up your data as you go along prevents** a backlog at the end of your study time.
- **You can show your work to a future employer** to prove you have developed the skills necessary for writing up properly; in industry, this is vital so that others in your team can interpret and develop your work.

Understanding what's expected – especially when taking notes for a lab-based practical, pay special attention to the aims and learning objectives (p. 30) of the session, as these will indicate the sorts of notes you should be taking, including content and diagrams, and the ways in which you should present these for assessment.

KEY POINT   A good set of lab notes should:
- outline the purpose of your experiment or observation;
- set down all the information required to describe your materials and methods;
- record all relevant information about your results or observations and provide a visual representation of the data;
- note your immediate conclusions and suggestions for further experiments.

Means of recording notes – the traditional paper-based bound notebook is still favoured by many, and has the advantage of reliability and portability. Increasingly, laptop and notebook computers are used, and an advantage here is that recorded data can be transferred easily between applications and associated files for analysis. In both cases, backup of data is essential.

Collecting and recording primary data

Individual observations (e.g. laboratory temperature) can be recorded in text form, but tables are the most convenient way to collect large amounts of information. When preparing a table for data collection, you should:

1. **Use a concise title or a numbered code for cross-referencing.**
2. **Decide on the number of variables to be measured and their relationship with each other and lay out the table appropriately:**
   - (a) The first column of your table should show values of the independent (controlled) variable, with subsequent columns for the individual (measured) values for each replicate or sample (Fig. 34.1(a)).
   - (b) If several variables are measured for the same organism or sample, each should be given a row (Fig. 34.1(b)).
   - (c) In time-course studies, put the replicates as columns grouped according to treatment, with the rows relating to different times (Fig. 34.1(c)).
3. **Make sure the arrangement reflects the order in which the values will be collected.** Your table should be designed to make the recording process as straightforward as possible, to minimise the possibility of mistakes. For final presentation, a different arrangement may be best (Chapter 73).
4. **Consider whether additional columns are required for subsequent calculations.** Create a separate column for each mathematical manipulation, so the step-by-step calculations are clearly visible.

Recording primary data – never be tempted to jot down data on scraps of paper: you are likely to lose them, or to forget what individual values mean.

Designing a table for data collection – use a spreadsheet or the table-creating facility in a word processor to create your table. This will allow you to reorganise it easily if required. Make sure there is sufficient space in each column for the values – if in doubt, err on the generous side.

Labelling your notes – always add a date and time to each of your primary record files. You may also wish to add details of the type of observation or experiment.

(a)

Level of controlled variable	Measured variable		
	Replicate 1	Replicate 2	Replicate 3
1			
2			
3			
4			

(b)

Treatment	Measured variable	
	Organism 1	Organism 2
A		
B		
C		
D		

(c)

Time	Measured variable	
	Control	Treated
1		
2		
3		
4		
5		
6		

Fig. 34.1 Example layouts of tables for recording data: (a) measurements on replicates at different levels of a controlled variable ('rep' being shorthand for replicate); (b) responses by two organisms to a range of treatments; (c) time course of responses in control and treated specimens. Note the grid-like nature of these tables used for ease of data entry compared with the 'horizontal lines only' style used for formal presentation of tables (Chapter 73).

Tips for recording notes on paper – bear in mind the following advice:

- Use a refill pad or spiral-bound notebook, so you can deal easily with mistakes by tearing out a page.
- If you need to use loose paper, make sure each sheet is dated and easily identifiable, and file these consecutively in a ring binder (the same applies to any traces, printouts and graphs produced by instruments).
- Write with a pencil, so that mistakes can be corrected easily.
- Write clearly, taking special care that individual numerals cannot be confused.
- Use a tally chart (p. 471) if recording events – this will also provide an 'instant' histogram of your results.

Electronic laboratory notebooks (ELNs) – a number of companies now offer software products for taking lab notes (e.g. LABtrak); however, these are mainly used by industrial companies to assist them to comply with regulations assuring quality. For student purposes a combination of 'office' products such as a word processsor, spreadsheet and possibly database should be satisfactory.

Use spreadsheet software (Chapter 13) if you are manipulating lots of data.

5. **Take sufficient time to record quantitative data accurately.**

6. **Record numerical data to an appropriate number of significant figures,** reflecting the accuracy and precision of your measurement (p. 497). Do not round off data values, as this might affect the subsequent analysis of your data.

7. **Prepare blank recording tables if your experiments or observations will be repeated.**

8. **Explain any unusual data values or observations in a footnote.** Do not rely on your memory.

Recording details of project work

The recommended system is one where you make a dual record.

Primary record

The primary record is made at the bench or in the field. In this, you must concentrate on the detail of materials, methods and results. Include information that would not be used elsewhere, but that might prove useful in error tracing: for example, if you note how a solution was made up (exact volumes and weights used rather than concentration alone), this could reveal whether a miscalculation had been the cause of a rogue result. Note the origin, type and state of the chemicals and organism(s) used. Make rough diagrams to show the arrangement of replicates, equipment, etc.

The basic order of the primary record should mirror that of a research report (see p. 115), including: the title and date; brief introduction; comprehensive materials and methods; the data and short conclusions.

Secondary record

You should make a secondary record which is better organised and presented. This will be used when discussing results with your supervisor, and when writing up a report or thesis, and may be part of your course assessment. Although these notes should retain the essential features of

Formal aspects of keeping a record – the diary aspect of the record can be used to establish precedence (e.g. for patentable research where it can be important to 'minute' where and when an idea arose and whose it was); for error tracing (e.g. you might be able to find patterns in the work affecting the results); or even for justifying your activities to a supervisor.

Analysing data as soon as possible – always analyse and think about data immediately after collecting them as this may influence your subsequent activities.

- A graphical indication of what has happened can be particularly valuable.
- Carry out statistical analyses before moving on to the next experiment because apparent differences among treatments may not turn out to be statistically significant when tested.
- Write down any conclusions you make while analysing your data: sometimes those that seem obvious at the time of doing the work are forgotten when the time comes to write up a report or thesis.
- Note ideas for further studies as they occur to you – these may prove valuable later. Even if your experiment appears to be a failure, suggestions as to the likely causes might prove useful.

the primary record, they should be more concise and the emphasis should move towards analysis of the experiment. Outline the aims more carefully at the start and link the experiment to others in a series (e.g. 'Following the results of Expt D24, I decided to test whether…'). You should present data in an easily digested form, e.g. as tables of means or as summary graphs. Use appropriate statistical tests (Chapter 76) to support your analysis of the results.

The dual method of recording outlined above deals with the inevitable lack of organisation notes taken at the bench or in the field; these often have to be made rapidly, in awkward positions and in a generally complex environment. Writing a second, neater version forces you to consider again details that might have been overlooked in the primary record and provides a duplicate in case of loss or damage.

If you find it difficult to decide on the amount of detail required in Materials and Methods, the basic ground rule is to record enough information to allow a reasonably competent scientist to repeat your work exactly. You must tread a line between the extremes of including pedantic, irrelevant detail and the omission of information essential for a proper interpretation of the data – better perhaps to err on the side of extra detail to begin with. An experienced worker can help you decide which subtle shifts in technique are important (e.g. batch numbers for an important chemical, or when a new stock solution is made up and used). Many scientific advances have been made because of careful observation and record taking or because coincident data were recorded that did not seem of immediate value.

When creating a primary record, take care not to lose any of the information content of the data: for instance, if you only write down means and not individual values, this may affect your ability to carry out subsequent statistical analyses.

There are numerous ways to reduce the labour of keeping a record. Do not repeat Materials and Methods for a series of similar experiments; use devices such as 'method as for Expt B4'. A photocopy might suffice if the method is derived from a text or article (check with supervisor). To save time, make up and copy a checklist in which details such as chemical batch numbers can be entered.

Special requirements for fieldwork

The main problems you will encounter in the field are the effects of the weather while taking a primary record and the distance you might be from a suitable place to make a neat secondary record. Wind, rain and cold temperatures are not conducive to neat note-taking and you should be prepared for the worst possible conditions at all times. Make sure your clothing allows you to feel comfortable while recording data.

The simplest method of protecting a field notebook is to enclose it in a clear polythene bag large enough for you to take notes inside (Fig. 34.2). Alternatively, you could use a clipboard with a waterproof cover to shield your notes or a special notebook with a waterproof cover. When selecting a field notebook, choose a small size – the dimensions of outside pockets may dictate the upper size limit.

Fig. 34.2 Simple but effective method for keeping notes dry in the field;

1. string attaching pencil to clip
2. pencil (not pen) writes on damp paper
3. transparent plastic bag
4. bulldog clip
5. hardboard (at least 31 × 22 cm for A4 paper)
6. record sheet.

Selecting a waterproof notebook – notebooks with waterproof paper are available (for example the 'Aquascribe' brand supplied by Hawkins and Mainwaring Ltd, Newark, Notts). These are relatively expensive, but you can reuse them by rubbing out.

SAFETY NOTE Maintaining and consulting communal lab records – these activities may form a part of the safety requirements for working in a laboratory.

Laptop- or notebook-based note-taking may be convenient in some field situations but may not be appropriate if poor weather is a possibility, or if there is no convenient method for charging the computer's battery. For these reasons, paper-based note-taking is usually preferred outdoors, in which case you should:

- **Use a pencil** as ink pens such as ballpoints smudge in wet conditions, are temperamental in the cold and may not work at awkward angles. Do not forget to take a sharpener.

- **Prepare well** to enhance the speed and quality of your field note-taking – the date and site details can be written down before setting out and tables can be made out ready for data entry.

- **Transcribe field notes to a duplicate primary record** at your base each time you return there. There is a very real risk of your losing or damaging a field notebook. Also, poor weather may prevent full note-taking and the necessary extra details should be written up while fresh in your memory.

- **Consider using a voice recorder** rather than a notebook, in which case voice transcription into written form should also take place while your memory is fresh in case the sound quality is poor.

- **Use photographs to set data in context,** when appropriate. Develop photographs as soon as possible to check their suitability. Consider using a digital camera when the suitability of a record must be guaranteed at the time of the visit (see Chapter 50).

Field data may be logged automatically, stored temporarily in the instrument's electronic memory, and downloaded to a portable computer ('data logger') when convenient. The information is then transferred to a data bank back at base. If you are using this system, make back-up copies of each period's data as soon as possible – if the recording instrument's memory is cleared or overwritten after reading, there may be no recourse if the logging machine fails.

Using communal records

If working with a research team, you may need to use their communal databases. These avoid duplication of effort and ensure uniformity in techniques. You will be expected to use the databases carefully and to contribute to them properly. They might include:

- **a shared file of common techniques** (e.g. how to make up media or solutions);

- **a set of simplified step-by-step instructions for use of equipment** – manuals are often complex and poorly written and it may help to redraft them, incorporating any differences in procedure adopted by the group;

- **an alphabetical list of suppliers of equipment and consumables;**

- **a list of chemicals required by the group** and where they are stored;

- **the risk-assessment sheets for dangerous procedures** (p. 136);

- **the record book detailing the use of radioisotopes and their disposal.**

Sources for further study

Anon. *Suggestions for Keeping Laboratory Notebooks.* Available: http://otl.stanford.edu/inventors/ resources/labnotebooks.html
Last accessed 18/02/15.
[A Stanford University website that looks at the laboratory notebook from the patenting perspective.]

Anon. *Aquascribe.* Available: http://www.aquascribe .co.uk/
Last accessed 18/02/15.
[Sales website for Hawkins and Manwaring Ltd, manufacturers of waterproof papers and notebooks suitable for field use.]

Kanare, H.M. (1985) *Writing the Laboratory Notebook.* American Chemical Society, Washington, DC.

Pechenik, J.A.A. (2015) *Short Guide to Writing About Biology,* 9th edn. Longman, Harlow.

Slingsby, D. and Cook, C. (1986) *Practical Ecology.* Macmillan Education, London.

STUDY EXERCISES

34.1 Design a primary data collection table for a behavioural study. Imagine you wish to describe the feeding behaviour of birds at their nest. You wish to study the nature of the food, the relative participation of both parents, and the frequency of visits during the daytime with a view to analysis using a spreadsheet.

34.2 Outline the advantages and problems that would be associated with using a tape recorder to record data. Consider the behavioural study described in Study exercise 34.1 as an example.

34.3 Design a secondary record table for the collection and analysis of a set of count data for the number of bacterial colonies developed in each of ten replicates for each of five different treatments. Assume that you need to calculate means, variances, etc., and then to compare the results for each treatment.

Answers to these study exercises are available at www.pearsoned.co.uk/practicalskills

35 Project work

Obtaining ethical approval – if any aspect of your project involves work with human or animal subjects, then you must obtain the necessary ethical clearance before you begin; consult your department's ethical committee for details.

The Internet as an information source – since many university departments have home pages on the Web, searches using relevant key words may indicate where research in your area is currently being carried out. Academics usually respond positively to emailed questions about their area of expertise.

Asking around – one of the best sources of information about supervisors, laboratories and projects is past students. Some of the postgraduates in your department may be products of your own system and they could provide an alternative source of advice.

Research projects are an important component of the final-year syllabus for most degree programmes in the life sciences, while shorter projects may also be carried out during courses in earlier years. Project work can be extremely rewarding, although it does present a number of challenges. The assessment of your project is likely to contribute significantly to your degree grade, so all aspects of this work should be approached in a thorough manner.

Deciding on a topic to study

Assuming you have a choice, this important decision should be researched carefully. Make appointments to visit possible supervisors and ask them for advice on topics that you find interesting. Use library texts and research papers to obtain further background information. Perhaps the most important criterion is whether the topic will sustain your interest over the whole period of the project. Other things to look for include:

- **Opportunities to learn new skills.** Ideally, you should attempt to gain experience and skills that you might be able to 'sell' to a potential employer.
- **Ease of obtaining valid results.** An ideal project provides a means to obtain 'guaranteed' data for your report, but also the chance to extend knowledge by doing genuinely novel research.
- **Assistance.** What help will be available to you during the project? A busy lab with many research students might provide a supportive environment should your potential supervisor be too busy to meet you often; on the other hand, a smaller lab may provide the opportunity for more personal interaction with your supervisor.
- **Impact.** Your project may result in publishable data: discuss this with your prospective supervisor.

Writing a proposal

You may be expected to submit a proposal for your project, especially where it is expected that you should define the precise area of study yourself. The structure for the proposal will probably be provided in the course handbook, or as a form you must complete. Box 35.1 outlines some common features of project proposals and what you should do to complete each part.

The aim of formulating a proposal is to ensure that:

- **the project has an appropriate theoretical background;**
- **the objectives you have set are achievable;**
- **the methods chosen are appropriate;**
- **safety and ethical issues have been considered;**
- **sufficient resources are available to complete the work,** such as matching students to available labs and supervisors;
- **you have set yourself a timetable with milestones** on which your progress can be judged;
- **you obtain feedback and suggestions about your plans.**

Box 35.1 How to write a project proposal

Below are listed some common elements of project proposals, with guidance regarding approach and content. Not every part may be included, and the titles for sections may differ – always follow the exact format specified for your course.

- **Contact details and suggested supervisor(s)** – you may be asked whether you have discussed the project with the named supervisor(s).

- **Proposed title** – this should be relatively short, following the style used in research papers. It may change for the final report.

- **Aims** – a general statement of what you plan to achieve.

- **Objectives** – a listing of specific outcomes you expect to fulfil. Typically you will have several specific objectives that all fit within the overall aim.

- **Brief description of the subject** – this section might have an alternative title such as Summary, Background, Review of Subject Area or Statement of the Problem to be Addressed. It should contain a brief synopsis of past work, a summary of current ideas and, if relevant, the hypothesis to be tested. In some cases it will consist of a mini-review of the subject area and act as a template for the introduction to your report.

- **Preliminary bibliography** – this will ensure (a) that you have read and understood relevant papers and (b) that you understand how to cite them properly (Chapter 9).

- **Research methods** – describes how you plan to carry out the investigation. Be quite specific, so the committee can arrive at a valid judgement. Quote references for methods and techniques, where available. Consider use of appropriate controls, experimental design and sampling procedures (Chapters 32 and 33). Check that the results obtained will tell you what you need to know to achieve your objectives and to test your hypothesis.

- **Resources required** – the samples, chemicals, instruments, etc. required to carry out your investigation. Listing these will require quite detailed consideration of the experimental design or field area, as relevant, and a thorough knowledge of methods, such as amounts or volumes of reagents used (Chapters 23–26).

- **Timetable/plan** – a realistic breakdown of the work required, with milestones leading to completion of the project (see Chapter 2 for advice on time management). Always allow time for the unexpected.

- **Statement or declaration in relation to safety and ethical rules** – this is to confirm that you have read and understood relevant issues, have completed relevant forms and processes and are in a position to proceed with the research. You may be required to attach copies of COSHH forms (Chapter 22) or ethics committee permission (Chapter 21).

Liaising with your supervisor(s) – this is essential if your work is to proceed efficiently. Specific meetings may be timetabled, e.g. to discuss a term's progress, review your work plan or consider a draft introduction. Most supervisors also have an 'open-door' policy, allowing you to air current problems or discuss your results. Prepare well for all meetings: have a list of questions ready before the meeting; provide results in an easily digestible form (but take your lab notebook along); be clear about your future plans for work.

Your proposal outline might be considered by your supervisor or by a formal committee before approval.

Planning your work

As with any lengthy exercise, planning is required to make the best use of the time allocated (Chapter 2). This is true on a daily basis as well as over the entire period of the project. It is especially important not to underestimate the time it will take to write and produce your thesis (see below). If you wish to benefit from feedback given by your supervisor, you should aim to have drafts in his/her hands in good time. Since a large proportion of marks will be allocated to the report, you should not rush its production.

If your department requires you to write an interim report, look on this positively as an opportunity to clarify your thoughts and get some of the time-consuming preparative work out of the way. If not, you should set your own deadlines for producing drafts of the introduction, materials and methods section, etc.

Fig. 35.1 Flowchart showing a recommended sequence of events in carrying out an undergraduate research project. Note the potentially iterative nature of this process, reflecting the scientific method discussed in Chapter 33.

> **SAFETY NOTE** It is essential that you follow all the safety rules applying to the laboratory or field site. Make sure you know all relevant procedures – normally there will be prominent warnings about these. If in doubt, ask.

> **KEY POINT** Project work can be very time-consuming at times. Try not to neglect other aspects of your course – make sure your lecture notes are up to date and collect relevant supporting information as you go along.

Getting started

Figure 35.1 is a flowchart illustrating how a project might proceed; at the start, do not spend too long reading the literature and working out a lengthy programme of research. Get stuck in and do an experiment. There is no substitute for 'getting your hands dirty' for stimulating new ideas:

- **even a 'failed' experiment will provide some useful information** that may allow you to create a new or modified hypothesis;
- **pilot experiments may point out deficiencies in experimental technique** that will need to be rectified;
- **the experience will help you create a realistic plan of work.**

Designing experiments or sampling procedures

Design of experiments is covered in Chapter 33, while sampling procedure is dealt with in Chapter 32. Avoid being too ambitious at the start of your work. It is generally best to work with a simple hypothesis and design your experiments or sampling around this. A small pilot experiment or test sample will highlight potential stumbling blocks, including resource limitations, whether in materials, or time, or both.

Working in a laboratory environment

During your time as a project student, you are effectively a guest in your supervisor's laboratory.

- **Be considerate** – keep your 'area' tidy and offer to do your share of lab duties such as calibrating the pH meter, replenishing stock solutions, distilled water, etc., maintaining cultures, tending plants or animals.
- **Use instruments carefully** – they could be worth more than you would think. Careless use may invalidate calibration settings and ruin other people's work as well as your own.
- **Do your homework on techniques you intend to use** – there is less chance of making costly mistakes if you have a good background understanding of the methods you will be using.
- **Always seek advice if you are unsure of what you are doing.**

Keeping notes and analysing your results

Tidy record keeping is often associated with good research, and you should follow the advice and hints given in Chapter 34. Try to keep copies of all files relating to your project. As you obtain results, you should always calculate, analyse and graph data as soon as you can (see Fig. 35.1). This can reveal aspects that may not be obvious in numerical or readout form. Do not be worried by negative results – these can sometimes be as useful as positive results if they allow you to eliminate hypotheses – and do not be too dispirited if things do not work first time. Thomas Edison's maxim 'Genius is one per cent inspiration and ninety-nine per cent perspiration' certainly applies to research work.

Writing your project report

The structure of scientific reports is dealt with in Chapter 18. The following advice concerns methods of accumulating relevant information.

Introduction

This is a big piece of writing that can be very time-consuming. Therefore, the more work you can do on it early on, the better. You should allocate some time at the start for library work (without neglecting field or bench work), so that you can build up a database of references (Chapter 9). Though photocopying can be expensive, you will find it valuable to have copies of key reviews and references handy when writing away from the library. Discuss proposals for content and structure with your supervisor to make sure your effort is relevant. Leave space at the end for a section on aims and objectives. This is important to orientate readers (including assessors), but you may prefer to finalise the content after the results have been analysed.

Materials and methods

You should note as many details as possible *when doing the experiment or making observations.* Do not rely on your memory or hope that the information will still be available when you come to write up. Even if it is, chasing these details can waste valuable time.

Results

Show your supervisor graphed and tabulated versions of your data promptly. These can easily be produced using a spreadsheet (Chapter 13), but you should seek your supervisor's advice on whether the design and print quality is appropriate to be included in your report. You may wish to access a specialist graphics program to produce publishable-quality graphs and charts: allow some time for learning its idiosyncrasies. If you are producing a poster for assessment (Chapter 14), be sure to mock up the design well in advance. Similarly, think ahead about your needs for any seminar or poster you will present.

Discussion

Because this comes at the end of your report, and some parts can be written only after you have all the results in place, the temptation is to leave the discussion to last. This means that it might be rushed – not a good idea because of the weight attached by assessors to your analysis of data and thoughts about future experiments. It will help greatly if you keep notes of aims, conclusions and ideas for future work as you go along (Fig. 35.1). Another useful tip is to make notes of comparable data and conclusions from the literature as you read papers and reviews.

Acknowledgements

Make a special place in your notebook for noting all those who have helped you carry out the work, for use when writing this section of the report.

Brushing up on your IT skills – word processors and spreadsheets are extremely useful when producing a thesis. Chapters 12 and 13 detail key features of these programs. You might benefit from attending courses on the relevant programs or studying manuals or texts so that you can use them more efficiently.

Using drawings and photographs –these can provide valuable records of sampling sites or experimental set-ups and could be useful in your report. Plan ahead and do the relevant work at the time of carrying out your research rather than afterwards. Refer to Chapters 30 and 50 for tips on technique.

References

Because of the complex formats involved (Chapter 9), these can be tricky to type. To save time, process them in batches as you go along.

> **KEY POINT** Make sure you are absolutely certain about the deadline for submitting your report and try to submit a few days before it. If you leave things until the last moment, you may find access to printers, photocopiers and binding machines is difficult.

Sources for further study

Luck, M. (1999) *Your Student Research Project.* Gower, London.

Marshall, P. (1997) *Research Methods: How to Design and Conduct a Successful Project.* How To Books, Plymouth.

McMillan, K.M. and Weyers, J.D.B. (2014) *How to Complete a Successful Research Project.* Pearson Education London.

STUDY EXERCISES

Note: These exercises assume that you have started a research project, or are about to start one, as part of your studies.

35.1 Prepare a project plan. Make a formal plan for your research project, incorporating any milestones dictated by your department, such as interim reports and final submission dates. Discuss your plan with your supervisor and incorporate his or her comments. Refer back to the plan frequently during your project, to see how well you are meeting your deadlines.

35.2 Resolve to write up your work as you go along. Each time you complete an experiment or observation,

write up the materials and methods, analyse the data and draw up the graphs as soon as you can. While you may reject or modify some of these drafts at a later stage, this approach will spread out the majority of the effort and allow time for critical thinking close to the final submission date.

35.3 Devise a computer database for keeping details of your references. Keeping these records up to date will save you a lot of time when writing up. You will need to decide on an appropriate referencing format, or find out about the one followed by your department (see Chapter 9).

Answers to these study exercises are available at www.pearsoned.co.uk/practicalskills

Obtaining and identifying specimens

Taking account of the environment – always remember that collecting is essentially a destructive process, so you should make maximum efforts to minimise impacts on the area/population/community being sampled.

Sites of Special Scientific Interest (SSSIs) – in the UK, these are areas of the environment that are protected by legislation to conserve particular habitats, geology, flora or fauna. It is a criminal offence to intentionally or recklessly damage or destroy any of the features of special interest of an SSSI, or to disturb wildlife for which the site was notified. Consent is required from the relevant conservation body (e.g. Natural England) for any collection activity on the site.

If you are required to collect material, you may need to use a formal sampling procedure (see Chapter 32) or simple qualitative collecting. Your choice of equipment will depend upon your objectives. Some of the main reasons for collecting include:

- **Obtaining specimens for subsequent laboratory experimentation or observation:** this requires collection of living material while causing minimum stress and damage to the organisms.
- **Making estimates of population and community parameters:** this can be destructive (requiring killing of specimens) or non-destructive, depending upon the objectives of the study and any requirements for subsequent laboratory work, e.g. sorting and identification.
- **Gathering specimens for museum-type collections:** here, the main objective is to obtain undamaged and representative specimens, usually in a preserved form (see Chapter 37).
- **Collecting for subsequent chemical/biochemical analysis:** this may require a formal protocol. This objective requires care in both the method of collection and subsequent storage to avoid inducing chemical changes that are artefacts of the collection and storage processes. Deep freezing is usually the preferred storage method when subsequent chemical analysis is likely.

The main rules for collecting are:

- **Collect only enough for your purposes.**
- **Treat animals with respect at all times:** do not cause unnecessary stress or suffering. There are formal rules for the handling of many vertebrate species but the same attitude should be taken towards all living organisms.
- **Minimise damage and stress during collection and transport:** stressed organisms are of little use for realistic experimentation.
- **Be aware of the limitations and bias of your collecting equipment:** this is particularly important for formal sampling procedures where collecting devices almost always have such problems. This may require specific testing for your particular usage.
- **Keep good records of collection details.**

KEY POINT   Check any legislation relating to the species or habitats you are intending to use. Obtain any permits required and strictly obey any regulations.

Equipment for collecting

There is an immense variety of collecting and sampling devices available; see, for example, Fig. 36.1. In general, the collection of remote and/or animal samples presents the greatest problems. The more remote the operator is from the point of sampling, the more difficult is the evaluation of the quality of the sample in terms of its representativeness. Animal collection can be difficult because of factors such as mobility and complications introduced when allowing for avoidance behaviour.

Fig. 36.1 Examples of devices used for collecting and sampling. (a) Small vertebrate trap – the animal walks up the ramp towards bait and releases the trap door by a simple trip mechanism. (b) Botanist's vasculum – used in the field for collecting and protecting plant, fungal and animal specimens. (c) Sweep net – used for catching flying insects or aquatic specimens. (d) Light trap – used for catching night-flying insects. (e) 'Pooter' – device for collecting small animals such as insects by suction. (f) Pitfall trap – animal falls into camouflaged trap and cannot escape.

Collecting formal samples

Here the objective is to obtain specimens that both qualitatively and quantitatively represent some well-defined habitat(s). Some of the more obvious practical considerations are:

- **Are there 'edge effects' associated with the sampling method?** Because of such effects you may need to adopt a special protocol, e.g. to determine whether or not a specimen falling on the boundary is included or excluded from the sample (p. 198).

- **Will the sample be uniform?** For example, remote sampling of marine sediments of different texture using grabs often results in 'bites' of different depth being taken.

- **Does the method sample all components of the biota equally well?** This may not be important as long as it adequately samples those components in which you are interested.

- **How accurately can the location of the sample be defined,** especially with regard to other samples intended as replicates? This is particularly important in remote sampling.

- **Is the size of the sampling unit adequate** for the size and distribution of the object, species or communities being sampled? In remote sampling, this can often be a problem and frequently the method used is a compromise.

- **Can interspecies interactions affect the integrity of the sample** after collection but before processing? Factors such as predation can be prevented by using an appropriate chemical fixative immediately upon collection (see Chapter 37). Freezing is a poor option

---

Definition

Edge effect – in this context, any phenomenon associated with the sampling procedure at the edge of a sampling device or area.

Defining location – for greatest precision, use the coordinates obtained from a large-scale map; for less precise measurements, coordinates given by satellite-based or radio-based position-fixing may be appropriate.

here since, upon thawing, many animal and plant tissues tend to disintegrate and the specimens are in poor condition for subsequent identification, etc.

General collecting strategies

For qualitative collecting, where a representative sample of the population or habitat is not the objective, the effectiveness of equipment may be of less significance. Here, suitability for capturing living and undamaged specimens may be the principal criterion in choosing your equipment. For mobile animals this often involves nets and traps of various kinds, combined with narcotising agents such as smoke for insects (see Chapter 37). If specimens can be killed before capture, then spray biocides can be a useful aid. Some useful general points about all collections are:

- **Think ahead and be prepared.** Your collecting equipment must be appropriate for the task.
- **Keep a good record of collection details:** this is particularly important if the collection is for museum or herbarium purposes.
- **Keep collected plants in a humid atmosphere** to ensure good condition. The vasculum originally designed for this purpose has largely been replaced by the polythene bag in most circumstances, except where mechanical damage is likely.
- **Keep all living animals in conditions as similar as possible to the environment from which they were collected.** Aquatic specimens are usually particularly temperature-sensitive and should be kept in a Thermos flask or 'Coolbox' to prevent rapid temperature changes.
- **Use appropriate collection vessels:** rigid containers are better than plastic bags for most purposes since they help to prevent damage to the specimens. Plastic containers should be chosen in preference to glass for most purposes. Make sure that the container seals properly to avoid the loss of specimens and to prevent loss of water.

Sources for further study

Anon. *List of Sites of Special Scientific Interest by Area of Search*. Available: http://en.wikipedia.org/wiki/List_of_Sites_of_Special_Scientific_Interest_by_Area_of_Search
Last accessed 27/03/15.

Bowles, J.M. *Guide to Plant Collection and Identification*. Available: http://www.queensu.ca/biology/facilities/herbarium/collecting.html
Last accessed 27/03/15.

Census of Marine Life. *Investigating Marine Life*. Available: http://www.coml.org/investigating/collecting/trawls
Last accessed 27/03/15.
[Contains a well-illustrated guide to marine collecting techniques.]

Lincoln, R.J. and Sheals, J.G. (1979) *Invertebrate Animals: Collection and Preservation*. Cambridge University Press, Cambridge.
[An excellent methodology source for all invertebrate groups. Note that some of the chemicals recommended have been superseded by other, usually less toxic, equivalents.]

Wolberg, D. and Reinard, P. (1997) *Collecting the Natural World: Legal Requirements and Personal Liability for Collecting Plants, Animals, Rocks, Minerals and Fossils*. Geoscience Press, Phoenix.
[Deals with US regulations; gives a good idea of what needs to be considered.]

STUDY EXERCISES

36.1 **Specify collection equipment.** What equipment would you use to make a collection of the following organisms and how would you transport them to the laboratory in a living condition?

(a) *Daphnia* (planktonic crustacean)

(b) earthworms

(c) bluebells

(d) ants

36.2 **Specify preservation methods.** How would you preserve organisms (a)–(d) in Study exercise 36.1 for subsequent examination?

36.3 **Specify labelling details.** What items would you include on the labels for preserved specimens of organisms (a)–(d) in Study exercise 36.1?

Answers to these study exercises are available at www.pearsoned.co.uk/practicalskills

Fixation is a chemical process that stops autolysis and stabilises protein and other major components of tissues so that during subsequent processing, the tissues retain as fully as possible the form they had in life. Preservation allows material to be stored indefinitely by destroying any bacteria, fungi, etc., that could degrade the specimen. Preservation and fixation used to be synonymous in that most of the commonly used preservatives also had a fixative action. This has changed with the introduction of 'phenoxetols'. These are good preservatives but since they do not arrest autolysis, their use must be preceded by treatment with a true fixative: for that reason, they are called post-fixation preservatives.

> **KEY POINT** Take care when fixing specimens – remember, anything that will fix your specimen is also capable of fixing you. Be very careful in your handling of these materials and obey all safety precautions.

Your choice of fixative will depend upon the material to be fixed and the purpose for which it is being fixed. Some of the most common fixatives (Table 37.1) may be used alone although more often they are used in mixtures, the object being to combine the virtues of the various ingredients.

Narcotisation

Narcotisation (relaxation) is usually advisable for animals since many are highly contractile and assume grossly distorted postures if placed straight into fixative; it is also more humane to narcotise before killing. Failure to narcotise may result in contortion, rupture of the body wall or evisceration and reduce the scientific value of the specimen. The need for narcotisation varies with the type of organism and the objectives of the study: some of the most widely used narcotising agents are given in Table 37.2.

Definitions

Autolysis – spontaneous breakdown of a specimen through the action of its own enzymes (i.e. self-digestion).

Fixation – treatment, usually by chemical, of cells, tissues or whole specimens, such that life processes (including autolysis) no longer occur (see also p. 265).

Narcotisation – anesthetisation: dulling of awareness, sense of pain and reduction of contortion associated with death.

Preservation – prevention of degradation of a specimen, usually by external agencies.

Fixing specimens for histological or cytological work – the procedures and chemicals used are critical to success. Consult specialist texts such as Kiernan (2008) or materials and methods sections of relevant research papers.

Acting humanely – always narcotise animals before carrying out any procedures that might cause pain.

Table 37.1 Some of the most widely used fixatives/preservatives and their properties. Note that there are significant safety issues with nearly all these substances (see notes in column 4)

Substance	Fixation	Usage	Notes
Formaldehyde	+	4% v/v	Comes as 40% v/v solution (= formalin): normal dilution 1 + 9. Make up with sea water for marine specimens; buffer for calcareous specimens. Use in a fume cupboard – health hazard
Ethanol	+	70% v/v aqueous	Highly volatile; inflammable; containers must be well sealed to prevent evaporation; causes shrinkage and decolorisation as well as loss of lipids
Acetic acid	+	In mixtures	Pungent vapour
Picric acid	+	In mixtures	Risk of explosion; detonates readily on contact with some metals. Not recommended for routine student use (significant health risk)
Mercuric chloride	+	In mixtures	Extremely poisonous; corrosive to metal implements; tissue will contain mercuric salt deposits
Osmium tetroxide	+	1% v/v in buffer or vapour	Both fluid and vapour highly toxic; use in fume cupboard only. Excellent for cytological detail but is expensive and can only be used for very small specimens due to poor penetration speed. Vapour good for protozoa
Propylene phenoxetol	–	1–2% v/v aqueous	Relatively expensive but innocuous and effective preservative: needs pre-fixation stage
Glutaraldehyde	+	2–4% v/v in buffer	Highly toxic and severe skin irritant. Must be used cold

Table 37.2 Narcotising agents and their characteristics

Agent	Usage	Notes
Cold (chilling)	Cold-blooded animals	Effective form of relaxing many animals such as tropical and subtropical invertebrates
Heat (warming)	Slow heat	Works for some animals. Start from ambient but keep time period as short as possible
Magnesium sulphate or Magnesium chloride	7% w/v in water 20% w/v in water	Quite effective for many invertebrates but beware of osmotic problems if made up in sea water: keep exposure times fairly short (1–2 h)
Menthol crystals	Float on water	Slow but effective for many aquatic animals
Chloral hydrate	1% w/v in sea water	General narcotising agent
Ethanol	10% v/v dropwise	Slow and rather tedious process for all but very small specimens
MS-222 (Tricaine)	Use as 0.05% w/v aqueous solution	Good for marine and freshwater fish: very rapid effect (15 s–1 min)
Chloretone	0.1–0.5% w/v in water	General narcotising agent
Ethyl acetate	Vapour	Effective for most insects (kills as well); highly volatile
Ether	Vapour	Effective for vertebrates
Chloroform	Vapour	Effective for vertebrates

> **SAFETY NOTE** Keep safety in mind when selecting a fixative or preservative – many fixatives and preservatives are toxic and a risk assessment should be made before they are used. Always use safety equipment when handling these potentially toxic chemicals, including fume hood, goggles, buttoned lab coat and latex gloves.

Selecting your fixative/preservative

The main factors you must consider in selecting a suitable preservative/fixative for your materials are:

- **Speed of penetration,** which determines the size of object that can be fixed. Some fixatives penetrate very slowly (e.g. osmium salts) and only very small pieces of tissue can be fixed. Others such as formaldehyde penetrate very rapidly and allow fixation of relatively large specimens, e.g. whole animals. It may be necessary to inject fixatives into the body cavities of large specimens to ensure adequate fixation.

- **Shrinkage:** fixatives such as ethanol and mercuric chloride cause tissues to shrink, thus distorting them significantly. The addition of glacial acetic acid is frequently used to reduce this problem.

- **Hardening:** ethanol is particularly liable to cause tissues to harden on prolonged exposure to concentrations above 70% v/v. For whole specimens, the addition of 3% v/v glycerol to 70% v/v alcohol reduces this, making it a useful medium for specimen storage (not to be used for histological studies). The time in hardening solutions must be carefully controlled for histological material.

- **Decolorisation:** for whole specimens, the loss of colour may be detrimental and solvents such as ethanol must be used with care since they readily remove many pigments.

- **Osmotic problems:** distortion of tissues by osmotic movements of water can be rapid and serious. Some fixatives are best made up in sea water if they are to be used for marine specimens to avoid osmotic swelling of tissues. Others such as osmium salts and glutaraldehyde usually need to be made up in a buffer solution isotonic with the tissues.

- **Decalcification:** acidic fixatives such as unbuffered formalin and those containing picric acid or acetic acid readily dissolve calcareous structures. For histological preparations this may be desirable, but for whole specimens it is undesirable. The acidic properties of formaldehyde can be overcome by neutralising with calcium salts.

- **Other chemical reactions:** fixatives containing mercuric chloride usually result in the deposition of mercuric precipitates in the tissues. For histological preparations, these must be removed by thorough washing and if necessary by post-treating with iodised alcohol. There is no good reason for using mercuric chloride fixatives other than for histological work and the benefits must be weighed against the dangers and difficulties. Note also that you must not use metal implements with such fixatives as they are extremely corrosive – use only wooden or glass implements.

Fig. 37.1 Method for constraining animal specimens liable to curl up or flex during wet preservation.

tape or elastic band

block of wood

Preservation

Wet preservation

This is used mainly for animals and is usually preceded by narcotisation. The process of fixation is then comparatively straightforward: sometimes it is necessary to arrange the body and appendages of the animal using tapes, elastic bands, etc. before fixation begins (see Fig. 37.1). The solutions most commonly used for wet preservation have been either:

- **a 5–10% v/v aqueous solution of formalin** (= 40% v/v aqueous formaldehyde), neutralised with calcium carbonate or some other agent to prevent any acidity in the solution resulting in the slow dissolution of calcareous structures (decalcification); or
- **50–70% v/v ethanol.**

Formaldehyde is cheap and non-inflammable but tends to stiffen and harden tissues on prolonged exposure: safety considerations make it problematical to use because of its noxious vapour. Ethanol is inflammable, highly volatile and tends to cause shrinkage and decolorisation in soft-bodied animals: pass them slowly through a graded series of concentrations to minimise this problem. Industrial methylated spirit is quite suitable for preservation if ethanol is unavailable.

A 1–2% v/v aqueous solution of propylene phenoxetol has been widely used for vertebrate and invertebrate preservation provided it is preceded by the use of a fixative: it preserves the natural colour well and leaves the material pliable. It is comparatively expensive but is non-flammable and non-volatile.

Wet preservation of plant material is relatively straightforward and generally has the advantage of conserving the form of the specimen providing it is done carefully. Herbaceous plants should be preserved in a solution with an alcohol base, while for succulents a formaldehyde base is better. Algae, fungi, ferns, lichens and seed plants can be stored wet in formalin acetic acid (FAA) solution: this is made from formalin, acetic acid and ethanol in the ratio 85 : 10 : 5. To retain the green coloration of ferns and seed plants, add a crystal of copper sulphate to the solution and incubate the specimens for 3–10 days, then transfer to ordinary FAA solution. Colours of flowers cannot easily be preserved using wet preservatives.

metal grids
(facilitate drying)

straps
(allow press to
be compressed)

Fig. 37.2 Typical plant press. Specimens are placed between absorbent paper sheets within the metal grids, then the straps are tightened.

Using a plant press – be sure to label each specimen carefully in a way that ensures that the label cannot be lost during paper changes.

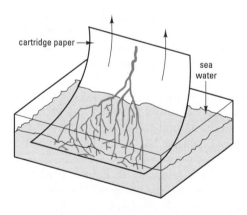

cartridge paper

sea water

Fig. 37.3 Method for 'floating out' algal specimens. Arrows show direction of lifting.

For marine specimens, make up a stock of 1 part propylene glycol with 1 part formalin. Use this in the ratio of 1 part stock to 8 parts sea water for fixation and 1 part stock to 9 parts sea water for subsequent preservation. This markedly reduces the formaldehyde concentration without loss of the fixative/preservative effect and has been shown to be satisfactory even for specimens with calcareous components.

Dry preservation

This is used mainly for vertebrate taxidermy, and for arthropod and plant material, although the development of freeze-drying has made this procedure applicable to almost any type of small or medium-sized organism. Arthropods and vascular plants, with their hard exoskeletons and vascular tissue respectively, are most easily dealt with since these rigid structures prevent collapse and loss of form upon drying. Similarly, some types of fungal fruiting bodies such as 'bracket fungi' can be preserved in a dry state.

Flowering plants, or parts of plants, mosses and liverworts, should be dried in a plant press (Fig. 37.2) between sheets of absorbent paper. Interpose a sheet of muslin or greaseproof paper between the paper and the specimen to prevent the specimen adhering to the paper. Great care must be exercised in the cleaning and arrangement of the plant so that essential features are not obscured. Change the drying sheets daily until dry. Speeding the drying by heating the press will help to preserve colour: use a low-temperature oven or place on a radiator.

Algae are usually dried in a press after floating them on to a sheet of cartridge paper that is immersed in sea water: this allows careful arrangement of the often delicate algal fronds (see Fig. 37.3). Once suitably arranged, the process is essentially the same as for terrestrial plants although regular changes of the absorbent paper layers are necessary and the drying procedure usually takes several days. Fungi are best preserved by drying without any attempt at pressing.

Storage

Careful maintenance of specimens after the fixation/preservation processes is very important. Specimens may be stored for long periods provided they are protected from pests such as mites: this may require the use of chemicals such as naphthalene in air-tight containers. Seeds are frequently stored after air-drying only, but they have a finite period of viability.

The storage conditions for all dried materials are critical; these include moderate to low temperatures and very low humidity. Storage of wet material should be in appropriate, vapour-tight vessels: even then the containers will need checking and topping up over time, particularly when volatile preservatives are used. Labelling must be comprehensive and should contain information on the fixation and preservation processes as well as ecological and taxonomic details.

Text reference

Kiernan, J.A. (2008) *Histological and Histochemical Methods: Theory and Practice,* 4th edn. Scion Publishing Ltd, Bloxham.

Sources for further study

Centre for Plant Diversity Center for Australian National Biodiversity Research. *An Introduction to Collecting Plants.* Available: http://www.anbg.gov.au/cpbr/herbarium/collecting/index.html Last accessed 27/03/15.

Lim, K.K.P. and Sivasoth, N. *A Guide to Methods of Preserving Animal Specimens in Liquid Preservatives.* Available: http://preserve.sivasothi.com Last accessed 27/03/15.

Lincoln, R.J. and Sheals, J.G. (1979) *Invertebrate Animals: Collection and Preservation.* Cambridge University Press, Cambridge.

Lo Bianco, S. *Classic Resources: The Methods Employed at the Naples Zoological Station for the Preservation of Marine Animals (Trans. E. O. Hovey).* Available: http://archive.org/details/methodsemployed00hoveygoog Last accessed 27/03/15.
[Web reprint of a 'classic' resource.]

Russell, H.D. *Notes on Methods for the Narcotization, Killing, Fixation, and Preservation of Marine Organisms.* Available: http://darchive.mblwhoilibrary.org/handle/1912/223 Last accessed 27/03/15.

Smalldon, G. and Lee, E.W. (1979) *A Synopsis of Methods for the Narcotisation of Marine Invertebrates.* Royal Scottish Museum, Edinburgh.

University of Hawaii Methods Manual for Managers: *Field Techniques for Algal Research.* Available: http://www.hawaii.edu/reefalage/publications/mthodsmanual/algalcollection.htm Last accessed 27/03/15.

STUDY EXERCISES

37.1 Write a protocol for collection, preservation and fixation of an animal specimen. This could be for any animal, but an illustrative answer is given for an earthworm specimen that is to be processed for microscopic sectioning.

37.2 Write a protocol for the collection and preservation of a plant specimen. This could be for any plant, but an illustrative answer is given for a marine seaweed to be used as a reference specimen for future identification purposes.

37.3 State the difficulties encountered when fixing materials for microscopic sectioning. Make a list of the safety problems associated with the use of five fixatives used in microscopy.

Answers to these study exercises are available at www.pearsoned.co.uk/practicalskills

Microbes can be studied by taking samples for analysis, usually in one of the following ways:

- **by direct examination of individual cells** of a particular microbe, e.g. using fluorescence microscopy (p. 262);
- **by isolating/purifying a particular species or related individuals** of a taxonomic group, e.g. the faecal indicator bacterium *Escherichia coli* in sea water;
- **by studying microbial processes,** rather than individual microbes, either *in situ,* or in the laboratory.

The sampling process

The main factors to be considered when deciding on a suitable sampling procedure are discussed in Chapter 32.

Sampling techniques include the use of swabs, sticky tape, strips and agar contact methods for sampling surfaces, bottles for aquatic habitats, plastic bags and corers for soils and sediments. A wide range of complex apparatus is available for accurately sampling water or soil at particular depths.

> **KEY POINT**   An important feature of all microbiological sampling protocols is that the sampling apparatus must be sterile; strict aseptic technique must be used throughout the sampling process (see Chapter 47).

The sampling method must minimise the chance of contamination with microbes from other sources, especially the exterior of the sampling apparatus and the operator. For example, if you are sampling an aquatic habitat, stand downstream of the sampling site. A portable Bunsen burner or spirit lamp can be used to assist sterile technique during field sampling, e.g. while flaming a loop (p. 285). Alternatively, use disposable plastic loops.

Process the sample as quickly as possible, to minimise any changes in microbiological status. As a general guideline, many procedures require samples to be analysed within six hours of collection. Changes in aeration, pH and water content may occur after collection. Some microbes are more susceptible to such effects, e.g. anaerobic bacteria may not survive if the sample is exposed to air. Sunlight can also inactivate bacteria; samples should be shielded from direct sunlight during collection and transport to the laboratory.

Soil and water samples are often kept cool (at 0–5 °C) during transport to the laboratory. In contrast, some microbes adapted to grow in association with warm-blooded animals may be damaged by low temperatures. An alternative approach is to keep the sample near the ambient sampling temperature using an insulated vessel (e.g. Thermos flask).

Isolation and purification techniques

Several different approaches may be used to obtain microbes in pure culture. The choice of method will depend upon the microbe to be isolated: some organisms are relatively easy to isolate, while others require more involved procedures.

Alternatives to traditional culture-based methods – microbes can now be studied by molecular methods, including the amplification of specific nucleic acid sequences by PCR (p. 403).

Avoiding contamination during sampling – always remember that you are the most important source of contamination of field samples: components of the oral or skin microflora are the most likely contaminants.

Subsampling – to minimise the effects of changes in temperature, aeration and water status during transportation, a primary sample may be returned to the laboratory, where the working sample (subsample) is then taken (e.g. from the centre of a large block of soil).

> **SAFETY NOTE**   When working with newly isolated microbes, you should always treat them as potentially harmful (p. 283) until they have been identified (Chapter 41).

Separation methods

Most microbial isolation procedures involve some form of separation to obtain individual microbial cells. The most common approach is to use an agar-based medium for primary isolation, with streak dilution, spread plating or pour plating to produce single colonies, each derived from a single type of microbe (pp. 286–7). It is often necessary to dilute samples before isolation, so that a small number of individual microbial cells is transferred to the growth medium. Strict serial dilution (pp. 148–9) of a known amount of sample is needed for quantitative work.

> **KEY POINT** If your aim is to isolate a particular microbe, perhaps for further investigation, you will need to subculture individual colonies from the primary isolation plate to establish a pure culture, also known as an axenic culture.

Pure cultures of most microbes can be maintained indefinitely, using sterile technique and microbial culture methods (Chapter 47).

Other separation techniques include:

- **Dilution to extinction.** This involves diluting the sample to such an extent that only one or two microbes are present per millilitre: small volumes of this dilution are then transferred to a liquid growth medium (broth). After incubation, most of the tubes will show no growth, but some tubes will show growth, having been inoculated with a single viable microbe at the outset. This should give a pure culture, though it is wasteful of resources.

- **Sonication/homogenisation.** This is useful for separating individual microbial cells from each other and from inert particles, prior to isolation. However, some decrease in viability is likely.

- **Filtration.** This can be useful where the number of microbes is low. Samples can be passed through a sterile cellulose ester filter (pore size 0.2μm), which is then incubated on the surface of an appropriate solidified medium. Sieving and filtration techniques (p. 153) are often used in soil microbiology to subdivide a sample on the basis of particle size.

- **Micromanipulation.** It may be possible to separate a microbe from contaminants using a micropipette and dissecting microscope (p. 139 and p. 275). The microbe can then be transferred to an appropriate growth medium, to give a pure culture. However, this is rarely an easy task for the novice.

- **Motility.** Phototactic microbes (including photosynthetic flagellates and motile cyanobacteria) will move towards a light source; heterotrophic flagellate bacteria will move through a filter of appropriate pore size into a nutrient solution, or away from unfavourable conditions (chemotaxis).

Selective and enrichment methods

Selective and enrichment techniques can be considered together, since they both enhance the growth of a particular microbe when compared with its competitors and they are often combined in specific media (Box 38.1).

The difference between selective and enrichment techniques is that the former use growth conditions unfavourable for competitors while the latter provide improved growth conditions for the chosen microbes.

Laboratory incubation under selective conditions will allow the particular microbes to be isolated in pure culture.

Obtaining a pure culture – if a single colony from a primary isolation medium is used to prepare a streak dilution plate and all the colonies on the second plate appear identical, then a pure culture has been established. Otherwise, you cannot assume that your culture is pure and you should repeat the subculture until you have a pure culture.

Using a sonicator – minimise heat damage with short treatment 'bursts' (typically up to 1 min), cooling the sample between bursts, e.g. using ice.

Definitions

Mesophile – a microbe with an optimum growth temperature of 20–45 °C (lit. 'middle-loving').

Psychrophile – a microbe with an optimum temperature for growth of <20 °C (lit. 'cold-loving').

Psychrotroph – a microbe with an optimum temperature for growth of ≥ 20 °C, but capable of growing at lower temperature, typically 0–5 °C (lit. 'cold-feeding').

Thermophile – a microbe with an optimum growth temperature of >45 °C (lit. 'heat-loving').

Box 38.1   Differential media for bacterial isolation: an example

MacConkey agar is both a selective and a differential medium, useful for the isolation and identification of intestinal Gram-negative bacteria. Each component in the medium has a particular role:

- **Peptone:** (a meat digest) provides a rich source of complex organic nutrients, to support the growth of non-exacting bacteria.
- **Bile salts:** toxic to most microbes apart from those growing in the intestinal tract (selective agent).
- **Lactose:** present as an additional, specific carbon source (enrichment agent).
- **Neutral red:** a pH indicator dye, to show the decrease in pH that accompanies the breakdown of lactose.
- **Crystal violet:** selectively inhibits the growth of Gram-positive bacteria. (This is only present in certain formations of MacConkey agar.)

Any intestinal Gram-negative bacterium capable of fermenting lactose will grow on MacConkey agar to produce large red-purple colonies, the red coloration being due to the neutral red indicator under acidic conditions while the purple coloration, often accompanied by a metallic sheen, is due to the precipitation of bile salts and crystal violet at low pH.

In contrast, enteric Gram-negative bacteria unable to metabolise lactose will give colonies with no obvious pigmentation. This differential medium has been particularly useful in medical microbiology, since many enteric bacteria are unable to ferment lactose (e.g. *Salmonella, Shigella*) while others metabolise this carbohydrate (e.g. *Escherichia coli, Klebsiella* spp.). Colonial morphology (Fig. 41.2) on such a medium can give an experienced bacteriologist important clues to the identity of an organism, e.g. capsulate *Klebsiella* spp. characteristically produce large, convex, mucoid colonies with a weak pink coloration, due to the fermentation of lactose, while *E. coli* produces smaller, flattened colonies with a stronger red coloration and a metallic sheen.

(a) obligate (strict) aerobe
(b) obligate (strict) anaerobe
(c) facultative anaerobe
(d) aerotolerant anaerobe
(e) microaerophile

Fig. 38.1   Agar shake tubes. Bacteria are suspended in molten agar at 45–50 °C and allowed to cool. The growth pattern after incubation reflects the atmospheric (oxygen) requirements of the bacterium.

KEY POINT Selective methods are based on the use of physicochemical conditions that will permit the growth of a particular group of microbes while inhibiting others. Enrichment techniques encourage the growth of certain bacteria, usually by providing additional nutrients in the growth medium.

Methods based on specific physical conditions include:

- **Temperature.** Psychrophilic and psychrotrophic microbes can be isolated by incubating the growth medium at 4 °C, while thermophilic microbes require temperatures above 45 °C for isolation. Short-term heat treatment of samples can be used to select for endospore-forming bacteria, e.g. 70–80 °C for 5–15 min, prior to isolation.

- **Atmosphere.** Many eukaryotic microbes are obligate aerobes, requiring an adequate supply of oxygen to grow. Bacteria vary in their responses to oxygen: obligate anaerobes are the most demanding, growing only under anaerobic conditions (e.g. in an anaerobic cabinet or jar). Oxygen requirements can be determined using the agar shake tube method as part of the isolation procedure (Fig. 38.1). Some pathogenic bacteria grow best in an atmosphere with a reduced oxygen status and increased CO_2 concentration: such carboxyphilic bacteria (capnophiles) are grown in an incubator where the gas composition can be adjusted.

- **Centrifugation.** This can be used to separate buoyant microbes from their non-buoyant counterparts – on centrifugation, such organisms will collect at the surface while the remaining microbes will sediment. Alternatively, density gradient methods may be used (p. 442). Centrifugation can be combined with repeated washing, to separate microbes from contaminants.

- **Ultraviolet irradiation.** Some microbes are tolerant of UV treatment and can be selected by exposing samples to UV light. However, the survivors may show a greater rate of mutation.

- **Illumination.** Samples may be enriched for cyanobacteria and microalgae by incubation under a suitable light regime. For dilute samples, where the number of photosynthetic microbes is too low to give the sample any visible green coloration, there is a risk of photoinhibition and loss of viability if the irradiance is too high. Such samples need shading during initial growth.

Chemical methods form the mainstay of bacteriological isolation techniques and various media have been developed for the isolation of specific groups of bacteria. The chemicals involved can be subdivided into the following groups:

- **Selectively toxic substances:** for example, salt-tolerant, Gram-positive cocci can be grown in a medium containing 7.5% w/v NaCl, which prevents the growth of most common heterotrophic bacteria. Several media include dyes as selective agents, particularly against Gram-positive bacteria.

- **Antibiotics:** for example, the use of antibacterial agents (e.g. penicillin, streptomycin, chloramphenicol) in media designed to isolate fungi, or the use of antifungal agents (e.g. cycloheximide, nystatin) in bacterial media. Some antibacterial agents show a narrow spectrum of toxicity and these can be incorporated into selective isolation media for resistant bacteria, e.g. metronidazole for anaerobic bacteria.

- **Nutrients that encourage the growth of certain microbes:** including the addition of a particular carbon source, or specific inorganic nutrients.

- **Substances that affect the pH of the medium:** for example, the use of alkaline peptone water at pH 8.6 for the isolation of *Vibrio* spp.

> **KEY POINT**   Note that subcultures from a primary isolation medium must be grown in a non-selective medium, to confirm the purity of the isolate.

Many of the selective and enrichment media used in bacteriology are able to distinguish between different types of bacteria: such media are termed differential media or diagnostic media and they are often used in the preliminary stages of an identification procedure. Box 38.1 gives details of the constituents of MacConkey medium, a selective, differential medium used in clinical microbiology (e.g. for the isolation of certain faecal bacteria), while Table 38.1 gives details of selective agents used. Further details on methods can be found in Collins *et al.* (2004). Note that isolation procedures for a particular microbe often combine several of the techniques described above. For instance, a protocol for isolating food-poisoning bacteria from a foodstuff might involve:

1. **homogenisation** of a known amount of sample in a suitable diluent;

2. **serial decimal dilution;**

3. **separation procedures** using spread or pour plates to quantify the number of bacteria of a particular type present in the foodstuff and provide a viable count (p. 297);

4. **selective/enrichment procedures,** e.g. specific media/temperatures/ atmospheric conditions, depending on the bacteria to be isolated;

5. **confirmation of identity:** any organism growing on a primary isolation medium would require subculture and further tests, to confirm the preliminary identification (Chapter 41).

Definitions

Aerotolerant anaerobe – a microbe that grows by fermentation, but which is insensitive to air/oxygen (in contrast to strict anaerobes, which are typically killed by exposure to air/oxygen).

Capnophile – a microbe that thrives in the presence of high levels of atmospheric carbon dioxide.

Facultative anaerobe – a microbe that grows by aerobic respiration when oxygen is present, switching to fermentation under anaerobic conditions.

Table 38.1 Selective agents in bacteriological media

Substance	Selective for
Azide salts	*Enterococcus* spp.
Bile salts	Intestinal bacteria
Brilliant green	Gram-negative bacteria
Gentian violet	Gram-negative bacteria
Lauryl sulphate	Gram-negative bacteria
Methyl violet	*Vibrio* spp.
Malachite green	*Mycobacterium*
Polymyxin	*Bacillus* spp.
Sodium selenite	*Salmonella* spp.
Sodium chloride	Halotolerant bacteria *Staphylococcus aureus*
Sodium tetrathionate	*Salmonella* spp.
Trypan blue	*Streptococcus* spp.
Tergitol/surfactant	Intestinal bacteria

Text reference

Collins, C.H., Lyne, P.M. and Grange, J.M. (2004) *Microbiological Methods,* 8th edn. Hodder-Arnold, London.

Sources for further study

Atlas, R.M. (2011) *Handbook of Microbiological Media,* 4th edn. CRC Press, Boca Raton.
[Gives details of culture media for a broad range of microbes.]

Eaton, A.D., Clesceri, L.S., Rice, E.W. and Greenberg, A.E. (2005) *Standard Methods for Examination of Water and Wastewater,* 21st edn. American Public Health Association, Washington, DC.
[Gives standard US protocols for a range of indicator bacteria.]

Hurst, C.J., Crawford, R.L., Garland, J.L., Lipson, D.A., Mills, A.L. and Stetzenbach, L.D. (2007) *Manual of Environmental Microbiology,* 3rd edn. American Society for Microbiology, Washington, DC.

Levett, P.N. (ed.) (1991) *Anaerobic Microbiology: A Practical Approach.* IRL Press, Oxford.
[Provides details of the methods used to isolate and culture anaerobic micro-organisms.]

Pepper, I.L. and Gerba, C.P. (2005) *Environmental Microbiology: A Laboratory Manual,* 2nd edn. Academic Press, New York.

Rochelle, P.A. (2001) *Environmental Molecular Microbiology: Protocols and Applications.* Horizon Scientific Press, Norwich.
[Covers molecular approaches to the detection and identification of microbes in environmental samples.]

STUDY EXERCISES

38.1 Plan a collection/sampling strategy for a target group of microbes. What approaches might you take in the following instances?

(a) Collecting representatives of the normal skin microflora.

(b) Sampling psychrotrophic anaerobes from the subsurface mud of an estuary.

(c) Collecting a sample of sea water to enumerate faecal indicator bacteria.

(d) Mapping the microflora on the surface of a leaf.

38.2 Decide on an appropriate isolation procedure for a particular microbe. How might you isolate the following microbes from a sample?

(a) Photosynthetic flagellate algae in a sample of pond water.

(b) Bacteria growing as a biofilm on the surface of sand particles.

(c) Faecal streptococci (enterococci) present at low density (<1 per mL) in a sample of river water.

(d) A distinctively shaped bacterium present at very low numbers in a sample containing a high number of other microbes of different shape but with similar nutritional requirements.

38.3 Investigate the selective basis of microbiological media. Using textbooks on bacteriological methods or the Web, research how each of the following media operates, in terms of their selective and diagnostic (differential) features:

(a) Mannitol salt agar for *Staphylococcus aureus.*

(b) Membrane lauryl sulphate broth for coliforms and *E. coli.*

(c) Slanetz and Bartley medium for faecal streptococci (enterococci).

(d) Mannitol pyruvate egg yolk polymyxin (MPYP) medium for *Bacillus cereus.*

Answers to these study exercises are available at www.pearsoned.co.uk/practicalskills

The use of scientific names is fundamental to all aspects of biological science since it aims to provide a system of identification that is precise, fixed and of universal application. Without such a system, comparative studies would be impossible.

There are two possible bases for such classification:

1. **Phenetic taxonomy,** which involves grouping on the basis of phenotypic similarity, frequently using complex statistical techniques to obtain objective measures of similarity. The characters used have been largely morphological and anatomical, but biochemical, cytological and other characters are increasingly used, especially for microbes (p. 248) where structural characters are few.

2. **Phylogenetic (=phyletic) taxonomy,** which involves grouping on the basis of presumed evolutionary, and therefore genetic, relationships.

These two systems are often broadly similar in outcome, since closely related organisms are usually fairly similar to each other and because judgements of evolutionary relationships are usually themselves based upon similarities. The situation is made more complex by phenomena such as convergent and divergent evolution; phyletic classifications are also liable to subjective bias.

New classifications have been proposed on the basis of studies of biomolecules, e.g. rRNA sequences, that are robust and objective. One such arrangement divides organisms into three major domains: Bacteria, Archaea and Eucarya (Fig. 39.1).

The hierarchical system

Another approach is to recognise six kingdoms if the viruses are included:

1. **Virales:** the viruses. Very simple in structure, obligate parasites of prokaryotic or eukaryotic cells.

2. **Prokaryotae (Monera):** the bacteria and cyanobacteria. Prokaryotic, non-nucleate and lacking membrane-bound organelles, such as: mitochondria, plastids, endoplasmic reticulum, Golgi apparatus.

3. **Protista:** includes algae, protozoa and slime moulds. Eukaryotic and mainly unicellular. A heterogeneous group.

4. **Plantae:** the plants. Eukaryotic, walled, mainly multi-cellular and usually photoautotrophs.

5. **Fungi:** eukaryotic, usually syncytial, walled and saprotrophic.

6. **Animalia:** the animals. Eukaryotic, multi-cellular and heterotrophs.

Six other levels of taxa are generally accepted: phylum, class, order, family, genus and species, although in botany and microbiology, division is used instead of phylum.

> **KEY POINT**   Whereas the levels of taxa above genus are rather subjective and vary among authorities, the use of genus and species names are governed by strict, internationally agreed conventions called Codes of Nomenclature.

Definitions

Classification (taxonomy) – the study of the theory and methods of organisation of taxa and, therefore, a part of systematics.

Identification – the placing of organisms into taxa (see Chapters 40 and 41).

Nomenclature – the allocation of names to taxa.

Systematics – the study of the diversity of living organisms and of the evolutionary relationships between them.

Taxon (plural taxa) – an assemblage of organisms sharing some basic features.

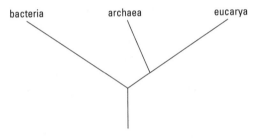

Fig. 39.1 Simplified diagram of the three major domains of the Tree of Life (based on the rRNA sequencing work of Carl Woese).

Definition

Saprotroph – a heterotrophic organism that feeds on dead organic matter. (Other nutritional terms are defined on p. 290.)

There are three major Codes of Nomenclature, the Botanical, Zoological and Bacteriological Codes, which operate on similar but not identical principles (Lapage *et al.*, 1992; Wakeham-Dawson *et al.*, 1999; McNeill *et al.*, 2006). Fungi are covered by the International Code of Botanical Nomenclature despite the fact that they are not plants.

The basis of classification

No simple definition of a species is possible, but there are two generally used definitions:

1. **A group of organisms capable of interbreeding and producing fertile offspring** – this, however, excludes all asexual, parthenogenetic and apomictic forms.

2. **A group of organisms showing a close similarity in phenotypic characteristics** – this would include morphological, anatomical, biochemical, ecological and life history characters.

> **KEY POINT**   The basic unit of classification is the species, which represents a group of recognisably similar individuals, clearly distinct from other such groups.

When species are compared, groups of species may show a number of features in common; they are then arranged into larger groupings known as genera (singular genus). This process can be repeated at each taxonomic stage to form a hierarchical system of classification whose different levels are known as taxonomic ranks. The number of levels in this system is arbitrary and based upon practical experience – the seven levels normally used have been found to be sufficient to accommodate the majority of the variation observed in nature.

When a generic or specific name is changed as a result of further study, the former name becomes a synonym; you should always try to use the latest name. Where a generic name has been changed recently, the old name is occasionally given in parentheses to allow easy reference to the old name. There are taxonomic reference works available for each discipline or subdiscipline, such as the *Flora Europea* (for plants) and the *Plymouth Marine Fauna* (for British marine animals); these often provide the current versions of a name and often its synonyms.

Nomenclature in practice

The scientific name of an organism is effectively a symbol or cipher that removes the need for repeated use of descriptions. It normally comprises two words and is, therefore, called a binomial term. The name of the genus is followed by a second term that identifies the species, e.g. *Quercus robur*, *Apis mellifera*, *Pseudomonas aeruginosa* or *Saccharomyces cerevisae* (Table 39.1).

DNA-based definition of a species – in microbiology, members of a bacterial species can be characterised by having a DNA similarity of ≥70%.

Example   When the cockle *Cardium edule* was renamed *Cerastoderma edule*, it was commonly referred to in textbooks as *Cerastoderma (Cardium) edule*: this is strictly not correct practice but can be helpful for non-specialists.

Writing taxonomic names – always underline or italicise generic and species names to avoid confusion: thus bacillus is a descriptive term for rod-shaped bacteria, while *Bacillus* is a generic name.

Table 39.1 Example of taxonomies for a plant, an animal, a bacterium and a fungus

Common name	English oak	Honey-bee	*Pseudomonas*	Baker's yeast
Kingdom	Plantae	Animalia	Monera	Fungi
Phylum/Division	Anthophyta	Arthropoda	Gracilicutes	Ascomycota
Class	Dicotyledonae	Insecta	Scotobacteria	Incertae sedis*
Order	Fagales	Hymenoptera	Pseudomonadales	Saccharomycetales
Family	Fagaceae	Apidae	Pseudomonadaceae	Saccharomycetaceae
Genus	*Quercus*	*Apis*	*Pseudomonas*	*Saccharomyces*
Species	*Q. robur*	*A. mellifera*	*P. aeruginosa*	*S. cerevisiae*

*Meaning = 'uncertain taxonomic position'

Box 39.1 Basic rules for the writing of taxonomic names

- Names of the seven levels of taxa should take lower-case initial letters, e.g. class Mollusca or kingdom Fungi.

- The Latin forms of all taxon names except the specific name take initial capital letters, e.g. 'the Arthropoda . . .' but anglicised versions do not, e.g. 'the arthropods . . .'.

- The names of the higher taxa are all plural, hence 'the Mollusca are. . .' while the singular of the anglicised version is used for a single member of that taxon, hence 'a mollusc is . . .'.

- The binomial system gives each species two names, the first being the generic name and the second the specific name, which must never be used by itself. The genus and species names are distinguished from the rest of the text either:
  (a) by being underlined (when handwritten), e.g. *Patella vulgata*; or
  (b) by being set in italics (in print or on a word processor), e.g. *Patella vulgata*.

- The generic name is singular and always takes an initial capital letter. If you use the generic name this implies that the point being made is a generic characteristic unless the specific name is present. Write the generic name in full when first used in a text, e.g. *Patella vulgata*, but subsequent references can be abbreviated to its initial letter, e.g. *P. vulgata*, unless this will result in confusion with another genus also being considered.

- The abbreviation 'sp.' should be used in place of the specific name if a single unspecified species of a genus is being referred to, e.g. *Patella* sp.; it is not underlined or italicised. If more than one unspecified species is meant, then the correct form is 'spp.', e.g. *Patella* spp.

- Common names should not normally be written with a capital letter, e.g. limpet.

- The name of each species should be followed by the authority: on first usage in formal reports and in titles, the name or names of the person(s) to whom that name is attributed and the date of that description should be quoted. These names may sometimes be abbreviated, e.g. L. for Linnaeus and standard abbreviations must be used. If the species was first described under its current generic name, the authority's name, often in abbreviated form is added, e.g. *Quercus robur* L. If, however, the species was first described under a different genus, the name of the author of the original description is presented in parentheses, e.g. *Escherichia coli* (Migula) Castellani and Chalmers. Note that in zoology, the date the description was published is also included, e.g. *Ischnochiton kermadecensis* Iredale 1914. The use of authorities should be confined to formal papers, final-year project reports, etc.; they would not normally be used in practical reports, short assignments or examinations.

Example The full, formal name of baker's yeast is *Saccharomyces cerevisiae* Meyen ex Hansen 1883. After first use, this would be abbreviated to *S. cerevisiae*.

Common names are often interesting, but totally unsatisfactory for use in biological nomenclature because of the lack of consistency in their use: the Codes of Nomenclature were established to prevent the ambiguities associated with informal/common names.

The Codes require that all scientific names are either Latin or treated as Latin, written in the Latin alphabet and subject to the rules of Latin grammar. Consequently, you must be very precise in your use of such names. In some cases, the Codes stipulate a standardised ending for the names of all taxa of a given taxonomic rank, e.g. names of all animal families must end in -idae while plant, fungal and bacterial families end in -aceae. When used in a formal scientific context, you should follow the specific name by the authority on which that name is based, i.e. the name of the person describing that species and the date of the description.

Box 39.1 summarises the basic rules for writing taxonomic names.

Taxa below the rank of species

Some use is made of taxa below the rank of species. Within zoology, this is confined to the term subspecies, so the names of subspecies have three components, e.g. *Mus musculus domesticus*, no rank term being necessary.

Applications of bacterial typing – typing methods are used widely in epidemiological studies, e.g. tracing a particular type of bacterium responsible for a food-poisoning outbreak, or tracking the development of a particular type of antibiotic-resistant bacterium in a hospital.

In bacteriology, the use of subspecies is again acceptable although a word indicating rank is usually inserted, e.g. *Bacillus subtilis* subsp. *niger.* The Bacteriological Code considers ranks from the level of subspecies up to, and including, class: the use of the term variety is discouraged, as the term is synonymous with subspecies. However, other terms are in widespread use for taxa below the species level, especially in medical microbiology and plant pathology, when a particular strain of bacterium has been identified (pp. 250–1). Subspecies identification is often referred to as typing and the following terms apply:

- **biovar, or biotype:** subdivided according to biochemical characteristics;
- **serovar or serotype:** subdivided by serological methods, using antibodies (see Chapter 59);
- **pathovar:** subdivided according to pathogenicity (ability to cause disease);
- **phagovar or phage type:** subdivided according to susceptibility to particular viruses.

Many micro-organisms are now referred to by their generic and specific names followed by a culture collection reference number, e.g. *Bacillus subtilis* NCTC 10400, where NCTC stands for the National Collection of Type Cultures and 10400 is the reference number of that strain in the collection.

Understanding microbiological terminology – the term *strain* is widely used, particularly in the context of the practice of lodging microbiological strains with culture collections, while the term *isolate* is often used for a pure culture derived from a natural (wild) population. Cell lines (Chapter 49) are often given code names and/or reference numbers.

In botany, several categories below the rank of species are recognised and a term of rank is used before the name, e.g. *Salix repens* var. *fusca*: the term var. is short for the Latin word *varietas* and is subordinate to the term subspecies in the Botanical Code. The term cultivar (cv.) is an important modern term frequently used in experimental work and refers to cultivated varieties of plants.

The special case of viruses

The classification and nomenclature of viruses are less advanced than for cellular organisms and the current nomenclature has been arrived at on a piecemeal, ad hoc basis. The International Committee for Virus Taxonomy proposed a unified classification system, dividing viruses into 50 families on the basis of:

- **host preference;**
- **nucleic acid type** (i.e. DNA or RNA);
- **whether the nucleic acid is single- or double-stranded;**
- **the presence or absence of a surrounding envelope.**

Virus family names end in -viridae and genus names in -virus. (Note that these names are *not* latinised and the genus-species binomial is not now approved.) However, this system has not yet been adopted universally and many viruses are still referred to by their trivial names or by code names (sigla), e.g. the bacterial viruses ϕX174, T4, etc. Many of the names used reflect the diseases caused by the virus. Often, a three-letter abbreviation is used, e.g. HIV (for human immunodeficiency virus), TMV (for tobacco mosaic virus).

Example   The virus that causes tobacco mosaic disease belongs to the genus *Tobamovirus* and can be referred to as tobacco mosaic tobamovirus, tobacco mosaic virus or TMV.

Text references

Lapage, S.P. *et al.* (eds) (1992) *International Code of Nomenclature of Bacteria (1990 revision)*. American Society for Microbiology, Washington, DC.

McNeill, J. *et al.* (2006) *International Code of Botanical Nomenclature: Vienna Code*. A.R.G. Ganter, Ruggell.

Wakeham-Dawson, A., *et al.* (eds) (1999) *International Code of Zoological Nomenclature*, 4th edn. The International Trust for Zoological Nomenclature, London.

Sources for further study

Anon. *International Code of Botanical Nomenclature (St Louis Code)*. Available: http://www.bgbm.fu-berlin.de/iapt/nomenclature/code/Saint Louis/0000st.luistitle.htm
Last accessed 28/03/15.
[Web version of Greuter *et al.* (2000).]

Buchen-Osmond, C. (2002) *Universal Virus Database of the International Committee on Taxonomy of Viruses*. Available: http://www.ncbi.nlm.nih.gov/ICTVdb/
Last accessed 28/03/15.
[Includes a catalogue of approved virus names, image gallery, descriptions, etc.]

Institute of Biology (2000) *Biological Nomenclature: Recommendations on Terms, Units and Symbols*. Institute of Biology, London.

[Includes sections on taxonomy and classification of organisms.]

Jeffrey, C. (1989) *Biological Nomenclature,* 3rd edn. Edward Arnold, London.

King A.M.Q. *et al.* (2011) *Virus Taxonomy: IXth Report of the International Committee on Taxonomy of Viruses*. Elsevier, London.

McNeill, J. *et al. International Code of Nomenclature for Algae, Fungi, and Plants (Melbourne Code)*. Available http://www.capt.toxon.org/nomen/main.php
Last accessed 28/03/15.

STUDY EXERCISES

39.1 Research full classifications. Provide the full classification of the following species, laid out as in the table in Study exercise 39.2.

(a) The limpet *Patella vulgata*

(b) The great white shark

(c) The giant redwood

(d) The earthworm *Lumbricus terrestris.*

39.2 What is wrong with the hierarchical classifications given in the table below?

Examples of hierarchical classifications

Nautilus	Edible mushroom	E. coli bacterium
Animalia	Fungi	Monera (Bacteria)
Mollusca	Basidiomycota	Proteobacteria
Cephalopoda	Basidiomycetes	Enterobacteriales
Nautilidae	Agaricales	Gamma subdivision
Nautilida	Agaricaceae	Enterobacteriaceae
Nautilus	*Agaricus*	*Escherichia*
pompilius	*bisporus*	*coli*

39.3 Compare a variety of current textbooks and Internet sites to discover alternative classification schemes at kingdom level. Make your own notes regarding the *evidence* used to support the alternative kingdom classifications.

Answers to these study exercises are available at www.pearsoned.co.uk/practicalskills

Fig. 40.1 Research scientists examining a rocky shore habitat at Orkney, Scotland, and making counts of the different species present, using quadrats. This image illustrates the problems facing a field biologist in identifying plants and animals. What are the species present here? How would you start to identify a specimen, say a mollusc or an alga, from this environment, if its identity were unknown to you? In the absence of specialised knowledge, the task is made far easier by using an identification guide (for flora and fauna) with a key that will guide you to descriptions of candidate species.

Out-of-date guides – taxonomists frequently change the names of taxa and update their classification. An up-to-date identification requires an up-to-date guide.

Observing rare specimens – take special care not to collect, disturb or destroy rare plants and animals in the course of your observations.

The normal way to identify plants or animals is to use identification guides. These consist of two parts:

1. **Written and pictorial descriptions of organisms,** which you compare with your unknown specimen to aid in its identification. Good descriptions direct you to the crucial diagnostic features for the relevant taxon, explain the range of variability found and point out biological and ecological characteristics of importance.

2. **Keys,** which help you find the likely description for your specimen rapidly and simply. Most keys are arranged to present you with a series of choices, usually dichotomous (dividing in two). The paired statements of each 'couplet' are framed to be contrasting and mutually exclusive. Each choice you make narrows down the possibilities for your specimen until you find the appropriate description.

The authors of identification guides assume that you have a live or preserved specimen to hand and the means to observe it closely and measure it (see Fig. 40.1). The terminology in guides is designed to combine precision with brevity. Guides for animal species are called faunas and those for plant species are called floras.

KEY POINT To use an identification guide properly, you need to know enough of the vocabulary to understand the choices presented to you, but all identification guides provide both a glossary and a list of abbreviations to help with this.

The best identification guides are those that lead you in the simplest way to a correct identification. If you need to choose one for your area of interest, think about the following questions:

- **What degree of prior knowledge is assumed?** Some guides are written for novices, whereas others assume an expert's command of terminology. If tempted to go for the former type, consider whether it will always be suitable for your needs.

- **What is the scope of the guide?** Guides may be restricted in the taxa they consider or in the geographical region that is covered; this will suit you if your interests are similarly narrow. However, if your interests are wide, the relevant guide may be so large as to be unwieldy in the field.

- **How well is the guide illustrated?** Good-quality illustrations enhance the ease of use of a guide – features can be shown pictorially that might otherwise involve an off-putting specialised vocabulary to describe. Accurately coloured illustrations can be helpful, but note that colour can be a variable character: look for good line diagrams that highlight the critical diagnostic points.

- **Is the guide divided into parts?** Good guides are arranged in short parts dealing with different levels of the taxonomic hierarchy. This speeds up identification by allowing you to skip initial material when you have a fair idea of the specimen's identity.

- **Do you like the style of the key?** As discussed below, there are several ways in which a key can be presented, one of which may suit you more than the others. If you cannot actually test out a key yourself on real specimens, the next best thing is to ask for the opinion of someone who has used it.

Types of key

Bracketed keys

Here, numbered pairs of adjacent lines in the key present you with a choice and either 'send' you to a new couplet or provide the tentative identification (Box 40.1).

capitulum

ligule

Fig. 40.2 Ragwort flower (*Senecio* spp.).
Adapted from figure from *New Flora of the British Isles*, 2nd edition, Cambridge University Press (Stace, C.A., 1997).

Box 40.1 **Example of a bracketed key**

Part of key to ragworts (Fig. 40.2), modified after Stace (1997):

1. Ligules <8 mm or 0; capitula cylindrical, about 2× as long as Wide 2
   Ligules >8 mm; capitula bell-shaped in flower, about 1.5× as long as wide 3
2. Ligules usually 0; achenes ≤ 2.5 mm   *Senecio vulgaris*
   Ligules usually present; achenes < 3 mm   *Senecio cambrensis*

If there are no ligules on your specimen or they are less than 8 mm in length, you should proceed to choice 2, but if they are present and greater than 8 mm in length, you should proceed to choice 3.

In this case, the choice at 2 is sufficient to pin down the species; sometimes quite early in a key a distinctive characteristic may allow the specimen to be 'identified' to species level, while for the other options the specimen's identity remains open. Note the use of more than one comparison in each couplet to provide confirmation.

Indented keys

In this method, the pairs of choices are indented and given the same number. They are separated by other choices further down the sequence. Having made a choice, you look at the next couplet below which will be one indent level further in. When a choice is sufficiently distinctive, the tentative identity of your specimen will be given (Box 40.2).

Box 40.2 **Example of an indented key**

Part of key for common species of true bumblebee (Fig. 40.3), modified after Prys-Jones and Corbet (1987):

1. Thorax with black area(s)
   2. Thorax all black
      3. Pollen baskets with red hairs   *Bombus rudarius*
      3. Pollen baskets with black hairs   *Bombus lapidarius*
   2. Thorax black with yellow or brown patches
      4. Tail white, buff or brown
         5. Scutellum black   *Bombus terrestris*
         5. Scutellum yellow   *Bombus hortorum*
      4. Tail red or orange   *Bombus pratorum*
1. Thorax without any black   *Bombus pascuorum*

If your bumblebee has a black thorax with yellow patches, proceed to choice 4; if its tail is brown, carry on to choice 5, etc.

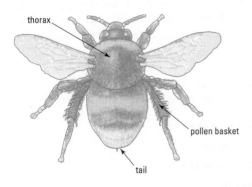

thorax

pollen basket

tail

Fig. 40.3 Bumblebees (*Bombus* spp., ♀).

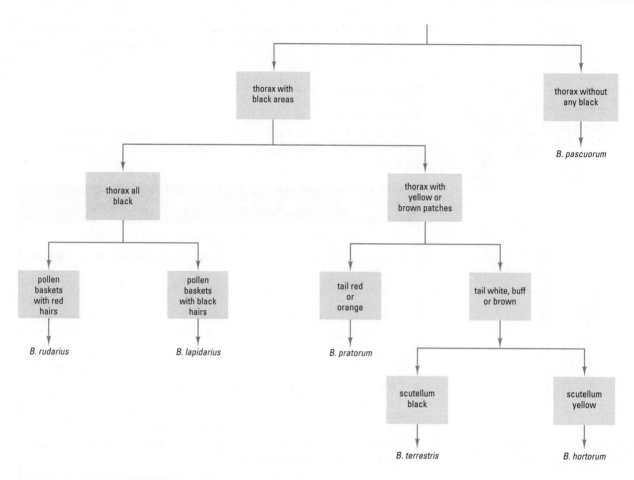

Fig. 40.4 A key for bumblebees *(Bombus* spp.) laid out in the form of a flowchart.

> **KEY POINT** Bracketed keys have the advantage that they keep the couplets close together for ready comparison, but indented keys show the relationships between taxa more clearly and allow you to backtrack more easily if an error has been made.

Flowchart keys

In this form of key, the choices are laid out in the form of a flowchart (Fig. 40.4), which allows easier cross-checking of options but is only feasible where there are a small number of choices. To use this type of key, follow the arrows after making each choice in the sequence; this will lead you on to another choice and eventually to the tentative identification.

Multi-access keys

These allow you to choose the characters used in the key according to the state of your specimen (Box 40.3). They are useful in situations where:

- **important characters are difficult to observe;**
- **characters are likely to be misinterpreted;**
- **a single character would be unreliable in isolation;**
- **a part is missing or seems abnormal.**

Species distribution – when using a key, it may be valuable to consider the known range of the species.

Box 40.3 **Example of a multi-access key**

Part of key to species of willowherbs, modified after Stace (1997):

Stigma 4-lobed	A
Stigma club-shaped	B
Seeds minutely uniformly papillose	C
Seeds with longitudinal papillose ridges	D
Stems erect or erect at apex	E
Stems trailing on ground	F

ACE	Petals 10–16 mm, purplish pink	*Epilobium hirsutum*
	Petals 5–9 mm, paler	*Epilobium parviflorum*
BCE	Petioles 4–15 mm, plant perennating by rosettes	*Epilobium roseum*
	Petioles ≤4 4 mm, plant perennating by stolons ending in tight bud	*Epilobium palustre*

If your specimen had a 4-lobed stigma and erect stems, but no seeds were available to examine, you could 'identify' it as either *E. hirsutum* or *E. parviflorum* and distinguish between these choices on the basis of petal size and colour.

Another way of presenting a multi-access key is in the form of a table. For instance, the taxa to be distinguished could make up the rows of the table and the relevant characteristics the columns (see Table 41.1). Like the flowchart, this type of key is limited to a small number of choices.

Computerised keys

These simplify the initial stages of identification by providing a series of menu-like choices for the user. They can rapidly provide a list of tentative identifications on even a few positive choices from these menus, ranking these in likelihood of being correct. You can then work down the list, comparing your specimen to a description of the proposed species (see below). At present, computerised keys are more likely to be used in a laboratory or field station than in the field.

Advice for using keys

- **Note down the route taken** (i.e. the numbering system for the decision tree): this makes it easier to trace back your path through the key.

- **At each step, read the full description for both choices** before arriving at a decision about which one to take.

- **Never guess if you do not know the precise meaning of the terms used** – consult the key's glossary and list of abbreviations. Where measurements are required, use a ruler – do not guess sizes.

- **If features are very small, use an appropriate lens to inspect them clearly.**

- **If the key is a multi-part one, look carefully at the descriptions for higher levels of taxa before progressing to the species key:** this not only acts as a check that you are correct up to this stage, but may also provide definitions of useful terminology.

- **If both of a pair of choices seem reasonable, try out each route** – one will usually prove to be unsuitable at a later stage.

Problems of identification – sometimes your best attempts to identify the specimen will be thwarted by the existence of sexual dimorphism or polymorphism, juvenile and adult phases (e.g. gametophyte and sporophyte phases for certain plants), local forms, non-native taxa, etc. A good guide will point out these problems where they occur.

KEY POINT When you arrive at the end of a key's path, do not simply accept this as a reliable identification of your specimen. Compare your specimen with the full description of the species.

Comparing specimens with descriptions

If the specimen does not fit the description properly, follow the instructions outlined below:

- **Compare the specimen with neighbouring descriptions:** in a well-organised guide, those of similar species will be together.

- **Go back along the path of the key and re-examine each decision you have made.** Try going down the alternative route for any that might have been questionable.

- **Check to see whether you inadvertently went down the wrong pathway** even though you made the correct diagnosis.

- **Bear in mind the possibility that your specimen is not typical.** A good key will use characteristics that are constant, but biological variation will often throw up an oddity to confuse you. Try to obtain another specimen, preferably not genetically related to the original.

- **Consider the possibility that the specimen could be outside its normal geographical range** or even new to science.

The ultimate check on an identification is a comparison of the specimen with an authentically named specimen in a museum or herbarium. The ultimate comparison would be with the type specimen, the specimen used when the species or subspecies was first described and named. If this is the only specimen collected by the author(s) who named the species, it is called a holotype. Other 'type' specimens include:

- **paratypes** – those other than the type specimen also used by the author(s) at the time of the original description;

- **syntypes** – a collection of specimens used for the original collection, but from which no one specimen was defined as the type specimen;

- **lectotypes** – a particular syntype subsequently chosen and designated through publication to act as the type specimen.

Text references

Prys-Jones, O.E. and Corbet, S.A. (1987) *Naturalist's Handbook 6. Bumblebees.* Cambridge University Press, Cambridge.

Stace, C.A. (1997) *New Flora of the British Isles,* 2nd edn. Cambridge University Press, Cambridge.

Sources for further study

Discover Life. *Identification Guides and Checklists of Plants and Animals.* Available: http://www.discoverlife.org/mp/20q
Last accessed 28/03/15
[An extensive set of guides to worldwide organisms.]

eNature Field Guides. Available: http://www.enature.com/fieldguides/
Last accessed 28/03/15
[Available online set of guides to UK species.]

Natural Pathways: *Recommended Field Guides, Plants Trees and Animals.* Available: http://www.natural-pathways.

co.uk/book-field-guides-bc-7.php
Last accessed 28/03/15.
[Recommended Guide Books on the UK flora and fauna.]

University Library, University of Illinois at Urbana-Champaign. *International Field Guides.* Available: http://www.library.illinois.edu/bix/fieldguides/main.htm
Last accessed 28/03/15.
[A searchable database of field guides for plants and animals in North America and around the world.]

STUDY EXERCISES

40.1 Devise a key. The table below gives details of some invertebrate phyla from freshwater habitats, together with illustrations of 'representative' animals. The taxa are presented in no particular order. Use this information to construct a dichotomous key to help with preliminary identification of specimens. Ask a colleague to test your key by correctly assigning the animals to phyla from their pictures alone. Refine the key if problems are encountered during testing.

Members of some invertebrate phyla

Taxonomic division and description	Representative animal
Phylum Mollusca, Class Bivalvia *Soft-bodied with hard shell in two parts*	Zebra mussel, *Dreissena polymorpha*
Phylum Arthropoda Subphylum Arachnida *Hard-bodied with four pairs of jointed legs*	Water spider, *Argyroneta aquatica*
Phylum Mollusca, Class Gastropoda *Soft-bodied with hard shell in one part*	River snail, *Viviparus viviparus*
Phylum Platyhelminthes *Soft-bodied, free moving, flattened and unsegmented, without legs*	Flatworm, *Dugesia lugubris*
Phylum Arthropoda Subphylum Insecta *Hard-bodied with three pairs of jointed legs*	Great diving beetle, *Dytiscus marginalis*

(continued)

STUDY EXERCISES (*Continued*)

Taxonomic division and description	Representative animal
Phylum Annelida Class Oligochaetae *Soft-bodied, free-moving, divided into 15 or more segments (without head capsule or leg-like appendages)*	 Water worms, various spp.
Phylum Nematoda *Soft-bodied, free-moving, thread-like and unsegmented, without legs*	 Nematodes, various spp.
Phylum Arthropoda Subphylum Crustacea *Hard-bodied, with four or more pairs of jointed legs*	 Freshwater shrimp, *Crangonyx pseudogracilis*
Phylum Coelenterata Class Hydrozoa *Soft-bodied, surface-attached, unsegmented and tube-like, without legs*	 e.g. *Hydra* spp.

40.2 Use a standard flora or fauna. Visit your university library and take out one of the standard keys for the local flora or fauna. Find a specimen plant or animal whose identity you do not know (take care not to kill or destroy any protected specimens) and use the key and its descriptions to identify it as best you can. Check your identification with a tutor.

40.3 Search for Web-based identification keys and test them. Hypertext (Chapter 11) lends itself well to dichotomous or multiple choices, taking you to new pages via links based on appropriate options. Search for a Web-based identification key using a search engine and try it out.

Answers to these study exercises are available at www.pearsoned.co.uk/practicalskills

41 Identifying microbes

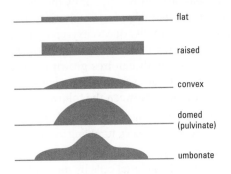

Fig. 41.1 Colony elevations (cross-sectional profile).

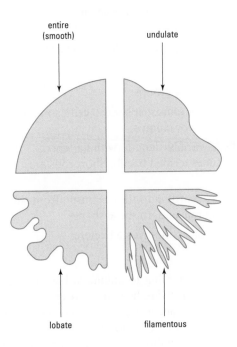

Fig. 41.2 Colony margins (surface view).

Most of the methods described in this chapter were developed for the identification of bacteria, and bacterial examples are used to illustrate the principles involved. While the basic techniques are equally applicable to other types of microbes, the identification systems for some protozoa, fungi and algae rely predominantly on microscopic appearance. Identification of viruses requires nucleic acid sequencing, electron microscopy or immunological techniques.

KEY POINT Identification of bacteria is often based on a combination of a number of different features, including growth characteristics, microscopic examination, physiological and biochemical characterisation, and, where necessary, immunological and nucleic acid tests.

Direct observation

Once a microbe has been isolated (Chapter 38) and cultured in the laboratory (Chapter 48), the visual appearance of individual colonies on the surface of a solidified medium may provide useful information. Bacteria typically produce smooth, glistening colonies, varying in diameter from <1 mm to >1 cm. Actinomycete colonies are often < 1 cm, with a shrivelled, powdery surface. Filamentous fungi usually grow as large, spreading mycelia composed of microscopic branching filaments (hyphae) with a matt appearance and are identified by microscopy, using the morphological characteristics of their reproductive structures. Yeasts produce small, glistening colonies; identification usually involves microscopy, combined with physiological and biochemical tests similar to those used for bacteria.

Colony characteristics

When measuring colony size, choose a typical colony, well spaced from any others, as colony size is affected by competition for nutrients. The characteristics of a microbial colony on a particular medium include:

- **Size:** some bacteria produce punctiform (pin-point) colonies, with a diameter of less than 1 mm, while motile bacteria may spread over the entire plate.
- **Form:** colonies may be circular, irregular, lenticular (spindle-shaped) or filamentous.
- **Elevation:** colonies may be flat, raised, convex, etc. (see Fig. 41.1).
- **Margin:** the edge of a colony may be entire (smooth) or more distinctive, e.g. undulate, lobate or filamentous (Fig. 41.2).
- **Consistency:** colonies may be viscous (or mucoid), butyrous (of similar consistency to butter) or friable (dry and granular), etc.
- **Colour:** some bacteria produce characteristic pigments. A few pigments are fluorescent under UV light.
- **Optical properties:** colonies may be translucent or opaque.
- **Haemolytic reactions on blood agar:** many pathogenic bacteria produce characteristic zones of haemolysis. Alpha haemolysis is a partial breakdown of the haemoglobin from the erythrocytes, producing a green zone around the colony, while beta haemolysis is the complete destruction of haemoglobin, producing a clear zone.

- **Odour:** some actinomycetes and cyanobacteria produce earthy odours, while certain bacteria and yeasts produce fruity or 'off' odours. However, odour is not a reliable characteristic in bacterial identification and its sensing is not recommended on safety grounds.

Microscopic examination

Bacteria are usually observed using an oil immersion objective at a total magnification of ×1000 (p. 274).

Motility

Wet mounts can be prepared by placing a small drop of bacterial suspension on a clean, degreased slide, adding a coverslip and examining the film without delay. For aerobes, areas near air bubbles or by the edge of the coverslip give best results, whereas anaerobes show greatest motility in the centre of the preparation, with rapid loss of motility due to oxygen toxicity.

Prepare wet mounts using young cultures in exponential growth in a liquid medium (p. 291). It is best to work with cultures grown at 20 or 25 °C, since those grown at 37 °C may not be actively motile on cooling to room temperature. It is essential to distinguish between the following:

- **True swimming motility,** due to the presence of flagella: bacteria dart around the field of view, changing direction in zigzag, tumbling movements.
- **Brownian motion:** non-motile bacteria show a localised, vibratory, random motion, due to bombardment of bacterial cells by molecules in the solution.
- **Passive motion,** due to currents within the suspension: all cells will be swept in the same direction at a similar rate of movement.
- **Gliding motility:** a slower, intermittent movement, parallel to the longitudinal axis of the cell, requiring contact with a solid surface.

Cell shape

Bacteria are subdivided into the following groups:

- **Cocci** (singular, coccus): spherical, or almost spherical, cells, sometimes growing in pairs (diplococci), chains or clumps.
- **Rods:** straight, cylindrical cells of variable length, with flattened, tapered or rounded ends; sometimes termed bacilli. Short rods are sometimes called cocco-bacilli.
- **Curved rods:** the curvature varies according to the organism, from short curved rods, sometimes tapered at one end, to spiral shapes.
- **Branched filaments:** characteristic of actinomycete bacteria.

Gram staining

This is the most important differential staining technique in bacteriology (Box 41.1 gives details). It enables us to divide bacteria into two distinct groups, Gram-positive and Gram-negative, according to a particular staining procedure (the technique is given a capital letter, since it is named after its originator, H.C. Gram). The basis of the staining reaction is the different structure of the cell walls of Gram-positive and Gram-negative bacteria. Heat fixation of air-dried bacteria causes some shrinkage, but cells retain their shape: to measure cell dimensions use a chemical fixative (p. 265).

Gram staining should be carried out using light smears of young, active cultures, since older cultures may give variable results. In particular, certain Gram-positive bacteria may stain Gram-negative if older cultures are used.

Using the hanging drop technique – place a drop of bacterial suspension on a coverslip and invert over a cavity slide so that the drop does not make contact with the slide: motile aerobes are best observed at the edge of the droplet, where oxygen is most abundant.

Assessing motility – if you have not seen bacterial motility before, it is worth comparing your unknown bacterium with a positive and a negative control.

Using cell shape in microbial identification – many bacteria are pleomorphic, varying in size and shape according to the growth conditions and the age of the culture: thus, other characteristics are required for identification.

SAFETY NOTE The Gram-staining procedure involves toxic dyes and flammable solvents: avoid skin contact and extinguish any naked flames (e.g. Bunsens).

Assessing the Gram status of an unknown bacterium – if a pure culture gives both Gram-positive and Gram-negative cells, identical in size and shape, it can be regarded as a Gram-positive organism that is demonstrating Gram-variability.

Box 41.1 Preparation of a heat-fixed, Gram-stained smear

Preparation of a heat-fixed smear

The following procedure will provide you with a thin film of bacteria on a microscope slide, for staining.

1. Take a clean microscope slide and pass it through a Bunsen flame twice, to ensure it is free of grease. Allow to cool.
2. Using a sterile inoculating loop, place a single drop of water in the centre of the slide and then mix in a small amount of sample from a single bacterial colony with the drop, until the suspension is slightly turbid. Smear the suspension over the central area of the slide, to form a thin film. For liquid cultures, use a single drop of culture fluid, spread in a similar manner.
3. Allow to air-dry at room temperature, or high above a Bunsen flame: air-drying must proceed gently, or the cells will shrink and become distorted.
4. Fix the air-dried film by passage through a Bunsen flame. Using a slide holder or forceps, pass the slide, film side up, rapidly through the hottest part of the flame (just above the blue cone). The temperature of the slide should be just too hot for comfort on the back of your hand: note that you must not overheat the slide or you may burn yourself (you will also ruin the preparation).
5. Allow to cool: the smear is now ready for staining.

Gram-staining procedure

The version given here is a modification of the Hucker method, since acetone is used to decolorise the smear. Note that some of the staining solutions used are flammable, especially the acetone decolorising solvent: you must make sure that all Bunsens are turned off during staining. The procedure should be carried out with the slides suspended over a sink, using a staining rack.

1. Flood a heat-fixed smear with 2% w/v crystal violet in 20% v/v ethanol : water and leave for 1 min.
2. Pour off the crystal violet and rinse briefly with tap water. Flood with Gram's iodine (2 g KI and 1 g I_2 in 300 mL water) for 1 min.
3. Rinse briefly with tap water and leave the tap running gently.
4. Tilt the slide and decolorise with acetone for 2-3 s: acetone should be added dropwise to the slide until no colour appears in the effluent. This step is critical, since acetone is a powerful decolorising solvent and must not be left in contact with the slide for too long.
5. Immediately immerse the smear in a gentle stream of tap water, to remove the acetone.
6. Pour off the water and counterstain for 10–15 s using 2.5% w/v safranin in 95% v/v ethanol: water.
7. Pour off the counterstain, rinse briefly with tap water, then dry the smear by blotting gently with absorbent paper: all traces of water must be removed before the stained smear is examined microscopically.
8. Place a small drop of immersion oil on the stained smear: examine directly (without a coverslip) using an oil-immersion objective (p. 274).

Gram-positive bacteria retain the crystal violet (primary stain) and appear purple while Gram-negative bacteria are decolorised by acetone and counterstained by the safranin, appearing pink or red when viewed microscopically.

Other decolorising solvents are sometimes used, including ethanol : water, ethyl ether: acetone and acetone : alcohol mixtures. The time of decolorisation must be adjusted, depending upon the strength of the solvents used, e.g. 95% v/v ethanol : water is less powerful than acetone, requiring around 30s to decolorise a smear.

Staining bacterial spores – these are relatively impermeable to dyes, and require heat treatment and specialised stains (p. 268).

This Gram-variability is due to autolytic changes in the cell wall of Gram-positive bacteria. Developing spores are often visible as unstained areas within older vegetative cells of *Bacillus* and *Clostridium*. Other stains are required to demonstrate sub-cellular structures such as, capsules or flagella (p. 268).

Basic laboratory tests

At least two simple biochemical tests are usually performed.

Oxidase test

Performing the oxidase test – never use a nichrome wire loop, as this will react with the oxidase reagent, giving a false positive result.

This identifies cytochrome *c* oxidase, an enzyme found in obligate aerobic bacteria. Soak a small piece of filter paper in a fresh solution of 1% (w/v) *N*-*N*-*N*′-*N*′-tetramethyl-*p*-phenylenediamine dihydrochloride on a clean microscope slide. Rub a small amount from the surface of a young, active colony on to the filter paper using a glass rod, a *plastic* loop or a wooden applicator stick: a purple-blue colour within 10 s is a positive result.

Avoiding false negatives – ensure you use sufficient material during oxidase and catalase testing, otherwise you may obtain a false negative result: a clearly visible 'clump' of bacteria should be used.

Carbohydrate utilisation tests and isolation media – many diagnostic agar-based media incorporate one or more specific carbohydrates and pH indicator dyes, thereby providing additional information as part of the isolation procedure.

Understanding biochemical identification – given the variability of different isolates of a single species, a large number of tests must be used to provide a high degree of probability of correctly identifying the isolate. This is often carried out in specialised labs using computer-aided analysis.

Catalase test

This identifies catalase, an enzyme found in obligate aerobes and in most facultative anaerobes, which catalyses the breakdown of hydrogen peroxide into water and oxygen ($2H_2O_2 \rightarrow 2H_2O + O_2$). Transfer a small sample of your unknown bacterium on to a coverslip using a disposable plastic loop or glass rod. Invert on to a drop of hydrogen peroxide: the appearance of bubbles within 30 s is a positive reaction. This method minimises the dangers from aerosols formed when gas bubbles burst.

The oxidase and catalase tests effectively allow us to subdivide bacteria on the basis of their oxygen requirements, without using agar shake cultures (p. 232) and overnight incubation, since, in broad terms:

- obligate aerobes will be oxidase-positive and catalase-positive;
- facultative anaerobes will be oxidase-negative and catalase-positive;
- microaerophilic bacteria, aerotolerant anaerobes and strict (obligate) anaerobes will be oxidase-negative and catalase-negative – the latter group will grow only under anaerobic conditions (p. 232).

Once you have reached this stage (colony characteristics, motility, shape, Gram reaction, oxidase and catalase status), it may be possible to make a tentative identification, at least for certain Gram-positive bacteria, at the generic level. To identify Gram-negative bacteria, particularly the oxidase-negative, catalase-positive rods, further tests are required.

Identification tables: further laboratory tests

Bacteria are asexual organisms and strains of the same species may give different results for individual biochemical/physiological tests. This variation is allowed for in identification tables (multi-access keys, pp. 242–3), based on the results of a large number of tests. Identification tables are often used for particular subgroups of bacteria, after Gram staining and basic laboratory tests have been performed: an example is shown in Table 41.1.

A large number of specific biochemical and physiological tests are used in bacterial identification including:

- **Carbohydrate utilisation tests.** Some bacteria can use a particular carbohydrate as a carbon and energy source. Acidic end products can be identified using a pH indicator dye (p. 166) while CO_2 is detected in

Table 41.1 Identification table for selected Gram-negative rods

Bacterium	Biochemical test								
	1	2	3	4	5	6	7	8	9
Escherichia coli	v	+	−	+	−	v	v	+	−
Proteus mirabilis	−	−	v	−	+	−	+	−	+
Morganella morganii	−	−	−	−	+	−	+	+	−
Vibrio parahaemolyticus	−	+	v	−	−	+	+	+	−
Salmonella spp.	−	+	v	−	−	+	+	−	+

Key to biochemical tests and symbols:

1. sucrose utilisation
2. mannitol utilisation
3. citrate utilisation
4. β-galactosidase activity
5. urease activity
6. lysine decarboxylase activity

7. ornithine decarboxylase activity
8. indole production
9. H_2S production
+, >90% of strains tested positive
−, <10% of strains tested positive
v, 10–90% of strains tested positive

Fig. 41.3 Durham tube in carbohydrate utilisation broth. Air within the Durham tube is replaced by broth during the autoclaving procedure.

liquid culture using an inverted small test tube (Durham tube, Fig. 41.3). Aerobic breakdown (via respiration) is termed oxidation, and anaerobic breakdown is known as fermentation. Identification tables usually incorporate tests for several different carbohydrates, e.g. Table 41.1.

- **Enzyme tests.** Most of these incorporate a substance that changes colour if the enzyme is present, e.g. a pH indicator, or a chromogenic substrate.
- **Tests for specific end products of metabolism,** e.g. the production of indole due to the metabolic breakdown of the amino acid tryptophan, or H_2S from sulphur-containing amino acids.

Identification kits

Some biochemical tests are now supplied in kit form, e.g. the API 20E system incorporates 20 tests within a sterile plastic strip (Fig. 41.4). After inoculation and overnight incubation, the results of the tests are converted into a seven-digit code, for comparison with known bacteria using either a computer program (the Analytical Profile Index), or a reference book. While kit identification systems save time and labour, they are more expensive and less flexible than conventional biochemical tests.

Immunological tests

Tests used in diagnostic microbiology include:

- **Agglutination tests:** based on the reaction between specific antibodies and a particular bacterium (p. 378). These tests are particularly useful for subdividing biochemically similar bacteria.
- **Fluorescent antibody tests:** the reaction between a labelled antibody and a particular bacterium can be visualised using UV microscopy. The direct fluorescent antibody test uses fluorescein isothiocyanate as the label.
- **Enzyme-linked immunoassay tests** using antibodies labelled with a particular enzyme, e.g. the double antibody sandwich ELISA or competitive ELISA tests (p. 381).

While such tests can give specific and accurate confirmation of the identity of a bacterium under controlled laboratory conditions, they are often too expensive and time-consuming for routine identification purposes, especially when large numbers of tests are required.

Typing methods

The identification of bacteria at subspecies level is known as typing: this is usually done in a specialist laboratory, e.g. as part of an epidemiological study to establish the source of an infection. Various methods are used:

- **Antigen typing** or serotyping is based on immunological tests.
- **Phage typing** is based on the susceptibility of different strains to certain bacterial viruses (phages).

Molecular approaches to microbial identification – several novel methods of detection and identification are based on nucleic acid techniques, including use of the PCR and Southern blotting to detect particular microbes (pp. 403–5).

Using antibiotic resistance to identify bacteria – patterns of resistance to antibacterial agents in lab tests can be used to generate an 'antibiogram', giving information that can aid identification, as well as providing data of clinical significance, e.g. for therapy of bacterial infections.

Practical applications of bacterial typing – *E. coli* O157:H7 is a serotype of this bacterium that is capable of causing severe human disease: it can be identified on the basis of an agglutination reaction with an appropriate antiserum. Typing is also discussed on p. 238.

Fig. 41.4 Example of a bacterial identification kit (API 20E). Property of bioMerieux S.A/Andrea Bannuscher.

- **Biotyping** is based on biochemical differences between different strains e.g. enzyme profiles or antibiotic resistance screening ('antibiograms').
- **Bacteriocin typing:** bacteriocins are proteins released by bacteria that inhibit the growth of other members of the same species.

Sources for further study

Barrow, G.I. and Feltham, R.K.A. (2004) *Cowan and Steel's Manual for the Identification of Medical Bacteria,* 3rd edn. Cambridge University Press, Cambridge. [A standard reference work for the identification of clinically important bacteria.]

Cullimore, D.R. (2010) *Practical Atlas for Bacterial Identification,* 2nd edn. CRC Press, Boca Raton. [Detailed coverage of practical methods applicable to a wide range of bacteria.]

Fox, A. (ed.) *Journal of Microbiological Methods,* Elsevier, London (available through Science Direct at: http://www.sciencedirect.com/). [Provides information on novel developments in all aspects of microbiological methods, including microbial culture/isolation and molecular approaches.]

Macfaddin, J.F. (2000) *Biochemical Tests for the Identification of Medical Bacteria,* 3rd edn. Williams & Wilkins, Philadelphia. [Explains the operating principles underlying most of the biochemical tests in routine use in diagnostic bacteriology.]

Richardson, M.D. and Warnock, D.W. (2012) *Fungal Infections: Diagnosis and Management*, 4th edn. Wiley, New York.

Spencer, J.F.T. and Spencer, A.L.R. (2010) *Environmental Microbiology: Methods and Protocols.* Humana Press, New York.

STUDY EXERCISES

41.1 Describe the colonial characteristics of selected microbes. Either research the features of the following microbes (e.g. via the Web), or look at well-isolated individual colonies in the laboratory, following overnight growth on a suitable medium:

(a) *Escherichia coli;*

(b) *Pseudomonas aerogenes;*

(c) *Staphylococcus aureus;*

(d) *Streptococcus pneumoniae;*

(e) *Bacillus subtilis;*

(f) *Saccharomyces cerevisiae.*

41.2 Find out how some of the tests used in microbial identification work. Research the operating principles that underpin the following tests, using microbiological textbooks or the Web, and prepare brief notes explaining how each of the following tests works:

(a) oxidase test;

(b) indole test;

(c) β-galactosidase test;

(d) urease test;

(e) lysine decarboxylase test;

(f) H_2S production.

41.3 Identify the following oxidase-negative, catalase-positive, Gram-negative rod-shaped bacteria, using Table 41.1.

(a) Positive for citrate, urease, ornithine decarboxylase and H_2S only.

(b) Positive for sucrose, mannitol, β-galacto- sidase, lysine decarboxylase and indole only.

(c) Positive for mannitol, citrate, lysine decarboxylase, ornithine decarboxylase and H_2S only.

(d) Positive for urease, ornithine decarboxylase and indole only.

(e) Positive for mannitol, β-galactosidase and indole only.

Answers to these study exercises are available at www.pearsoned.co.uk/practicalskills

Examining specimens and samples

42 The purpose and practice of dissection

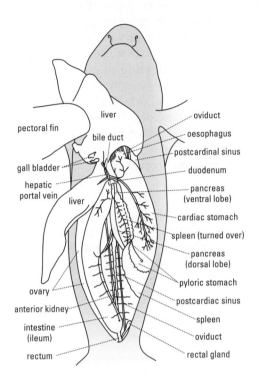

liver
pectoral fin
bile duct
gall bladder
hepatic portal vein
liver
ovary
anterior kidney
intestine (ileum)
rectum

oviduct
oesophagus
postcardinal sinus
duodenum
pancreas (ventral lobe)
cardiac stomach
spleen (turned over)
pancreas (dorsal lobe)
pyloric stomach
postcardiac sinus
spleen
oviduct
rectal gland

Fig. 42.1 Labelled diagram of abdomen of young female dogfish. Good drawings are essential to get the most out of dissections (see Chapter 30).

Dissection in zoology involves the display or removal of parts of any dead animal, whereas vivisection is an operation on a living animal. Dissection is usually associated with teaching animal structure, but the skills involved are used widely within the life sciences for:

- **investigating anatomy and morphology;**
- **making physiological preparations** of nerves, muscles and other organs;
- **investigating parasites** in various body organs;
- **removing specific body organs/tissues for chemical analysis;**
- **investigating reproductive status;**
- **removing organs/tissues for histological/histochemical investigation;**
- **manipulating living material**, as in grafting processes and preparing diagrams to show selected features (Fig. 42.1).

There is considerable debate about the ethics of using animals for dissection (p. 134) but this must be distinguished from that of vivisection (Chapter 21). It is part of a complex and emotive issue, but one viewpoint is that dissection is required for effective zoology teaching: experience of dissection at an *appropriate* stage of the curriculum is enlightening and also teaches an essential technical skill. However, if dissection is definitely not for you, it is best to discover this as early as possible in your career.

> **KEY POINT** The primary objectives of dissection are:
> - personal exploration of animal structure and function;
> - development of manipulative skills;
> - production of reports based upon personal observation.

The ground rules of dissection

There are some important rules to be considered if dissection is to be an acceptable procedure in zoological teaching.

Humane treatment

The use of any animal, whatever its level of organisation, for experimentation or dissection must be a considered act with due regard for humane treatment and killing (see Chapter 21). Remember there are very specific regulations for the use of vertebrate species and some higher invertebrates: check with your supervisor.

Maximum benefit

Any animal should be used for as many investigations as possible.

Preparation

Prepare for the exercise by ensuring that practical schedules and relevant texts are consulted before attempting a dissection. If you have to miss a practical, inform the organiser in advance to prevent animals being killed or prepared unnecessarily.

Your responsibility – because of the ethical aspect to dissection, you have a particular responsibility to attend to and make careful and effective use of a dissection specimen. Chapter 21 considers the broader aspects of the use of animals in scientific research and study.

Types of dissection

Dissection can be carried out at three levels of sophistication, related directly to the size of the organism or structure being investigated. Gross dissection of large organisms requires equipment very different from that used for 'normal' dissection in that large knives replace scalpels, etc. Normal-scale dissection involves creatures from a dog down to an earthworm, and requires equipment typical of commercial dissection kits/instruments. Fine-scale dissection for the removal or display of organs, glands, etc., from small animals usually requires only mounted needles, fractured glass edges for cutting and a dissecting microscope or magnifier. Fine-scale dissection requires extensive practice, and the preparation of material (relaxation and fixation) is much more critical than for larger specimens.

Equipment

Table 42.1 lists basic equipment required for normal dissection: some comments on use are given below (see also Fig. 42.2). Commercial dissection kits often contain inappropriate components and you should buy your instruments individually from specialist suppliers, if possible. A set of equipment for fine-scale dissections can be assembled easily to your own specifications.

Table 42.1 List of basic equipment recommended for dissection

Quantity	Description
1	All-metal scalpel, stainless steel, 45 mm blade
1 each	Swann–Morton scalpel handles, sizes 3 and 4
1 each	Swann–Morton blades, packets of nos. 10, 11, 12, 15, 22 and 24
2	Dissecting needles, straight, stainless steel
2	Blunt seekers, stainless steel: metal handles
1	Fine forceps, blunt points, stainless steel, 112 mm length
1	Coarse forceps, blunt points, stainless steel, 112 mm length
1	Coarse scissors, open shanks, straight points, stainless steel, 150 mm length
1	Fine scissors, open shanks, straight fine points, stainless steel, 110 mm length
1	Section lifter

Scalpels

There are two basic types of scalpels, fixed blade and replaceable blade. Buy at least one 45-mm blade length solid forged scalpel, preferably made of stainless steel: use this for coarser cutting procedures as it can be resharpened using an oil-stone. The Swann–Morton scalpel comprises a handle and a disposable blade: blades come in a variety of shapes and sizes (Fig. 42.2), curved-edge ones being the most useful. Blades fit only specific handle sizes so make sure that they match.

Forceps

Buy at least one each of coarse and fine stainless steel forceps. The latter are very delicate and you should check that the points meet precisely before purchasing. Look after them very carefully as the points are easily damaged: use only on soft tissues. Use large and small forceps for general purposes but keep fine ones for delicate work only.

SAFETY NOTE Taking care with dissection equipment – sharp blades should be used with extreme caution and disposed of immediately after use in a sharps container.

v. fine   fine   coarse   curved

forceps types

Swann–Morton scalpel blade types

Fig. 42.2 Variations in design of dissection equipment.

Dissecting scissors

Buy two pairs of stainless steel, pointed scissors, one medium-large for coarse work and cutting small bones and one fine pair for delicate work. Points and edges are easily damaged and require professional sharpening when repair is needed, so use carefully. A pair of bone cutters is optional but do *not* use scissors to cut large bones.

Dissecting (mounted) needles

Buy at least two with metal handles and protect the points; use them for dissecting membranes in areas where damage is acceptable. Needles have many non-dissection functions associated with other fine manipulative techniques.

Dissecting seekers

These have blunt points and should have metal handles. Use for breaking down membranes holding delicate organs together. Again, they have multiple uses.

Recommended accessory equipment

A teat (Pasteur) pipette is valuable for washing delicate organs and tissues. A camel-hair brush is useful for removing material from delicate structures. Pins and awls are usually provided by your department and are used for fixing the dissection specimen to the board (awls) or wax dish (pins): small pins are invaluable for pinning organs aside for display purposes. For measurement, and as an aid to drawing, a pair of dividers is very useful, as is a small steel ruler.

Carrying out the dissection

The basic sequence of steps in a typical dissection is shown in Box 42.1 and Fig. 42.3, while Box 42.2 gives tips to help you improve your dissection technique.

Virtual dissections

For both practical and ethical reasons, a number of virtual dissections have been developed in recent years. Examples include the 'virtual earthworm' dissection published by McGraw-Hill; the 'virtual frog dissection' by Robertson *et al.;* and the 'virtual pig' dissection by Fleck *et al.* These allow you to view an animal in the process of dissection (e.g. Fig. 42.4), with access to a range of images and substitute techniques. Although they can never replicate exactly the experience and skills required for 'real' dissection, these programs do offer some advantages:

- **Everyone sees the same specimen.**
- **There is no need to breed or kill additional animals.**
- **They are suitable for large classes, where resources might be limiting.**
- **A range of images can be included, including diagrams, photographs of real dissections and microscopy.**
- **It is possible to 'undissect' the specimen and 'redo' the dissection.**
- **Key organs and other parts can be highlighted and labelled.**

- **Links can be made to tutorials and explanations about functions of key parts.**
- **Online quizzes can be provided to check understanding.**

Virtual dissections may be used as a substitute for dissection or as preparation to ensure you get the most out of a real dissection.

Box 42.1 **Basic stages of an animal dissection**

Fig. 42.3 Flowchart for dissection.

The basic sequence of steps is outlined in Fig. 42.3.

1. **The animal should have been killed as humanely and as recently as possible:** this will probably be done for you by the class supervisor or technician according to the rules appropriate to the type of animal. The method of killing should be chosen carefully to keep the specimen relaxed. If preserved material is used, wash out excess fixative thoroughly before dissecting.

2. **Orient the specimen carefully;** determine the dorsal/ventral, anterior/posterior or other oral/aboral axes (see Fig. 46.4) and work out the correct orientation of the specimen for dissection. Invertebrates are usually dissected from the dorsal surface (the nerve cord being ventral in position) and vertebrates from the ventral surface (the nerve cord being dorsal in position): however, special objectives may require a different orientation.

3. **Open the body cavity carefully.** This is usually done using forceps to lift the skin away from underlying organs while using a fine pair of scissors to make an initial opening; the scissors are then used to extend this opening in anterior–posterior and lateral directions until the skin flaps can be pinned back. Pin out by placing tension on the skin flaps, breaking down any restricting membranes using a seeker.

4. **Subsequent dissection procedure depends upon your objectives.** To display the system being investigated as clearly and neatly as possible requires:

   (a) identification of organs, blood vessels and nerves initially visible;

   (b) separation of these structures from the membranes that hold them in place. This is best achieved using a blunt instrument such as a seeker to avoid damage;

   (c) removal or displacement of organs that obscure parts of the system you wish to display. Displacement using pins to hold the organ in position is the preferred option when possible;

   (d) if your objective is to display a blood system or a nervous system, you must follow individual vessels/nerves from their point of origin to their destination organ; do this using a blunt seeker to remove covering membranes by working carefully along the structure. Do not pull sideways during this procedure as this often results in breakages; such tissues are usually much stronger in the direction of their length than when pulled laterally.

   (e) tidy up loose pieces and wash away residual blood, etc.; use more pins to finalise the display of the system and then make notes and drawings as necessary (see e.g. Figs 30.3 and 42.1).

| Box 42.2 | Tips for improving dissection technique |

The following tips should help you develop your dissection technique:

- Use a blunt seeker and controlled tension on the tissue to remove the connective tissue that binds tissues and organs together: use scalpels sparingly.
- Insert pins/awls obliquely so that they do not interfere with further dissection. In segmented animals such as earthworms, fine pins can be used to mark the position of specific segments.
- Dissect most invertebrates under water: the water buoys up the tissues/organs and assists dissection. Change the water if it becomes clouded but do not allow flowing water to run directly on to the specimen as it will damage delicate structures.
- Dissect vertebrates and larger invertebrates in air on a dissection board, using cotton-wool swabs to remove excess blood and other body fluids.
- Keep tissues under tension while dissecting, but avoid damaging them with forceps/fingers.
- Cut away from the organs when using scalpels or scissors and keep scissor points away from structures you are attempting to free. This is essential when opening the body cavities of small animals.

- Use appropriately sized equipment: e.g. do not attempt to use large scissors to open an earthworm. Never use delicate instruments for coarse work.
- *Never* remove anything until you know what you are removing.
- Dissect along structures and not across them, particularly for tubular structures such as nerves and blood vessels.
- Use fresh material whenever possible. Fixatives make tissues more brittle and inelastic: alcohol storage tends to harden skin, muscle and connective tissue.
- Keep your instruments clean and sharp: you cannot dissect with blunt or dirty equipment. Dry instruments after washing and wipe with an oily cloth. Cut nothing but tissues with scalpels and scissors. Do not sharpen pencils with scalpels or stick mounted needles into the bench or the dissecting board.
- Be hygienic. Wear rubber gloves if you have cuts or lesions on your hands. Wash hands and equipment thoroughly when finished. Dispose of animal remains carefully as instructed.

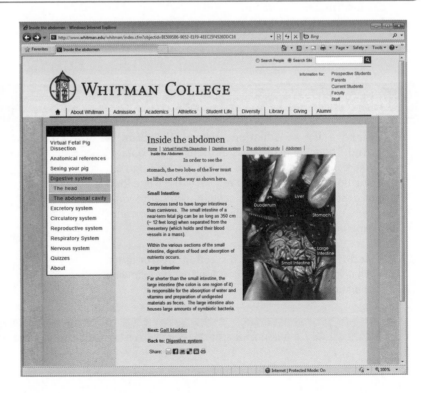

Fig. 42.4 Screenshot from the 'virtual fetal pig' dissection at http://www.whitman.edu/content/virtualpig. Reproduced with permission.

Text references

Fleck, E.W. *et al. Virtual Fetal Pig Dissection.* Available: http://www.whitman.edu/biology/virtualpig Last accessed 30/03/15.

McGraw-Hill Biology: *Virtual Lab Earthworm Dissection.* Available: http://www.mhhe.com/biosci/genbio/virtual_labs/BL_14/BL_14.html Last accessed 30/03/15.

Sources for further study

About Inc. *About Biology: Online dissections.* Available: http://biology.about.com/od/onlinedissections/Online_ Dissections.htm Last accessed 30/03/15.
[A compendium of Web-based dissections.]

Fishbeck, D. and Sebastini, A. (2008) *Comparative Anatomy: Manual of Vertebrate Dissection,* 2nd edn. Morton Publishing Co., Englewood, NJ.

Kardong, K.V. and Zalisko, E.J. (2005) *Comparative Vertebrate Anatomy: A Laboratory Dissection Guide*, 4th edn. McGraw-Hill, Boston.

Microscopy UK. *Earthworm Anatomy and Dissection.* Available: http://www.bideggcorner.com/worksheets/earthworm_dissection.html Last accessed 30/03/15.

Robertson, D., Johnston, W. and Nip, W. *Virtual Frog Dissection: Interactive 3D Graphics Via the Web.* Available: http://froggy.lbl.gov/papers/WWW.94/paper.html Last accessed 31/10/11.

Walker, W.F. and Homberger, D. (1999) *Anatomy and Dissection of the Rat,* 3rd edn. W.H. Freeman, New York.

Whitehouse, R.H. and Grove, A.J. (1947) *The Dissection of the Crayfish.* University Tutorial Press, London.
[Part of a classic series including manuals on frog, rabbit, dogfish, earthworm and cockroach. Includes lots of information on how to carry out detailed dissections.]

STUDY EXERCISES

42.1 Make a list of situations where dissection is used. There are many areas of zoology (and botany) where dissection technique is required. List up to ten situations where you would require to use some form of dissection technique to acquire material to work with, or to carry out manipulations before or after experiments.

42.2 Decide on appropriate dissection instruments. What instruments should you use for the following operations:

(a) following a nerve or blood vessel from source to ending?

(b) opening up the body cavity of the earthworm?

(c) removing or severing muscle tissue in the body wall of a vertebrate?

42.3 Research general rules of dissection. What is the general rule for deciding on which side a general dissection should be commenced: (a) for invertebrates; and (b) for vertebrates?

Answers to these study exercises are available at www.pearsoned.co.uk/practicalskills

43 Introduction to microscopy

Many features of interest in biological systems are too small to be seen by the naked eye and can only be observed with a microscope. All microscopes consist of a coordinated system of lenses arranged so that a magnified image of a specimen is seen by the viewer (Figs 43.1 and 43.2). The main differences are the wavelengths of electromagnetic radiation used to produce the image, the nature and arrangement of the lens systems and the methods used to view the image.

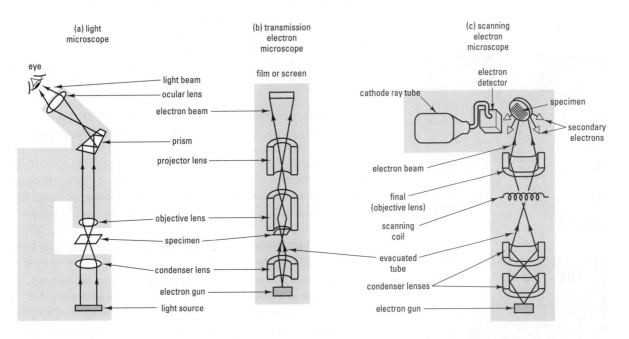

Fig. 43.1 Simplified diagrams of light and electron microscopes. Note that the electron microscopes are drawn upside-down to aid comparison with the light microscope.

Microscopes allow objects to be viewed with increased resolution and contrast. Resolution is the ability to distinguish between two points on the specimen – the better the resolution, the 'sharper' the image. This is affected by lens design and inversely related to the wavelength of radiation used. Contrast is the difference in intensity perceived between different parts of an image. This can be enhanced (a) by the use of stains, and (b) by adjusting microscope settings, usually at the expense of resolution.

The three main forms of microscopy are light microscopy, transmission electron microscopy (TEM) and scanning electron microscopy (SEM). Their main properties are compared in Table 43.1 and their suitability for observing cells and organelles is shown in Table 43.2.

Light microscopy

Two forms of the standard light microscope, the binocular (compound) microscope and the dissecting microscope, are described in detail in Chapter 45. These are the instruments most likely to be used in routine practical work. Figure 43.2(a) shows a typical image from a light microscope. In more advanced project work, you may use one or more

Table 43.1 Comparison of microscope types. Resolution is that obtained by a skilled user. LM, light microscope; SEM, scanning electron microscope; TEM, transmission electron microscope

Property	Type of microscope		
	LM	TEM	SEM
Resolution	200 nm	1 nm	10nm
Depth of focus	Low	Medium	High
Field of view	Good	Limited	Good
Specimen preparation (ease)	Easy	Skilled	Easy
Specimen preparation (speed)	Rapid	Slow	Quite rapid
Relative cost of instrument	Low	High	High

Fig. 43.2 Examples of images of a similar specimen (the stomatal complex of *Commelina communis* L.) obtained using different microscopic techniques: (a) light microscopy (surface view); (b) transmission electron microscopy (transverse section through guard cell pair at mid pore); and (c) scanning electron microscopy (surface view). As an indication of scale, the width of a single guard cell is about 10 μm.

of the following more sophisticated variants of light microscopy to improve image quality:

- **Dark-field illumination** involves a special condenser that causes reflected and diffracted light from the specimen to be seen against a dark background. The method is particularly useful for near-transparent specimens and for delicate structures like flagella. Care must be taken with the thickness of slides used – air bubbles and dust must be avoided and immersion oil must be used between the dark-field condenser and the underside of the slide.

- **Ultraviolet microscopy** uses short-wavelength UV light to increase resolution. Fluorescence microscopy uses radiation at UV wavelengths to make certain fluorescent substances (e.g. chlorophyll or fluorescent dyes that bind to specific cell components) emit light of visible wavelengths. Special light sources, lenses and mountants are required for UV and fluorescence microscopy, and filters must be used to prevent damage to users' eyes.

- **Phase-contrast microscopy** is useful for increasing contrast when viewing transparent specimens. It is superior to dark-field microscopy because a better image of the interior of specimens is obtained. Phase contrast operates by causing constructive and destructive interference effects in the image, visible as increased contrast. Adjustments must be made, using a phase telescope in place of the eyepiece, for each objective lens and a matching phase condenser, and the microscope must be set up carefully to give optimal results.

- **Nomarski or differential interference contrast (DIC) microscopy** gives an image with a three-dimensional quality. However, the relief seen is optical rather than morphological, and care should be taken in interpreting the result. One of the advantages of the technique is the extremely limited depth of focus that results: this allows 'optical sectioning' of a specimen.

- **Polarised-light microscopy** can be used to reveal the presence and orientation of optically active components within specimens (e.g. starch grains, cellulose fibres), showing them brightly against a dark background.

- **Confocal microscopy** allows three-dimensional views of cells or thick sections. A finely focused laser is used to create electronic images of layered horizontal 'slices', usually after fluorescent staining. Images can be viewed individually or reconstructed to provide a 3-D computer-generated image of the whole specimen.

Electron microscopes

Electron microscopes offer an image resolution up to 200 times better than light microscopes (Table 43.1) because they utilise radiation of shorter wavelength in the form of an electron beam. The electrons are produced by a tungsten filament operating in a vacuum and are focused by electromagnets. TEM and SEM differ in the way in which the electron beam interacts with the specimen: in TEM, the beam passes through the specimen (Fig. 43.1(b)), while in SEM the beam is scanned across the specimen and is reflected from the surface (Fig. 43.1(c)). In both cases, the beam must fall on a fluorescent screen before the image can be seen. Permanent images ('electron micrographs') are produced after focusing the beam on photographic film (Figs 43.2(b) and (c)).

Table 43.2 Dimensions of some typical cells and organelles with an indication of suitable forms of microscopy for observing them. LM, light microscope; SEM, scanning electron microscope; TEM, transmission electron microscope. Column 2 data after Rubbi (1994)

Cell or organelle	Approximate diameter or width (µm)	Suitable form of microscopy
Prokaryote cell	0.15–5	LM, SEM, TEM
Eukaryote cell	10–100	LM, SEM, TEM
Fungal hypha	5–20	LM, SEM, TEM
Nucleus	5–25	LM, SEM, TEM
Mitochondrion	1–10	SEM, TEM
Chloroplast	2–8	LM, SEM, TEM
Golgi apparatus	1	SEM, TEM
Lysosome/ peroxisome	0.2–0.5	SEM, TEM
Plant cell wall	0.1–10	LM, SEM, TEM

You are unlikely to use either type of electron microscope as part of undergraduate practical work because of the time required for specimen preparation and the need for detailed training before these complex machines can be operated correctly. However, electron microscopy is extremely important in understanding cellular and subcellular structures, and you may be shown electron micrographs with one or more of the following objectives:

- **to demonstrate cell ultrastructure (TEM);**
- **to show surface features of organisms,** e.g. when surface features are coated or when cells are 'freeze-fractured', then coated **(SEM);**
- **to investigate changes in the number, size, shape and condition of cells and organelles (TEM);**
- **to carry out quantitative studies of cell and organelle disposition (TEM).**

Aspects of the interpretation of electron micrographs are dealt with in Chapter 46.

Preparative procedures

Without careful preparation of the material being studied, the biological structures viewed with any microscope can be rendered meaningless due to artefacts. Figure 43.3 summarises the processes involved for the main types of microscopy discussed above. The processes involved in preparing material for light microscopy are outlined in Chapter 44.

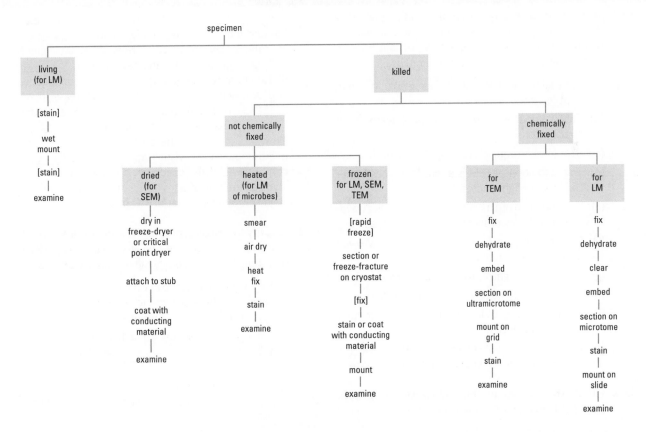

Fig. 43.3 Flowchart of procedures necessary to prepare specimens for different forms of microscopy. Steps enclosed in brackets are optional. LM, light microscope; SEM, scanning electron microscope; TEM, transmission electron microscope.

Text reference

Rubbi, C.P. (1994) *Light Microscopy Essential Data.* Wiley, Chichester.

Sources for further study

Bradbury, S. (1989) *An Introduction to the Optical Microscope*, 2nd edn. Bios Scientific Publishers, Abingdon.

Davidson, M.W. and Abramowitz, M. *Molecular Expressions. Exploring the World of Optics and Microscopy.* Available: http://micro.magnet.fsu.edu/ Last accessed 19/02/15.
[Covers many areas of basic knowledge underlying microscopy. Includes a microscopy primer.]

John Innes Centre *About Microscopy.* Available: https://www.jic.ac.uk/Microscopy/about.html Last accessed 19/02/15
[Discusses light microscopy, electron microscopy and scale and provides links to a range of other websites.]

Mertz, J. (2009) *Introduction to Optical Microscopy.* Roberts and Co., Greenwood Village.

Murphy, D.B. (2012) *Fundamentals of Light Microscopy and Electronic Imaging*, 2nd edn. Wiley- Liss, New York.

van Egmond, W. and Walker, D. *A Virtual Pond-dip.* Available: http://www.microscopy-uk.org.uk/index.html? http//www.microscopy.uk.org.uk/ponddip Last accessed 19/02/15.
[Introduction to things you can see in pond water using a standard light microscope.]

STUDY EXERCISES

43.1 Test your microscopy knowledge. Indicate whether the following statements about light microscopy, scanning electron microscopy (SEM) or transmission electron microscopy (TEM) are true or false.

(a) TEM allows you to see at finer resolution than light microscopy.

(b) TEM allows you to see surface features of specimens.

(c) SEM always requires staining of specimens.

(d) The resolution of TEM is about 200 times better than that of light microscopy.

(e) The resolution of a microscope is linked to the wavelength of electromagnetic radiation employed.

(f) The specimen in both TEM and SEM is viewed under near-vacuum conditions.

(g) Specimens for light microscopy can be living or dead.

(h) SEM provides better resolution than TEM.

(i) The depth of focus in light microscopy is greater than that in SEM.

(j) Light microscopy, SEM and TEM all involve the use of a condenser lens within the microscope.

43.2 Fill in the blanks in the following paragraph. Dark-field microscopy involves shining reflected and _____ light on the specimen against a dark background. It is particularly useful for _____ specimens. UV microscopy uses short-wavelength UV light in order to increase image _____. Phase-contrast microscopy utilises constructive and destructive _____ effects to increase image _____. Nomarski microscopy provides a pseudo _____ image, with a very small depth of _____, allowing _____ to be carried out. _____ light microscopy allows visualisation of optically active components in the specimen. Confocal microscopy involves the use of a _____ light source and can yield computer-generated 3-D images.

43.3 Identify the missing preparative procedures. In each sequence below, one or two steps have been missed out. Using Fig. 43.3, identify the missing procedures.

a) For light microscopy on a killed and fixed specimen: fix – dehydrate – clear – _____ – section – _____ – mount – examine.

b) For light microscopy on a heat-fixed microbial specimen: smear – _____ – heat fix – _____ – examine.

c) For TEM on a killed and fixed specimen: fix – _____ – embed – section – mount stain – examine.

Answers to these study exercises are available at www.pearsoned.co.uk/practicalskills

44 Preparing specimens for light microscopy

Preparative techniques are crucial to successful microscopical investigation because the chemical and physical processes involved have the potential for making the material difficult to work with and for producing artefacts. The basic steps (outlined in Fig. 43.3) are similar in most cases, but the exact details (e.g. timing, chemicals used and their concentrations) differ according to the material being examined and the purpose of the investigation. It is usually best to follow a recipe that has worked in the past for your material (see Grimstone and Skaer, 1972; Kiernan, 2008).

Chemical fixation

The main purpose of fixation is to preserve material in a lifelike manner. The process of fixation for microscopy is much more critical than for whole specimens (see Chapter 37) and only small pieces of tissue should be used. The fixation solutions used for microscopy are intended to:

- **penetrate rapidly** to prevent post-mortem changes in the cells;
- **coagulate the cell contents** into insoluble substances;
- **protect tissues against shrinkage and distortion** during subsequent processing;
- **allow cell parts to become selectively and clearly visible** when stained.

Fixative solutions are usually mixtures of chemicals selected for their combined properties (see Chapter 37). Your choice of fixative from the numerous recipes available in reference texts will depend upon both the type of investigation and the nature of the material.

Fixing specimens – tissues must be fixed as soon as possible after death. If this is not possible, specimens should be stored at low temperature (4 °C) and for as short a time as possible.

> **KEY POINT** Poor fixation can produce artefacts, particularly where coagulant fixatives are used.

When using a fixative for microscopy, observe the following points:

- **Use fresh solutions:** some of the fluids are unstable and do not keep well. Do not reuse fixative.
- **Always use plenty of fixing fluid** compared with the volume of material to be fixed (not less than a 10 : 1 fixative: sample volume ratio).
- **Avoid underfixation or overfixation:** in general the optimum time will be a function of several factors including:
  - (a) The penetration capacity of the fixative.
  - (b) The size of the piece of tissue: this should always be small and have as large a surface : volume ratio as possible.
  - (c) The type of tissue to be fixed: uniform tissues fix more quickly than complex tissues, where one component may form a barrier to others. The presence of a chitin exoskeleton usually means a slow rate of penetration. Tissues filled with air can be difficult to submerge and infiltrate; this can be overcome by fixation in a partial vacuum.
  - (d) The temperature: increased temperature results in increased penetration rate, but also tends to make tissue brittle.
- **Wash the specimen thoroughly after fixation:** residues of fixative can interfere with subsequent processes. The washing may be in water or another appropriate solution.

Decalcifying specimens – this may be necessary if calcareous structures remain after fixation: this is usually done using a 5–10% solution of EDTA followed by thorough washing.

> Allowing time for dehydration – never be tempted to rush, because incomplete dehydration will do more than anything else to ruin a preparation.

> **SAFETY NOTE** The sharp blades used in hand sectioning are an obvious hazard. Identify beforehand the location of the nearest first-aid kit. Take care to carry out such procedures with care and away from others who may inadvertently brush against you.

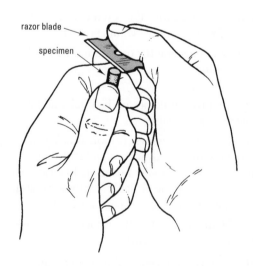

razor blade

specimen

Fig. 44.1 Preparation of a hand-cut section.

Dehydration and clearing

A high water content in tissues will usually hinder subsequent processing so they must be dehydrated. This is done with an organic solvent using a series of solutions graded from pure water to pure solvent. Ethanol/water mixtures are often used in histology. Dehydration must be carried out carefully, using prescribed time schedules to avoid distortion and hardening. Protect delicate specimens by mixing solutions in the container rather than by transferring the specimen as this is when damage could occur.

The chemicals used for dehydration are not usually soluble in the embedding medium and tissues must therefore be infiltrated with an intermediate fluid, miscible with the waxes and resins normally used: such fluids make the tissues transparent and are termed clearing agents. The most widely used ones are hydrocarbons such as xylene; many are volatile, pose significant health risks, and tend to harden tissues rapidly. Clearing oils such as clove oil and cedarwood oil are safer, but slower in action. Terpineol is useful since it does not require such complete dehydration and can be used straight from 90% v/v ethanol. Note that every trace of oil must be removed during the infiltration (embedding) stage, or specimens will not embed properly.

Embedding and sectioning

Embedding involves infiltrating the specimen with a medium that will solidify and support it when sectioned. The specimen is either passed through a series of gradually increasing concentrations of the embedding material (e.g. wax or epoxy resin) dissolved in the dehydrating or clearing agent, or placed straight into pure embedding agent. Several changes of embedding material are made to remove the last traces of solvent, then the embedding agent is solidified by cooling or by a polymerisation treatment. Detailed protocols can be found in specialist books (e.g. Kiernan, 2008). The specimen must be orientated carefully during the solidification process to aid section-cutting in known planes.

The aim of sectioning is to provide a thin slice through the tissues to allow observation of cellular details at maximum resolution. After trimming, blocks of embedded material are sectioned using a microtome. Section-cutting techniques are complex and require time for familiarisation. Sections are attached to slides after flattening the sections by a flotation procedure carried out either directly on the glass slide or in a water bath. The sections are attached to the slide by coating the latter with a very thin layer of albumen before drying them down on to this sticky surface.

For some procedures, specimens are frozen in isopentane, freon or dry-ice/ethanol mixtures and then maintained below −20 °C until sectioned. Sectioning is performed in a cold chamber (cryostat) where there is a microtome; sections are removed directly to glass slides. If storage is necessary, slides are placed over a desiccant in a refrigerator, but this is possible only for short periods.

Hand-cut sections can be made through certain relatively stiff materials, notably stems and roots of herbaceous plants:

1. **Grasp the object firmly** between your index finger and thumb.
2. **Brace your elbows against your ribs** to steady your hands.
3. **Rest the side of a fresh and sharp razor blade on the index finger** holding the specimen (Fig. 44.1).

(a)

(b)

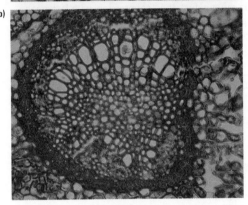

Fig. 44.2 Simulation of the introduction of contrast into a biological specimen by staining. Leaf midrib transverse section showing in greyscale the effects 'before' (a) and 'after' (b) staining xylem tissue red with phloroglucinol. In colour, the effects of staining are even more distinct.

Definitions

Counterstains – stains that apply a background colour to contrast with stained structures.

Metachromic stains – have the capacity to stain different structures different colours.

Mordants – chemicals (salts and hydroxides of divalent and trivalent metals) that increase the efficiency of stains usually by forming complexes with the stain.

Negative staining – where the background is stained rather than the structure of interest.

Orthochromic stains – never change colour whatever they stain.

---

4. **Pull the blade towards your body** so that you cut the specimen with a slicing action.

5. **Repeat this action quickly,** pushing the object slowly upwards with your thumb.

6. **Float off the cut sections** on to a watch glass of water.

This is a relatively tricky procedure, requiring practice. You may find it helps to lubricate the razor and object by wetting them before cutting. Thin material, such as a leaf, is best sectioned when supported between the halves of a longitudinally split cylinder of pith or fresh carrot. Always cut lots of sections, because only a small proportion will be thin enough to use – the best ones are often wedge-shaped, tapering off to thin edges.

Staining

The purpose of staining in microscopy is to:

- **add contrast to the image;**
- **identify chemical components of interest;**
- **locate particular tissues, cells or organelles.**

This is achieved in different ways for different types of microscopy. In standard light microscopy, contrast is achieved by staining the structure of interest with a coloured dye (e.g. Fig. 44.2); in UV microscopy, contrast is obtained using fluorescent stains. Physico-chemical properties of the stain cause it to attach to certain structures preferentially or be taken up across cell membranes.

Stains for light microscopy are categorised according to the charge on the dye molecule. Stains like haematoxylin, whose coloured part is a cation (i.e. basic dyes), stain acidic, anionic substances like nucleic acids: such structures are termed basophilic. Stains like eosin, whose coloured part is an anion (i.e. acid dyes), stain basic, cationic substances: such structures are termed acidophilic. Acid dyes tend to stain all tissue components, especially at low pH, and are much used as counterstains. Staining is progressive if it results in some structures taking up the dye preferentially. Staining is regressive if it involves initial over-staining followed by decolorisation (differentiation) of those structures that do not bind the dye tightly (e.g. Gram staining, pp. 248–9).

Certain 'vital' stains (e.g. neutral red) are used to determine cell viability or the pH of cell compartments such as plant vacuoles. 'Mortal' stains (e.g. Evans' blue) are excluded from living cells but diffuse into dead ones and are used to assay cell mortality.

Stains and staining procedures

There is a huge range of stains for light microscopy but the features of some used commonly for cytology in botany, microbiology and zoology are given in Table 44.1. Results depend on technique: follow the recommended procedures carefully. There are many stains used in histochemistry for identifying various classes of macromolecules such as DNA, RNA, proteins, lipids, carbohydrates (chitin, cellulose, starch, callose, pectins, glycogen) and heteropolymers (e.g. lipopolysaccharides, peptido-glycans, proteoglycans). Consult appropriate texts for full details of (a) how to make up stains and (b) the protocol to use when applying them (e.g. Grimstone and Skaer, 1972; Horobin and Kiernan, 2002).

Table 44.1 A selection of stains for light microscopy of sections

	Stain	What it stains	Comments
Plant cells	Chlorazol black	Cell walls: black Nuclei: black, yellow or green Suberin: amber	The solvent used (70% ethanol in water or water alone) affects colours developed
	Neutral red	Living cells: pink (pH < 7)	A 'vital' stain used to determine cell viability or to visualise plant protoplasts in plasmolysis experiments; best used at neutral external pH
	Phloroglucinol/HCl	Lignified cell walls: red	Care is required because the acid may damage microscope lenses
	Ruthenium red	Pectins: red	Shows up the middle lamella
	Safranin + Fast green	Nuclei, chromosomes, cuticle and lignin: red Other components: green	Stain in safranin first, then counterstain with fast green (light green will substitute). A differentiation step is required
	Toluidine blue	Lignified cell walls: blue Cellulose cell walls: purple	Best to apply dilute and allow progressive staining to occur
Fungi and bacteria	Giemsa	Bacterial chromosome: purple Bacterial cytoplasm: colourless	Also used in zoology to stain protozoa
	Gram	Gram-positive bacteria: violet/purple Gram-negative bacteria: red/pink Yeasts: violet/purple	See p. 249 for procedure
	Gray	Bacterial flagella: red	Uses toxic chemicals: mercuric chloride and formaldehyde. Leifson's stain is an alternative
	Lactophenol cotton blue	Fungal cytoplasm: blue (hyphal wall unstained)	Shrinkage may occur
	Nigrosin or India ink	Background: grey–black	Negative stains for visualisation of capsules: requires a very thin film
	Shaeffer and Fulton	Bacterial endospores: green Vegetative cells: pink/red	Malachite green is primary stain, heated for 5 min. Counterstained with safranin
	Ziehl–Neelsen	Actinomycetes: red Bacterial endospores: red Other microbes: blue	Requires heat treatment of fuchsin primary stain, decolorisation with ethanol–HCl and a methylene blue counterstain (acid-fast structures remain red)
Animal cells	Azure A/eosin B	Nuclei, RNA: blue Basophilic cells: blue-violet Most other cells: pale blue Muscle cells: pink Necrosing cells: pink Cartilage matrix: red-violet Bone: pink Red blood cells: orange-red Mucins: green-blue/blue-violet	Used in pathology – shows up bacteria as blue; must be fresh; care required over pH: Mann's methyl blue/eosin gives similar results
	Chlorazol black	Chitin: greenish-black Nuclei: black, yellow or green Glycogen: pink or red	Solvent (70% v/v ethanol in water or water alone) affects colours formed
	Iron haematoxylin	Nuclei, chromosomes and red blood cells: black Other structures: grey or blue-black	Good for resolving fine detail; iron alum used as mordant before haematoxylin to differentiate
	Mallory	Nuclei: red Nucleoli: yellow Collagen, mucus: blue Red blood cells: yellow Cytoplasm: pink or yellow	Simple, one-stage stain; fades within a year; not to be used with osmium-containing fixatives. Heidenhain's azan gives similar results but does not fade. Cason's one-step Mallory is a rapidly applied stain that is particularly good for connective tissue
	Masson's trichrome	Collagen, mucus: green Cytoplasm: orange or pink	Used as a counterstain after, for example, iron haematoxylin, which will have stained nuclei black. Not to be used after osmium fixation
	Mayer's (haematoxylin and eosin; 'H&E'; haemalum)	Nuclei: blue/purple Cytoplasm: pink	Alum used as mordant for haematoxylin; eosin is the counterstain. To show up collagen, use van Gieson's stain as counterstain

Fig. 44.3 How to irrigate a section with stain by drawing it through with a piece of filter paper.

Stains for light microscopy are normally applied by one of four methods:

- **Floating the sections on the stain.**
- **Applying the stain to a smear fixed on to a slide,** e.g. when staining bacteria and blood.
- **Drawing the stain through under a coverslip** as shown in Fig. 44.3.
- **Immersing slides with sections attached into a staining trough.** This is best for bulk staining.

Most stains need to act in an aqueous medium, so sections that have been embedded in wax must be rehydrated before staining. The wax is dissolved, e.g. in Histo-clear for 1–3 min, the Histo-clear replaced by 100% ethanol (1 min) then 70% (v/v) ethanol: water for 1 min, finally transferring to distilled water. Fresh sections of plants or heat-fixed smears of microbes generally require no pretreatment before staining if they are not to be retained after examination.

Mounting sections

Wet mounts

These are used for observing fresh specimens. The following steps are involved:

1. **Isolate the specimen.**
2. **Place the specimen in a small droplet of the relevant fluid** (fresh water, sea water, etc.) on a microscope slide.
3. **Gently lower a coverslip on to the droplet,** using forceps or two needles and avoiding bubbles.
4. **Remove any excess water** on or around the coverslip with absorbent paper.

Entire specimens can be examined under the light microscope providing they are small enough to be mounted on a glass slide. They may be mounted in cavity slides or by using ring mounts.

Temporary mounts

These essentially involve wet mounting in a mountant with a short useful life, e.g. for identification purposes. It may be desirable to clear the specimen first and a dual-purpose substance such as lactophenol, which will clear from 70% (v/v) ethanol, is recommended.

Permanent mounts

These protect sections during examination and allow storage without deterioration. A permanent mount involves sealing your section under a coverslip in a mountant. The mountants used are clear resins dissolved in a slowly evaporating solvent. A good mountant has a similar refractive index to the tissue being mounted, remains clear through time, is chemically inert and will harden quickly. Natural resins like Canada balsam take a long time to dry, are variable in quality and tend to colour up and crack in time. The newer synthetic resins and plastics such as DPX mountant

Transferring sections – a camel-hair paintbrush is useful for transferring sections between staining solutions and on to a slide.

Using coverslips – all wet specimens for microscopic examination must be covered with a coverslip to protect the objective lens from water, stains and dirt.

Using mountants – most mountants require that all water is removed from the section by transfer through increasing concentrations of ethanol until 100% ethanol is reached.

mountant on underside of coverslip

Fig. 44.4 How to lower a coverslip painted with mountant on its underside on to sections on a slide.

Making squash preparations – these may be required for any type of mount. The smallest specimens can be squashed after mounting by applying gentle pressure on the coverslip with your forceps. Larger specimens can be squashed between two slides before fixing and mounting – this ensures that higher pressures are applied evenly.

are superior: they dry quickly, are available in a range of refractive indices, and do not yellow with age. For tissue components soluble in organic solvents, aqueous mounting media based on, for example, gelatine or glycerol should be used. Find out which mountant is recommended for your particular sections.

The recommended procedure when mounting sections on a slide is:

1. **Apply a little mountant to a coverslip** of appropriate size.
2. **Turn the coverslip over and place on its edge** to one side of the sections as in Fig. 44.4.
3. **Lower the coverslip slowly down on to the sections** so as to displace all the air and sandwich the sections between the slide and the coverslip.
4. **Press firmly from the centre outwards** to distribute the mounting medium evenly.
5. **Allow the solvent to evaporate** – best results come from slow drying when time allows, but many synthetic mountants will tolerate brief heating when speed is essential.

Always label slides at the time of preparation – include date, material, section type and stain. Store them in specialised trays or containers, dust free and avoiding contact with each other.

Text references

Grimstone, A.V. and Skaer, R.J. (1972) *A Guidebook to Microscopical Methods*. Cambridge University Press, Cambridge.

Horobin, R.W. and Kiernan, J.A. (eds) (2002) *Conn's Biological Stains. A Handbook of Dyes, Stains and Fluorochromes for use in Biology and Medicine*, 10th edn. Bios Scientific Publishers, Oxford.

Kiernan, J.A. (2008) *Histological and Histochemical Methods: Theory and Practice*, 4th edn. Scion Publishing, Bloxham.

Source for further study

Davidson, M.W. and Abramowitz, M. *Molecular Expressions. Exploring the World of Optics and Microscopy*. Available: http://micro.magnet.fsu.edu/index.html

Last accessed 19/02/15.
[Covers many areas of basic knowledge underlying microscopy, including preparative procedures.]

STUDY EXERCISES

44.1 Select appropriate stains. From Table 44.1, identify a stain you could use to help indicate the presence of the following: (a) glycogen in a liver section; (b) woody (lignified) cells in a plant stem section; (c) a fungal pathogen in a leaf section; (d) living cells in an onion epidermal peel; (e) mucus in a lung section; (f) bacteria in a food sample.

44.2 Explain the processes involved in preparing a section for light microscopy. Prepare one-sentence summaries of the reasons for carrying out the following procedures: (a) fixation; (b) dehydration; (c) embedding; (d) sectioning; (e) staining; (f) permanent mounting.

44.3 Investigate toxicity and safety aspects of some of the staining reagents used for light microscopy. Many chemicals used in staining are toxic or otherwise dangerous. As examples, investigate the specific hazards associated with the use of the following stains: (a) chlorazol black; (b) neutral red; (c) malachite green; (d) safranin.

Answers to these study exercises are available at www.pearsoned.co.uk/practicalskills

Fig. 45.1 Standard features of an objective lens. Most lenses are inscribed to show the details labelled above. The numerical aperture is a measure of the light-gathering power of the lens.

Using binocular eyepieces – if you do not know your interpupillary distance, ask someone to measure it with a ruler. You should stare at a fixed point in the distance while the measurement is taken (in millimeters). Take a note of the value for future use.

Issues for spectacle and contact lens wearers – those who wear glasses can remove them for viewing, as microscope adjustments will accommodate most deficiencies in eyesight (except astigmatism). This is more comfortable and stops the spectacle lenses being scratched by the eyepiece holders. However, it may create difficulties in focusing when drawing diagrams. Those with contact lenses should simply wear them as normal for viewing.

The light microscope is one of the most important instruments used in biology practicals and knowing how to use one correctly is a basic and essential skill of biology. A standard undergraduate binocular microscope consists of three main types of optical unit (Fig. 43.1): eyepieces (two), an objective (Fig. 45.1) and a condenser. These are attached to a stand that holds the specimen on a stage (Fig. 45.2). A monocular microscope is constructed similarly but has one eyepiece lens rather than two.

Setting up a binocular light microscope

Before using any microscope, familiarise yourself with its component parts.

> **KEY POINT** Never assume that the previous person to use your microscope has left it set up correctly: apart from taking account of differences in users' eyes, the microscope needs to be properly set up for each lens combination used.

The procedures outlined below are simplified to allow you to set up microscopes like those of the Olympus CX series (Fig. 45.2). For monocular microscopes, disregard instructions for adjusting eyepiece lenses (3) and (5).

1. **Place the microscope at a convenient position on the bench.** Adjust your seating so that you are comfortable operating the focus and stage controls. Unwind the power cable, plug in and switch on after first ensuring that the lamp setting is at a minimum. Then adjust the lamp setting to about two-thirds of the maximum.

2. **Select a low-power (e.g. ×10) objective.** Make sure that the lens clicks home.

3. **Set the eyepiece (ocular) lenses to your interpupillary distance;** this can usually be read off a scale on the turret. You should now see a single circular field of vision. If you do not, try adjusting in either direction.

4. **Put a prepared slide on the stage.** Examine it first against a light source and note the position, colour and rough size of the specimen. Place the slide on the stage (coverslip up!) using the slide holder and, viewing from the side, position it with the stage adjustment controls so that the specimen is illuminated.

5. **Focus the image of the specimen** using first the coarse and then the fine focusing controls (Fig. 45.3). The image will be reversed and upside down compared with that seen when viewing the slide directly.

   (a) If both eyepiece lenses are adjustable, set your interpupillary distance on the scale on each lens. Close your left eye, look through the right eyepiece with your right eye and focus the image with the normal controls. Now close your right eye, look through the left eyepiece with your left eye and focus the image by rotating the eyepiece holder. Take a note of the setting for future use.

   (b) If only the left eyepiece is adjustable, close your left eye, look with the right eye through the static right eyepiece and focus the image

Fig. 45.2 Diagram of the Olympus binocular microscope model CX41.

- The lamp in the base of the stand (1) supplies light; its brightness is controlled by an on–off switch and voltage control (2). Never use maximum voltage or the life of the bulb will be reduced – a setting two-thirds to three-quarters of maximum should be adequate for most specimens. A field–iris diaphragm may be fitted close to the lamp to control the area of illumination (3).
- The condenser control focuses light from the condenser lens system (4) on to the specimen and projects the specimen's image on to the front lens of the objective. Correctly used, it ensures optimal resolution.
- The condenser–iris diaphragm (5) controls the amount of light entering and leaving the condenser; its aperture can be adjusted using the condenser–iris diaphragm lever (6). Use this to reduce glare and enhance image contrast by cutting down the amount of stray light reaching the objective lens.
- The specimen (normally mounted on a slide) is fitted to a mechanical stage or slide holder using a spring mechanism. Two controls allow you to move the slide in x and y planes. Vernier scales (see p. 151) on the slide holder can be used to return to the same place on a slide. The fine and coarse focus controls adjust the height of the stage relative to the lens systems. Take care when adjusting the focus controls to avoid hitting the lenses with the stage or slide.
- The objective lens (9) supplies the initial magnified image; it is the most important component of any microscope because its qualities determine resolution, depth of field and optical aberrations. The objective lenses are attached to a revolving nosepiece (10). Take care not to jam the longer lenses on the stage or slide as you rotate the nosepiece. You should feel a distinct click as each lens is moved into position. The magnification of each objective is written on its side; a normal complement would be ×4, ×10, ×40 and ×100 (oil immersion).
- The eyepiece lens (11) is used to further magnify the image from the objective and to put it in a form and position suitable for viewing. Its magnification is written on the holder (normally ×10). By twisting the holder for one or both of the eyepiece lenses you can adjust their relative heights to take account of optical differences between your eyes. The interpupillary distance scale (12) and adjustment knob allow compensation to be made for differences in the distance between users' pupils.

From a photograph of Olympus Model CX41 binocular microscope, supplied by Olympus Microscopes, Olympus Optical Company (UK) Ltd, published courtesy of Olympus Optical Company (UK) Ltd.

Fig. 45.3 Importance of correct focus in light microscopy. Stomatal complex of *Commelina communis* L., a specimen that is a monolayer of cells approximately 30–50 μm thick. (a) Focal plane is on 'internal' walls of the cells; (b) focal plane is on the 'external' walls and stomatal pore. The two images are different, and while it would not be possible to measure the stomatal pore in (a), it would not be possible to see the vacuolar crystals in (b). When looking at specimens, always use the fine-focus control to view different focal planes.

with the normal controls. Now close your right eye, look through the left eyepiece with your left eye and focus the image by rotating the eyepiece holder. Take a note of the setting for future use.

6. **Close the condenser–iris diaphragm** (aperture–iris diaphragm), then open it to a position such that further opening has no effect on the brightness of the image (the 'threshold of darkening'). The edge of the diaphragm should not be in view. Turn down the lamp if it is too bright.

7. **Focus the condenser.** Place an opaque pointed object (the tip of a mounted needle or a sharp pencil point) on the centre of the light source. Adjust the condenser setting until both the specimen and needle tip/pencil point are in focus together. Check that the condenser–iris diaphragm is just outside the field of view.

8. **For higher magnifications, swing in the relevant objective** (e.g. ×40), carefully checking that there is space for it. Adjust the focus using the fine control only. If the object you wish to view is in the centre of the field with the ×10 objective, it should remain in view (magnified, of course) with the ×40. Adjust the condenser–iris diaphragm and condenser as before – the correct setting for each lens will be different.

9. **When you have finished using the microscope, remove the last slide and clean the stage if necessary.** Turn down the lamp setting to its minimum, then switch off. Clean the eyepiece lenses with lens tissue. Check that the objectives are clean. Unplug the microscope from the mains and wind the cable round the stand and under the stage. Replace the dust cover.

If you have problems in obtaining a satisfactory image, refer to Box 45.1; if this does not help, refer the problem to the class supervisor.

Procedure for observing transparent specimens

Some stained preparations and all colourless objects are difficult to see when the microscope is adjusted as below (Fig. 45.4). Contrast can be improved by closing down the condenser–iris diaphragm. Note that when you do this, diffraction haloes appear round the edges of objects. These

---

Adjusting a microscope with a field–iris diaphragm – adjust this before the condenser–iris diaphragm: close it until its image appears in view as a circle of light, if necessary focusing on the edge of the circle with the condenser controls and centring it with the centring screws. Now open it so the whole field is just illuminated.

High-power objectives – never remove a slide while a high power objective lens (i.e. ×40 or ×100) is in position. Always turn back to the ×10 first. Having done this, lower the stage and remove the slide.

Box 45.1	Problems in light microscopy and possible solutions

No image; very dark image; image dark and illuminated irregularly

- Microscope not switched on (check plug and base)
- Illumination control at low setting or off
- Objective nosepiece not clicked into place over a lens
- Diaphragm closed down too much or off-centre
- Lamp failure

Image visible and focused but pale and indistinct

- Diaphragm needs to be closed down further (see Fig. 45.4)
- Condenser requires adjustment

Image blurred and cannot be focused

- Dirty objective
- Dirty slide
- Slide upside down
- Slide not completely flat on stage
- Eyepiece lenses not set up properly for user's eyes
- Fine focus at end of travel
- Oil-immersion objective in use, without oil

Dust and dirt in field of view

- Eyepiece lenses dirty
- Objective lens dirty
- Slide dirty
- Dirt on lamp glass or upper condenser lens

(a)

(b)

Fig. 45.4 Effect of closing the condenser–iris diaphragm on contrast. Head of human head louse, *Pediculus humanus capitus* DeGeer; (a) with condenser–iris diaphragm open; (b) with condenser–iris diaphragm closed (all other settings the same). Note difference in detail that can be seen in (b), but also that image (b) is darker – when using the condenser–iris diaphragm in this way you may need to compensate by increasing the light setting.

> **SAFETY NOTE**   Take care when moving microscopes, not only because of the cost of replacement if damaged, but also because they weigh several kilograms and could cause injury if dropped. Always carry a microscope using two hands.

obscure the image of the true structure of the specimen and may result in loss of resolution. Nevertheless, an image with increased contrast may be easier to interpret.

Procedure for oil-immersion objectives

These provide the highest resolution of which the light microscope is capable. They must be used with immersion oil filling the space between the objective lens and the top of the slide. The oil has the same refractive index as the glass lenses, so loss of light by reflection and refraction at the glass/air interface is reduced. This increases the resolution, brightness and clarity of the image and reduces aberration. Use oil-immersion objective(s) as follows:

1. **Check that the object of interest is in the field of view** using, for example, the ×10 or ×40 objective.

2. **Apply a single small droplet of immersion oil** to the illuminated spot on the top of the slide, having first swung the ×40 objective away. Never use too much oil: it can run off the slide and mess up the microscope.

3. **Move the high-power (×100) oil-immersion objective into position carefully,** checking first that there is space for it. Focus on the specimen using the fine control only. You may need a higher brightness setting.

4. **Perform condenser–iris diaphragm and condenser focusing adjustments** as for the other lenses.

5. **When finished, clean the oil-immersion lens** by gently wiping it with clean lens tissue. If the slide is a prepared one, wipe the oil off with lens tissue.

You should take great care when working with oil-immersion lenses as they are the most expensive to replace. Because the working distance between the lens and coverslip is so short (less than 2 mm), it is easy to damage the lens surface by inadvertently hitting the slide or coverslip surface. You must also remember that they need oil to work properly. If working with an unfamiliar microscope, you can easily recognise oil-immersion lenses. Look for a white or black ring on the lens barrel, near the lens, or for 'oil' clearly marked on the barrel.

Care and maintenance of your microscope

Microscopes are delicate precision instruments. Handle them with care and never force any of the controls. Never touch any of the glass surfaces with anything other than clean, dry lens tissue. Bear in mind that a replacement would be very expensive.

If moving a microscope, hold the stand above the stage with one hand and rest the base of the stand on your other hand. Always keep the microscope vertical (or the eyepieces may fall out). Put the microscope down gently.

Clean lenses by gently wiping with clean, dry lens tissue. Use each piece of tissue once only. Try not to touch lenses with your fingers as oily fingerprints are difficult to clean off. Do not allow any solvent (including water) to come into contact with a lens; sea water is particularly damaging.

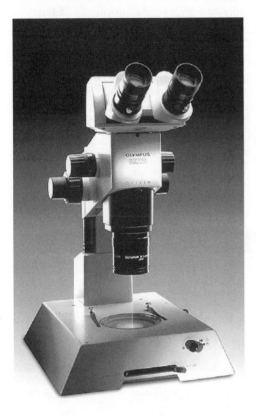

Measuring specimens using a dissecting microscope – because of the low magnification, sizes can generally be estimated by comparison with a ruler placed alongside the specimen. If accurate measurements are required, eyepiece graticules can be used (Fig. 46.2).

The dissecting (stereoscopic) microscope

The dissecting microscope (Fig. 45.5) is a form of stereoscopic microscope used for observations at low total magnification ($\times 4$ to $\times 50$) where a large working distance between objectives and stage is required, perhaps because the specimen is not flat or dissecting instruments are to be used. A stereoscopic microscope essentially consists of two separate lens systems, one for each eye. Some instruments incorporate zoom objectives. The eyepiece–objective combinations are inclined at about 15° to each other and the brain resolves the compound image in three dimensions as it does for normal vision. The image is right side up and not reversed, which is ideal for dissections. Specimens are often viewed in a fresh state and need not be placed on a slide – they might be in a Petri dish or on a white tile. Illumination can be from above or below the specimen, as desired.

Most of the instructions for the binocular microscope given above apply equally well to dissecting microscopes, although the latter do not normally have adjustable condensers or diaphragms. With stereoscopic microscopes, make specially sure to adjust the eyepiece lenses to suit your eyes so that you can take full advantage of the stereoscopic effect.

Fig. 45.5 From a photograph of Olympus Model SZX12 dissecting microscope, supplied by Olympus Microscopes, Olympus Optical Company (UK) Ltd, published courtesy of Olympus Optical Company (UK) Ltd.

Sources for further study

Bradbury, S. and Bracegirdle, B. (1998) *Introduction to Light Microscopy.* Bios Scientific Publishers, Oxford.

Davidson, M.W. *Nikon MicroscopyU: Introduction to Microscope Objectives.* Available: http://www.microscopyu.com/articles/optics/objectiveintro.html.
Last accessed 19/02/15.
[Covers the optics of objective lenses.]

Fellers, T.J. and Davidson, M.W. *Olympus Microscopy Resource Centre: Microscopy Primer.* Available: http://www.olympusmicro.com/primer
Last accessed 19/02/15.
[A primer on optical microscopy, including FAQs and links to further resources.]

45.1 Test your knowledge of the parts of a binocular light microscope. Cover up the legend on the left of Fig. 45.2 with a piece of paper or card. Now identify the parts of the light microscope numbered on the diagram. Check your answers from the legend.

45.2 Identify roles of parts of a binocular light microscope. State briefly the *primary* role of each component of a standard binocular light microscope:

(a) condenser

(b) objective lens

(c) condenser–iris diaphragm

(d) interpupillary distance scale

(e) Vernier scales on the mechanical stage.

45.3 Identify the correct sequence of adjustments when setting up a light microscope.

(a) Focus specimen – set interpupillary distance – adjust condenser–iris diaphragm – make individual eyepiece adjustment – focus condenser – focus specimen.

(b) Make individual eyepiece adjustment – set interpupillary distance – focus condenser – focus specimen – adjust condenser–iris diaphragm – focus specimen.

(c) Set interpupillary distance – focus specimen – make individual eyepiece adjustment – adjust condenser–iris diaphragm – focus condenser – focus specimen.

(d) Make individual eyepiece adjustment – set interpupillary distance – focus specimen – adjust condenser–iris diaphragm – focus condenser – focus specimen.

(e) Focus specimen – make individual eyepiece adjustment – set interpupillary distance – focus condenser – adjust condenser–iris diaphragm.

Microscope images, whether viewed directly or as photomicrographs, need to be interpreted with extreme care. This is not only because of the potentially damaging procedures involved in specimen preparation, but also because scale, section orientation and staining combine to create images that can be misinterpreted by the human mind.

Eliminating non-biological artefacts

There are several types of non-biological artefacts (e.g. those shown in Fig. 46.1). It is only through experience that you can learn to recognise these. Sometimes they are mistaken for specimens by novices, because the specimen itself is hard to see. A useful tip is to look at your slide carefully before placing it on the stage; if you do this against a light background, you may be able to see the specimen and judge its location as you move the slide under the objective. For slides where specimens are tiny and spread out, such as blood or bacterial smears, it can be useful to focus on the edge of the coverslip (as in Fig. 46.1(a)) to determine the correct focal plane before searching. Correctly setting up the microscope is crucial in such cases.

Establishing scale and measuring objects

Working out the length of objects on photomicrographs

On a micrograph, the scale will usually be provided as a magnification factor (e.g. ×500) or in the form of a bar of defined length (e.g. 100 nm). If you need to estimate the dimensions of an object in the micrograph, follow the steps below:

1. **Measure the object as it appears on the micrograph** with a ruler or set of Vernier calipers (see p. 151). It may be difficult to decide exactly where the boundary of a structure lies; rather than a discrete boundary, you may be dealing with shades of grey. It is essential to be consistent.

2. **If the scale is given as a bar, measure the bar too and find the object's size by proportion.** For example, if the object measures 32 mm and a bar representing 100 nm is 20 mm long, then the size of the object is 32/20 × 100 nm = 160 nm.

3. **If the scale is given as a magnification factor, divide your measurement by this number to obtain the object's size,** taking care to enter the correct units in your calculator. For example, if the object measures 32 mm and the print magnification is stated as ×200 000, then its size is 32×10^{-3} m/200 000 = 1.6×10^{-7} m = 160 nm.

Avoid putting too many significant figures in any estimates of dimensions: there may be quite large errors in estimating print magnifications that could make the implied accuracy meaningless (p. 497).

Fig. 46.1 Artefacts in microscopy. Do not mistake any of the following for biological specimens: (a) edge of coverslip; (b) dust on slide; (c) shard of glass; (d) air bubble; (e) grease spots; (f) cellulose fibre.

Fig. 46.2 Calibrating an eyepiece micrometer (graticule). Align the two scales and read the number of stage micrometer divisions (shown in colour) for a particular number of eyepiece micro-meter divisions (shown in black). In this case 26.5 scale divisions of 0.01 mm (= 2.65 mm) are equivalent to 100 eyepiece divisions, so each eyepiece division 0.265 ÷ 100 = 0.00265 mm = 2.65 μm. This is a typical value for a total magnification of ×400 (e.g. ×40 objective and ×10 eyepiece).

Using an eyepiece graticule – choose the eyepiece lens corresponding to your stronger eye and check that you have made the correct adjustments to the eyepiece lenses as detailed on p. 271.

Examples An object occupying one-third of the area of a field of diameter 0.45 mm would have a cross-sectional area of 0.159/3 (= 0.053 mm²).

Seven objects in a field of area 0.159 mm² would be present at a density of 7 ÷ 0.159 (= 44 objects mm⁻²).

Example For a field area of 0.159 mm², a droplet volume of 30 mm³ and a coverslip of area 484 mm², the field volume would be 30 × (0.159 ÷ 484) = 0.01 mm³, so if you saw 25 objects per field, their density would be 25 ÷ 0.01 = 2500 mm⁻³.

Adding linear scales to your own drawings

The magnification of a light microscope image is calculated by multiplying the objective magnification by the eyepiece magnification. However, the magnification of the image bears no certain relation to the magnification of any drawing of the image – you may equally well choose to draw the same image 10 mm or 10 cm long. For this reason, *it is essential to add a scale to all your diagrams.* You can provide either a bar giving the estimated size of an object of interest, or a bar of defined length (e.g. 100 μm).

The simplest method of estimating linear dimensions is to compare the size of the image to the diameter of the field of view. You can make a rough estimate of the field diameter by focusing on the millimetre scale of a transparent ruler using the lowest power objective. Estimate the diameter of this field directly, then use the information to work out the field diameters at the higher powers pro rata. For example, if the field at an overall magnification of ×40 is 4 mm, at an overall magnification of ×100 it will be: 40/100 × 4 mm = 1.6 mm (1600 μm).

Greater accuracy can be obtained if an eyepiece micrometer (graticule) is used. This carries a fine scale and fits inside an eyepiece lens. The eyepiece micrometer is calibrated using a stage micrometer, basically a slide with a fine scale on it. Figure 46.2 shows how to calibrate an eyepiece micrometer, along with a worked example. Once you have calibrated your eyepiece micrometer for each objective lens used, you can use it to measure objects: in the example shown in Fig. 46.2, the scale reading is multiplied by 2.65 μm to give the value in micrometres. So, if you measured the width of a human hair at 34 eyepiece micrometer units, then this will be equal to 34 × 2.65=90.1 μm. An alternative approach is to put a scale bar on a diagram, e.g. a 100 μm scale bar would be equivalent to the length of almost 38 eyepiece micrometer divisions.

Estimating the area and area density of objects

You can calculate the area of the field at each magnification using a field diameter estimate obtained as above (obtain the area from πr^2, but do not forget that radius r = diameter ÷ 2). The units you use to measure the diameter will be squared; thus, a field diameter of 0.45 mm converts to a field area of 0.159 mm². Use the field area to estimate the cross-sectional area of a specimen by proportion.

If several objects appear in a field, you can express their frequency on an area basis by dividing the number seen by the field area. It is nearly always best to take an average from several fields of view before calculating.

Volume–density estimates

The density of particles in a liquid suspension can be *roughly* estimated as follows: place a drop of well-mixed suspension on to a slide with a Pasteur pipette. You can estimate the drop volume by weighing drops of water and assuming 1 mg = 1 mm³. Alternatively, use a pipettor (see p. 140). Quickly cover the drop with a coverslip (see p. 270), avoiding air bubbles – a 22 × 22 mm coverslip should allow a 30 mm³ drop to spread out completely. Now count the numbers of the object of interest within each of several fields and calculate a mean value. The volume of each field will be:

$$(\text{drop volume} \times \text{field area}) \div \text{coverslip area} \qquad [46.1]$$

and the density will be:

$$(\text{mean no. of objects per field}) \div \text{field volume} \qquad [46.2]$$

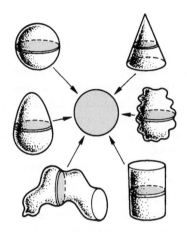

Fig. 46.3 A two-dimensional section may be representative of many different three-dimensional structures.

Where accurate estimates of volume–density are required, you should use a haemocytometer (see p. 294).

Determining the position and orientation of the section within the specimen

When interpreting microscope images, you need to determine and take into account what position and orientation the section was taken from.

Always remember that a section is a two-dimensional slice through a three-dimensional structure. Thus, a circular profile in two dimensions could arise from sectioning numerous three-dimensional shapes (Fig. 46.3); you cannot determine which shape without further information (e.g. from serial sections or sections taken in another plane). Likewise, the size of a profile in a section may reveal little about the size of the structure from which it arose. Also bear in mind that very thin sections (say 75 nm thick) may not provide a representative sample of all the organelles present in a cell (which may have a depth of 1000–100 000 nm).

Always relate what you see to what you know; at a minimum, you should take account of your basic knowledge of animal, microbial or plant cell structure. Note that, in the absence of a defining characteristic, you cannot be sure about the nature of a specimen, because it may simply be due to chance that the section does not include the feature.

The terminology used for sections taken in different planes is shown in Figs 46.4 and 46.5.

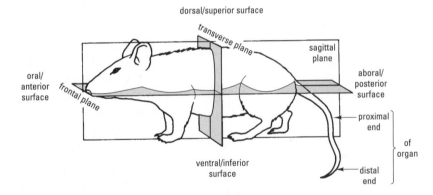

Fig. 46.4 The terms used for various parts and sections through the animal body.

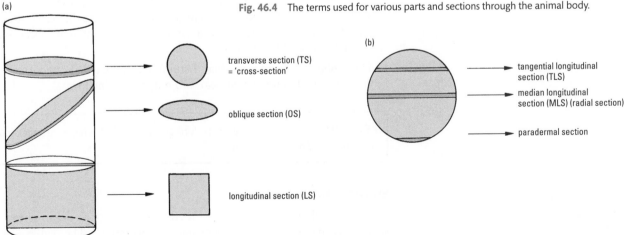

Fig. 46.5 The main section planes: (a) sections through a cylindrical object; (b) various types of longitudinal sections as seen in transverse orientation.

Table 46.1 The main stains used in electron microscopy

Stain	What it stains
Lead citrate	Lipoproteins Ribonucleoproteins Glycogen
Osmium tetroxide	Lipids and proteins
Phosphotungstic acid	Proteins (at pH >3) Polysaccharides Glycoproteins (at pH <3)
Uranyl acetate	Proteins Nucleic acids (at pH 3.5)

Comparing methods – it is informative to compare the same tissue prepared and viewed by different methods.

Accounting for methodology

As discussed on pp. 263, 265–70, a great deal of preparative work is required before specimens can be viewed under LM or TEM. Under normal circumstances, it is hoped that the stain has the greatest effect on the image. The colours produced in various cell types and structures by the main stains used in light microscopy are tabulated on p. 268. You should bear in mind that colours are subjective and that dyes may exhibit different colours in different chemical environments (e.g. at different pH values).

Stains for TEM are generally compounds or ions containing elements of high atomic number, such as the heavy metals, because these scatter electrons strongly. On a positive micrograph from a positively stained specimen, dark regions show where electrons have been scattered, i.e. where the stain has attached. Light regions show where electrons have passed relatively unaffected through the specimen, i.e. where the stain has not been taken up. Table 46.1 shows the main stains used in TEM and the specific cell components to which they attach.

Images from SEM are relatively easy to interpret because the brain is used to interpreting surface features. However, it is possible for optical illusions to occur – for instance, it may be difficult to decide whether a shadowed feature projects out of a surface or represents a depression in it (Chapter 29).

Artefacts

The following is a list of the main types of artefact occurring during the preparation process:

- **Fixation:** swelling or shrinkage may occur, often due to the use of hypotonic or hypertonic fixative solutions. Glutaraldehyde-fixed tissue is particularly likely to shrink if this fixative is used at high concentration.

- **Dehydration:** shrinkage may occur as water is withdrawn from the tissue. Lipids may dissolve in the solvent used and be lost from the tissue.

- **Embedding:** difficulties in sectioning may be caused by incorrect hardness of embedding resins.

- **Sectioning:** too thick a section can result in a blurred image. Compression effects, tearing and knife chattering may occur.

- **Mounting:** folds may be introduced into the section. These can be detected by the presence of discontinuities in features that cross the fold. In TEM, the pattern of the metal grid may sometimes obscure the image (as a solid black area).

- **The stain(s) used on the section:** if staining is not carried out correctly, precipitates of stain may appear, often in crystalline form.

KEY POINT No preparative procedure gives a 'perfect' image. Artefacts can be introduced at many stages.

Stereological studies

Stereology allows the study of the three-dimensional organisation of specimens based on two-dimensional information. The subject is a complex one, relying on geometric and statistical principles, and specialist texts should be consulted for detailed background. There are three main approaches:

1. **Cells or organelles are assumed to have a regular geometric shape,** whose surface area or volume can be estimated from measurements of the principal axes (see Table 74.3). Cell and organelle shapes are generally so irregular as to make such estimates very approximate.

2. **A physical or computer-generated model of structures in three dimensions is built up** from a set of serial sections. Considerable expertise is required to obtain the basic information required.

3. **Representative sections are sampled and information derived from the profiles seen.** This method can be based on simple count data and is surprisingly informative. Three parameters of interest are:

   (a) Numerical density, N_V (the numbers of a component found per unit volume). For spherical components, this can be estimated from:

   $$N_V = N_A/\overline{D} \qquad [46.3]$$

   where N_A is the number of components per unit area and $\overline{D}$ is their mean diameter. N_A is simply obtained by counting component profiles in defined areas. The mean diameter (of spheres) will be approximately equal to the largest diameters seen in the sample, or may be calculated from $\overline{D} = 4/\pi\overline{d}$, where $\overline{d}$ is the mean component diameter of the sample. The situation for non-spherical components is more complex.

   (b) Surface density, S_V (the surface area of the component per unit containing volume). For any shape of object, this can be estimated from:

   $$S_V = 4/\pi B_A \qquad [46.4]$$

   where B_A is the length of profile boundaries per unit area of section. Profile boundary lengths can be measured by fitting a thread or by using a map-measuring wheel or a planimeter. An even simpler method is to count the number of intersections of the bounding membrane I_i with test lines of total length L_T on an overlay (Fig. 46.6). Then S_V is obtained from:

   $$S_V = 2I_i / L_T \qquad [46.5]$$

   Note that this method depends on the components being randomly orientated with respect to the lines on the overlay; this can be tested by comparing results with the lines placed in different orientations.

   (c) Volume density, V_V (the volume of the component as a proportion of the total volume). This is numerically equivalent to A_A, the component density on the test area. A_A can be most simply estimated:

   (i) Destructively, by the cut-and-weigh method. Here, the total area to be considered is cut out, weighed, then the component profiles are cut out and weighed: the ratio of weights equals A_A.

   (ii) Non-destructively, by point counting, where a patterned overlay or lattice (e.g. Fig. 46.6) is placed over the micrograph and counts made of the number of points overlaying the feature of interest and the total area. A_A is estimated by the ratio of these point counts.

Fig. 46.6 Weibel's multi-purpose test overlay for stereological studies. The number of intersections of a membrane with the solid lines can be used when estimating surface densities, while point counts for estimates of profile density on test areas can be made using the ends of the lines.

Both of these methods of estimating A_A require that a reasonably large representative area is sampled. Other methods for estimating A_A include planimetry and linear integration.

Sources for further study

Crang, R.F.E. and Klomparens, K.L. (1988) *Artefacts in Biological Electron Microscopy.* Plenum Press, New York.

Russ, J.C. and Dehoff, R.T. (2000) *Practical Stereology,* 2nd edn. Plenum Publishing Corporation, New York.

STUDY EXERCISES

46.1 Calculate the size of an object seen down a microscope using graticule readings. At ×100 overall magnification, you find that 1 stage micrometer unit of 0.1 mm = 11 eyepiece micrometer units. The length of a parasitic worm on a slide viewed at this magnification appears to be 25 eyepiece units. What is its estimated true length to the nearest µm?

46.2 Calculate the area density of objects seen down a microscope. A light microscope at ×400 overall magnification has a field diameter estimated as 0.425 mm. Observing a sample of leaf surface at this magnification, you count 25 hairs in a single field of view. What is your estimate of the area density (frequency) of hairs, expressed on a per mm² basis, to three significant figures?

46.3 Calculate the volume density of objects seen down a microscope. A diluted blood sample viewed at the same magnification as in Study exercise 46.2 gives a mean count of 32 white blood cells per field of view. If the thickness of the liquid layer below the coverslip is 0.1 mm, what is the volume density of white blood cells (express on a per mm³ basis, to four significant figures)?

46.4 Use point count data to estimate relative volumes of cellular components. A research team fixed and embedded leaf mesophyll cells, sectioned them randomly, stained the sections appropriately and

used a TEM to obtain photomicrographs. Then, using an overlay like the one illustrated in Fig. 46.6, they made point counts within each of 25 cell outlines and took a mean of the results (see table below). From their data, answer the following questions, to two decimal places:

(a) What percentage of the cell volume did the cell wall account for?

(b) What percentage of the cell volume was taken up by the cytoplasm?

(c) What is the estimated cyoplasm: vacuole ratio for this cell?

(d) What percentage of the cytoplasm was taken up by chloroplasts?

Point counts over random mesophyll sections

Organelle	Mean point count
Cell wall	34.0
Cytosol	35.0
Nucleus	16.9
Mitochondria	5.1
Chloroplasts	160.8
Other cytoplasmic bodies	2.6
Vacuole	264.7
Total	519.1

Answers to these study exercises are available at www.pearsoned.co.uk/practicalskills

47 Sterile technique

Sterile technique (aseptic technique) is the name given to the procedures used in cell culture. Although the same general principles apply to all cell types, you are most likely to learn the basic procedures using bacteria and most of the examples given in this section refer to bacterial culture.

Sterile technique serves two main purposes:

1. **To prevent accidental contamination of laboratory cultures** due to microbes from external sources, e.g. skin, clothing or the surrounding environment.

2. **To prevent microbial contamination of laboratory workers**, in this instance you and your fellow students.

> **KEY POINT** *All* microbial cell cultures should be treated as if they contained potentially harmful organisms. Sterile technique forms an important part of safety procedures, and must be followed whenever cell cultures are handled in the laboratory. Avoid any 'hand-to-mouth' operations at all times.

Care is required because:

- **You may accidentally isolate a harmful microbe as a contaminant** when culturing a relatively harmless strain.
- **Some individuals are more susceptible to infection and disease than others** – not everyone exposed to a particular microbe will become ill.
- **Laboratory culture involves purifying and growing large numbers of microbial cells** – this represents a greater risk than small numbers of the original microbe.
- **A microbe may change its characteristics,** perhaps as a result of gene exchange or mutation.

The international biohazard symbol, shown in Fig. 47.1, is used to indicate a significant risk due to a pathogenic microbe (p. 286).

Sterilisation procedures

Given the ubiquity of microbes, the only way to achieve a sterile state is by their destruction or removal. Several methods can be used to achieve this objective:

Heat treatment

This is the most widespread form of sterilisation and is used in several basic laboratory procedures including the following:

- **Red-heat sterilisation.** Achieved by heating metal inoculating loops, forceps, needles, etc. in a Bunsen flame (Fig. 47.2). This is a simple and effective form of sterilisation as no microbe will survive even a brief exposure to a naked flame. Flame sterilisation using alcohol is used for glass rods and spreaders (see below).
- **Dry-heat sterilisation.** Here, a hot-air oven is used at a temperature of at least 160 °C for at least 2 h. This method is used for the routine sterilisation of laboratory glassware. Dry-heat procedures are of little value for items requiring repeated sterilisation during use.

Achieving a sterile state – you should assume that all items of laboratory equipment have contaminating microbes on their surfaces, unless they have been destroyed by some form of sterilisation. Such items will only remain sterile if they do not come into contact with the non-sterile environment.

Fig. 47.1 International symbol for a biohazard. Usually red or black on a yellow background, or black on a red background.

loop handle

wire

blue 'cone' (unburnt gas)

air inlet fully open

Fig. 47.2 'Flaming' a wire loop. Keep the loop in the hottest part of the Bunsen flame (just outside the blue 'cone') until the wire is red-hot.

Fig. 47.3 Autoclave tape – the bottom sample is untreated while the upper sample (with dark diagonal lines) has been autoclaved.

Using a sterile filter – most filters are supplied as pre-sterilised items. Make sure you follow a procedure that does not contaminate the filter on removal from its protective wrapping.

SAFETY NOTE When working with biocides, take care to avoid skin contact or ingestion, as most are toxic and irritant. If you spill biocide on your skin, rinse with plenty of water. If ingested, seek immediate medical help.

Using molten agar – a water bath (at 45–50 °C) can be used to keep an agar-based medium in its molten state after autoclaving. Always dry the outside of the container on removal from the water bath, to reduce the risk of contamination from microbes in the water, e.g. during pour plating (p. 287).

- **Moist-heat sterilisation.** This is the method of choice for many laboratory items, including most fluids, apart from heat-sensitive media. It is also used to decontaminate liquid media and glassware after use. The laboratory autoclave is used for these purposes. Typically, most items will be sterile after 15 min at 121 °C, although large items may require a longer period. The rapid killing action results from the latent heat of condensation of the pressurised steam, released on contact with cool materials in the autoclave. Although special heat-sensitive tape (Fig. 47.3) is sometimes used to check that the autoclave is operating correctly, a better approach is to use spores of *Bacillus stearothermophilus*.

Radiation

Many disposable plastic items used in microbiology and cell biology are sterilised by exposure to UV or ionising radiation. They are supplied commercially in sterile packages, ready for use. Ultraviolet radiation has limited use in the laboratory, while ionising radiation (e.g. γ-rays) requires industrial facilities and cannot be operated on a laboratory scale.

Filtration

Heat-labile solutions (e.g. complex macromolecules, including proteins, antibiotics, serum) are particularly suited to this form of sterilisation. The filters come in a variety of shapes, sizes and materials, usually with a pore size of either 0.2 µm or 0.45 µm. The filtration apparatus and associated equipment is usually sterilised by autoclaving, or by dry heat. Passage of liquid through a sterile filter of pore size 0.2 µm into a sterile vessel is usually sufficient to remove bacteria but not viruses, so filtered liquids are not necessarily virus-free.

Chemical agents

These are known as disinfectants, or biocides, and are most often used for the disposal of contaminated items following laboratory use, e.g. glass slides and pipettes. They are also used to treat spillages. The term 'disinfection' implies destruction of disease-causing bacterial cells, although spores and viruses may not always be destroyed. Remember that disinfectants require time to exert their killing effect – any spillage should be covered with an appropriate disinfectant and left for at least 10 min before mopping up.

Use of laboratory equipment

Working area

One of the most important aspects of good sterile technique is to keep your working area as clean and tidy as possible. Start by clearing all items from your working surface, wipe the bench down with disinfectant and then arrange the items you need for a particular procedure so that they are close at hand, leaving a clear working space in the centre of your bench.

Media

Cells may be cultured in either a liquid medium (broth), or a solidified medium (p. 290). The gelling agent used in most solidified media is agar, a

Agar – a Malay word meaning 'jelly' (originally agar-agar).

Working with plastic disposable loops – these are used in many research laboratories: pre-sterilised and suitable for single use, they avoid the hazards of naked flames and the risk of aerosol formation during heating. Discard into a disinfectant solution after use.

Using a Bunsen burner to reduce airborne contamination – working close to the updraught created by a Bunsen flame reduces the likelihood of particles falling from the air into an open vessel.

Using glass pipettes – these are plugged with cotton wool at the top before being autoclaved inside a metal can. Flame the open end of the can on removal of a pipette, to prevent contamination of the remaining pipettes. Autopipettors and sterile disposable tips (p. 140) offer an alternative approach.

Table 47.1 Classification of microbes on the basis of hazard. The following categories are recommended by the UK Advisory Committee on Dangerous Pathogens (ACDP). A similar approach is followed by the World Health Organization

Hazard group	Comments
1	Unlikely to cause human disease
2	May cause disease: possible hazard to laboratory workers, minimal hazard to community
3	May cause severe disease: may be a serious hazard to laboratory workers, may spread to community
4	Causes severe disease: is a serious hazard to laboratory workers, high risk to community

complex polysaccharide from red algae that produces a stiff transparent gel when used at 1–2% (w/v). Agar is used because it is relatively resistant to degradation by most bacteria and because of its rheological properties – an agar medium melts at 98 °C, remaining solid at all temperatures used for routine laboratory culture. Once melted however, it does not solidify again until the temperature falls to about 44 °C. This means that heat-sensitive constituents (vitamins, blood, cells, etc.) can be added aseptically to the medium after autoclaving.

Inoculating loops

The initial isolation and subsequent transfer of microbes between containers can be achieved by using a sterile inoculating loop. Most teaching laboratories use nichrome wire loops in a metal handle. A wire loop can be repeatedly sterilised by heating the wire, loop downwards and almost vertical, in the hottest part of a Bunsen flame until the whole wire becomes red-hot. Then the loop is removed from the flame to minimise heat transfer to the handle. After cooling for 8–10 s (without touching any other object), it is ready for use.

When re-sterilising a contaminated wire loop in a Bunsen flame after use, do not heat the loop too rapidly, as the sample may spatter, creating an aerosol: it is better to soak the loop for a few minutes in disinfectant than to risk heating a fully charged (contaminated) inoculating loop.

Containers

There is a risk of contamination whenever a sterile bottle, flask or test tube is opened. One method that reduces the chance of airborne contamination is quickly to pass the open mouth of the glass vessel through a flame. This destroys any microbes on the outer surfaces nearest to the mouth of the vessel. In addition, by heating the air within the neck of the vessel, an outwardly directed air flow is established, reducing the likelihood of microbial contamination.

You should flame the mouth of each vessel immediately after opening and then repeat the procedure just before replacing the top. Caps, lids and cotton wool plugs must not be placed on the bench during flaming and sampling: they should be removed and held using the smallest finger of one hand, to minimise the risk of contamination. This also leaves your remaining fingers free to carry out other manipulations. With practice, you will be able to remove the tops from two tubes, flame each tube and transfer material from one to the other while holding one top in each hand.

Laminar-flow cabinets

These are designed to prevent airborne contamination, e.g. when preparing media or subculturing microbes or tissue cultures. Sterile air is produced by passage through a high-efficiency particulate air (HEPA) filter: this is then directed over the working area, either horizontally (towards the operator) or downwards. The operator handles specimens, media, etc., through an opening at the front of the cabinet. Note that standard laminar-flow cabinets do *not* protect the worker from contamination and must not be used with pathogenic microbes: special safety cabinets are used for work with ACDP hazard group 3 and 4 microbes (Table 47.1) and for samples that might contain such pathogens.

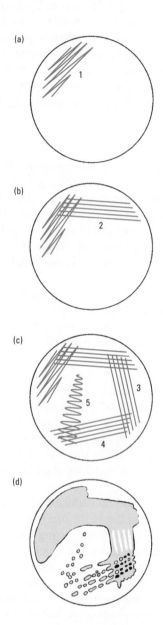

Fig. 47.4 Preparation of a streak plate for single colonies. (a) Using a sterile metal loop, take a small sample of the material to be streaked. Distribute the sample over a small sector of the plate (area 1), then flame the loop and allow to cool (approximately 8–10 s). (b) Make several small streaks from the initial sector into the adjacent sector (area 2), taking care not to allow the streaks to overlap. Flame the loop and allow to cool. (c) Repeat the procedure for areas 3 and 4, re-sterilising the loop between each step. Finally, make a single, long streak, as shown for area 5. (d) The expected result after incubation at the appropriate temperature (e.g. 37 °C for 24 h): each step should have diluted the inoculum, giving individual colonies within one or more sectors on the plate. Further subculture of an individual colony should give a pure (clonal) culture.

Microbiological hazards

> **KEY POINT** The most obvious risks when handling microbial cultures are those due to ingestion or entry via a cut in the skin – all cuts should be covered with a plaster or disposable plastic gloves. A less obvious source of hazard is the formation of aerosols of liquid droplets from microbial suspensions, with the risk of inhalation, or surface contamination of other objects.

The following steps will minimise the risk of aerosol formation:

- **Use stoppered tubes when shaking, centrifuging or mixing microbial suspensions.**
- **Pour solutions gently, keeping the difference in height to a minimum.**
- **Discharge pipettes on to the side of the container.**

Other general rules that apply in microbiology laboratories include:

- **Take care with sharp instruments,** including needles and glass Pasteur pipettes.
- **Avoid any 'hand-to-mouth' actions,** e.g. chewing the end of your pencil while thinking.
- **Do not pour waste cultures down the sink** – they must be autoclaved.
- **Put other contaminated items (e.g. slides, pipettes) into disinfectant after use.**
- **Wipe down your bench with disinfectant when practical work is complete.**
- **Always wash your hands before leaving the laboratory.**

Plating methods

Many culture methods make use of a solidified medium within a Petri plate (= Petri dish). A variety of techniques can be used to transfer and distribute the organisms prior to incubation. The three most important procedures are described below.

Streak-dilution plate

Streaking a plate for single colonies is one of the most important basic skills in microbiology, since it is used in the initial isolation of a cell culture and in maintaining stock cultures, where a streak-dilution plate with single colonies all of the same type confirms the purity of the strain. A sterile inoculating loop is used to streak the organisms over the surface of the medium, thereby diluting the sample (Fig. 47.4). The aim is to achieve single colonies at some point on the plate: ideally, such colonies are derived from single cells (e.g. in the case of unicellular bacteria, animal and plant cell lines) or from groups of cells of the same species (e.g. in filamentous or colonial forms).

> **KEY POINT** Single colonies, containing cells of a single species and derived from a single parental cell, form the basis of all pure culture methods.

Note the following:

- **Keep the lid of the Petri plate as close to the base as possible** to reduce the risk of aerial contamination.

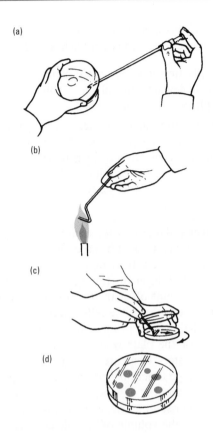

(a)

(b)

(c)

(d)

Fig. 47.5 Preparation of a spread plate. (a) Transfer a small volume of cell suspension (0.05–0.5 mL) to the surface of a solidified medium in a Petri plate. (b) Flame sterilise a glass spreader and allow to cool (8–10 s). (c) Distribute the liquid over the surface of the plate using the sterile spreader. Ensure an even coverage by rotating the plate as you spread: allow the liquid to be absorbed into the agar medium. Incubate under suitable conditions. (d) After incubation, the microbial colonies should be distributed over the surface of the plate.

- **Allow the loop to glide over the surface of the medium.** Hold the handle at the balance point (near the centre) and use light, sweeping movements, as the agar surface is easily damaged and torn.

- **Work quickly, but carefully.** Do not breathe directly over the exposed agar surface and replace the lid as soon as possible.

Spread plate

This method is used with cells in suspension, either in a liquid growth medium or in an appropriate sterile diluent. It is one method of quantifying the number of viable cells (or colony-forming units) in a sample, after appropriate dilution (pp. 148–50).

An L-shaped glass spreader is sterilised by dipping the end of the spreader in a beaker containing a small amount of 70% v/v alcohol, allowing the excess to drain from the spreader and then igniting the remainder in a Bunsen flame. After cooling, the spreader is used to distribute a known volume of cell suspension across the plate (Fig. 47.5). *There is a significant fire risk associated with this technique,* so take care not to ignite the alcohol in the beaker, e.g. by returning an overheated glass rod to the beaker. The alcohol will burn with a pale blue flame that may be difficult to see, but will readily ignite other materials (e.g. a laboratory coat). Another source of risk comes from small droplets of flaming alcohol shed by an overloaded spreader on to the bench and this is why you *must* drain excess alcohol from the spreader *before* flaming. Many laboratories now provide plastic disposable spreaders for student use, to avoid this risk.

Pour plate

This procedure also uses cells in suspension, but requires molten agar medium, usually in screw-capped bottles containing sufficient medium to prepare a single Petri plate (i.e. 15–20 mL), maintained in a water bath at 45–50 °C. A known volume of cell suspension is mixed with this molten agar, distributing the cells throughout the medium. This is then poured without delay into an empty sterile Petri plate and incubated, giving widely spaced colonies (Fig. 47.6). Furthermore, as most of the colonies are formed within the medium, they are far smaller than those of the surface streak method, allowing higher cell numbers to be counted (e.g. up to 1000 colonies per plate): some workers pour a thin layer of molten agar on to the surface of a pour plate after it has set, to ensure that no surface colonies are produced. Most bacteria and fungi are not killed by brief exposure to temperatures of 45–50 °C, though the procedure may be more damaging to microbes from low temperature conditions, e.g. psychrophilic bacteria.

One disadvantage of the pour-plate method is that the typical colony morphology seen in surface-grown cultures will not be observed for those colonies that develop within the agar medium. A further disadvantage is that some of the suspension will be left behind in the screw-capped bottle. This can be avoided by transferring the suspension to the Petri plate, adding the molten agar, then swirling the plate to mix the two liquids. However, even when the plate is swirled repeatedly and in several directions, the liquids are not mixed as evenly as in the former procedure.

Working with phages

Bacterial viruses ('bacteriophages', or simply 'phages') are often used to illustrate the general principles involved in the detection and enumeration of viruses. They also have a role in genome mapping of bacteria (p. 401).

Fig. 47.6 Preparation of a pour plate. (a) Add a known volume of cell suspension (0.05–1.0 mL) to a small bottle of molten agar medium from a 45 °C water bath. (b) Mix thoroughly, by rotating between the palms of the hands: do not shake or this will cause frothing of the medium. (c) Pour the mixture into an empty, sterile Petri plate and allow to set. Incubate under suitable conditions. (d) After incubation, the microbial colonies will be distributed throughout the medium: any cells deposited at the surface will give larger, spreading colonies.

Individual phage particles (virions) are too small to be seen by light microscopy, but are detected by their different effects on susceptible host cells:

- **Virulent phages** will infect and replicate within actively growing host cells, causing cell lysis and releasing new infective phages – this 'lytic cycle' takes ≈30 min for T-even phages of *E. coli,* e.g. T4.
- **Temperate phages** are a specialised group, capable either of lytic growth or an alternative cycle, termed lysogeny – the phage becomes latent within a host cell (lysogen), typically by insertion of its genetic information into the host cell genome, becoming a 'prophage'. At a later stage, termed induction, the prophage may enter the lytic cycle. A widely used example is λ phage of *E. coli.*

The lytic cycle can be used to detect and quantify the number of phages in a sample. A known volume of sample is mixed with susceptible bacterial cells in molten soft agar medium (45–50 °C), then poured on top of a plate of the same medium, creating a thin layer of 'top agar'. The upper layer contains only half the normal amount of agar, to allow phages to diffuse through the medium and attach to susceptible cells. On incubation, the bacteria will grow throughout the agar to produce a homogeneous 'lawn' of cells, except in those parts of the plate where a phage particle has infected and lysed the cells to create a clear area, termed a plaque (Fig. 47.7). Each plaque is due to a single functional phage (i.e. a plaque-forming unit, or PFU). A count of the number of plaques can be used to give the number of phages in a particular sample (e.g. as PFU mL^{-1}), with appropriate correction for dilution and the volume of sample counted in an analogous manner to a bacterial plate count (p. 295). When counting plaques in phage assays you should view them against a black background to make them easier to see: mark each plaque with a spirit-based marker to ensure an accurate count. Temperate phages often produce cloudy plaques, because many of the infected cells will be lysogenised rather than lysed, creating turbidity within the plaque. Samples of material from within the plaque can be used to subculture the phage for further study, perhaps in a broth culture where the phages will cause widespread cell lysis and a decrease in turbidity. Alternatively, phages can be stored by adding chloroform to aqueous suspensions – this will prevent contamination by cellular microorganisms. A similar approach can be used to detect and count animal or human viruses, using a monolayer of susceptible host cells.

Electron microscopy (EM, pp. 262–3) provides an alternative approach to the detection of viruses, avoiding the requirement for culture of infected

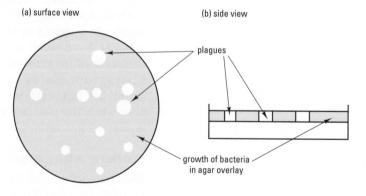

Fig. 47.7 Phage plaques in a 'lawn' of susceptible host bacterium.

host cells, and giving a faster result. However, it requires specialised equipment and expertise. EM counts are often higher than culture-based methods, for similar reasons to those described for bacteria (p. 293).

Labelling your plates and cultures

Petri plates should always be labelled on the *base,* rather than the lid. Restrict your labelling to the outermost region of the plate, to avoid problems when counting colonies, assessing growth, etc. After labelling, Petri plates usually are incubated upside down in a temperature-controlled incubator (often at 37 °C) for an appropriate period (usually 18–72 h). Plates are also usually kept upside down on the lab bench – following incubation, the base (containing medium and microbes) can then be lifted from the lid and examined.

Labelling Petri plates – the following information should be recorded on the *base*:
- date
- the growth medium
- your name or initials
- brief details of the experimental treatment.

Sources for further study

Anon. (1995) *Advisory Committee on Dangerous Pathogens: Categorisation of Biological Agents According to Hazard and Categories of Containment*, 4th edn. HSE Books, London.

Anon. (2013) *Advisory Committee on Dangerous Pathogens: Approved List of Biological Agents.* Available: http://www.hse.gov.uk/pubns/misc208.pdf Last accessed 13/07/15.

Collins, C.H., Lyne, P.M., Grange, J.M. and Falkinham, J. (2004) *Collins and Lyne's Microbiological Methods,* 8th edn. Hodder-Arnold, London.

Hawkey. P. and Lewis, D. (2004) *Medical Bacteriology: A Practical Approach,* 2nd edn. Oxford University Press, Oxford.

World Health Organization (2004) *Laboratory Biosafety Manual,* 3rd edn. Available: http://www.who.int/csr/resources/publications/biosafety/en/Biosafety7.pdf Last accessed 13/07/15.

STUDY EXERCISES

47.1 Decide on the best method of sterilisation. What would be the most appropriate method of sterilisation for the following items?

(a) A box of 100 plastic tips to be used with a pipettor.

(b) A 50 mL batch of blood, for use in 5% v/v blood agar plates.

(c) A 1 L batch of MacConkey agar.

(d) Ten 5 mL glass pipettes.

(e) A microbiological wire, used for 'stab' cultures.

(f) A 10 mL sample of a heat-sensitive solution of an antibiotic, to be used as a component of a selective isolation medium.

47.2 Find out the biohazard classification (UK ACDP categorisation – Table 47.1) for the following microbes:

(a) *Salmonella typhimurium*

(b) *Leptospira interrogans*

(c) *Shigella dysenteriae* (type 1)

(d) *Escherichia coli* K12

(e) *Escherichia coli* O157

(f) Human immunodeficiency virus (HIV)

(g) *Cryptococcus neoformans var. neoformans*

(h) *Mycobacterium tuberculosis*

(i) *Lactobacillus plantarum*

(j) Marburg virus.

47.3 Consider the advantages and disadvantages of spread-plating and pour-plating methods. Having read through this chapter, list up to six pros/cons of each plating method and compare your answers either with the list we have provided on the website, or with those of other students, as a group exercise.

Answers to these study exercises are available at www.pearsoned.co.uk/practicalskills

48 Cell culture

vented cap

flat base, for culture of cells as a monolayer

Fig. 48.1 Plastic flask for animal cell culture – this design provides a large surface area for growth of an adherent monolayer of cells.

Harvesting bacteria from an agar plate – colonies can be harvested using a sterile loop, providing large numbers of cells without the need for centrifugation. The cells are relatively free from components of the growth medium; this is useful if the medium contains substances that interfere with subsequent procedures.

Subculturing – when subculturing microbes from a colony on an agar medium, take your sample from the growing edge, so that viable cells are transferred.

Microbial, animal and plant cell culture methods are based on the same general principles, requiring:

- **a pure culture** (also known as an axenic culture), perhaps isolated as part of an earlier procedure, or from a culture collection;
- **a suitable nutrient medium** to provide the necessary components for growth. This medium must be sterilised before use;
- **satisfactory growth conditions** including temperature, pH, atmospheric requirements, ionic and osmotic conditions;
- **sterile technique** (Chapter 47) to maintain the culture in pure form.

Heterotrophic animal cells, fungi and many bacteria require appropriate organic compounds as sources of carbon and energy. Non-exacting bacteria can utilise a wide range of compounds and they are often grown in media containing complex natural substances (including meat extract, yeast extract, soil, blood). Animal cells have more stringent growth requirements (pp. 298–9).

Photoautotrophic bacteria, cyanobacteria and algae are grown in a mineral medium containing inorganic ions including chelated iron, with a light source and CO_2 supply. Plant cells may require additional vitamins and hormones (p. 299). For chemoautotrophic bacteria, the light source is replaced by a suitable inorganic energy source, e.g. H_2S for sulphur-oxidising bacteria, NH_3/NH_4^+ for nitrifying bacteria, etc.

Growth on solidified media

Many organisms can be cultured on an agar-based medium (pp. 284–5).

KEY POINT An important benefit of agar-based culture systems is that an individual cell inoculated onto the surface can develop to form a visible colony: this is the basis of most microbial isolation and purification methods, including the streak dilution, spread-plate and pour-plate procedures (pp. 286–7).

Animal cells are often grown as an adherent monolayer on the surface of a plastic or glass culture vessel (Fig. 48.1), rather than on an agar-based medium.

Several types of culture vessel are used:

- **Petri plates** (Petri dishes): usually the pre-sterilised, disposable plastic type, providing a large surface area for growth.
- **Glass bottles or test tubes:** these provide sufficient depth of agar medium for prolonged growth of bacterial and fungal cultures, avoiding problems of dehydration and salt crystallisation. Inoculate aerobes on the surface and anaerobes by stabbing down the centre, into the base (stab culture).
- **Flat-sided bottles:** these are used for animal cell culture, to provide an increased surface area for attachment and allow growth of cells as a surface monolayer. Usually plastic and disposable.

The dynamics of growth are usually studied in liquid culture, apart from certain rapidly growing filamentous fungi, where increases in colony diameter can be measured accurately, e.g. using Vernier calipers (p. 151).

Growth in liquid media

Many cells, apart from primary cultures of animal cells, can be grown as a homogeneous unicellular suspension in a suitable liquid medium, where growth is usually considered in terms of cell number (population growth) rather than cell size. Most liquid culture systems need agitation, to ensure adequate mixing and to keep the cells in suspension. An Erlenmeyer flask of 100–2000 mL capacity (Fig. 48.2) can be used to grow a batch culture on an orbital shaker, operating at 20–250 cycles per minute. For aerobic organisms, the surface area of such a culture should be as large as possible: restrict the volume of medium to not more than 20% of the flask volume. Larger cultures may need to be gassed with sterile air and mixed using a magnetic stirrer rather than an orbital shaker. The simplest method of air sterilisation is filtration, using glass wool, non-absorbent cotton wool or a commercial filter unit of appropriate pore size (usually 0.2 μm). Air is introduced via a sparger (a glass tube with many small holes, so that small bubbles are produced) near the bottom of the culture vessel to increase the surface area and enhance gas exchange. More complex systems have baffles and paddles to further improve mixing and gas exchange.

Liquid culture systems may be subdivided under two broad headings: batch culture and continuous culture.

Batch culture

This is the most common approach for routine liquid culture. Cells are inoculated into a sterile vessel containing a fixed amount of growth medium. Your choice of vessel will depend upon the volume of culture required: larger-scale vessels (e.g. 1 L and above) are often called 'fermenters' or 'bioreactors', particularly in biotechnology. Growth within the vessel usually follows a predictable S-shaped (sinusoidal) curve when plotted in log–linear format (Fig. 48.3), divided into four components:

1. **Lag phase:** the initial period when no increase in cell number is seen. The larger the inoculum of active cells the shorter the lag phase will be, provided the cells are transferred from similar growth conditions.

2. **Log phase,** or exponential phase: where cells are growing at their maximum rate. This may be quantified by the growth rate constant, or specific growth rate (μ, or k), where:

$$\mu = \frac{2.303(\ \log N_x -\ \log N_0)}{(\ t_x - t_0)} \qquad [48.1]$$

where N_0 is the initial number of cells at time t_0 and N_x is the number of cells at time t_x. For times specified in hours, μ is expressed as h^{-1}.

Prokaryotes grow by binary fission, whereas eukaryotes grow by mitotic cell division; in both cases each cell divides to give two identical offspring. Consequently, the doubling time or generation time (g, or T_2) is:

$$g = \frac{0.301(t_x - t_0)}{\log N_x - \log N_0} \qquad [48.2]$$

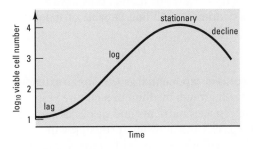

mouth of flask plugged with cotton wool bung to prevent contamination

wide base provides stability and large surface area for mixing

Fig. 48.2 Conical (Erlenmeyer) flask.

Fig. 48.3 Population growth curve for cells in batch culture (liquid medium).

Example Suppose you counted 2×10^3 cells ($\log_{10} = 3.30$) per unit volume at t_0 and 6.3×10^4 cells ($\log_{10} = 4.80$) after 2 h (t_x).

Substitution into eqn [48.1] gives $[2.303(4.8 - 3.3)] \div 2 = 1.727\ h^{-1}$ (or $0.0288\ min^{-1}$).

Substituting the same values into eqn [48.2] gives:
$[0.301 \times 2] \div [4.8 - 3.3] = 0.40\ h$ (or 24 min).

Cells grow at different rates, with doubling times ranging from under 20 min for some bacteria to 24 h or more for animal and plant cells. Exponential phase cells are often used in laboratory experiments, since growth and metabolism are nearly uniform.

3. **Stationary phase:** growth decreases as nutrients are depleted and waste products accumulate. Any increase in cell number is offset by death. This phase is usually termed the 'plateau' in animal cell culture.

4. **Decline phase,** or death phase: this is the result of prolonged starvation and toxicity, unless the cells are subcultured. Like growth, death often shows an exponential relationship with time, which can be characterised by a rate constant (death rate constant), equivalent to that used to express growth or, more often, as the decimal reduction time (d, or T_{90}), the time required to reduce the population by 90%:

$$d = \frac{t_x - t_0}{\log N_x - \log N_0} \qquad [48.3]$$

> **Example** Suppose you counted 5.2×10^5 cells ($\log_{10} = 5.716$) per unit volume at t_0 and 3.7×10^3 cells ($\log_{10} = 3.568$) after 60 min (t_x). Substitution into eqn [48.3] gives $60 \div [2.148] = 27.9$ min. To the nearest minute, this gives a value for d of 28 min.

Some cells undergo rapid autolysis at the end of the stationary period while others show a slower decline.

Batch-culture methods can be used to maintain stocks of particular organisms; cells are subcultured on to fresh medium before they enter the decline phase. However, primary cultures of animal cells have a finite life unless transformed to give a continuous cell line, capable of indefinite growth (p. 300).

Continuous culture

This is a method of maintaining cells in exponential growth for an extended period by continuously adding fresh growth medium to a culture vessel of fixed capacity. The new medium replaces nutrients and displaces some of the culture, diluting the remaining cells and allowing further growth.

After inoculating the vessel, the culture is allowed to grow for a short time as a batch culture, until a suitable population size is reached. Then medium is pumped into the vessel: the system is usually set up so that any increase in cell number due to growth will be offset by an equivalent loss due to dilution, i.e. the cell number within the vessel is maintained at a steady state. The cells will be growing at a particular rate (μ), counterbalanced by dilution at an equivalent rate (D):

Fig. 48.4 Example of a two-dimensional lab equipment diagram of components of a chemostat.

$$D = \frac{\text{flow rate}}{\text{vessel volume}} \qquad [48.4]$$

where D is expressed per unit time (e.g. h^{-1}). In a chemostat, the growth rate is limited by the availability of some nutrient in the inflowing medium, usually either carbon or nitrogen (see Fig. 48.4). In a turbidostat, the input of medium is controlled by the turbidity of the culture, measured using a photocell. A turbidostat is more complex than a chemostat, with additional equipment and controls.

To determine the specific growth rate (μ, or k) of a continuous culture:

1. **Measure the flow of medium through the vessel over a known time interval** (e.g. connect a sterile measuring cylinder or similar volumetric device to the outlet), to calculate the flow rate.

> **Example** Suppose a continuous culture system of 2000 mL volume had a flow of 600 mL over a period of 40 min (flow rate $600 \div 40 = 15$ mL min^{-1}. Substitution into eqn [48.4] gives a dilution rate D of $15 \div 2000 = 0.00075$ min^{-1} or $0.00075 \times 60 = 0.45$ h^{-1}.

Example Suppose you wanted to convert a doubling time of 20 min to a specific growth rate. Rearrangement of eqn [48.5] gives $\mu = 0.693 \div 20 = 0.03465$ min^{-1} ($= 2.08$ h^{-1}). For the example given on page 292, with a growth rate of 0.45 h^{-1}, substitution into eqn [48.4] gives $0.693/0.45 = 1.54$ h (approx. 92 min).

2. **Divide the flow rate by the vessel volume** (eqn [48.4]) to give the dilution rate (D).

3. **This equals the specific growth rate,** since $D = \mu$ at steady state.

4. **If you want to know the doubling time (g), calculate using the relationship:**

$$g = \frac{0.693}{\mu} \qquad\qquad [48.5]$$

(Note that eqn [48.5] also applies to exponential phase cells in batch culture and is useful for interconverting g and μ.)

Continuous culture systems are more complex to set up than batch cultures. They are prone to contamination, having additional vessels for fresh medium and waste culture: strict aseptic technique is necessary when the medium reservoir is replaced, and during sampling and harvesting. However, they offer several advantages over batch cultures, including the following:

- **The physiological state of the cells is more clearly defined,** since actively growing cells at the same stage of growth are provided over an extended time period. This is useful for biochemical and physiological studies.
- **Monitoring and control can be automated and computerised.**
- **Modelling can be carried out for biotechnology/fermentation technology.**

Measuring growth in cell cultures

The most widely used methods of measuring growth are based on cell number.

Direct microscopic counts

One of the simplest methods is to count the cells in a known volume of medium using a microscope and a counting chamber or haemocytometer (Box 48.1). While this gives a rapid assessment of the total cell number, it does not discriminate between living and dead cells. It is also time-consuming as a large number of cells must be counted for accurate measurement. It may be difficult to distinguish individual cells, e.g. for cells growing as clumps.

Electronic particle counters

These instruments can be used to give a direct (total) count of a suspension of microbial cells. The Coulter counter detects particles due to change in electrical resistance when they pass through a small aperture in a glass tube (Fig. 48.5). It gives a rapid count based on a larger number of cells than direct microscopy. It is well suited for repeat measurements or large sample numbers and can be linked to a microcomputer for data processing. If correctly calibrated, the counter can also measure cell sizes. A major limitation of electronic counters is the lack of discrimination between living cells, dead cells, cell clumps and other particles (e.g. dust). In addition, the instrument must be set up and calibrated by trained personnel. Flow cytometry is a more specialised alternative, since particles can be sorted as well as counted (see Macey, 2010).

to vacuum pump

pulse measurement system

sample vessel

internal electrode

external electrode

aperture

cell suspension

Fig. 48.5 Components of an electronic particle counter. During operation, the cell suspension is drawn through the aperture by the vacuum, creating a 'pulse' of resistance between the two electrodes as each cell passes through the aperture.

Alternative approaches for measuring growth – these include biomass, dry weight, turbidity (light scattering: nephelometry), uptake of radiolabelled substrates, absorbance or assay of any major cellular component, e.g. protein, nucleic acid, ATP, etc.

Box 48.1 How to use a counting chamber or haemocytometer

A counting chamber is a specially designed slide containing a chamber of known depth with a grid etched on to its lower surface. When a flat coverslip is placed over the chamber, the depth is uniform. Use as follows:

1. Place the special coverslip over the chamber. Press the edges firmly, to ensure that the coverslip makes contact with the surface of the slide, but take care that you do not break the slide or coverslip by using too much force. When correctly positioned, you should be able to see interference rings (Newton's rings) at the edge of the coverslip.

2. Add a small amount of your cell suspension to fill the central space above the grid. Place on the microscope stage and allow the cells to settle (2-3 min).

3. Examine the grid microscopically. Using the ×10 objective lens first, since the counting chamber is far thicker than a standard microscope slide. Then switch to the ×40 objective: take care not to scratch the surface of the objective lens, as the special coverslip is thicker than a normal coverslip. For a dense culture, the small squares are used, while the larger squares are used for dilute suspensions. You may need to dilute your suspension if it contains more than 30 cells per small square.

4. Count the number of cells in several squares: at least 600 cells should be counted for accurate measurements. Include those cells that cross the upper and left-hand boundaries, but not those that cross the lower or right-hand rulings (p. 198). A hand tally may be used to aid counting. Motile cells must be immobilised prior to counting (e.g. by killing with a suitable biocide).

5. Divide the total number of cells (C) by the number of squares counted (S), to give the mean cell count per square.

6. Determine the volume (in mL) of liquid corresponding to a single square (V), e.g. a Petroff-Hausser chamber has small squares of linear dimension 0.2 mm, giving an area of 0.04 mm²; since the depth of the chamber is 0.02 mm, the volume is 0.04 × 0.02 = 0.0008 mm³; as there are 1000 mm³ in 1 mL, the volume of a small square is 8 × 10^{-7} mL; similarly, the volume of a large square (equal to 25 small squares) is 2 × 10^{-5} mL. Note that other types of counting chamber will have different volumes: check the manufacturer's instructions. For example, the improved Neubauer chamber (Fig. 48.6) has small squares of volume 0.00025 mm³ = 2.5 × 10^{-7} mL.

7. Calculate the cell number per mL by dividing the mean cell count per square by the volume of a single square (in mL).

Fig. 48.6 Haemocytometer grid (improved Neubauer rulings) viewed microscopically. The large square (delimited by triple etched lines) has a volume of 1/250th mm³ (0.04 mm³ = 4 nL) while each small square (16 contained within the large square) has a volume of 1/4000 mm³ (0.00025 mm³ = 0.25 nL). Note that the boundary line for squares delimited by triple-etched lines is the *middle* line, so this line must be used when counting (see Fig. 32.4, for how to deal with objects straddling the gridlines).

8. Remember to take account of any dilution of your original suspension in your final calculation by multiplying by the reciprocal of the dilution (M), e.g. if you counted a 1 in 20 dilution of your sample, multiply by 20, or if you diluted to 10^{-5}, multiply by 10^5.

The complete equation for calculating the total microscopic count is:

$$\text{Total cell count (per mL)} = (C \div S \div V) \times M \quad \quad [48.6]$$

For example, if the mean cell count for a hundred–fold dilution of a cell suspension, counted using a Petroff–Hausser chamber, was 12.4 cells in ten small squares, the total count would be

$$(12.4 \div 10 \div 8 \times 10^{-7}) \times 10^2 = 1.55 \times 10^8 \text{ mL}^{-1}$$

A simpler, less accurate approach is to use a known volume of sample under a coverslip of known area on a standard glass slide, counting the number of cells per field of view using a calibrated microscope of known field diameter, then multiplying up to give the cell number per mL (see p. 278).

---

1. Prepare serial decimal dilutions of the sample in a sterile diluent (pp. 148–9). The most widely used diluents are 0.1% w/v peptone water or 0.9% w/v NaCl, buffered at pH 7.3. Take care that you mix each dilution before making the next one. For soil, food, or other solid samples, make the initial decimal dilution by taking 1 g of sample and making this up to 10 mL using a suitable diluent. Gentle shaking or homogenisation may be required for organisms growing in clumps. The number of decimal dilutions required for a particular sample will be governed by your expected count: dilute until the expected number of viable cells is around 100–1000 mL^{-1}.

2. Transfer an appropriate volume (e.g. 0.05–0.5 mL) of the lowest dilution to an agar plate using either the spread-plate method or the pour-plate procedure (p. 287). At least two, and preferably more, replicate plates should be prepared for each sample. You may also wish to prepare plates for more than one dilution, if you are unsure of the expected number of viable cells.

3. Incubate under suitable conditions for 18–72 h, then count the number of colonies on each replicate plate at the most appropriate dilution. The most accurate results will be obtained for plates containing 30–300 colonies. Mark the base of the plate with a spirit-based pen each time you count a colony. Determine the mean colony count per plate at this dilution (C).

4. Calculate the colony count per mL of that particular dilution by dividing by the volume (in mL) of liquid transferred to each plate (V).

5. Now calculate the count per mL of the original sample by multiplying by the reciprocal of the dilution: this is the multiplication factor (M); e.g. for a dilution of 10^{-3}, the multiplication factor would be 10^3. For soil, food or other solid samples, the count should be expressed per g of sample.

The complete equation for calculating the viable count is:

$$\text{Count per mL (or per g)} = (C \div V) \times M \quad [48.7]$$

For example, for a sample with a mean colony count of 5.5 colonies per plate for a volume of 0.05 mL at a dilution of 10^{-7}, the count would be:

$$(5.5 \div 0.05) \times 10^7 = 1.1 \times 10^9 \text{ CFU mL}^{-1}$$

The count should be reported as colony-forming units (CFU) per mL, rather than as cells per mL, since a colony may be the product of more than one cell, particularly in filamentous microbes or in organisms with a tendency to aggregate. You should also be aware of the problems associated with counts of zero – these are best recorded as '<1', and you should then apply the appropriate correction factors for dilution and volume to obtain the detection limit. For example, a zero count (<1) of 100 μL of a five-fold dilution gives a detection limit of (<1 ÷ 0.1) × 5 = <50 CFU mL^{-1}.

Using culture-based methods and selective media – injured cells and stressed microbes may not be able to grow on selective media under standard aerobic conditions unless a non-selective resuscitation stage is added. Alternative approaches involve the addition of scavengers of reactive oxygen species (ROS), e.g. pyruvate.

Definition

CFU – colony-forming unit: a cell or group of cells giving rise to a single colony on a solidified medium.

Culture-based counting methods

A variety of culture-based techniques can be used to determine the number of microbes in a sample. A major assumption of such methods is that, under suitable conditions, an individual viable microbial cell will be able to multiply and grow to give a visible change in the growth medium, i.e. a colony on an agar-based medium, or turbidity ('cloudiness') in a liquid medium. You are most likely to gain practical experience using bacterial cultures, counted by one or more of the following methods:

- **Spread- or pour-plate methods** ('plate counts', p. 287). The most widespread approach is to transfer a suitable amount of the sample to an agar medium, incubate under appropriate conditions and then count the resulting colonies (Box 48.2).

- **Membrane filtration.** For bacterial samples where the expected cell number is lower than 10 CFU mL^{-1}, pass the sample through a sterile filter (pore size 0.2 μm or 0.45 μm). The filter is then incubated on a suitable medium until colonies are produced, giving a count by dividing the mean colony count per filter by the volume of sample filtered.

Alternative approaches in plate counting – when large numbers of samples have to be counted, a single agar plate can be divided into segments and a single droplet of each dilution placed into the appropriate segment ('Miles and Misra' droplet counting).

Counting injured or stressed microbes – a resuscitation stage may be required, to allow cells to grow under selective conditions, Box 38.1.

Molecular approaches to enumerating microbes – DNA-based methods can be used, including real-time quantitative polymerase chain reaction (RT-qPCR, Reed et al., 2016).

● **Multiple tube count, or most probable number (MPN). A** bacteriological technique where the sample is diluted and known volumes are transferred to several tubes of liquid medium (typically, five tubes at three volumes), chosen so that there is a low probability of the smallest volumes containing a viable cell. After incubation, the number of tubes showing growth (turbidity) is compared to tabulated values to give the most probable number (MPN per mL).

The principal advantage of culture-based counting procedures is that dead cells will not be counted. However, for such techniques, the incubation conditions and media used may not allow growth of all cells, underestimating the true viable count. Further problems are caused by cell clumping and dilution errors. In addition, such methods require sterile apparatus and media and the incubation period is lengthy before results are obtained. An alternative approach is to use direct microscopy, combined with 'vital' or 'mortal' staining. For example, the direct epifluorescence technique (DEFT) uses acridine orange and UV epifluorescence microscopy to separate living and dead bacteria, while neutral red is a vital stain used for plant cells. Chapter 44 gives examples of vital/mortal stains for other cell types. A further approach is to use DNA-based methods, such as the polymerase chain reaction (PCR, pp. 403–4).

Text references

Macey, M.H. (2010) *Flow Cytometry: Principles and Applications.* Humana Press, London.

Reed, R.H., Holmes, D., Weyers, J.D.B. and Jones, A.M. (2016) *Practical Skills in Biomolecular Sciences,* 5th edn. Pearson, Harlow.

Sources for further study

Cann, A.J. (2000) *Virus Culture: A Practical Approach.* Oxford University Press, Oxford.

Davis, J.M. (2002) *Basic Cell Culture: A Practical Approach,* 2nd edn. Oxford University Press, Oxford.

Helgason, C.D. and Miller, C. (2013), *Basic Cell Culture Protocols,* 4th edn. Human Press, London.

STUDY EXERCISES

48.1 **Calculate the specific growth rate and doubling time of cells in culture.** What are the specific growth rates and doubling times of the following (give all answers to three significant figures)?

(a) A broth culture of the yeast *Saccharomyces cerevisiae* growing in exponential phase and containing 5.2×10^4 cells at 10 a.m. and 3.4×10^6 cells at 7 p.m.

(b) A log phase culture of *Escherichia coli* containing 3.0×10^4 CFU mL^{-1} at the start of the experiment and 6.7×10^7 CFU mL^{-1} at the end of the experiment, 200 min later.

(c) An actively growing culture of *Bacillus subtilis* containing 32 bacteria in 25 squares of a haemo-

cytometer chamber at 13.00 h and 250 bacteria in 25 squares of the same haemocytometer chamber at 15.30 h.

48.2 **Practise the calculations involved in using a haemocytometer to make a direct microscopic count.** Express your answers to three significant figures in all cases.

(a) The mean cell count of a yeast suspension per small square of an improved Neubauer counting chamber was 6.42. If the volume of a small square is 2.5×10^{-7} mL, what is the cell count per mL?

(continued)

STUDY EXERCISES (continued)

(b) The following counts were obtained for bacteria in 20 individual small squares of a Petroff–Hauser counting chamber: 26, 36, 42, 35, 27, 16, 29, 50, 24, 43, 41, 35, 18, 36, 33, 47, 25, 46, 32, 57. If the volume of each small square is 8×10^{-7} mL, what is the cell count per mL?

(c) A 10^{-2} dilution of a dense suspension of *E. coli* was examined microscopically using an improved Neubauer counting chamber, giving a total of 78 cells in a total of 25 small squares. If the volume of each small square is 2.5×10^{-7} mL, what is the cell count per mL of the original (undiluted) suspension.

(d) A dilute suspension of yeast cells was concentrated ten-fold by centrifuging 10 mL and resuspending the pellet in 1 mL. The mean cell count of the concentrated yeast suspension per large square of a Petroff–Hauser counting chamber was 6.8. If the volume of the large square is 2×10^{-5} mL, what is the cell count per mL of the original suspension?

48.3 **Practise the calculations involved in making a plate count.** Express your answers to three significant figures.

(a) The mean spread-plate count for 100 μL of a 10^{-5} dilution of a culture of *E. coli* was 54.4 CFU. What is the mean plate count per mL of the original suspension?

(b) Three replicate plates of nutrient agar were each spread with 200 μL of a 10^{-3} dilution of a bacterial suspension, giving colony counts of 34, 40 and 37 after 24 h incubation. What is the mean plate count per mL of the original suspension?

(c) A twenty-fold dilution of a yeast suspension was used to prepare four replicate pour plates, each containing 500 μL of this dilution and giving counts of 211, 186, 194 and 202 after incubation. What is the mean plate count per mL of the original suspension?

(d) A sample of 50 g of raw seafood was homogenised and diluted to 2% w/v in sterile saline solution. Three replicate pour plates were prepared using 1 mL of the diluted sample, giving counts for *E. coli* of 35, 41 and 32. What is the average count of *E. coli* per 100 g of raw seafood?

(e) Duplicate samples of 250 mL of river water were filtered through separate sterile membranes of pore size 0.2 μm. The membranes were then transferred to the surface of an agar-based medium and incubated for 48 h at 37 °C, giving colony counts of 45 and 32 respectively. What is the mean count per 100 mL of river water?

(f) A sample of bottled drinking water was processed by serial decimal dilution and 500 μL samples of each dilution were pour-plated and incubated at 22 °C for 72 h. The colony counts obtained for the three replicate plates of the 10^{-1} dilution were 28, 32 and 39. Does this water meet the EU regulations for mineral water, which specify a maximum average plate count at 22 °C of 100 CFU mL^{-1}?

Answers to these study exercises are available at www.pearsoned.co.uk/practicalskills

Although the aim of many studies is to isolate, quantify or characterise individual molecules from a biological system, e.g. the purification of a particular enzyme (p. 390), this is not always the most appropriate course of action. Depending on the purpose of the investigation, it may be more relevant to study the functioning of biomolecules within more complex systems, in order to understand their role in a particular biological process. At one extreme this may be carried out *in vivo*, using whole multi-cellular organisms (e.g. individual animals or plants), while the other extreme is represented by *in vitro* studies, using subcellular 'cell-free' extracts. Between these two extremes, a range of tissue and cell culture techniques offers some of the biological complexity of the intact organism combined with a degree of experimental control that may not be obtainable *in vivo*. The ethics and costs of whole animal experimentation have provided an additional stimulus to the development of *in vitro* methods, e.g. toxicity tests using mammalian cell culture, while developments in molecular genetics have led to further applications of cell and tissue culture.

Animal tissues and organs

Physiological experiments are carried out using either whole organisms, or a range of animal organs and tissues, including heart, liver, muscle, etc.

> **KEY POINT** A major practical consideration is that the tissue should be studied as soon as possible after the death of the animal, typically under laboratory conditions that mimic the *in vivo* environment as closely as possible.

In most instances, the experiments are relatively short term (<24 h) and the aim is to maintain the tissue in a physiological state similar to that within the living organism. For metabolic studies, the whole organ or a tissue slice (typically 1–10 mm thick) will be bathed in an appropriate perfusion fluid, supplied either by gravity or by peristaltic pump. Practical considerations include:

- **Inorganic solute requirements** – the chemical composition of the perfusion fluid is usually chosen to reflect the major inorganic ion requirements of the tissue. For short-term studies, a number of so-called 'physiological salt solutions' may be used, e.g. Ringer's solution, one formulation of which is given in Table 49.1.

- **Oxygen requirements** – it may be necessary to increase the O_2 content of the perfusion fluid by bubbling with air, in order to meet the oxygen demand of the innermost parts of the tissue. However, this can lead to oxygen toxicity in the outermost parts and an alternative approach is simply to increase the rate of perfusion.

- **Physico-chemical conditions** – including temperature (usually controlled to ± 1 °C of normal body temperature), water status (the perfusion fluid and the tissue should be isotonic, p. 161), pH and buffering capacity (e.g. some perfusion fluids have elevated an $NaHCO_3$ concentration, to mimic the buffering capacity of mammalian serum).

Definitions

In silico – occurring within a computer system (literally 'in silicon').

In vitro – occurring outside a living organism, in an artificial environment (literally 'in glass').

In vivo – occurring within a living organism (literally 'in life').

Working with vertebrate animals and their organs/tissues – remember that procedures must be consistent with the law, i.e. in the UK, the *Animals Scientific Procedures Act 1986*.

Using human tissues – in the UK, the *Human Tissue Act 2004* applies and work is licensed and overseen by the Human Tissue Authority. Note that established cell lines (p. 300) are excluded from the regulations.

Table 49.1 Composition of Ringer's solution (simplified formulation, for amphibians, etc., pH 7.4–7.6)

Compound	Concentration (g L^{-1})
NaCl	6.0
KCl	0.075
CaCl$_2$	0.1
NaHCO$_3$	0.1

Working with plant tissues and organs – the ethical problems associated with animal and human tissues are avoided, and there is a decreased risk of infection to the laboratory worker; practical manipulations are often carried out in horizontal laminar-flow cabinets.

Table 49.2 Components of Long Ashton medium (nitrate version)

Stock solution: mass of component required per litre of solution (g)	Volume of stock solution to make 1 L of medium (mL)
Major nutrients	
KNO_3: 50.60	8
$Ca(NO_3)_2$: 80.25	8
$MgSO_4.7H_2O$: 46.00	8
$NaH_2PO_4.2H_2O$: 52.00	4
Micronutrients	
FeKEDTA: 3.30	5
$MnSO_4.4H_2O$: 2.23	1
$ZnSO_4.7H_2O$: 0.29	1
$CuSO_4.5H_2O$: 0.25	1
H_3BO_3: 3.10	1
$Na_2MoO_4.2H_2O$: 0.12	1
NaCl: 5.85	1
$CoSO_4.7H_2nO$: 0.056	1

Effects of humans on plants – remember that your exhaled breath will be nearly saturated with water vapour and will contain CO_2 at 3–4% v/v, some 100 times more concentrated than atmospheric CO_2: your breath can thus affect rates of transpiration and photosynthesis.

- **Organic nutrient requirements** – for longer-term studies, suitable nutrients will be required: these may be chemically defined additives, e.g. vitamins, amino acids, proteins, etc., or biological fluids such as plasma or serum. Glucose is often added as a carbon and energy source (Freshney, 2010).

Plant tissues and organs

Individual plant components (e.g. leaves, leaf slices and epidermal strips) can be isolated from the main plant body for study under controlled conditions. Since photosynthetic plant parts are autotrophic, they may be maintained *in vitro* for longer than animal organs, given adequate light and CO_2. However, most plant cells are joined via plasmodesmata, and the separation of such connections when the component is removed from the plant often leads to death when these connections are broken.

KEY POINT Plants show wound responses that may affect the metabolic processes under study. For these reasons, the most suitable systems for longer-term studies are often whole organs, e.g. whole leaves or entire root systems.

Water culture (hydroponics), e.g. in Long Ashton medium (Table 49.2), is an alternative approach, offering greater control over the root environment (Smith, 2000).

It is best to use vigorous, healthy stock plants and to follow a well-established procedure, taking account of the following:

- **Sterility** – strict attention to sterile technique can be essential to the success of many longer-term experiments. Decontamination of plant organs may be especially difficult where specimens are obtained from soil: to achieve this, use a surface wash with disinfectant (e.g. 10% w/v sodium hypochlorite), followed by several rinses with sterile water.

- **Gaseous environment** – in general, the experimental system should be well ventilated. Actively photosynthetic tissues will rapidly deplete the atmospheric CO_2 in a closed vessel: plant parts may also produce physiologically active gases, such as ethylene, especially at wound sites. Turgor loss may occur in isolated plant parts unless a high humidity is maintained.

- **Nutrition** – plant tissues may benefit from a supply of inorganic ions, including K^+, SO_4^{2-}, etc., and may require certain vitamins, micronutrients and plant hormones for prolonged studies.

- **Physico-chemical conditions** – light is the most important environmental requirement for green plant parts. At atmospheric CO_2 concentrations, the light compensation point (p. 433) is about 5–10 µmol photons PAR $m^{-2}s^{-1}$ and photosynthesis is usually saturated between 500 and 2000 µmol photons PAR $m^{-2}s^{-1}$, depending on the plant type. Light quality and photoperiod (daylength) are also important: fluorescent tubes and incandescent bulbs that mimic the photosynthetic spectrum of sunlight are available (see p. 373), while photoperiod can be controlled using a timer. The water potential of aqueous media can be adjusted using an impermeant osmoticum such as mannitol, and pH values are often kept close to those of the apoplasm ($\approx$pH 6) using appropriate buffers, if necessary (pp. 167–8).

Definitions

Cell line – a cell culture derived by passage of a primary culture.

Clone – a population of cells derived from a single original cell, i.e. sharing the same genotype.

Confluence/confluent growth – merging of individual cells to form a continuous layer.

Continuous cell line – a culture with the capacity for unlimited multiplication *in vitro*. Sometimes termed an established cell line.

Finite cell line – a culture with a limited capacity for growth *in vitro* (maximum number of cell doublings).

Hybridoma – a hybrid cell produced by fusion of a tumour (myeloma) cell and an antibody-producing B lymphocyte (p. 377). A cell line derived from a single hybridoma will produce a single antibody type (monoclonal).

Immortalisation – conversion of a finite cell line into a continuous cell line.

Passage – an alternative term for subculture.

Primary culture – a cell culture derived from tissue or organ fragments (explants). Primary culture ends on first subculture.

Senescence – the end point in the limited lifespan of a finite cell line, characterised by the lack of proliferation. Conversion to an established (continuous) cell line requires 'escape' from senescence.

Stem cell – an undifferentiated animal cell that can generate new stem cells through mitosis, or can develop and differentiate into a particular specialised cell type.

Transformation – a permanent alteration in the growth characteristics of a finite cell line that may include (i) changes in morphology, (ii) an increased growth rate and/or (iii) the acquisition of an infinite lifespan, often termed immortalisation. Transformation may be spontaneous or may be induced by chemical agents or viruses, and often involves a change in chromosome number.

Cell and tissue cultures

Many of the basic principles involved in culturing animal and plant cells are broadly similar to those described for microbial cell culture (p. 290). Definitions for several key terms are given in the margin.

Applications of cell and tissue culture

The main uses of animal and plant cell culture systems include:

- **Experimental model systems in biochemistry, pharmacology and physiology:** cell culture offers certain advantages over whole organism studies, with greater control over environmental conditions and biological variability. The use of genetically defined clones of cells may simplify the analysis of experimental data. Conversely, results obtained with specialised cell-based systems might be unrepresentative of a broader range of cell types and may be more difficult to interpret in terms of the whole organism.

- **Studies of the growth requirements of particular cells:** including studies of the positive effects of growth factors or growth-promoting substances, and the negative effects of xenobiotics or cytotoxic compounds. The use of cell culture in bioassays and mutagenicity testing is considered in Chapter 51.

- **Studies of cell development and differentiation:** including aspects of the cell cycle and gene expression. Cell cultures retaining their ability to differentiate *in vitro* are particularly interesting to researchers, while the lack of differentiation and unlimited growth of many animal cell lines makes them useful models of tumour development.

- **Pathological studies:** including culture of foetal cells for karyotyping and detection of genetic abnormalities, e.g. trisomy.

- **Genetic manipulation:** cell culture techniques have played an essential role in the development of molecular biology, including the production of transgenic animals and plants by techniques such as transfection.

- **Biotechnology:** including the industrial production of therapeutic proteins, vaccines and monoclonal antibodies using large-scale hybridoma culture techniques similar to those used in microbiology (pp. 291–3).

- **Stem cell technology:** increasingly important in biomedical research and development.

> **KEY POINT** One of the main differences between organ and tissue incubation techniques and those used in cell/tissue culture is that the former aim simply to maintain metabolic and physiological activity for a limited period, while the latter provide conditions suitable for cell growth, division and development *in vitro* over an extended timescale, from a few days to several months.

Animal cell culture systems

These may be established either from whole organisms (e.g. chick embryo), discrete organs (e.g. rat liver) or from blood (e.g. lymphocytes), typically using *mild* tissue disruption techniques, e.g. enzyme treatment, where necessary.

KEY POINT Although, in theory, it is possible to culture nucleated cells from virtually any source, in practice, the highest rates of success are most often achieved with young, actively growing tissues.

The principal considerations in animal cell culture are:

SAFETY NOTE When working with tissue cultures safely, do not confuse laminar-flow hoods with biosafety cabinets (p. 285).

SAFETY NOTE When sterilising during tissue culture (Box 49.1), take care when using 70% v/v alcohol near a naked flame (e.g. Bunsen or spirit lamp) – it is easily ignited and burns with a weakly visible flame.

Example Dulbecco's modified Eagle's medium contains Ca^{2+}, Fe^{3+}, Mg^{2+}, K^+, Na^+, Cl^-, SO_4^{2-}, PO_4^{2-}, glucose, 20 amino acids, 10 vitamins, inositol and glutathione. Foetal calf serum is usually added at up to 20% v/v.

Using antibiotics in cell culture – a typical antimicrobial supplement might include antibacterial agents, e.g. penicillin and streptomycin, an antifungal agent, e.g. griseofulvin, and an antimycoplasmal agent, e.g. gentamicin. Use sparingly to treat contamination and discontinue once the contaminant has been eradicated.

- **Safety:** it is important to be aware of the potential dangers of infection from cell cultures. Although avian and rodent cells present a reduced risk of disease transmission compared with human cells, all cell cultures must be regarded as a potential source of pathogenic microbes, and appropriate sterile technique (Box 49.1) must be used at all times (see also pp. 283–4). Work involving human tissues and cell cultures must be carried out in a biosafety cabinet by trained, experienced personnel – you are unlikely to gain practical experience with such cultures in the early stages of your course. For other cell cultures, you will need to follow the code of safe practice of your department – consult your departmental safety officer if you have any doubts about safe working procedures.

- **Whether to use a primary culture or a cell line:** freshly isolated cells are more likely to reflect the biochemical activities of cells *in vivo,* though they will have a limited lifespan in culture, requiring repeated isolation for longer-term projects. Continuous cell lines are more easily cultured and offer the advantage that their growth requirements in culture may be known in some detail, especially for the more widely used cell lines (e.g. BHK, HeLa).

- **The requirements for a solid substratum:** some cells must be attached to a solid surface in order to grow. Anchorage dependence is a typical feature of primary cultures and finite cell lines – such cultures show density-dependent growth inhibition once the cells have formed a confluent monolayer on the surface of the substratum. An alternative approach is to grow such cells on a particulate support using 'micro-carrier' beads. In contrast, many continuous cell lines can be maintained in suspension culture, as individual cells or aggregates.

- **The physico-chemical conditions,** including pH (typically 7.2–7.5) and buffering capacity, osmolality (usually 300 ± 20 mosmol kg^{-1}) and temperature (e.g. 35–37 °C for mammalian cells).

- **The requirements of the culture medium:** these will include the provision of inorganic ions (as a balanced salt solution), a carbon/energy source plus other organic nutrients, and in some cases a supplement containing antimicrobial agents to counter the risks of contamination. For example, Dulbecco's modified Eagle's medium is used for many mammalian cell types that grow as adherent monolayers, while suspension cultures of continuous cell lines can be maintained using less stringent media. To support growth, the basal medium is usually supplemented with serum (usually foetal calf serum, at up to 20% v/v), or a chemically defined serum-like supplement containing a mixture of proteins, polypeptides, hormones, lipids and trace components. The *in vitro* level of CO_2 and O_2 must also be considered: many cell cultures are buffered using bicarbonate, and must be maintained in an atmosphere of elevated CO_2, either in a sealed culture vessel or in a CO_2 incubator, to maintain pH balance.

Box 49.1 Sterile technique and its application to animal and plant cell culture

Although Chapter 47 gives general advice on the basic principles of sterile technique with microbial cultures, animal and plant cell cultures need additional precautions due to the complexity of some of the procedures and the likelihood of rapid overgrowth of any contaminating microbes. For routine work on the open bench, you should be aware of the following points:

- **Consider personal clothing and hygiene:** long hair should be tied back, or retained by a net/cap. Avoid pendulous earrings and similar items. Wash your hands at the outset, to remove any loose skin flakes: surgical gloves may be worn, though there is loss of tactile sensitivity and comfort. Hands/gloves should be swabbed with 70% v/v alcohol to further reduce the risk of contamination. If you have a cold/cough, consider wearing a face mask.

- **Work in a designated quiet area:** there should be no air currents (avoid open windows) and no 'through traffic' or other activity that might give rise to contamination (e.g. microbes should not be cultured in the same location). The term 'quiet area' also highlights that no talking should occur, to reduce the risk of aerosol contamination from the oral microflora.

- **Organise your work surface:** clear everything from the bench, swab with 70% v/v alcohol, then position all items around a central area, so you can reach every item without reaching across anything. A well-organised workspace reduces the risk of accidental contact between sterile and non-sterile items. Swab bottles and flasks with 70% v/v alcohol before positioning them at the edge of your workspace. Also, swab your work surface between procedures and at the end of the session.

- **Work close to a Bunsen flame,** positioned centrally within your workspace where convection currents create an upward airflow, reducing the likelihood of particles falling from the air into an open vessel. Always flame the tops of glass bottles (p. 285), but not plastic items, for 2–3 s both *before* and *after* opening and closing, rotating the bottles during exposure to the flame. Flame glass pipettes, as described on p. 285.

- **Tilt flasks and bottles during use:** uncapped culture flasks and media bottles are best kept at a shallow angle, to minimise the risk of airborne contamination.

- **Work without delay, but do not hurry:** always keep in mind that the air contains contaminant microbes and the longer you leave a vessel open to the air, the more likely it is that it will become contaminated.

When preparing to work in a laminar-flow cabinet, the following additional aspects should be considered:

- **Note whether the airflow is horizontal or vertical:** horizontal flow gives greatest protection to the work area and provides the most stable air flow, while vertical-flow cabinets reduce exposure of the operator. A biosafety cabinet should be used for work with potentially hazardous cultures (e.g. human cells/tissues, or with any cells known to be infected with a virus).

- **Prepare the cabinet:** switch on the cabinet and leave running for at least 10 min, then swab the work surface and other interior surfaces with 70% v/v alcohol. Swab the outsides of bottles, flasks, etc., before bringing them into the cabinet. Arrange items around a crescent-shaped central work area.

- **Carry out your work with due regard for the airflow within the cabinet:** use of a Bunsen burner is often discouraged, as it disrupts the correct airflow, creates heat and can ignite flammable items within the restricted interior of the cabinet. Always try to keep your hands/arms further away from the sterile airstream than the items within the cabinet, e.g. avoid working *behind* an open vessel in a horizontal-flow cabinet and *above* an open vessel in a vertical-flow cabinet.

- **Take particular care when using pipettes:** a laminar-flow hood will have a restricted interior volume, and it is easy for the novice to touch a pipette tip on the hood, or on another item during use. When removing a pipette from its packaging and connecting to a pipette filler (p. 140), you should point the tip of the pipette away from the user and into the air flow, holding it well above the graduated scale markings to avoid contamination. The pipette and filler are then held horizontally until required: as this can be a tricky procedure for the novice, you should not take a pipette out of its container until you are ready to use it.

- **Work carefully with open bottles and flasks:** caps can be either placed top-down on the work surface, or held in a crooked finger (Fig. 47.6(a)). Since the airflow is sterile, there is less need to work swiftly when compared to the open bench, but you should still recap all containers as soon as you have finished a particular procedure.

Definitions

Callus – an aggregation of undifferentiated plant cells in culture.

Embryogenic callus – tissue with the capacity to differentiate under defined laboratory conditions, typically in response to plant growth regulators in the medium.

Explant – a fragment of tissue used to initiate a culture (the term is also used in animal culture).

Protoplasts – cells lacking their cell walls.

Sphaeroplasts (US, spheroplasts) – cells with attached fragments of their cell walls: osmotically sensitive.

Totipotency – the ability (of any plant cell) to de-differentiate and re-differentiate into any of the cell types found in the mature plant.

In some cases, a pH-sensitive dye (e.g. phenol red, p. 166) may be incorporated into the growth medium, to provide a visual check on pH status during growth – a colour change indicates acidification of the medium and the growth medium should be renewed.

- **The equipment required:** this may include a laminar-flow hood or biosafety cabinet to reduce the possibility of microbial contamination, suitable culture vessels (typically pre-sterilised, disposable polystyrene dishes, bottles and flasks, treated to create a negatively charged, hydrophilic surface), a supply of high-purity water (typically distilled, deionised and carbon-filtered), a suitable incubator with temperature control of $\pm 0.5\,°C$ or better, often with CO_2 control and mechanical mixing, and an inverted microscope to examine adherent cell monolayers during growth.

The successive stages of isolation of animal components are shown in Fig. 49.1(a). Box 49.1 gives advice on key aspects of sterile technique in cell culture; additional guidance on practical procedures in animal cell culture is given by Reed *et al.* (2016).

Plant tissue and cell culture systems

Plant tissue cultures can be established by growing explants of sterilised tissue on the surface of an agar-based growth medium to give a callus of undifferentiated cells. Initial sterilisation is usually achieved by incubation for 15–20 min in 10% w/v sodium hypochlorite.

(a)

whole animal

perfused tissue slice

continuous cell line in suspension culture

primary culture on solid substratum

(b)

whole plant

detached leaf in aqueous solution

leaf slice in incubation medium

isolated cell in suspension culture

protoplast in suitable osmoticum

Fig. 49.1 Isolation of animal (a) and plant (b) components for *in vitro* study (note decreasing scale, from organisms to cells).

KEY POINT For most plants, cell cultures can be established from a broad range of tissue types, reflecting the totipotency of many plant cells.

Using an orbital incubator – continued rapid agitation (>100 r.p.m.) maintains a homogeneous suspension and encourages gas exchange between the culture medium and the atmosphere.

Example Murashige and Skoog's medium contains a balanced mixture of the principal inorganic ions, plus seven trace element compounds, three vitamins, inositol, glycine and sucrose (at 30 g/L) as the major carbon source.

Working with protoplasts – remember that all solutions must contain a suitable osmoticum, to prevent bursting. Mannitol is widely used at ≈400 mmol L^{-1}.

Assessing cell and protoplast viability – cytochemical techniques are often used, e.g. exclusion of the mortal stain Trypan blue or Evan's blue. The fluorogenic vital stain fluorescein diacetate is an alternative: it is cleaved by esterases within living cells, liberating fluorescein and giving green-yellow fluorescent cells when viewed by UV fluorescence microscopy.

Embryogenic callus may be induced to differentiate, forming tissues and organs on a medium containing appropriate plant hormones: in many cases, these cultures will develop to form plantlets that can be grown on to mature plants, or encapsulated to produce so-called 'artificial seeds'. This approach can be used to study the conditions necessary for differentiation and development, or to propagate rare plants and other valuable stock (e.g. virus-free stock, or genetically altered plants). Callus derived from anthers can be used to provide haploid cell cultures and haploid plants – these are often useful for experimental genetics and breeding purposes.

Plant cell suspension cultures (Fig. 49.1(b)) are usually obtained by transferring fragments of actively growing callus to a 'shake' flask containing liquid medium in an orbital incubator: gentle agitation causes fragmentation of the callus tissue, to give a suspension culture that will contain individual plant cells and cell aggregates. This culture can be maintained by repeated subculturing of material from the upper layers of the liquid, encouraging the growth of small aggregates. However, in contrast to animal cell suspension cultures, it is rare for plant suspension cultures to be entirely unicellular, due to the presence of plasmodesmata and plant cell walls. Suspension cultures often require a minimum inoculum size on subculture – an inoculum volume of 10% v/v may be necessary to ensure successful subculture. It may also be helpful to add a small amount of 'conditioned' medium from a previous culture. Commercial-scale suspension cultures are used to produce certain plant pigments and secondary metabolites such as flavourings and high-value pharmaceutical compounds.

The growth media used for callus and suspension cultures are more complex than those required for intact plant organs and tissues, with organic nutrient supplements in addition to a balanced salt solution. Such organic supplements often include a major carbon and energy source, e.g. sucrose, plus various vitamins and growth regulators (typically, at least one auxin and a cytokinin), together with undefined components such as yeast extract and hydrolysed casein in some instances. Otherwise, the techniques are broadly similar to those described for microbial systems (Chapters 38 and 48).

Plant protoplasts

Some experimental procedures using plant cells require the enzymatic removal of their cell walls, creating protoplasts that can be manipulated *in vitro*, then maintained under conditions that allow the regeneration of cell walls, with subsequent growth and differentiation to give genetically modified plants. Protoplast isolation often involves pretreatment in a concentrated osmoticum (e.g. sucrose or mannitol, at 300–500 mmol L^{-1}) to plasmolyse the cells, weakening the linkage between cell wall and plasma membrane. Enzymatic treatment is often prolonged, taking several hours in a suitable mixture of enzymes, e.g. cellulase and Macerozyme. The resulting material can be sieved through fine nylon mesh, centrifuged at low speed and then resuspended, to remove subcellular debris and cell aggregates. The protoplast preparation can be checked for fragments of cell wall using a suitable stain, e.g. 0.1% w/v calcofluor white and a fluorescence microscope.

The fusion of protoplasts from different plants can be used to produce a somatic hybrid: this process can be used to circumvent interspecies reproductive barriers, creating novel plants. Protoplast fusion can be induced by chemical 'fusogens', e.g. using polyethylene glycol (PEG) at high concentration, or by electrofusion (incubation under low alternating current to encourage aggregation, then brief exposure to a high-voltage electrical field – typically 1000 Vcm^{-1} for 1–2 ms – causing protoplast fusion). Similar techniques can be used with animal cells. The successive stages of isolation of plant components are illustrated in Fig. 49.1(b).

Text references

Freshney, R.I. (2010) *Culture of Animal Cells: A Manual of Basic Technique and Specialised Applications,* 6th edn. Wiley, New York.

Reed, R.H., Holmes, D., Weyers, J.D.B. and Jones, A.M. (2016) *Practical Skills in Biomolecular Sciences,* 5th edn. Prentice Hall, Harlow.

Smith, R.H. (2000) *Plant Tissue Culture,* Academic Press, London.

Sources for further study

Anon. (2006) *Fundamental Techniques in Cell Culture – A Laboratory Handbook.* Sigma-Aldrich/European Collection of Cell Cultures. Available: http://www.sigmaaldrich.com/life-science/cell-culture/learning-center/ecacc-handbook.html. Last accessed 14/07/15.

Butler, M. (2003) *Animal Cell Culture and Technology: The Basics,* 2nd edn. Garland Science Press, Oxford.

Davey, M.R. and Anthony, P. (2010) *Plant Cell Culture: Essential Methods.* Wiley, New York.

Davis, J.M. (2002) *Basic Cell Culture: A Practical Approach,* 2nd edn. Oxford University Press, Oxford.

Davis, J. (2010) *Animal Cell Culture: Essential Methods.* Wiley, New York.

Loyola-Vargas, V. M. and Vazques-Flota, F. (2005) *Plant Cell Culture Protocols,* 2nd edn. Humana Press, New Jersey.

Neumann, K.H., Kumar, A. and Imani, J. (2009) *Plant Cell and Tissue Culture: A Tool in Biotechnology – Basics and Applications.* Springer, New York.

Reinart, J. and Yeoman, M.M. (2012) *Plant Cell and Tissue Culture: A Laboratory Manual.* Springer, New York.

STUDY EXERCISES

49.1 Calculate the osmolarity of Ringer's solution from its individual constituents. Use the data in Table 49.1 and the information in Chapter 25, and assume that each constituent behaves according to ideal thermodynamic principles. Note the following M_r values: NaCl 58.44; KCl 74.55; CaCl$_2$ 110.99; NaHCO$_3$ 84.01, required to convert the data in Table 49.1 to molar concentrations (see p. 158 if you are unsure about this conversion). Give your final answer in mosmol kg^{-1} to one decimal place.

49.2 Find out about the availability of cell cultures using the Web. Research the culture collections of particular countries via their websites. How useful is each site in terms of features such as: ease of use; access to catalogues; information on individual strains, e.g. culture media; details of costs?

49.3 Research the origins of the code names used for individual cell lines. In many instances, the code names given to individual cell lines are derived from their original source. Use the Web to find out the origin of the following descriptors for cell lines: (a) HeLa; (b) Vero; (c) BHK; (d) CHO; (e) TBY-2.

Answers to these study exercises are available at www.pearsoned.co.uk/practicalskills

Photography and imaging techniques are used primarily for:

- **documentation and validation of observations** for projects or coursework;
- **recording visual information for later measurement,** counting and/or analysis;
- **preparation of material for illustrating talks;**
- **preparation of poster displays;**
- **producing 'hard copy' of microscope images.**

Professional technical assistance may be available for this type of work but there is an important role for you in defining your requirements precisely. A basic understanding of the key factors that affect the final product will help you do this.

Types of photography

Photographs can be taken using two main systems: the digital camera and the film camera. Although the basic operation of both types is similar, they differ in their light-capturing and image-recording systems.

Digital photography and imaging

This is the production of an electronically stored image, using either a digital camera (still or video) or a scanning device from a pre-existing image. The image is held as a series of discrete picture elements called pixels (Fig. 50.1); the greater the number of pixels per unit surface area, the better the potential resolution of the image and the greater the amount of detail captured. The image may be in black/white, greyscale or colour. Image resolution is constrained by the nature of the sensor array and the amount of memory available for recording, while size and quality will also depend on the display or printer used.

Film photography

Black-and-white or colour-print photography involves capturing images on a film of plastic coated with light-sensitive silver halide salts. When exposed to light and subsequently chemically treated (developed), silver grains are produced to form a negative image, later converted to a positive image during printing. Transparency (slide or positive) films are based on colour dyes and are effectively grainless. Table 50.1 summarises the relationship between speed, grain and resolution for print films. The choice of film is a compromise between speed and resolution. Slower film is best for fine detail, but requires high light levels for effective results. Faster film is more suitable for moving subjects or when light levels are low, but there is a loss of contrast and detail.

Comparing digital and film systems

Despite advances in digital image technology and reductions in the price of digital cameras and associated memory systems, even the best digital images will always be of somewhat lower quality (resolution) than the highest-quality film systems. However, for practical purposes, and

---

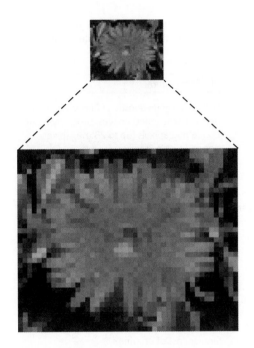

Fig. 50.1 Pixel structure in a digital image.

Table 50.1   The speed and grain relationship in film

Speed of film	ISO number	DIN number	Grain	Resolution
Slow	25–64	15–19	Very fine	Very sharp
Medium	64–200	19–24	Fine	Sharp
Fast	200–400	24–27	Medium	Medium
Extra-fast	400–1600	27–32	Coarse	Poor to medium

especially for undergraduate studies, digital photography has a significant number of advantages, including:

- **Immediate review of images.** There is no wait for film to be developed. If there is a problem with a photograph, you can take another picture instantly.
- **Photographic flexibility.** Settings are easy to adjust and may be programmed automatically. For example, ISO speed settings can be set to adapt during shooting when the weather changes from bright sunlight to cloudy. In film photography, the current film would need to be unloaded and a new film with the desired ISO speed loaded.
- **Convenient selection, manipulation and printing of images.** Photos can be copied from one digital medium to another without any deterioration in the quality of the image. Images can be refined using image editors such as *Photoshop* or *PaintShop Pro* and then printed using a computer and consumer-grade printer.
- **Additional features.** Some cameras have potentially valuable functions such as antishake, rapid multi-shot, and display and recording of information from a global positioning system (GPS).
- **Capacity to record digital video as well as still images.** This would require separate cameras and media if using film photography.
- **Metadata are stored with the image file.** This feature can simplify data recording for your records. Film cameras have a very limited ability to handle metadata.
- **Minimal costs and simplicity of image capture and storage.** Many hundreds of photographs can be captured and stored on the same media device (depending on capacity and image quality). By contrast, a film camera requires regular changing of film (typically, after every 24 or 36 shots).

Digital photography does have some disadvantages that you should consider:

- **Digital cameras are entirely dependent on an electrical supply.** This is usually batteries but sometimes a power cord when in 'tethered' mode. Without backup, factors affecting battery life, such as cold, can be important. Spare batteries are required when working in the field.
- **Image quality can be suspect in certain circumstances.** For example:
  - (a) when highlights burn out, they burn to white without details, while film cameras retain a reduced level of detail;
  - (b) image 'noise' may manifest as multi-coloured speckles in digital images taken at high ISO values, similar to the 'grain' of high-ISO film;
  - (c) patterns or artefacts can result from electronic processing.

Manipulation of images – the ease with which digital images can be adjusted can lead to problems because the final image may not represent the true and original situation. For scientific work, you should employ the original, unmanipulated image whenever possible – although some changes to image quality such as contrast may be valuable and acceptable.

Definition

Metadata – for digital photographs, this information can include: date and time; place (for cameras with GPS facility); keywords; photographer; details of the camera and its settings.

Capturing moving images – this may be valuable in biology, for example, to record animal behaviour, speed up plant movements or study fungal growth patterns.

Table 50.2 A comparison of digital and film photography. This table assumes you would be using a 12 megapixel SLR (single-lens reflex) digital camera and a 35 mm SLR film camera with 100 ASA film. Although it is not the only factor determining image quality, results would in general be poorer for a compact digital camera or one offering fewer megapixels

Aspect	SLR digital camera	SLR film camera
Resolution	Primarily limited by number of pixels, i.e. 12–18 megapixels.	Limited by grain density of film and generally better. For scanned images, probably higher than 6 megapixels (up to 25 megapixels, depending on scanner).
Image quality	Better for shadow detail. Long-exposure images may be of poor quality. Delays in image capture may make it difficult to capture moving subjects. Depth of field longer with compact digital cameras.	Better for highlights. Grain is present, and will show if exposure time is long and/ or image is magnified. Good-quality long exposures are possible. Easier to control depth of field.
Convenience	Easier to take repeated shots of same subject; easier to send images to others digitally; easier to incorporate into word-processed documents.	Results not known until film is developed – hence use of bracketing, which is wasteful of film. Unless scanned, less easy to send or incorporate into word-processed documents.
Costs	Costs of equipment, operation and printing similar. No film or processing costs, but memory card required.	Costs of equipment, operation and printing similar, but film must be bought and subsequently processed, with additional cost. Similar cost of printing.
Post-exposure image manipulation	Easy using commercially available software.	Possible at developing stage but only after scanning thereafter.
Storage and copying of images	Electronic. Digital images can be stored without loss of quality and occupy little physical space. Copies of files are exact facsimiles. Back-up easier.	Physical (hard copy). Colour-film images may fade through time and negatives and prints occupy significant space. There is always loss of quality when they are copied. Back-up less easy, unless scanned.

Understanding the light sensitivity of photographic systems – both digital and film media are rated in terms of their 'speed', measured in either ISO (= ASA) or DIN units. A good range for general purposes is 100–160 ISO (21-23 DIN). Film of 1600 ISO is available, as are digital cameras with settings as high as 3200 ISO. These are useful for low light situations, but with some loss of image quality.

Focusing images correctly – the autofocus feature on cameras may result in lack of focus on the parts of the image you are interested in because the algorithm used (generally spot or area) may not be suitable for your purpose. Better results may be obtained using manual focusing.

- **Digital file formats may cause difficulty.** They may be incompatible with certain software, or become obsolete or replaced. Conversion programs are usually available, but they can result in loss of image quality.
- **Dust particles can adhere to the cover glass of the image sensor.** This can result in persistent dust spots on images. It is easier to clean the mechanical systems of a film camera.

Important aspects of photography using digital and film systems are further compared in Table 50.2.

Capturing images

Camera and lens types

For both digital and film systems, the primary difference is whether you preview the image through the lens (TTL), often referred to as single-lens reflex (SLR), or whether a separate viewfinder is used. Digital cameras also incorporate a monitor that allows you to preview and review images. Lenses are either fixed, i.e. have a set focal length, or zoom, allowing a range of focal lengths, and hence magnification and angle of view. In general, higher-quality images can be taken with an SLR camera; these systems have better-quality lenses and allow greater control over the final image.

Tips for better photography:

- Use your camera on a tripod whenever possible, and use high shutter speed to minimise the effect of movements. Otherwise, adopt a posture with your elbows firmly locked against your body to reduce camera shake.
- Use electronic flash wherever possible as it provides uniformity of colour balance. Unless you want shadowless lighting, do not place the flash on the camera (hot-shoe) connection.
- Take care when using flash as it can result in shadowing and overexposure (especially when photographing wet materials). Use a ring-flash system to counteract shadowing, or a tripod with slower exposure speed under natural light to avoid reflection.

Storing colour film – colour film is highly sensitive to environmental factors such as heat and humidity, which cause changes in film speed and colour rendition. Make sure that your film has a sufficiently long expiry date. Store film for extended periods in a refrigerator or freezer: if stored in a freezer, allow a 24 h thawing period before use.

Setting up for photography

In the biosciences, this is usually carried out in one of three general situations:

1. **Laboratory or studio environments** – where conditions are under your control and problems are relatively easy to overcome.
2. **Field situations** – where many factors may be difficult to control, and you may need to use artificial means to enhance the shot (e.g. flash photography, or fast film speed).
3. **Working with specialised equipment** – such as photo-microscopes or electron microscopes, where control of the photographic process is largely managed by the instrument's operating system.

The amount and type of light available to create an image are very important, especially for colour photography. Colour-image creation requires the appropriate colour temperature balance to be selected. Colour temperature is measured in kelvin (K: see Fig. 50.2). In digital cameras this is usually automatic, but can be manually controlled via the 'white balance' settings. Filters are used with film cameras, to control overall colour balance, but the most important factor is the film type selected:

- **daylight film** is balanced for daylight conditions (5400 K) and electronic flash;
- **artificial light film** types A (3400 K) or B (3200 K) are designed for studio lighting conditions.

If using a digital camera, remember to adjust settings for the appropriate light temperature: this usually requires you to set for tungsten (indoor) lighting or outdoor light. Digital images are also modifiable after being taken, unlike film images where minimal adjustment is possible after taking the photograph.

The quantity of light can be measured by a photographic light meter, which may be external or built-in. The more light there is, the smaller the lens aperture you can select (larger f-number), and the greater will be the depth of field (= depth of focus). With the camera on a tripod, you can select slow shutter speeds and larger aperture settings (f-numbers), to maximise depth of focus. Where an electronic flash system is employed to provide some or all of the lighting this is even easier, since the effective shutter speed with electronic flash is extremely short (1 ms). Remember that your choice of lighting arrangement will affect the quality of the picture: shadowless lighting is appropriate in some situations, but often shadows help to give three-dimensional form to subjects. In general, the use of more than one light source is advisable to prevent hard shadows.

KEY POINT To use an SLR camera properly, you need to understand the relationship between aperture (f-number), shutter speed and depth of field. For moving objects, give priority to fast shutter speed by opening the lens aperture (small f-number). Where depth of field is required, close the lens aperture (high f-number); this will result in slower shutter speeds, so take care to avoid camera shake. Bracket your exposures (make extra exposures at plus and minus 1 f-number at same shutter speed) to ensure good results.

Fig. 50.2 Colour temperature and the kelvin scale in relation to sensitivity of film type.

Backing up digital image files – since no physical copy exists, you must remember to download and back up digital image files to appropriate media (hard drive, network drive, CD or USB memory stick) as soon as possible.

Using scales with images – these are vital to the interpretation of any biological image (see Chapter 46 for information on different types and how to calculate them).

Scanning

Scanners can be useful to digitise a printed image. The quality of the scanned image is defined in terms of the number of dots per inch (dpi). Dedicated facilities may be available in your department; however, many modern printers also have a scan feature, though often at lower resolution. A scanner with a value of 600 dpi is adequate for most purposes.

Image processing

Images are recorded in digital cameras as files on various forms of memory cards and the data are later transferred either to a computer for manipulation or directly to a printer. An appropriate card reader or USB cable may be required. For critical work, the rule is to capture images using the largest possible file size, as reductions (e.g. compression of file size) of digital information are possible, but you can never add information.

Digital image files can be stored in various formats (e.g. .bmp, .jpg, .tiff) and can be manipulated using programs such as Adobe *Photoshop* or *PaintShop Pro*. Aspects of image quality such as contrast and colour balance can easily be adjusted and images can be cropped and resized to suit your application (e.g. small file sizes are recommended for websites). Digital images are valuable for use with word processors and Web-authoring software and can be transferred readily between such programs.

Film images require chemical development, usually a professional task. A film image may be impaired irreversibly by faulty development and the resulting image can only be modified to a limited extent during subsequent printing. This is a key reason for selecting digital image systems.

Photomicrography

This requires the use of a camera system mounted on a microscope. The important operations for successful photomicrography are outlined below:

1. **Carefully prepare the specimen:** ensure thorough cleanliness of any slides or coverslips used.

2. **Choose the correct settings** and any filters required for alteration of colour balance, depending upon the type of light source available.

3. **Decide on the magnification to be used:** make sure that you know how this relates to the magnification of the negative/transparency/digital image. Always include a photograph of a stage micrometer (p. 278) so the final magnification can be calculated and given in the legend to the photograph.

4. **Carefully focus the specimen on to the image plane:** there will be a system-specific method for doing this (consult the manual, as this is a critical aspect).

5. **Bracket extra exposures** above and below the automatic settings, especially for colour photography using film.

Printing images

The most appropriate choice of print type will depend on your purpose. Glossy-finish prints generally appear sharper than matt or other finishes. Moreover, the addition of lettering and scales is often only possible on the smooth surface of a glossy print. However, when preparing prints for

Fig. 50.3 Use of white areas for labelling dark prints, illustrated by an SEM micrograph of a stellate hair on a leaf surface.

display on a poster, matt/velvet finishes are often preferable, as there is less glare for the viewer. Printing from computers using ink-jet and similar printers requires care, since colours viewed on monitors do not always reproduce on paper as expected. You may need to experiment with settings and paper quality to obtain the appropriate results.

Adding scales and labels to images

Having acquired a suitable print, it is often necessary to add information to it:

- When digital images are copied into word-processor or presentation software, additional information such as lettering, scales and labels can be added by overlaying a 'text box', or equivalent. You may need to make the background transparent, e.g. using the 'no fill' option in *PowerPoint*. Lettering can be added to printed media using transfer letters (e.g. Letraset).
- Choose a simple font type that will not distract and a size that is legible but not too dominant, e.g. 12 point in final reproduction. Always use the same font and type size for a related sets of images.
- Choose an appropriate part of the image for the lettering, i.e. a black area for a white letter. If the background is variable, create a coloured or white area and place the letters on this (see Fig. 50.3).

Storing and mounting images

Digital images

Indexing software can be used to organise and catalogue your images (e.g. *Picasa,* available for free download at: http://picasa.google.com/). These programs allow you to add appropriate information and store this electronically along with the image, facilitating viewing and selection when your collection of images is large.

Prints and transparencies

Prints, negatives and transparencies are easily damaged and should be stored with care:

- **Prints should be stored flat** in boxes or in albums. When used for display, mount on stiff board either with modern photographic adhesives or dry-mounting tissue.
- **Negative strips should be enclosed** in transparent or translucent paper filing sheets. Obtain a set of contact prints for each film and store this with the negatives for easy reference.
- **Transparencies (slides) should be mounted** in plastic mounts but not between glass, as this often causes more problems than it solves. Labelling of transparencies is best done on the mount, using small, self-adhesive labels.

Beware of cheap plastic filing materials as some contain residual chemicals that can cause damage over time. Avoid dampness: silica gel desiccant can be used where necessary. Both physical copies and records should be well organised, including all potentially relevant information.

Recording information about images – this should include subject, location, image-capture details and processing procedures. Bear in mind that these data may be of use when the image is used in a different context from the one you initially planned.

Sources for further study

Adobe Systems Inc. *Digital Imaging and Photo Editing Software for Digital Photography*. Available: http://www.adobe.com/digitalimag/main.html Last accessed 30/03/15.

Ang, T. (2009) *Digital Photographer's Handbook,* 4th edn. Dorling Kindersley, London.

Gookin, D. (2000) *Digital Scanning and Photography.* Microsoft Press International, Redmond, Washington.

McCartney, S. (2001) *Mastering the Basics of Photography.* Allworth Press, New York.

Oldfield, R., Dickes, F.W. and Rost, F.W.D. (2000) *Photography with a Microscope.* Cambridge University Press, Cambridge.

STUDY EXERCISES

50.1 Choose an appropriate image-recording method. You wish to record a stage in an experiment as an image. What are the relative merits of doing this using a film camera relative to using a digital camera? What problems might you encounter?

50.2 Make a list of situations where an image is often required. When in biology project work might you require to use images?

50.3 Choose an appropriate combination of film/lighting/image for specific purposes. What film, lighting and image types would you select to record the following items *in situ*?

(a) a flower in a field

(b) a field site

(c) plants in a greenhouse

(d) mice in a laboratory experiment.

Answers to these study exercises are available at www.pearsoned.co.uk/practicalskills

The ability of organisms to grow, develop and respond to stimuli is fundamental to life. The aims of studies concerned with these phenomena include:

- **investigating developmental processes;**
- **comparing developmental patterns;**
- **understanding behavioural responses;**
- **characterising physiological processes;**
- **measuring the levels of a particular compound (bioassays);**
- **assessing sensitivity to a stimulus.**

> **KEY POINT**   The growth and development of a multi-cellular organism are complex phenomena that cannot be easily measured by a single variable. An overall picture emerges from changes in size, weight, number, complexity and function evaluated at cell, tissue and organ levels.

Growth and development

Length and area are relatively easy to measure (see p. 151 for techniques) though curved organisms may present difficulties. Volume measurements can be made from the displacement of water employing Archimedes' principle, or derived from fresh weight if the density of the organism can be estimated. Growth may not occur smoothly through time (e.g. plants increase in height mainly at night, when cell turgor is highest). For rapidly growing organisms it is desirable to have a continuous photographic or video record, though this may only be feasible with sedentary organisms. Care is required to ensure that focus is maintained on the points that will be measured, or your data will be inaccurate. If you are interested in growth of different parts of the organism, use reference points, e.g. recognisable morphological features such as hairs or marks made with ink or paint.

Fresh weight might at first sight appear a useful measure of growth: however, water content may vary according to availability and consumption. Dry weight (p. 152) does not have this disadvantage, but it may decrease when respiration dominates metabolism (e.g. during seed germination), despite the fact that growth is obviously occurring. Protein or nitrogen content may be better correlated with 'mass of protoplasm', but they are both complex to measure and are 'destructive' measurements.

Whichever of the above variables is measured, growth generally shows a sigmoidal increase in time (Fig. 51.1(a)) that is reminiscent of the population growth curve of cells in batch culture (p. 291). The slope of this curve at any given time represents the absolute growth rate, with units of increment in size per unit time. Plotting this against time generally shows a bell-shaped curve (Fig. 51.1(b)). The relative growth rate or rate of increase in size per unit of size measures the 'efficiency' of growth (Fig. 51.1(c)).

Cell number is a common measure of the growth of unicellular suspensions (p. 295). For multi-cellular organisms, this can be estimated from measurements of average cell size and total volume. Numbers of organs (e.g. leaves), specialised cells (e.g. hairs) or organelles (e.g. mitochondria) can indicate the progress of differentiation and development,

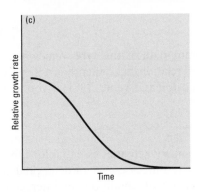

Fig. 51.1 Different ways of analysing growth: (a) simple plot of size against time; (b) plot of absolute growth rate for s the same data; (c) plot of relative growth rate for the same data.

but in general these are difficult processes to characterise quantitatively. Changes in numbers can also be expressed as absolute and relative rates, as above.

Display and analysis of growth data

Care should be taken when plotting a growth curve. When plotting change in size against time for a single organism, a smooth curve passing through all measurement points is appropriate (i.e. measurement error is assumed to be negligible). If dealing with a sample of organisms, the approach depends on whether (a) the same sample is measured each time (as with non-destructive harvesting, e.g. for linear measurement), when the curve should pass through each mean point, or (b) the samples are independent (as with destructive harvesting, e.g. for dry-weight measurement), when a smooth trend line should be drawn that need not pass through all of the points. Regarding statistical analysis, note that time-series measurements can only be treated as independent if the experimental design is as in (b) above; otherwise, in comparing changes in size with time, you should use tests for paired data. Note that variance may increase with size, so non-parametric tests (Chapter 76) are likely to be appropriate.

When presenting growth rates, use histograms if the data are obtained from differences in mean values divided by the time interval between measurements. A smooth curve can be used if it represents the slope (i.e. first derivative) of a function that has been fitted to the growth curve. Growth data are often log-normally distributed, so curves should be fitted to logarithmically transformed data. Fitting functions to growth curves is complex and specialist advice should be sought.

Allometry

This is the study of relationships between the dimensions of organisms. It can be used for describing changes in the allocation of resources within a single species (e.g. root-to-shoot ratio in plants; Fig. 51.2) or among species (e.g. brain size in relation to body mass). The underlying relationship between dimensions x and y is generally:

$$y = ax^b \qquad [51.1]$$

where a is a positive constant. The coefficients a and b can be estimated from a log–log plot (e.g. Fig. 51.2), which may show one or more linear phases of the form:

$$\log y = \log a + b \log x \qquad [51.2]$$

Care should be taken in interpreting data of this type: remember that eqn [51.1] is an experimentally derived relationship rather than a theoretical one. Consult a specialist text (e.g. Causton and Venus, 1981) if you wish to fit a line or lines to such data.

Responses to stimuli

Responses to stimuli are involved in many activities essential for life. For example, they are required to capture resources, escape or deter predators, move towards or away from a given set of conditions or interact with other individuals in a community. Understanding these responses requires answers to the following questions:

- **What is the precise response and what causes it?**
- **How does the response develop?**

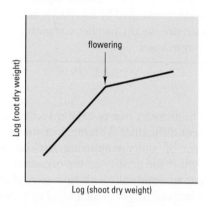

Fig. 51.2 Example showing changes in an allometric relationship based on a published analysis of the growth of a grass. Both root and shoot dry weight increase as the plant ages. The onset of flowering causes a clear change in their relationship.

Analysing allometric data – when fitting eqn [51.2] to allometric data, you should use Model II regression as the variables are interdependent (see p. 525). Note that for monophasic log–log relationships, a can be determined from the intercept ($\log a$) and b from the slope ($\log b$).

- **What is the function of the response?**
- **How might the response have evolved?**
- **How much of the response is inherited and how much learned?**

The first stage when investigating a response is to describe it fully (preferably under conditions as near to those in nature as possible) and determine the precise stimuli that cause it. For instance, a study of plant phototropism might start with research into which cells perceive the light stimulus and which respond; effective wavelengths and doses of light might then be investigated. When working with animals, avoid anthropomorphism and describe objectively what happens. All circumstances of potential relevance (e.g. weather, time of day) should be recorded until a clear pattern emerges. A full description of most responses obviously requires physiological and biochemical techniques whose scope is too wide to be covered here.

You may wish to follow an individual animal in space or time and this will require either recognising its features or marking it in some way that does not interfere with the response (e.g. leg-ringing birds or painting symbols on insects). When recording the elements of a behavioural pattern, a coding system may be helpful (e.g. 1 = flaps wings, 2 = drops neck, etc.). As with growth, photography and video techniques are good aids for recording movements and analysing them later. Tape recorders can be used to record a description of responses so long as your voice does not interfere with the response. They can also record the sound component of a response and might be used later to elicit a response under controlled conditions (e.g. for a suspected alarm cry).

The second question listed above requires a study of the response throughout the developmental stages of the organism. The third and fourth questions require a certain amount of interpretive skills but, as with the fifth question, are amenable to experimentation. You may be presented with a natural experiment if conditions change in the right way in the course of your observations. Experiments in a laboratory give you control over conditions but this may well be at the expense of realism.

In designing experiments to investigate animal behaviour, you must take account of the fact that an animal's behaviour may change according to its experience. If not actually investigating learning per se, you need to ensure that the opportunity for a subject to learn is the same for each treatment, or bias will occur. Ways round this problem are either to use naive animals for every replicate or to adopt a Latin square design (p. 206) so that experiences are shared evenly among treatments (fully randomized or randomised block designs may not achieve this).

Bioassays and their applications

A bioassay is a method of quantifying a chemical substance (analyte) by measuring its effect on a biological system under controlled conditions. The hypothetical underlying phenomena are summarised by the relationship:

$$A + Rec \rightleftharpoons ARec \rightarrow \rightarrow R \qquad [51.3]$$

where A is the analyte, Rec the receptor, ARec the analyte–receptor complex and R the response. This relation is analogous to the formation of product from an enzyme–substrate complex and, using similar mathematical arguments to those of enzyme kinetics (p. 390), it can be shown that the expected relationship between [A] and rate of response is

---

Definition

Anthropomorphism – attributing human characteristics to animals.

Distinguishing between innate and learned behaviour – innate behaviour is genetically determined; it is always executed perfectly and in its entirety the first and every subsequent time the stimulus occurs. Learned behaviour is modified by experience.

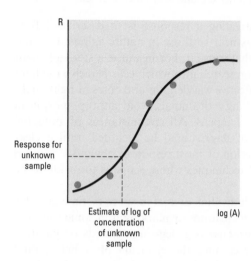

Fig. 51.3 Typical bioassay response curve, showing estimation of an unknown sample. Closed circles represent responses with standard samples. R = response; A = analyte.

Examples The following are typical bioassays:

- measuring the amount of a plant growth substance in a plant extract using the coleoptile straight growth assay;
- measuring the effect of a drug on the dilation of a person's pupils;
- measuring the mutagenic properties of a chemical compound in the Ames test (see Box 51.1).

Legal use of bioassays – in the UK, where bioassays involving 'higher' (vertebrate) animals are controlled by the *Animals Scientific Procedures Act (1986)*, they can only be carried out under the direct supervision of a scientist licensed by the Home Office.

hyperbolic (sigmoidal in a log–linear plot like Fig. 51.3). This pattern of response is usually observed in practice if a wide enough range of [A] is tested.

To carry out the assay, the response elicited by the unknown sample is compared with the response obtained for differing concentrations of the substance, as shown in Fig. 51.3. When fitting a curve to standard points and estimating unknowns, the available methods, in order of increasing accuracy, are:

1. **fitting by eye;**
2. **using linear regression** on a restricted 'quasi-linear' portion of the assay curve;
3. **linearisation followed by regression** (e.g. by probit transformation);
4. **non-linear regression** (e.g. to the Morgan–Mercer–Flodin equation).

In general, bioassay techniques have more potential faults than physicochemical assay techniques. These may include the following:

- **A greater level of variability:** error in the estimate of the unknown compound will result because no two organisms will respond in exactly the same way. Assay curves vary through time, and, because they are non-linear, a full standard curve is required each time the assay is carried out.
- **Lack of chemical information:** bioassays provide information about *biological* activity; they say little about the chemical structure of an unknown compound. The presence of a specific compound may need to be confirmed by a physico-chemical method (e.g. mass spectrometry).
- **Possibility of interference:** though many bioassays are very specific, it is possible that different chemicals in the extract may influence the results.

Despite these problems, bioassays are still much used. They are 'low tech' and generally cheap to set up. They often allow detection at very low concentrations. Bioassays also provide the means to assess the biological activity of chemicals and to study changes in sensitivity to a chemical, which physico-chemical techniques cannot do. Changes in sensitivity may be evident in the shape of the dose–response curve and its position on the concentration axis.

Bioassays can involve responses of whole organisms or parts of organisms. 'Isolated' responding systems (e.g. excised tissues or cells, Chapter 49) decrease the possibility of interference from other parts of the organism. Disadvantages include disruption of nutrition and wound damage during excision. Isolation can continue down to the molecular level, as in immunoassays (Chapter 53).

Bioassays are the basis for characterising the efficacy of drugs and the toxicity of chemicals. Here, response is often treated as a quantal (all-or-nothing) event. The E_{50} is defined as that concentration of a compound causing 50% of the organisms to respond. Where death is the observed response, the LD_{50} describes the concentration of a chemical that would cause 50% of the test organisms to die within a specified period under a specified set of conditions. Box 51.1 presents details of the Ames test, a widely used bioassay to assess the mutagenicity of chemicals.

Box 51.1 Mutagenicity testing using the Ames test – an example of a widely used bioassay

Chemical carcinogens can be identified by the formation of tumours in laboratory animals exposed to the compound under controlled conditions. However, such animal bio-assays are time-consuming and expensive. Dr B.N. Ames and co-workers have shown that most carcinogens are also mutagens, i.e. they will induce mutational changes in DNA. The Ames test makes use of this correlation to pro-vide a simple, rapid and inexpensive bioassay for the initial screening of potential carcinogens. The test makes use of particular strains of *Salmonella typhimurium* with the fol-lowing characteristics:

- histidine auxotrophy – the tester strains are unable to grow on a minimal medium without added histidine: this characteristic is the result of specific mutational changes to the DNA of these strains, including base substitutions and frame shifts;

- increased cell envelope permeability, to permit access of the test compound to the cell interior;

- defects in excision repair systems and enhanced error-prone repair systems, to reduce the likelihood of DNA repair after treatment with a potential mutagen.

When grown in the presence of a chemical mutagen, the bacteria may revert to prototrophy as a result of back muta-tions that restore the wild-type phenotype: such revertants grow independently of external histidine and are able to form colonies on minimal medium, unlike the original test strains. The extent of reversion can be used to assess the mutagenic potential of a particular chemical compound. Since many chemicals must be activated *in vivo*, the test in-corporates a rat liver homogenate (so-called 'S-9 activator', containing microsomal enzymes) to simulate the metabolic events within the liver. The tester strains are mixed with S-9 activator and a small amount of molten soft agar, then poured as a thin agar overlay (top agar) on a minimal medi-um plate. The top agar layer contains a very small amount of histidine, to allow the tester strains to divide a few times and express any mutational changes (i.e. prototrophy).

The test can be performed in one of two ways:

1. Spot test: a concentrated drop of the test compound is placed at the centre of the plate, either directly on the agar surface, or on a small filter paper disc. The test compound

will diffuse into the agar and revertants appear as a ring of colonies around the site of inoculation, as shown in Fig. 51.4. The distance of the ring of colonies from this site provides a measure of the toxicity of the compound, while the number of colonies within the ring gives an indication of the relative mutagenicity of the test substance. The spot test is often carried out in a simplified form, without added S-9 activator, as a rapid preliminary test prior to quantitative analysis.

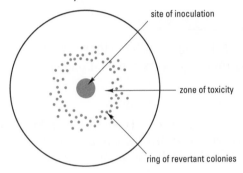

Fig. 51.4 Typical outcome of spot test (Ames test).

2. Agar incorporation test: known amounts of the test compound are mixed separately with molten top agar and the other constituents and poured on to separate minimal agar plates. After incubation for 48 h, revertants will appear as evenly dispersed colonies throughout the agar overlay. The number of colonies reflects the relative mutagenicity of the test compounds, with a direct relationship between colony count and the amount of mutagen. Agar incorporation tests can be used to generate dose–response curves similar to that shown in Fig. 51.3. The simplicity, sensitivity and reproducibility of the Ames test has resulted in its widespread use for screening potential carcinogens in many countries, though it is not an infallible test for carcinogenicity.

SAFETY NOTE Correct handling procedures **must** be followed at all times, as the test substances may be carcinogenic – testers should wear gloves and avoid skin contact or ingestion.

Considerations when setting up a bioassay

- **The response should be easily measured** and as metabolically 'close' to the initial binding event as possible.

- **The experimental conditions should mimic the *in vivo* environment.**

- **The standards should be chemically identical to the compound being measured** and spread over the expected concentration range being tested.

• **The samples should be purified if interfering compounds are present** and diluted so the response will be on the 'linear' portion of the assay curve.

To check for interference, the bioassay may be standardised against another method (preferably physico-chemical). Related compounds known to be present in the analyte solution should be shown to have minimal activity in the bioassay. If an interfering compound is present, this may show up if a known amount of standard is added to sample vials – the result will not be the sum of independently determined results for the standard and sample.

Text reference

Causton, D.R. and Venus, J.C. (1981) *The Biometry of Plant Growth.* Edward Arnold, London.

Sources for further study

Drickamer, L.C., Vessey, S.H. and Jakob, E.M. (2001) *Animal Behavior: Mechanisms, Ecology, Evolution.* McGraw-Hill, Dubuque, Iowa.

Hunt, R. (1990) *Basic Growth Analysis.* Unwin Hyman, London.

Martin, P. and Bateson, P. (2007) *Measuring Behaviour: An Introductory Guide,* 3rd edn. Cambridge University Press, Cambridge.

Price, E.O. and Stokes, A.W. (eds) (1975) *Animal Behavior in Laboratory and Field,* 2nd edn. W.H. Freeman, San Francisco.

Salanki, J., Jeffrey, D. and Hughes, G.M. (1994) *Biological Monitoring of the Environment: A Manual of Methods.* International Union of Biological Sciences Methodology Series, CAB International, Oxford.

STUDY EXERCISES

51.1 Estimate relative growth rate. A baby hamster weighed at the start of a week has a mass of 35 g. By the end of one week it has a mass of 38.5 g.

(a) What is its absolute growth rate on a daily basis, expressed in mg day^{-1}?

(b) What is its relative growth rate on a weekly basis, expressed as a percentage?

51.2 Use bioassay data to estimate the amount of analyte in unknown samples. Bioassay data obtained by exposing water fleas (*Daphnia* sp.) to a purified algal toxin are shown in the table opposite.

(a) Graph the data and use the graph to estimate the toxin contents of two samples obtained from two lochs that gave (i) a heart rate of 130 beats per minute and (ii) a heart rate of 25 beats per minute.

(b) Which one of the two estimates would be the less reliable and why?

(c) How could you improve the reliability of this estimate, if you had more of the sample?

(d) What unmeasured factors might limit the usefulness of this assay?

Responses of *Daphnia* to toxin

Toxin concentration (mmol L^{-1})	Mean heart rate of Daphnia sp. (beats per minute)
1.00×10^{-3}	20
3.16×10^{-4}	23
1.00×10^{-4}	26
3.16×10^{-5}	50
1.00×10^{-5}	180
3.16×10^{-6}	192
1.00×10^{-6}	195

(continued)

51.3 Use agar diffusion bioassay data to quantify a growth inhibitor in a sample. Agar diffusion bioassay is a simple means of estimating the amount of an inhibitory substance in a test sample, by adding the substance, in aqueous solution, to wells cut in a Petri plate of agar medium seeded with a microbe sensitive to the substance and then incubating the plate at an appropriate growth temperature. Diffusion of the inhibitory substance gives a zone of growth inhibition proportional to $\log_{10}$ of the concentration of substance. In the following example, the amount of the antibiotic nisin in a test sample can be determined by comparing the degree of growth inhibition of *Bacillus subtilis* for the test solution and a series of standards containing known concentrations of nisin.

Prepare a calibration curve and determine the nisin concentration of solution 5 (unknown). Express your answer in $\mu g \ mL^{-1}$, to one decimal place.

Growth inhibition of *B. subtilis* in solutions containing different amounts of nisin

Solution	Nisin content ($\mu g \ mL^{-1}$)	Zone of growth inhibition (mm)
1	40	26.5
2	20	23
3	10	19.5
4	5	17
5	unknown (test solution)	22.5

Answers to these study exercises are available at www.pearsoned.co.uk/practicalskills

Many biological studies require measurement of chemical components, ranging from elements to complex organic molecules. This usually involves sampling, purification and measurement stages. Most analytical methods rely on one or more chemical or physical properties of the test substance (the analyte) for detection and/or measurement, with two main approaches:

1. **Qualitative analysis** – where a sample is assayed to determine whether a biomolecule is present or absent. As an example, a blood sample might be analysed for a particular drug or a specific antibody (Chapter 59), or a bacterial cell might be probed for a nucleic acid sequence (Chapter 62).

2. **Quantitative analysis** – where the quantity of a particular biomolecule in a sample is determined, either as an amount (e.g. as g, or mol) or in terms of its concentration in the sample (e.g. as gL^{-1}, or $mol\ m^{-3}$). For example, a blood sample might be analysed to determine its pH ($-\log_{10}[H^+]$), alcohol concentration in $mg\ mL^{-1}$, or glucose concentration in $mmol\ L^{-1}$. Skoog *et al.* (2006) give details of methods.

Analysing biological and environmental materials

Figure 52.1 presents a summary of the typical stages of a chemical analysis. Broadly, these include:

- **sampling** (see, for example, Chapters 55–57);
- **preservation and pretreatment** (Chapter 37);
- **sample preparation** (e.g. separation and dilution/concentration);

Fig 52.1 The stages of a chemical analysis (after Schwedt, 1997).

- **identification and quantitative measurement of the analyte** (technique-specific, see Table 52.1 and Chapters 63–70);
- **expressing results in appropriate units** (Chapters 71–76).

The choice of approach will be determined by the purpose of the investigation and by the level of accuracy and precision required (p. 175). Many of the basic quantitative methods described in Chapters 59–70 rely on chemical reactions of the analyte and involve assumptions about the nature of the test substance and the lack of interfering compounds in the sample: such assumptions are unlikely to be wholly valid at all times.

> **KEY POINT** In general, you should aim to use the simplest procedure that satisfies the purpose of your investigation – there is little value in using a complex, time-consuming or costly analytical procedure in a study where a high degree of accuracy is not required.

If you need to make more definitive measurements of a particular analyte, it may be necessary to separate it from the other components in the sample, e.g. using chromatography (Chapter 69), centrifugation (Chapter 67), or electrophoresis (Chapter 70), and then identify the separated components, e.g. using spectroscopic methods (Chapter 68). However, each stage in the separation and purification procedure may introduce further errors and/or loss of sample.

Identification of a chemical substance requires specific information about its molecular structure. For some chemicals and techniques, this is straightforward, as very specific signals are produced, e.g. absorption spectra of metal ions. In cases where the probability of interference by other compounds is minimal, a purification step may not be necessary. In other cases, there may be difficulties due to the possibility of isomers providing similar signals to the analyte of interest. An additional complication in biological and environmental samples is that these compounds are always present in complex mixtures, so purification is generally essential. Consequently, an element of uncertainty is present in any analytical technique.

Choosing an appropriate analytical technique

Several criteria influence the selection of a particular technique:

- **accuracy and precision;**
- **sensitivity and detection limit** in relation to the expected concentration range in the samples;
- **amount of available sample;**
- **the selectivity required;**

Selecting a method – Most routine methods of chemical analysis are destructive, since the analyte is usually converted to another substance which is then assayed, e.g. in colorimetric assays (p. 448). In contrast, analytical methods based on physical properties are often non-destructive (e.g. the intrinsic absorption spectrophotometry, p. 449, or nuclear magnetic resonance techniques, p. 412). Non-destructive methods are often preferred, as they allow further analysis.

Recognising the limitation of certain techniques – complex and expensive techniques may not be available to undergraduates in practicals or project work. In some cases, you will carry out simpler, 'proxy' measurements, such as conductivity (Chapter 55) as a measure of total dissolved salts, or colorimetric assays (Chapter 68) as a measure of pigment production. You should be aware of the limitations of these methods and discuss these in your report.

- **nature of the sample:** whether solid, liquid or gas phase;
- **whether the method is destructive or non-destructive;**
- **reliability of the procedure and instrument;**
- **simplicity:** the level of skill and practice required;
- **number of samples to be analysed;**
- **speed or capacity of analysis;**
- **cost and convenience.**

Table 52.1 provides examples of techniques for specific analytes.

Sample preparation

Once a sample is obtained, a range of techniques may be used to prepare the material for analysis. Examples include:

- **dissolving in a suitable solvent** (Chapters 25 and 26);
- **ashing and acid digestion** (Chapter 56);
- **centrifuging solidus filtering, to remove particles** (Chapters 24 and 67);
- **enzymic digestion;**
- **chromatographic techniques** (Chapter 69).

Table 52.1 Examples of techniques typically selected for specific analytes and sample types

Analyte and type of sample	Technique	Comment
Metal, e.g. Cu^{2+} in water sample from former industrial site	Atomic absorption spectrometry (Chapter 68)	Signal highly specific to analyte; purification rarely needed
Atmospheric gas, e.g. CO_2 produced by leaf during photosynthesis	Infrared gas analysis (Chapter 66)	Signal not specific to analyte; some purification required
Dissolved gas, e.g. O_2, in aquatic environment, such as a pond	Oxygen electrode (Chapter 65)	Underlying chemical reaction is specific; purification rarely needed
Sugar, e.g. glucose produced by a microbe degrading a specific polysaccharide	Gas chromatography (Chapter 69)	Derivatisation of compound may be required; 'identification' relies on retention time, and identity of analyte should be confirmed using method such as mass spectroscopy (Chapter 68)
Biomolecule, e.g. the plant hormone GA_3 in a sample derived from leaves	Immunoassay (Chapter 59)	Signal reasonably specific, but other compounds in sample may interfere
Amino acid, e.g. serine in serum sample as part of nutritional study	High-performance liquid chromatography (Chapter 69)	Measurement often involves mixture of similar compounds; 'identification' relies on retention time, and identity of analyte should be confirmed using a method such as mass spectroscopy
Macromolecule, e.g. protein in molecular study	Chromatography (Chapter 69) coupled with specific methods such as electrophoresis (Chapter 70)	'Identification' involves describing class characteristics, such as M_r, overall charge, constituent amino acids. Purification and separation can be complex, involving several stages
Genetic material, e.g. DNA in forensic or genetic study	Polymerase chain reaction (Chapter 62) followed by electrophoresis (Chapter 70)	Identification involves determination of the sequence of nucleotide bases

It may be essential to carry out pretreatments on samples as soon as they are obtained, to prevent enzymic or microbial decomposition or chemical changes. Commonly used techniques include:

- **cooling or freezing** to slow the rate of chemical reactions and microbial activity, e.g. placing pond water samples in an ice bath during transport to the laboratory;
- **addition of a biocide** to minimise biological activity, e.g. addition of sodium azide to prevent microbial action;
- **acidification** to prevent metal precipitation or adsorption reactions, e.g. adding nitric acid to a soil sample;
- **adding reducing agents** to prevent oxidation, e.g. dithiothreitol to a tissue extract;
- **adding agents to 'fix' the level of the analyte**, e.g. iodide and Mn^{2+} to precipitate free oxygen in water samples (p. 429).

Calculating an appropriate dilution – follow the guidance in Chapters 24 and 25.

Seek advice from your supervisor if developing a pretreatment procedure as part of project work.

In most cases, it will be necessary to dilute or concentrate your samples before analysis, so that analyte concentrations fall within the measurement range of the chosen technique. Representative sample concentration procedures include:

- **simple evaporation** facilitated by heating or vacuum;
- **methods involving extraction into different phases**, such as extraction from aqueous solution into an organic medium, followed by evaporation;
- **Simple chromatography**, such as use of ion-exchange columns (p. 458).

Using reagent blanks – these should be included in each analysis to account for interference deriving from the chemicals used.

You should take care to prevent concentration of a sample due to evaporation of solvent, or dilution of an analyte due to its volatility – check original sample volumes are maintained, e.g. by weighing, and use sample vials with tightly fitting stoppers.

Validity and quantitative analysis

Before using a particular procedure, you should consider its possible limitations in terms of:

Evaluating a new Method – results from a novel technique can be compared with an established 'standardised' technique by measuring the same samples by both methods and analysing the results by correlation (p 523).

- **measurement errors, and their likely magnitude:** these might include processing errors (e.g. in preparing solutions and making dilutions), instrument errors (e.g. a pH electrode that has not been set up correctly), calibration errors (e.g. converting a digital readout analyte concentration) and errors due to the presence of interfering substances;
- **sampling errors:** these may occur if the material used for analysis is not representative, e.g. due to biological differences between the individual organisms used in the sampling procedure (Chapter 32).

Replication will allow you to make quantitative estimates of several potential sources of error: for example, repeated measurements of the same sample can provide information on the precision of the analytical method, e.g. by calculating the coefficient of variation (p. 509), while measurements of several different samples can provide information on biological and sample variability, e.g. by calculating the standard deviation (Chapter 76). Analyses of different sets of samples at different times (e.g. on different days) can provide information on 'between batch' variability, as opposed to 'within batch' variability (based on a single set of analytical data).

'Standard' analytes (substances with known identity), sometimes called 'controls', are often used to calibrate methods and may assist in identification. These are often prepared in the laboratory by adding a known amount of analyte to a real sample (this is often termed 'spiking' a sample), or by preparing an artificial sample containing a known amount of analyte along with other relevant components (e.g. the major sample constituents and possible interfering substances). In many instances, several standards (including a 'blank' or 'zero') are assayed to construct a 'standard curve', which is then used to convert sample measurements to amounts of analyte (see Chapter 53 for details). Such standard curves form the basis of many routine laboratory assays: while hand-drawn linear calibration curves are sufficient for basic assays, more complex mathematical functions are often fitted to data sets using a computer program (e.g. bioassays, p. 315). Standards can also be used to check the calibration of a particular method: a mean value based on repeated measurements of an individual 'standard' can be compared with the true value using a modified t-test (p. 521, eqn [76.1]), in which there is only one standard error term, i.e. that associated with the measured values.

Validation of a particular method can be important in certain circumstances, e.g. in a forensic science or a clinical biochemistry laboratory, where particular results can have important implications. Such laboratories operate strict quality assurance and quality control procedures, including: (i) adherence to standard operating procedures for each analytical method; (ii) calibration of assays using certified reference materials containing known amount of analyte and traceable to a national reference laboratory; (iii) effective systems for internal quality control and external quality assurance; (iv) detailed record-keeping, covering all aspects of the analysis and recording of results. Although such rigour is not required for routine analysis, the general principles of standardisation, calibration, assessment of performance and record-keeping are equally valid for all analytical work.

Example When measuring a sugar in a cell extract in a formal research investigation, it would be insufficient to regard the retention time of a peak on a gas chromatography trace as providing sufficient information to identify the compound, even when compared with selected standards, as in practice many other compounds could have similar retention times. In such cases, gas chromatography may be used as a separation method coupled with mass spectroscopy (Chapter 68) to provide further information about the molecule in question.

Interpreting results from 'spiked' samples – remember that such procedures tell you nothing about the extraction efficiency of biomolecules from a particular sample, e.g. during homogenisation (Chapter 24).

Text references

Schwedt, G. (1997) *The Essential Guide to Analytical Chemistry.* Wiley, Chichester.

Skoog, D.A., Holler, F.J. and Crouch, S.R. (2006) *Instrumental Analysis Principles.* Brooks-Cole, Pacific Grove.

Sources for further study

Hage, D.S. and Carr, J.R. (2010) *Analytical Chemistry and Quantitative Analysis.* Pearson Education, Harlow.

Venn, R.F. (ed.) (2008) *Principles and Practice of Bioanalysis,* 2nd edn. CRC Press, Boca Raton.

STUDY EXERCISES

52.1 Check your understanding of the fundamental principles of chemical analysis. Distinguish between each of the following pairs of terms:

(a) qualitative analysis and quantitative analysis;

(b) sensitivity and selectivity;

(c) accuracy and precision;

(d) validation and replication;

(e) noise and drift.

52.2 Research sources of calibration standards and reference materials. Using the Web, find a supplier of a validated reference standard for the following:

(a) cortisol in serum;

(b) cotinine in urine.

52.3 Compare equipment specifications. The table (alongside) gives operational details for two techniques for measuring glucose, A and B.

Which technique has:

(a) the lower detection limit?

(b) the wider linear dynamic range?

(c) the greater accuracy?

(d) the better level of precision?

(e) the smaller sample volume?

Performance of two glucose measurement techniques

Aspect	Technique A	Technique B
Units of measurement	mmol L^{-1}	mg dL^{-1}
Detection limit	0.10	1.00
Linearity	to 20.0	to 250.0
Accuracy	±0.050	±1.50
Relative standard deviation (%)	3.5	2.5
Sample volume	40 µL	0.05 mL

Answers to these study exercises are available at www.pearsoned.co.uk/practicalskills

Understanding quantitative measurement – Chapters 27 and 52 contain details of the basic principles of valid measurement, while Chapters 59–70 deal with some of the specific analytical techniques used in biology. The use of internal standards is covered on pp. 459–60.

Calibrating laboratory apparatus – this is important in relation to validation of equipment, e.g. when determining the accuracy and precision of a pipettor by the weighing method: see p. 141.

There are many instances where it is necessary to measure the quantity of a test substance using a calibrated procedure. You are most likely to encounter this approach in one or more of the following practical exercises:

- **Quantitative spectrophotometric assay of biomolecules** (Chapter 68).
- **Flame or atomic absorption spectroscopic analysis** of metal ions in biological solutions (pp. 450–2).
- **Using a chromogenic or fluorogenic substrate** to determine the activity of an enzyme (p. 387).
- **Quantitative chromatographic analysis,** e.g. GC or HPLC (Chapter 69).
- **Using a bioassay system** (pp. 315–18) to quantify a test substance: examples include immunodiffusion (pp. 378–9) and radioimmuno-assay (pp. 380–1).

> **KEY POINT** In most instances, calibration involves the establishment of a relationship between the measured response (the 'signal') and one or more 'standards' containing a known amount of substance.

In some instances, you can measure a signal due to an inherent property of the substance, e.g. the absorption of UV light by nucleic acids (p. 404), whereas in other cases you will need to react it with another substance to see the result (e.g. molecular weight measurements of DNA fragments after electrophoresis, visualised using ethidium bromide, p. 403), or to produce a measurable response (e.g. the reaction of cupric ions and peptide bonds in the Biuret assay for proteins, p. 388).

The different types of calibration curve

By preparing a set of solutions (termed 'standards'), each containing either (i) a known *amount* or (ii) a specific *concentration* of the substance, and then measuring the response of each standard solution, the underlying relationship can be established in graphical form as a 'calibration curve', or 'standard curve'. This can then be used to determine either (i) the amount or (ii) the concentration of the substance in one or more test samples. Alternatively, the response can be expressed solely in mathematical terms: an example of this approach is the determination of chlorophyll pigments in plant extracts by measuring absorption at particular wavelengths, and then applying a formula based on previous (published) measurements for purified pigments (p. 450).

There are various types of standard curve: in the simplest cases, the relationship between signal and substance will be linear, or nearly so, and the calibration will be represented best by a straight-line graph (see Box 53.1). In some instances (Fig. 53.1(a)), you will need to transform either the *x* values or *y* values (Chapter 71), in order to produce a linear graph (e.g. in radial immunodiffusion bioassay, where the *y* values are squared, p. 379. In other instances, the straight-line relationship may only hold up to a certain value (the 'linear dynamic range') and beyond this point the graph may curve (e.g. in quantitative spectrophotometry,

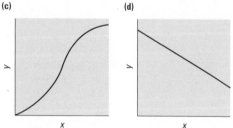

Fig. 53.1 Calibration curves: (a) log-linear; (b) curvilinear; (c) sigmoid, or S-shaped; (d) inverse.

Box 53.1   The stages involved in preparing and using a calibration curve

1. **Decide on an appropriate test method** – for example, in a project, you may need to research the best approach to the analysis of a particular metabolite in your biological material.

2. **Select either (a) amount or (b) concentration, and an appropriate range and number of standards** – in practical classes, this may be given in your schedule, along with detailed instructions on how to make up the standard solutions. In other cases, you may be expected to work this out from first principles (Chapters 23–25 give worked examples) – aim to have evenly spaced values along the *x* axis.

3. **Prepare your standards very carefully** – due attention to detail is required: for example, you should ensure that you check the calibration of pipettors beforehand, using the weighing method (p. 141). Do not forget the 'zero standard' plus any other controls required, e.g. to test for interference due to other chemical substances. Your standards should cover the range of values expected in your test samples.

4. **Assay the standards and the unknown (test) samples** – preferably all at the same time, to avoid introducing error due to changes in the sensitivity or drift in the zero setting of the instrument with time. It is a good idea to measure all of your standard solutions at the outset, and then measure your test solutions, checking that the 'zero standard' and 'top standard' give the same values after, say, every six test measurements. If the remeasured standards do not fall within a reasonable margin of the previous value, then you will have to go back and recalibrate the instrument, and repeat the last six test measurements. If your test samples lie outside the range of your standards, you may need to repeat the assay using diluted test samples (extrapolation of your curve may not be valid, see pp. 479–84).

5. **Draw the standard curve, or determine the underlying relationship** – Figure 53.2 gives an example of a typical linear calibration curve, where the spectrophotometric absorbance of a series of standard solutions is related to the amount of substance. When using a spreadsheet (Box 53.2) or graphic package (Chapter 13), it is often

Fig. 53.2 Typical calibration curve for spectrophotometric analysis.

appropriate to use Model I linear regression (p. 525) to produce a linear trend line (also termed the 'line of best fit') and you can then quote the value of r^2, which is a measure of the 'fit' of the measurements to the line (see pp. 523–6). However, you should take care not to use a linear plot when the underlying relationship is clearly nonlinear (Fig. 72.19) and you must consider whether the assumptions of the regression analysis are valid (e.g. for transformed data, Chapter 71).

6. **Determine the amount or concentration in each unknown sample** – either by reading the appropriate value from the calibration curve, or by using the underlying mathematical relationship, i.e. $y = a + bx$ (pp. 500–1). Make sure you draw any horizontal and vertical construction lines very carefully – students often lose marks unnecessarily by submitting poorly drawn construction lines within practical reports.

7. **Correct for dilution or concentration, where appropriate** – for example, if you diluted each test sample by ten-fold, then you would need to multiply by 10 to determine the value for the undiluted test sample. As another example, if you assayed 0.2 mL of test sample, you would need to multiply the value obtained from the calibration curve by 5, to give the value per mL.

8. **Quote your test results to an appropriate number of significant figures** – this should reflect the accuracy of the method used (see p. 497), not the size of your calculator's display.

the Beer–Lambert relationship often becomes invalid at high absorbance, giving a curve, Fig. 53.1(b)). Some calibration curves are sigmoid (Fig. 53.1(c)). Finally, the signal may *decrease* in response to an increase in the substance (Fig. 53.1(d)), e.g. radioimmunoassay, where an inverse sigmoid calibration curve is obtained (p. 380). In some practical classes, you may be told that the relationship is expected to be linear, curvilinear, or whatever, while in others you may be expected to decide the form of the standard curve as part of the exercise.

Practical considerations

Amount or concentration?

This first step is often the most confusing for new students. It is vital that you understand the difference between *amount* of substance (e.g. mg, ng, etc.), and *concentration* (the amount of substance per unit volume, e.g. mmol L^{-1}, mol m^{-3}, % w/v, etc.) before you begin your practical work.

> **KEY POINT** Essentially, you have to choose whether to work in terms of either (i) the total amount of substance in your assay vessel (e.g. test tube, or cuvette) or (ii) the final concentration of the substance in your assay vessel.

The interconversion of amount and concentration is covered in more detail on p. 147. Either way, this is usually plotted on the x (horizontal) axis and the measured response on the y (vertical) axis.

Choice of standards

In your early practical classes, you may be provided with a stock solution (pp. 147–8), from which you then have to prepare a specified number of standard solutions. In such cases, you will need to understand how to use dilutions to achieve the required amounts or concentrations (p. 148). In later work and projects, you may need to prepare your standards from chemical reagents in solid form, where the important considerations are purity and solubility (p. 145). For professional analysis (e.g. in forensic science or clinical biochemistry), it is often important to be able to trace the original standard or stock solution back to national or international standards or to certified reference materials.

How many standards are required?

This may be given in your practical schedule, or you may have to decide what is appropriate (e.g. in project work and research). If the form and working range of the standard curve are known in advance, these may influence your choice – for example, linear calibration curves can be established with fewer standards than curvilinear relationships. In some instances, analytical instruments can be calibrated using a single standard solution, often termed a 'single point calibrator'. Replication of each standard solution is a good idea, since it will give you some information on the variability involved in preparing and assaying the standards. Consider whether you should plot mean values on your standard curve, or whether it is better to plot the individual values (if one value appears to be well off the line, you have made an error, and you may need to check and repeat the measurement).

Example A test tube containing 8 mL of water plus 2 mL of 1% w/v NaCl ($M_r = 58.44$) would have a *mass concentration* of 0.2% w/v NaCl (a five-fold dilution of the original NaCl solution), which can also be expressed as 2 g L^{-1}; in terms of *molar concentration* (pp. 155–6), this would be equivalent to $2 \div 58.44 = 0.0342$ mol L^{-1} (to three significant figures). Expressed in terms of the *amount* of NaCl in the test tube, this would be 0.02 g in *mass*, or $0.02 \div 58.44 = 3.42 \times 10^{-4}$ mol (342 µmol) in *moles*.

Organisations providing national/international standards – these include:
- Laboratory of the Government Chemist (http://www.lgc.co.uk/), for UK standards and European Reference Materials.
- National Institute for Science and Technology (http://www.cstl.nist.gov/), for biochemical/chemical standards in the USA.
- Institute for Reference Materials and Measurements (http://www.irmm.jrc.be/), for European standards, including BCR and ERM materials.
- OIE Biological Standards Commission (http://www.oie.int/international-standard-setting/overview/), for international animal standards.

Plotting a standard curve – do not force your calibration line to pass through zero if it clearly does not. There is no reason to assume that the zero value is any more accurate than any other reading you have made.

Box 53.2 How to use a spreadsheet (e.g. Microsoft *Excel*) to produce a linear regression plot

Two approaches are possible: either using the trendline feature or using the regression analysis function within the data analysis tool pack. Both make the assumption that the criteria for Model I linear regression are met (see p. 525). In the example shown below, the following simple data set has been used:

Amount (ng)	Absorbance
0	0.00
10	0.19
20	0.37
30	0.56
40	0.63
50	0.78

Using the *Trendline* feature

This quick method provides a line of best fit on an *Excel* chart and can also provide a set of equation values for predictive purposes.

1. **Create a graph** *(chart)* **of your data.** Enter the data in two columns within your spreadsheet, select the data array (highlight using left mouse button) and then, using the *Chart Wizard* icon, select *Chart type > XY (Scatter)*.

2. **Add a trendline.** Right-click on any of the data points on your graph, and select the *Add Trendline* menu. Choose the *Linear* trendline option, but do not click *OK* at this stage. Rather, from the *Options* menu, select: (i) *Display equation on chart* and (ii) *Display R-squared value on chart*. Now click *OK*. The equation (shown in the form $y = bx + a$) gives the slope and intercept of the line of best fit, while the R-squared value (coefficient of determination, p. 526) gives the proportional fit to the line (the closer this value is to 1, the better the fit of the data to the trendline).

3. **Modify the graph to improve its effectiveness.** For a graph that is to be used elsewhere (e.g. in a lab write-up or project report), adjust the display to remove the default background and gridlines and change the symbol shape (see Box 72.2 for more advice and examples). Right-click on the trend line and use the *Format Trendline* menu to adjust the *Line Style > Width* of the line to make it thinner. Drag and move the equation panel if you would like to alter its location on the chart, or delete it, having noted the values. Figure 53.3 shows a calibration curve produced in this way for the data presented above.

Fig. 53.3 Calibration curve showing line of best fit and details of linear regression equation.

4. **Use the regression equation to estimate unknown** (test) **samples.** By rearranging the equation for a straight line and substituting a particular *y*-value, you can predict the amount/concentration of substance (*x*-value) in a test sample. This is more precise than simply reading the values from the graph using construction lines (Box 53.1). If you are carrying out multiple calculations, the appropriate equation, $x = (y - a)/b$, can be entered into a spreadsheet, for convenience.

Using the Regression Analysis tool

This requires the 'data analysis tool pack' to be loaded beforehand (using *Excel Options* under the Microsoft *Office* button in the top left corner, then select *Add-Ins > Analysis ToolPak)* and provides summary output that contains details of slope, intercept and coefficient of determination along with an analysis of variance (ANOVA) table (Chapter 76 gives further details of these aspects, including an example of output in Study exercise 76.4).

Note: instructions illustrated here may vary among the different versions of Microsoft *Office* Programs. Check the exact functions and syntax using the *Insert function* menu option (fx button in some versions).

Dealing with interfering substances – one approach is to use the method of 'standard additions', where the standards all contain a fixed additional amount of the sample (for more detail of this approach, see e.g. Dean, 1997). Internal standards (pp. 459–60) can also be used to detect such problems.

Preparing your standards

It is extremely important to take the greatest care to measure out all chemicals and liquids very accurately, to achieve the best possible standard curve. The grade of volumetric flask used and temperature of the solution also affect accuracy (grade A apparatus is best). You may also consider what other additives might be required in your standard solutions. For example, do your test samples have high levels of potentially interfering substances, and should these also be added to your standards? Also consider what controls and blank solutions to prepare.

> **KEY POINT** The validity of your standard curve depends upon careful preparation of standards, especially in relation to accurate dispensing of the volumes of any stock solution and the diluting liquid (diluent) – the results for your test samples can only be as good as your standard curve.

Preparing the calibration curve and determinining the amount of the unknown (test) sample(s)

This is described in stepwise fashion within Box 53.1. Check you understand the requirements of graph drawing, especially in relation to plotted curves (pp. 477–84) and the mathematics of straight-line graphs (pp. 500–1). Spreadsheet programs such as Microsoft *Excel* can be used to produce a regression line for a straight-line calibration plot (p. 525). Examples of how to do this are provided in Box 53.2.

Text reference

Dean, J.D. (1997) *Atomic Absorption and Plasma Spectroscopy,* 2nd edn. Wiley, Chichester. [Chapter 1 deals with calibration, and covers the principle of standard additions.]

Sources for further study

Cable, M. (2005) *Calibration: A Technician's Guide.* ISA, Raleigh.

Harris, D.C. (2010) *Quantitative Chemical Analysis,* 8th edn. Freeman, New York.

Miller, J.N. and Miller, J.C. (2010) *Statistics and Chemometrics for Analytical Chemistry*, 6th edn. Prentice Hall, Harlow. [Gives detailed coverage of calibration methods and the validity of analytical measurements.]

STUDY EXERCISES

53.1 Determine unknowns from a hand-drawn calibration curve. The following data are for a set of calibration standards for Zn, measured by atomic absorption spectrophotometry.

Absorbance measurements for a series of standard solutions containing different amounts of zinc

Zinc concentration (µg mL^{-1})	Absorbance
0	0.000
1	0.082
2	0.174
3	0.257
4	0.340
5	0.408
6	0.463
7	0.511
8	0.543
9	0.561
10	0.575

Draw a calibration curve by hand using graph paper and estimate the concentration of zinc in the following water samples:

(a) an undiluted sample, giving an absorbance of 0.157

(b) a twenty-fold dilution, giving an absorbance of 0.304

(c) a five-fold dilution, giving an absorbance of 0.550.

Give your answer to three significant figures in each case.

53.2 Determine unknowns from a calibration curve produced in *Excel*. The following data are for a set of calibration standards for protein content, measured by the Lowry (Folin–Ciocalteau) method (p. 388).

Absorbance measurements for a series of protein standards

Protein (mg per tube)	Absorbance
0.00	0.000
0.02	0.161
0.04	0.284
0.06	0.438
0.08	0.572
0.10	0.762

Using PC-based software (e.g. *Excel*), fit a trend line (linear regression) and determine the protein content of solutions with the following absorbances:

(a) 0.225   (b) 0.465   (c) 0.682

Give your answers to three significant figures in each case.

53.3 Identify the errors in a calibration curve. The figure shows a calibration curve of the type that might be submitted in a practical write-up. List the errors and compare your observations with the list given on the website.

Calibration graph

Answers to these study exercises are available at www.pearsoned.co.uk/practicalskills

Environmental investigations

Community – a number of populations living and interacting in the same geographical area.

Ecology – the study of the distribution and abundance of organisms and their interactions with each other and the physical environment.

Ecosystem – the combination of communities of living organisms and their physical envronment.

Individual – a single representative of a population or sample.

Population – a group of individuals existing within a specified time or space (see also p. 196).

Sample – a subset of a biological population, whose properties may be studied to gain information about the whole.

Definition

Modular organism – one which develops by repeatedly producing subunits (modules) rather than developing into a complete (unitary) organism with fixed number of parts. Examples include many plants, fungi and colonial animals such as corals.

Definitions

Abundance – the number of individuals present in a geographical location.

Cover – the proportion of ground that is occupied by a species or community. For plant populations, this includes the perpendicular projection of the aerial parts of individuals down on to the substrate, and is usually expressed as a percentage.

The study of ecology deals with the collection, presentation, analysis and classification of data relating to properties of communities and populations of organisms. This includes investigating species composition and changes in spatial and temporal organisation as a result of the processes of birth, growth, ageing, death and migration. The aim is usually to determine the significance of the measured values within the natural environment.

Commonly measured ecological features include:

- **species composition, abundance and cover;**
- **dispersion within populations;**
- **life and fecundity tables;**
- **age and size distribution;**
- **population growth rates.**

Investigating species composition, abundance and cover

Determining the ecological characteristics of populations and communities will nearly always involve sampling (Chapter 32) to estimate the true values, since most communities and populations are too large to enumerate all component individuals.

Methods for analysing immobile animal and plant populations are numerous and require careful planning and statistical treatments. Commonly used methods include the use of: quadrats, sample plots, belt transects, line intercepts and point-frequency intercepts (see Chapter 32). Techniques used depend on such aspects as the frequency, size and layering of organisms and the nature of boundaries between communities (see, for example, Rodwell, 2006). Description of communities involves species identification (Chapter 40) and measures of abundance and cover. Seasonality in distribution or cover is a factor to consider (for example, deciduous or annual plants and hibernating or migrating animals).

Abundance estimates quantify the number of individuals present, but for modular organisms, number rarely reflects dominance or cover. Although separate and distinct properties, cover and abundance may be combined in a qualitative description such as the Domin scale (Table 54.1), but this is subjective at low cover densities and as a result the data are difficult to analyse statistically.

Table 54.1 The Domin scale (after Rodwell, 2006)

Domin scale value	Description
10	Cover of 91–100%
9	Cover of 76–90%
8	Cover of 51–75%
7	Cover of 34–50%
6	Cover of 26–33%
5	Cover of 11–25%
4	Cover of 4–10%
3	Cover of <4% with many individuals
2	Cover of <4% with several individuals
1	Cover of <4% with few individuals

Fig. 54.1 Example of standard record sheet as used in the UK Joint Nature Conservation Committee National Vegetation Survey. Note data recorded to specify the site and its location; the use of a diagram to illustrate patterns of plant distribution; and the species list with associated Domin value (Table 54.1). Reproduced with permission.

Table 54.2 The major vegetation community types in the UK with codes

Code	Vegetation type
A	Aquatic communities
CG	Calcicolous grasslands
H	Heaths
M	Mires
MC	Maritime cliff communities
MG	Mesotrophic grasslands
OV	Vegetation of open habitats
S	Swamps and tall-herb fens
SD	Shingle, strandline and sand-dune communities
SM	Salt-marsh communities
U	Calcifugous grasslands and montane communities
W	Woodlands and scrub

Fig. 54.2 Grid map showing distribution of *Lepus timidus*, the mountain hare, in Great Britain and Ireland (based on sightings in 10 km squares).
NERC 100017897 2004, reproduced with permission (Crown Copyright).

Description of communities involves a combination of qualitative and quantitative features. An example is the UK National Vegetation Survey, which uses a standard record sheet (Fig. 54.1, opposite) to note relevant information. Communities can be classified following computerised data analysis (e.g. Table 54.2) and subsequently are coded, numbered and named after predominant species, for example, SM16, the *Festuca rubra* salt marsh community.

At a national level, the distribution of animal or plant species can be mapped using coarse-scale information about their occurrence within individual areas of a country, typically 10 km grid squares. An example of this approach is shown in Fig. 54.2. Representative uses of such grid-based species range maps include:

- **studying changes in the distribution of a species over time;**
- **comparing the distribution of a species with similar-scale meteorological or habitat data;**
- **as an aid to identification (Chapter 40), by considering likelihood of occurrence in a specific location;**
- **investigating the spatial coincidence of two or more species;**
- **highlighting the scarcity of rare or threatened species.**

The study of animal populations is generally more difficult than the study of plants, since animals are relatively rare, hidden and are not usually stationary. When choosing a method for recording and collection (Chapter 36) you should bear in mind that sampling bias can occur both due to operator effects and the nature and behaviour of the animal. Approaches fall into three main categories:

1. **The true census:** this requires a direct count of all individuals in a given area and is similar in principle to the human population census. It is very difficult to achieve for most wild species but can be used where areas of concentration occur, e.g. waterfowl on wintering grounds, rookeries, roosts and breeding colonies of birds and mammals. Photography and subsequent image analysis are often appropriate for such studies.

2. **The sampling estimate:** derived from counts on sample plots. Sampling estimates (see Chapter 32) make two basic assumptions:

   (a) mortality and recruitment during the sampling period are negligible, or can be accounted for;

   (b) all members of the population have an equal chance of being sampled, i.e. there is no bias in the method being used.

   It is important, therefore, that the sampling procedure is appropriate for the species, time, place and purpose of the study.

3. **Indirect indices:** these include counts of faecal materials, counts of calls of vocal species, and measures of other animal signs, such as territorial marks and songs. All data are relative and must be compared with data from other areas or other time periods. The results do not give estimates of absolute populations but can indicate trends.

The main methods of obtaining the raw data for such approaches are direct sampling methods such as trapping (Chapter 36), mark–recapture methods (Box 54.1) and photography (Chapter 50).

Box 54.1 Marking animals to study ecological populations

Marking is necessary when you need to identify certain individuals at some future date, either for studies of behaviour or movement or to estimate populations by the mark–recapture method. In the mark-recapture method a proportion of the population is marked in some way, returned to the original population and then, after mixing (usually assumed to be 'complete'), a second sample is taken. The method works on the principle that the number of marked individuals in the second sample will have the same ratio to the total number in that sample as the total of marked individuals originally released would have to the total population. Since the first three quantities are known, the total population N can be estimated from:

$$N = MC/R \qquad [54.1]$$

where M = total number of animals captured and marked on the first visit; C = total number of animals captured on the second visit; and R = number of animals captured on the first visit (i.e. marked) that were then recaptured on the second visit. More complex formulae may be used to remove bias and sophisticated statistical analysis is possible.

The mark–recapture method has the advantage over fixed-unit sampling methods in that its accuracy does not depend on an assessment of the number of sampling units available within the habitat. A basic prerequisite for this method is that there is a method of marking individuals so that they can be released unharmed and unaffected back into the wild and can be recognised upon recapture. Such marked individuals can also be used for a variety of studies such as dispersal, longevity and growth.

The most widely used marking methods include:

- **Recognition of individuals with identifiable natural markings:** this has been used, for example, with many whale species and with apes, and is usually supported by a photographic record.

- **Tagging labels:** common for birds, which are normally labelled by serially numbered aluminium bands (bird rings), often bearing colours which facilitate individual recognition in the field. Fish are also marked by tagging, typically through a fin.

- **Paints and dyes:** useful for animals such as terrestrial snails which are easily marked with quickdrying cellulose paint. Aquatic species can be marked with marine paints. Mammalian fur and bird plumage can be marked by coloured dyes to permit recognition of individuals at a distance. In powder form, certain dyes and fluorescent substances may be appropriate for hairy insects, but they tend to wear off relatively easily.

- **Modification of body parts:** this is carried out with some small mammals, frogs, toads, salamanders and lizards. Individuals can be identified via a system of toe clipping (removal of the distal part of one or more toes). Mammals may also be marked by notching of ears. Snakes and lizards can be marked by removing scales or patches of scales in particular combinations. Crabs and other crustaceans can be identified by modifying exoskeletal components.

- **Radioactive isotopes:** these can be used to mark animals that are secretive in habit, live in dense cover, spend part or all of their life underground or have radically different phases in their life cycle. Animals are fed small amounts of gamma-emitting radioactive material which becomes incorporated into the tissues. However, use of radioisotopes requires particular care and appropriate permissions (Chapter 64).

Note that arthropods, snakes, etc. shed their skin (moult) periodically and that this may affect the permanence of any external markings.

Dispersion within populations

The dispersion (distribution) of individuals within populations may be uniform, random or clumped (Chapter 31) – characteristics that may be of ecological interest. Data collected from quadrats, point quadrats and other methods may be used to quantify the nature of dispersion (Chapter 32). However, you should take account of the fact that the analysis can be affected by the size of the sampling unit you select. Where the density of individuals is low relative to the available surface area or volume, the Poisson distribution usually applies, and can be used to test for randomness of distribution (pp. 517–18).

Life and fecundity tables

A life table (e.g. Table 54.3) provides a clear and systematic picture of mortality and survival, according to age. The concept was developed from studies of human populations by insurance companies. It consists of a series of columns, each of which describes mortality statistics for members of a population, such as daily mortality rate or life expectation. To construct a life table, you must be able to determine the age of individuals and then allocate them to age intervals, the length of which must be appropriate to the longevity of the organism. Information is then needed on survival or mortality for each age class.

There are three main types of life table:

1. **Horizontal (dynamic) life tables:** these are constructed by following a cohort of individuals all born within a short time period from birth to the death of the last individual (e.g. Table 54.3).

2. **Vertical (static) life tables:** these are created by sampling the population and ageing the individuals to obtain a distribution of size classes at a particular moment. This type of table assumes that each age class is sampled in proportion to its numbers in the population, that the birth and death rates are constant and the population size is static. In practice, these assumptions rarely apply, and data may require adjustment or smoothing to minimise the resulting effects.

3. **Dynamic-composite life tables:** these record the same information as the dynamic (horizontal) tables but they take as the cohort a composite of a number of animals marked over a period of years rather than at just one birth period. An example would be the marking of each batch

> Using life tables for organisms which have distinct stages to their life cycles – for insects and other similar animals, life tables are best divided into developmental stages rather than time intervals.

Table 54.3 Life table for *Phlox drummondii*, annual phlox, at Nixon, Texas (source: Leverlich and Levin, 1979)

Age interval (days)	Survivors at end of period	Survivorship	No. dying during interval	Average mortality rate per day	Mean expectation of life (days)
0–63	996	1.0000	328	0.0052	122.87
63–124	668	0.6707	373	0.0092	104.73
124–184	295	0.2962	105	0.0059	137.59
184–215	190	0.1908	14	0.0024	137.05
215–231	176	0.1767	2	0.0007	115.72
231–247	174	0.1747	1	0.0004	100.96
247–264	173	0.1737	1	0.0003	85.49
264–271	172	0.1727	2	0.0017	68.94
271–278	170	0.1707	3	0.0025	62.71
278–285	167	0.1677	2	0.0017	56.78
285–292	165	0.1657	6	0.0052	50.42
292–299	159	0.1596	1	0.0009	45.19
299–306	158	0.1586	4	0.0036	38.46
306–313	154	0.1546	3	0.0028	32.36
313–320	151	0.1516	4	0.0038	25.94
320–327	147	0.1476	11	0.0107	19.55
327–334	136	0.1365	31	0.0325	13.85
334–341	105	0.1054	31	0.0422	9.90
341–348	74	0.0743	52	0.1004	5.58
348–355	22	0.0221	22	0.1428	3.50
355–362	0	0.0000			

Table 54.4 Fecundity table for *Phlox drummondii*, annual phlox, at Nixon, Texas, based on seed production (source: Leverlich and Levin, 1979)

Age interval (days)	Survivors at end of period	Survivorship	Average no. of seeds per individual during interval	Contribution to net reproductive rate during interval
0–299	996	1.0000	0.0000	0.0000
299–306	158	0.1586	0.3394	0.0532
306–313	154	0.1546	0.7963	0.1231
313–320	151	0.1516	2.3995	0.3638
320–327	147	0.1476	3.1904	0.4589
327–334	136	0.1365	2.5411	0.3470
334–341	105	0.1054	3.1589	0.3330
341–348	74	0.0743	8.6625	0.6436
348–355	22	0.0221	4.3072	0.0951
355–362	0	0.0000	0.0000	0.0000
				Total = 2.4177

of newly hatched birds over a period of years and following the fate of each year's group. Data are then pooled and all marked animals are treated as a single cohort.

Both static and dynamic-composite tables are inevitably inaccurate because the assumptions of constancy from year to year are rarely valid. In spite of this, such tables may present a reasonable description of average population characteristics.

A fecundity table (e.g. Table 54.4) can be constructed if you know the reproductive productivity of each age class of females in a population. The productivity can be determined by counts of litter size, brood counts, young fledged or egg/seed production, depending upon the organism being investigated.

Age and size structure

The age structure of plant, animal and human populations determines, in part, population reproductive rates, death rates, vigour, survival and other demographic attributes. The structure of populations can be characterised by:

- **specific age categories,** such as years or months;
- **life history stages,** for example, eggs, pupae, larvae and instars in insects;
- **size classes,** perhaps where age is difficult to determine.

The use of age structure in studies of organisms such as plants is complicated by their modular nature and their extensive use of asexual reproduction. The use of size classes is therefore often substituted. This is because each module has its own age which is independent of the age of the genet.

> **Life tables in botany** – plant ecologists have adapted the use of life tables to study the dynamics of plant modules such as leaves that exhibit age-specific mortality.

Definitions

Genet – the term given to the genetic individual that results from the sexual reproduction of a modular species.

Zygote – the product of fertilisation in a sexually reproducing species.

KEY POINT Modular species are composed of the genet (e.g. a tree), which is the product of the original zygote, and the modules (leaves, flowers, roots) – units of construction, each of which has its own birth, life and death, usually independently of the genet.

Some methods of determining the age of organisms are discussed in Box 54.2.

Box 54.2   Measuring age in organisms

Plants

Determining age in plants usually involves following a co-hort of marked individuals over time or taking a sample and determining age by use of growth rings, bud scans or other measurements. Cambial growth rings (Fig. 54.3) are usually laid down as a result of the interruption of growth by seasonal changes in temperature and/or light. They should be interpreted with care: they may fail to be laid down or multiple rings may be produced by short-term changes in cambial activity during the growing season.

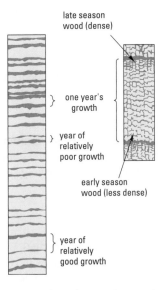

Fig. 54.3 Transverse section through a woody stem, showing features of annual rings.

late season wood (dense)

one year's growth

year of relatively poor growth

early season wood (less dense)

year of relatively good growth

Animals

Ageing wild animals is not easy, although a number of techniques have been developed:

- **Juvenile/adult characteristics:** ageing techniques for birds and mammals may include the use of specific characteristics of body parts. For birds, plumage development is often used and for mammals, tooth characteristics can be useful.

- **Growth rings:** some fish can be aged using scales or ear stones (otoliths), both of which lay down growth rings which vary in spacing or density (Fig. 54.4). Any bony or calcareous structure which is laid down from the outside has the potential for possessing growth rings. Such structures exist in echinoderm skeletal elements, molluscan and barnacle shells and the vertebrae of fish.

Fig. 54.4 Structures used to age teleost fish: (a) scale of salmonid fish showing circuli; (b) otolith of flatfish showing annual rings.

- **Length-frequency distribution:** this can be useful in species with a polymodal population distribution, particularly where there are very distinct and short reproductive and settlement phases. The statistical analysis of polymodal frequency distributions has been refined from commercial fish population studies but works well for other polymodal species, such as mussels.

Table 54.5 Types of interactions between species. These may vary in strength and could include as examples, symbiosis (mutualism), parasitism (antagonism), contests for mutual prey (competition).

Type of interaction	Effect on A	Effect on B
Mutualism – both species benefit	+	+
Commensalism – one species benefits	+	0
Neutralism – neither species benefits	0	0
Antagonism – one species benefits, other harmed	+	−
Amensalism – one species harmed	0	−
Competition – both species harmed	−	−

Measuring community parameters

Some of the key parameters analysed for community structures are described below.

Interspecific association

Some species in a community may occur together more frequently than predicted by chance. Measurement of this association provides an objective way of recognising natural groupings of species. Associations may be positive or negative, each providing information on the nature of interactions between the organisms considered (Table 54.5), although the explanations for such associations may be complex. A commonly used measure is that of the coefficient of association (C) which varies from +1.0 for a maximum positive association to −1.0 for a maximum negative association: a value of 0 indicates no statistical association. Data sets noting presence or absence in a defined area are all that is required and testing of the significance of a relationship requires the application of the Chi^2 test (see Chapters 61 and 76).

Community similarity

It may be necessary to compare the species composition of communities over space or time. The similarity of communities can be measured from data as basic as presence/absence of species or from more complex data sets such as those based on density or frequency. A variety of measures exist but you are advised to consult a specialist text such as Fowler *et al.* (1998) if required to use particular statistical parameters and tests: the former include similarity coefficients, percentage similarity and the coefficient of community.

Community ordination

Ordination is the technique of arranging components in unidimensional or multi-dimensional order in such a manner that the position of each component along the axis or axes conveys information about its composition or relationship to the other components.

Ordination is founded on the assumption that community composition varies gradually over a continuum of environmental conditions. It may be accomplished by two different approaches to deriving the axes:

1. **changes in environmental conditions;**
2. **changes in composition of the community.**

When the axes are based on environmental changes, the relative position of the communities along the axes reflects changes in composition caused by the environment (gradient analysis). Figure 54.5 is a simplified generic example to illustrate the concept. When based on community composition, their relative positions are based on similarities of species composition. The approaches used are various and complex, though they have been simplified by the existence of computer programs. If you wish to carry out a community ordination, you should consult relevant texts, such as Whittaker (1982).

Quantifying species diversity

Species richness, S, is a simple measure of the number of species present in a community, and hence diversity. However, this value provides little

Fig. 54.5 A much-simplified example of an ordination for two species showing their abundance relative to a selected environmental variable, indicating differing preferences for specific values of the variable and potentially explaining their observed spatial distribution.

Example Beals *et al.* (2000) provide a worked calculation of Shannon–Weiner *H* and *E* for a study of ant diversity in Costa Rican primary forest and banana plantations. The species richness *S* of ants in these habitats was 37 and 14, respectively. The corresponding values of *H* were 3.182 and 0.832, whereas those of *E* were 0.881 and 0.315. The relatively low *E* value for the plantations (lack of species evenness) reflects the domination in that habitat of a single species (83% of all ants sampled), while the most abundant forest species only made up 12.5% of the ants sampled.

Examples Ficetola *et al.* (2008) studied amphibian activity by examining mitochondrial DNA in filtrates from different ponds. The authors concluded that the technique allowed the reliable detection of secretive organisms such as bullfrogs in wetlands without the need for time-consuming observation.

de Vere *et al.* (2012) described standard DNA 'barcodes' for formally identified specimens of the complete Welsh flora, allowing the identification of plant species from material which in the past would have been difficult or impossible. This database has been used to investigate the role of indigenous hoverflies in pollination by studying the DNA present in the pollen on their bodies.

information about the relative abundance of each species present, which is an important feature. The most widely used diversity indices are:

- **Simpson's Index, *D*:** this measures the probability that two individuals randomly selected from a sample will belong to the same species (or some category other than species). Summing over all the species identified in an area, it is calculated as:

$$D = \sum \left[\frac{n(n-1)}{N(N-1)} \right]^2 \qquad [54.2]$$

- where *n* = the total number of organisms of a particular species and *N* = the total number of organisms of all species. *D* ranges between 0 and 1 and the bigger the value of *D*, the lower the diversity. *D* is often subtracted from 1 to give Simpson's Index of Diversity, a more intuitive value which increases from 0 to 1 with increasing diversity.

- **The Shannon-Weiner diversity index, *H* and evenness index, *E*:** this diversity index accounts for both abundance and evenness of the species present. Summing over all the species present in the area, *S*, it is calculated as:

$$H = \sum p \cdot \ln p \qquad [54.3]$$

- where *p* is the proportion of numbers of a particular species relative to the total number of individuals present. The value of the index typically ranges from 1.5 (low species richness and evenness) to 3.5 (high species evenness and richness). The Shannon–Weiner evenness index is calculated from:

$$E = H/\ln S \qquad [54.4]$$

E takes values between 0 and 1, with 1 indicating complete evenness.

Measuring environmental nucleic acids

DNA and RNA are present in many environmental samples and provide a useful indirect method for characterising and tracking organisms. This type of study is variously called metagenomics, environmental genomics, ecogenomics and community genomics. The approach is well established for examining the diversity of microbes in ecosystems and has revealed that in many cases standard microbial isolation techniques based on laboratory culture (Chapter 38) greatly underestimate the range of species present. Similar methods can also be used to study the distribution of viruses in nature. This is a particularly useful approach as PCR amplification of virus DNA avoids the need for replication in a suitable host cell, thereby enabling the study of viruses not currently known from cell-based replication studies (Suttle, 2007). Yet another application is to study the distribution of eukaryote organisms, allowing their presence to be inferred from their DNA 'fingerprint' rather than direct observation.

DNA is the nucleic acid of choice for most analyses, including for sequences coding for RNA. Typically, for DNA analysis, the technique might involve:

1. **Sampling** (following the principles outlined in Chapter 32) and stabilisation
2. **Filtering and centrifugation,** to precipitate cellular remains, containing DNA
3. **DNA extraction** (e.g. using a commercial kit, Chapter 62)
4. **Polymerase chain reaction (PCR) amplification of the DNA** (Chapter 62) with relevant primers (e.g. segments of DNA coding for

ribosomal RNA, rRNA, in bacteria and archaea (p. 235), and segments of mitochondrial cytochrome-b in eukaryotes)

5. **Analysis and sequencing of PCR-amplified DNA** (e.g. hypervariable rRNA regions) from the samples.

Text references

Beals, M., Gross, L. and Harrell, S. (2000) *Diversity Indices: Shannon's H and E.* Available: http://www.tiem.utk.edu/~gross/bioed/bealsmodules/shannonDI.html Last accessed 03/03/15.

de Vere, N., Rich, T.C.G., Ford, C.R., Trinder, S.A. *et al.* (2012) *DNA Barcoding the Native Flowering Plants and Conifers of Wales.* PLoS ONE 7(6): e37945.doi:10.1371/journal.pone.0037945#

Ficetola, G.F., Miaud, C., Pompanon, F. and Taberlet, P. (2008) Species detection using environmental DNA from water samples. *Biology Letters,* **4**, 423–5. Available: http://rsbl.royalsocietypublishing.org/content/4/4/423. Last accessed 03/03/15.

Fowler, J., Cohen, L. and Jarvis, P. (1998) *Practical Statistics for Field Biology,* 2nd edn. John Wiley, Chichester.

Leverlich, W.J. and Levin, D.A. (1979) Age specific survivorship and reproduction in *Phlox drummondi. The American Naturalist,* **113**, 881–903.

Rodwell, J.C. (2006) *National Vegetation Classification: Users' handbook* (Joint Nature Conservation Committee, Peterborough). Available: www.ukgbc.org/site/document/download/?document_id=556 Last accessed 14/07/15.

Suttle, C.A. (2007) Marine viruses – major players in the global ecosystem. *Nature Reviews Microbiology* **5**, 801–812.

Whittaker, R.H. (ed.) (1982) *Ordination of Plant Communities.* Junk, The Hague.

Sources for further study

Krebs, C.J. (2009) *Ecology,* 6th edn. Benjamin Cummings, San Francisco.

Molles, M.C. Jr (2005) *Ecology: Concepts and Applications,* 3rd edn. McGraw-Hill, New York.

National Biodiversity Network. *NBN Gateway.* Available: https://data.nbn.org.uk/ Last accessed 14/07/15.

New, T.R. (1998) *Invertebrate Surveys for Conservation.* Oxford University Press, New York.

Smith, R.L. and Smith, T.M. (2001) *Ecology and Field Biology*, 6th edn. Benjamin Cummins, New York. [Especially Appendix B on 'measuring community and population structure'.]

STUDY EXERCISES

54.1 Carry out some research on indirect indices as a measure of animal populations. In each case, try to find out: what measure was used; how it was used to quantify the population; and what problems were identified by the researchers in obtaining accurate results.

54.2 Some terrapins have been accidentally released into a pond in a botanic garden and are eating a collection of rare aquatic plants. Before a concerted attempt to capture all the animals, the staff would like to estimate how many there are in the pond. They catch a sample of 17 terrapins, mark their backs with paint, then release them back into the pond. Ten days later they return and catch 19 animals, of which 6 have paint markings. What is their best estimate of the total number of terrapins in the pond)? [Hint: use eqn [54.1], Box 54.1.]

54.3 How would you determine/estimate age in the following organisms: (a) a perennial plant species; (b) a cockle (Mollusca: Bivalvia: *Cerastoderma edule*); (c) a blowfly or related Dipteran insect; (d) a reptile; (e) a bird species; (f) and a marine mammal?

Answers to these study exercises are available at www.pearsoned.co.uk/practicalskills

A simple three-fold classification of aquatic environments can be made on the basis of salinity:

- **marine** (salinity relatively constant at ca. 35% w/w);
- **estuarine** (variable salinity between 0 and 35% w/w);
- **freshwater** (zero or very low salinity).

The freshwater class may be further subdivided into:

- **rivers and streams** (characterised by flowing water);
- **lakes and ponds** (characterised by relatively stationary water).

Many analytical methods are common to all of the above environments (e.g. determination of suspended sediment concentration or pH measurement) while others (e.g. measurement of river discharge) are environment-specific. For an introduction to the principles of chemical analysis, refer to Chapter 52.

> **KEY POINT** The sampling approach (Chapter 32) you choose when analysing aquatic environments will depend on the method selected for the analysis.

Salinity, conductivity and temperature

Salinity is of considerable ecological importance in many habitats and it affects the types of organisms that live in a body of water, or on land supplied by a specific water source. The term applies to total salts – usually sodium chloride is most abundant, but in some cases such as inland saline lakes, Mg and Ca sulphates and bicarbonates are important.

A simple, though not very accurate, way to measure NaCl salinity is by evaporating a water sample of known mass to dryness, weighing the amount of salts that remain and expressing the latter as a ratio of the former. The customary laboratory procedure, however, involves the titration of a water sample with silver nitrate ($AgNO_3$), whereby:

$$NaCl(aq) + AgNO_3\,(aq) \rightarrow NaNO_3\,(aq) + AgCl(s) \qquad [55.1]$$

The mass of halide halogen ion precipitated as AgCl from the sample is then compared with the precipitate from a Standard Seawater sample.

You are most likely to measure salinity using some form of field probe, often known as a salinometer (Fig. 55.1). Such probes measure the electrical conductivity of the water and convert this to a salinity reading. The most versatile and recommended instruments measure electrical conductivity, salinity and water temperature, and can be used in both fresh and salt waters. Probes can be lowered over the side of a boat permitting measurements of these three parameters simultaneously at various depths (determined using a graduated cable) in the water column.

Definition

Salinity – traditionally defined as the mass ratio between dissolved salts and water. It is often expressed as practical salinity units (psu), where 1 psu = 1 part per thousand (ppt or %, w/w). Practical salinity (S) is usually calculated from the ratio of the conductivity of a water sample to that of International Association for Physical Sciences of the Ocean (IAPSO) Standard Seawater, batches of which are calibrated against a defined KCl solution. Since S is a ratio, it is dimensionless.

Categorisation of tolerance of aquatic organisms to salt – the following terms are frequently used:

- **Halophile** – an organism (often a microbe) able to survive in very salty conditions (hence, halophyte, a plant adapted to saline conditions).
- **Euryhaline** – an organism able to withstand a wide range of salinities.
- **Stenohaline** – an organism able to survive in a narrow range of salinities.

Definition

Electrical conductivity – often abbreviated to conductivity, this is a measure of a solution's ability to conduct electricity. Electrical conductivity is the reciprocal of electrical resistivity and its units were formerly based on 'reciprocal ohms' (mho m^{-1}). The correct SI units are S m^{-1}, where 1 S (siemen) = 1 mho. The conductivity of sea water is close to 54 μSm^{-1} For measurements in freshwater, the practical units are thus also μSm^{-1}. The relationship between electrical conductivity, temperature and salinity is complex (see e.g. Howard, 1998).

Fig. 55.1 An example of a hand-held salinometer. Image courtesy of David Bryson/Clandonia Resources.

Looking after field probes – make certain that these are well maintained and calibrated in accordance with manufacturers' instructions prior to use. The accuracy and precision of your readings will depend on this.

SAFETY TIP as with all operations carried out on water bodies or beside or in running water, extreme care should be taken. Never work alone. Wear appropriate clothing, footwear (i.e. waders) and life jackets.

Using a Secchi disk – an individual observer should always take Secchi depth measurements in the same manner each time (e.g. standing, sitting, with spectacles, without spectacles, etc.). To further standardise procedures, you should always take measurements on the shady side of a boat and use a rope that will neither stretch nor shrink in contact with water.

pH

There are many types of test kits and pH meters available for field and laboratory use. Meters are recommended due to their greater accuracy. Those designed for field operation tend to have more robust probes compared with the lab equivalents, but care must still be taken not to damage them. In the field, they can be deployed in collected water samples (e.g. from specific depths in a lake) or directly into the water (e.g. the surface water of a lake). The general procedure for measuring pH is described in Chapter 26.

Water transparency

Water transparency is an important environmental factor as it affects both the potential survival and growth of photosynthetic organisms and can be used an indicator of the trophic status of water bodies. The main methods of measuring transparency are the Secchi disk and electrical instruments that measure either the transmittance or scattering of light.

The Secchi disk

The Secchi disk provides an easy to use and inexpensive means of measuring water transparency or clarity. It consists of a circular disk, 20–30 cm in diameter, that is either white or divided into alternating black and white quadrants. White Secchi disks are conventionally used for marine investigations, whereas black and white disks are usually used in studies of lake water quality. The disk is attached to a graduated rope and has a weight at its base to ensure vertical descent through a water column. The procedure for use is as follows (see also Fig. 55.2):

1. **Slowly lower the disk into the water** from the side of the boat until it disappears from view to the observer above.
2. **Record the depth from the graduated rope.**
3. **Carefully raise the disk until it just reappears in view.**
4. **Record the depth.**
5. **Work out the average of the two depth readings** – this is known as the Secchi disk depth, often shortened to Secchi depth.

disk raised slowly to point where it reappears

Secchi depth disk is midway between the two points

disk lowered slowly until it disappears from view

Fig. 55.2 Field determination of Secchi disk depth.

The value recorded is a complex function of many factors:

- **the ambient light level;**
- **the reflection of light from the water surface;**
- **the light absorption characteristics of the water;**
- **water colour;**
- **suspended sediment concentration;**
- **concentration of algae, plankton, etc.**

Nevertheless, readings of Secchi disk depth provide a simple and effective way of comparing the clarity of different waters or variations in clarity within the surface waters of a water body (e.g. a lake). They can also provide ground measurements to correlate with remote sensing observations of water bodies.

Optical instruments

Various optical methods are available to provide information on the characteristics and concentrations of materials suspended in water. These work either on the principle of light beam attenuation or light beam scattering.

- **Light attenuation instruments (transmissometers)** – a beam of light is shone directly at a sensor; the greater the attenuation of the light, the higher the concentration of suspended materials. Note that one very large particle in the light path (e.g. a leaf fragment) could produce the same attenuation as hundreds of suspended grains of silt.
- **Light scatterance instruments/turbidimeters** – a sensor measures the light scattered (i.e. reflected and refracted) by particles suspended in the water column to provide a measure of scatterance (also called turbidity and cloudiness). The higher the scatterance, the greater the number of particles in suspension. Note, however, that scatterance readings are a complex function of particle size, particle shape and the reflective and refractive properties of the particles. The units recorded are 'notional turbidity units' (NTUs), and hence non-SI.

Always record averages of several readings and ensure that water samples that have been static for any length of time are well mixed before taking readings.

Water flow velocity

The easiest way to measure the velocity of flowing waters is to record, with a stop watch, the time it takes for a float to travel between two known points. Select, for example, a straight river reach with a minimum length of 100 m. To obtain an accurate flow velocity, the minimum duration of float movement should be 20–25 s. The use of several floats across a river enables variations in flow velocity with distance from the banks to be determined, together with horizontal flow patterns. It is, however, important to note that this method measures the velocity of surface waters only.

To measure water flow velocity at depth below the surface, it is necessary to use some form of current meter. The simplest type consists of a propeller, mounted at the upstream end of a streamlined, torpedo-shaped

Long-term optical monitoring – when deploying optical instruments to monitor materials suspended in water over long periods (as for example in a river), it is important to clean the covers on the sensor and light beam source regularly to prevent algal growth, which could erroneously be interpreted as high concentrations of suspended matter.

Floats for flow rate measurement – oranges or grapefruit can be used as practical partially submerged floats and can be identified with an indelible felt pen where necessary.

Fig. 55.3 A propeller-type current meter mounted on a rod for measurement of flow velocity in a river.

Average flow velocity – in a river this occurs at approximately 0.6 × the water depth, measured from the surface downwards. The velocity of flow of surface waters, as determined by float tracking, is, as a general guide, approximately equal to 1.25 × the average flow velocity of the river.

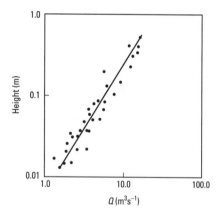

Fig. 55.4 Example of a stream flow rating curve. Note that the x-axis is logarithmic.

weighted body (Fig. 55.3). The number of revolutions of the propeller per unit time increases with increasing speed of water flow. Rotations are recorded, over fixed periods, either electronically or mechanically by a control box that can be installed on a boat for marine use or hand-held by the operator for river measurements. An instrument calibration curve enables the number of revolutions over a fixed time period to be converted to a flow speed. Current meters for marine use include a compass that enables flow direction (and therefore velocity) to be determined.

For river measurements, current meters are usually mounted on to a hand-held graduated rod that has a plate at its base. It may be necessary to wade into the river and measure the current speed at known distances from the bank and known depths. Depth may be recorded by reference to the graduated rod and distance from the bank by stretching a tape across the stream. In this way variations in flow rate with depth and distance from the banks may be determined. Stand as far as possible downstream of the current meter propeller so as to cause minimal turbulence of the water. On very large rivers, a current meter is deployed from a cable stretched across the river from which the meter can be lowered to chosen depths at specific distances.

River discharge measurements

Discharge measurements are used in ecology to characterise and compare river and stream habitats and to calculate parameters such as sediment movement and productivity.

To determine the volume of water per unit time (or discharge) passing through a particular cross-section of a river, it is necessary to know the water velocity and area of cross-section. This is usually achieved at gauging stations where a river has been engineered to flow through a well-defined structure, for example, a concrete flume, a rectangular weir or a V-notch weir. This enables cross-sectional area variations to be more accurately determined. The water level can be measured from a ruler or by using a depth recording instrument, such as an echo sounder.

The product of the flow velocity (ms^{-1}) and the cross-sectional area (m^2) at any time gives the instantaneous discharge (Q, $m^3\ s^{-1}$). In practice, a series of discharge measurements is made at various flow conditions over a period of years. This enables the construction of a stage–discharge relationship, or streamflow rating curve, where stage (h) is the river water level in m. A rating curve usually approximates to a straight line on a log–log scale (Fig. 55.4); thus the relationship is of the power function form:

$$Q = K(h)^n \qquad [55.2]$$

where K is a coefficient and n is an exponent.

At least 30 data points are needed to establish a good rating curve. Once this has been established, discharge can simply be determined by monitoring the water level and reading the corresponding value of discharge from the stage–discharge curve.

If you want to select your own site for river gauging, it is important that:

- **the reach is straight and has relatively uniform cross-sectional area and slope angle;**
- **the water depth is sufficient** for the immersion of a current meter;

- **the channel is free from weed growth and other obstacles** such as large boulders or trees;
- **the site chosen is not close to a tributary confluence;**
- **the banks and channel bed are stable** and not subject to change during measurements.

Computation of cross-sectional area where the bed is irregular is problematic. You will find that it is necessary to measure water depth at several points across the channel using a graduated rod. The position and spacing between the points should be chosen so as to define the bed profile as accurately as possible.

A combination of water discharge measurements and determinations of suspended sediment concentration, as determined from the filtration of water samples (see Box 55.1), enables you to calculate the instantaneous suspended sediment discharge of a river as mass per unit time. Thus:

$$\text{suspended sediment discharge} = \text{water discharge} \times \text{suspended sediment concentration} \quad [55.3]$$

Collection of large numbers of samples over long periods of time permits a suspended sediment rating curve to be established, in which water discharge is plotted against suspended sediment discharge on a log–log scale. Thus, for a given water discharge rate, the suspended sediment discharge of a particular stream can be estimated.

Suspended biological materials – plankton

The collection of plankton, the assemblage of organisms that live suspended in the open waters of lakes and the sea, is usually qualitative. In general, a plankton net consists of a cone of nylon mesh with a collecting vessel at the pointed end to retain the concentrated sample (Fig. 55.5). At the broad end of the net, the nylon is attached to a bridle hoop which has three attachment points to a rope. Nets of different mesh sizes (conventionally quoted in pre-metric units as number of mesh per inch) are used to collect different types of plankton (see Table 55.1).

Tows can be horizontal or vertical. Always tow the plankton net slowly through the water. The ideal tow speed is directly related to the gauge of the net mesh; the finer the gauge, the more slowly it must be towed. The duration of tow depends on the concentration of plankton in the water column and the size of sample you hope to collect. If you want to obtain quantitative plankton samples, you should collect (see Chapter 36) and filter known volumes of water through meshes of the same gauge as those

> **Calculating cross-sectional area** – plot the cross-section of the stream to scale on graph paper. Work out the area represented by each 1 mm × 1 mm square. By counting the total number of squares, you can estimate the cross-sectional area.

> **Units for sediment discharge** – be careful not to confuse units when calculating suspended sediment discharge. Water discharge is conventionally measured in $m^3\,s^{-1}$ and suspended sediment concentration in $mg\,L^{-1}$. For example, to express suspended sediment discharge in $kg\,day^{-1}$, it is necessary to convert mg to kg, litres to m^3 (see Chapter 28) and s to days.

Fig. 55.5 A plankton net.

Table 55.1 Plankton net mesh sizes

Mesh size (mesh per inch)	Type of plankton collected
60	Coarse planktonic organisms
100	Zooplankton
200	Phytoplankton
up to 500	Ultraplankton

Box 55.1 How to determine suspended sediment concentration

1. Either:

   (a) **Write numbers on a series of filter papers.** This should be done neatly at the margin (i.e. so that the number will not become obscured by sediment after filtration) using a drawing pen with a fine nib (0.2 mm) filled with waterproof ink. Allow the ink to dry.

   Or:

   (b) **Use unlabelled filter papers in individually numbered Petri dishes.** Note: filter papers should be handled using flat-bladed tweezers at all stages.

   It is normal to use 0.45 mm pore diameter filter paper (e.g. Whatman GF/C glass fibre filter papers of 47 mm diameter). The choice of pore diameter is somewhat arbitrary but it is now widely accepted that 0.45 mm marks the division between solid and dissolved materials.

2. **Pre-weigh the filter papers** using a balance capable of a weighing accuracy of 10μg. This must be done after any numbers have been added to take account of the dry mass of the ink.

3. **Record the initial mass, W_1.**

4. **Set up the vacuum filtration system.** Place the filter papers on to filter holders; attach the water containers above the filter holders and mount the assembly on to vacuum flasks attached to a vacuum pump (Fig. 55.6). It is a good idea to include a water trap or safety reservoir between the vacuum line and the pump to avoid water being sucked back into the pump.

5. **Agitate each water sample** to ensure good dispersion of the particles prior to pouring them carefully into the containers. Take care not to spill any of the samples.

> **KEY POINT** The amount of water to be filtered depends to a large extent on its suspended sediment concentration. For very low concentrations, it may be necessary to filter several litres. For high concentrations, less than 1 L may prove sufficient. However, the volume of water filtered (V) must be accurately known. It is important that you do not try to filter so much sediment that the paper becomes clogged. This may cause a strain on the pump.

6. **Filter the water samples.** It is usual to run several vacuum flasks from one pump.

7. **When all the water has been filtered, turn off the pump.**

8. **Carefully remove the wet filter papers from the holders,** taking care not to tear them or to lose any of the collected sediment.

9. **Dry the filter papers** on trays covered with aluminium foil in an oven at 105 °C, or air dry in individual Petri dishes.

10. **Record the final mass, W_2.** Note: before reweighing the filter papers, it is necessary to allow them to equilibrate with the atmosphere. Many clay minerals and organic matter are hygroscopic and so absorb moisture from the air. If you reweigh before equilibration is complete, you will notice the mass of the filter paper plus sediment increasing as you observe at the balance readout.

11. **Determine the suspended sediment concentration as**

$$(W_2 - W_1)/V \qquad\qquad [55.4]$$

and express in units of mg L^{-1}.

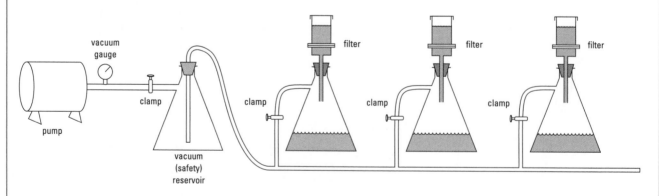

Fig. 55.6 A vacuum filtration system for the determination of suspended sediment concentration in water samples.

of the different plankton nets described in Table 55.1. Use water from a wash bottle to prevent plankton from adhering to the sides of the sample container. Such samples may require subsampling prior to identification and counting.

Free-swimming biological materials – nekton

Nets of various mesh apertures, towed behind a boat, are the conventional means of collection for nektonic species, such as fish. However, specialised echo sounders operating at high frequencies, that receive echoes from the bodies of fish, can be used to detect the presence of shoals in the water column. These provide a means of locating large numbers of fish and therefore can determine points at which it is best to sample. Typical analyses of nekton might include:

* **percentages of different species** within a total catch;
* **body lengths** (a measure of age) of individuals of the same species;
* **stomach contents,** as a means of determining predator–prey relationships.

Bottom-dwelling biological materials – benthos

Benthic organisms are usually sampled by means of various forms of grab (e.g. Fig. 55.7). Such samplers will collect both organisms living in the bottom sediment and the sediment itself. It is therefore necessary to separate the organisms from the sediment particles, prior to counting and identification. A series of sieves is generally used. This process can be done either on the boat or immediately after the samples have been returned to the laboratory, as follows:

1. **Pass the sediment sample or a subsample through a stainless steel sieve** (typically of 2 mm aperture) held over a bucket. Note that if you wish to collect a quantitative sample, you must first measure the volume of sediment (e.g. using a measuring cylinder; this is best done in the laboratory).
2. **Rinse with plenty of the relevant kind of water from a wash bottle.** Take care not to lose any of the sample at this stage.
3. **Place the sieve on a large white tray, half filled with water, and gently agitate.**
4. **Carefully remove any particles from the sieve mesh** using tweezers. These may include large pieces of inorganic material (e.g. pebbles) or dead organic material (e.g. shell fragments, twigs and leaves).
5. **Search for benthic organisms retained on the sieve with the naked eye.** These should be transferred, using tweezers, to a numbered collection jar containing 4% v/v buffered formaldehyde.
6. **Discard the remaining sediment on the sieve,** once all the organisms have been removed.

The material retained in the bucket can be similarly passed through a sieve of smaller aperture if you wish to obtain and analyse smaller benthic organisms.

(a)

(b)

Fig. 55.7 An example of a specific type of sediment sampling grab, the Van Veen type: (a) open position; (b) closed position.

Definitions

Biological oxygen demand (BOD) – the amount of oxygen (typically measured in mg L^{-1}) required for the oxidation of organic matter by biological action under specific standard test conditions. This is sometimes called **biochemical oxygen demand**, also abbreviated BOD.

Chemical oxygen demand (COD) – the amount of oxygen (typically measured in mg L^{-1}) consumed under specific conditions in the oxidation of organic and oxidisable inorganic matter contained in water.

Dissolved oxygen (DO) – the amount of gaseous oxygen (typically measured in mg L^{-1} or as percentage saturation) that is dissolved in an aqueous solution.

Basis of the Winkler method – iodine substitutes for the oxygen dissolved in a 'fixed' aqueous sample. The iodine is then titrated with sodium thiosulfate to a clear end point. The amount of sodium thiosulfate required can be used to calculate the original dissolved oxygen concentration. See Box 65.3 for further details.

Dissolved oxygen, biological (or biochemical) oxygen demand and chemical oxygen demand

Dissolved oxygen

Chapter 65 details several laboratory methods for determining dissolved oxygen (DO) but for environmental studies you are most likely to require field measurements. If you do return samples to the laboratory for DO analysis they should be chemically fixed on collection, typically using 1 mL each of 'standard' manganous sulphate solution and alkaline iodide solution, as a precursor to the iodometric or Winkler titration method of analysis (Box 65.3). The Winkler titration method carried out under laboratory conditions is the most reliable method for the determination of DO in such 'fixed' field samples, but an alternative is to measure DO in the field using one of the many DO kits commercially available.

DO can also be measured immediately in the field using various types of DO probe which measure the rate of oxygen diffusion across a membrane (see Chapter 65). DO probes may be placed directly in water or into a water sample collected by bucket or water sampler from various depths in the water column. It is important not to get any air bubbles in the sample. Follow the general procedure outlined in Box 65.1, recording the DO concentration to the nearest 0.1 mg L^{-1} or nearest 0.1% saturation. Some DO probes include a thermistor which permits the simultaneous reading of DO and temperature. Many contain permanent membranes, eliminating the need for membrane replacement and recalibration. DO electrodes are not usually suitable for analysing industrial and domestic wastewater samples.

BOD and COD

The determination of biological oxygen demand (BOD) is usually carried out on a replicate water sample after an initial determination of DO on the first sample. The procedure is described in Box 55.2. You are less likely to measure chemical oxygen demand (COD) than BOD since COD is usually determined in specialist laboratories. COD is, however, commonly measured by adding water samples (typically 2 mL in volume) to small test tubes containing a pre-measured amount of a strong oxidising reagent, such as potassium dichromate. The tubes are sealed and heated until digestion is complete (typically 2 h at 150 °C) and then cooled to room temperature. The COD level is then determined using either a colorimeter or a spectrophotometer (see Chapter 68).

Box 55.2 How to determine the BOD of a water sample

1. Collect two replicate water samples.
2. Measure the DO of one sample immediately, e.g. using a DO probe or test kit in the field. Note that readings from DO probes require adjustment for temperature and salinity. Some field probes have built-in temperature compensation.
3. Incubate the second sample in complete darkness for 5 days at 20 °C. If you do not have access to an incubator, wrap the sample bottles in aluminium foil and store at room temperature in complete darkness.
4. After 5 days measure the DO of the incubated water sample using the same technique/equipment as previously.
5. Calculate the BOD (mg L^{-1} by subtracting the second DO reading (day 6 reading) from the initial one (day 1 reading).

Text reference

Howard, A.G. (1998) *Aquatic Environmental Chemistry.* Oxford University Press, Oxford.

Sources for further study

Langmuir, D. (1996) *Aqueous Environmental Chemistry.* Prentice Hall, New Jersey.

Manahan, S.E. (2009) *Fundamentals of Environmental Chemistry,* 3rd edn. Lewis Publishers/CRC Press, Boca Raton.
[Chapter 11 is particularly relevant.]

STUDY EXERCISES

55.1 What factors influence the level of oxygen found in an aquatic system? Make a list for (a) a stream system and (b) a shallow sea system.

55.2 How would you determine the amount of planktonic material in a sample from a lake ecosystem? How would this sample have to have been treated to have any value and why?

55.3 Calculate a discharge rate from raw data. A river is flowing through a 10 m wide rectangular weir and the level of water is 16 cm. A flow meter is used to estimate the flow velocity at nine evenly spaced points across the weir and gives the following readings (ms^{-1}): 1.8, 2.1, 2.2, 2.4, 2.7, 2.3, 2.5, 2.1, 1.9. What is best estimate of the instantaneous discharge for this river?

Answers to these study exercises are available at www.pearsoned.co.uk/practicalskills

(a)

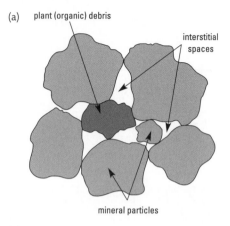

plant (organic) debris

interstitial spaces

mineral particles

(b)

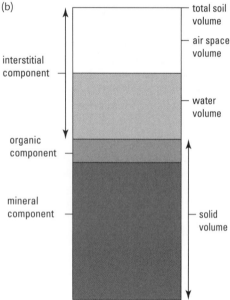

total soil volume

air space volume

interstitial component

water volume

organic component

mineral component

solid volume

Fig. 56.1 Two representations of soils: (a) diagrammatic cross-section of a mixed grade sediment/ soil showing interstitial (pore) spaces; (b) histogram showing the subdivision into solid, liquid and gaseous components. A typical average volume composition might be: minerals, 45%; organic material, 5%; water, 25%; and air, 25%.

Soil colour – many soils exhibit characteristic colouring due to minerals or organic material present. Red soils are typically high in Fe oxides, for example. This characteristic is sometimes used in soil classification systems but is never wholly diagnostic.

Soils and sediments are important components of most ecosystems. They form a habitat, substrate and nutritional source for plants, microbes and animals. It is therefore important to be able to characterise and classify the different types of soils and sediments, and to follow changes in their spatial or temporal composition.

Soils consist of solid, liquid and gaseous phases, and their chief components are often considered under the following headings:

- **the mineral component** – the dominant particulate constituent;
- **the organic component** – living or dead materials;
- **the interstitial component(s)** – air or water or both.

Figure 56.1 provides two simple illustrations of the basic composition of soils.

Soil classification

The wide range of soil and sediment types arises from:

- **differences in the source material;**
- **the age of the soil;**
- **the past climatic history of the area,** especially temperature and rainfall;
- **the influence of plants, microbes and animals living in the soil;**
- **aspects of topography,** such as height (or depth) and slope;
- **human influences.**

Although soils can be classified on the basis of physical and chemical properties, in practice it is difficult to identify particular samples because there is continuous variation in soil properties across the landscape and up and down the soil profile, or pedon. Zones with different properties in soil profiles are known as horizons. The upper horizons are affected most by weathering, addition of organic material and human activities such as farming. The lower horizons are closer in properties to the source rock or sediment. Figure 56.2 illustrates a typical soil profile and indicates the major properties and processes defining each horizon.

Two extensively used soil classifications are those published by the US Natural Resources Conservation Service and the UN bodies FAO and UNESCO. These are broadly similar and an abbreviated version based on USDA-NRCS (1999) is shown in Table 56.1. This taxonomy classifies soil types within 12 orders, according to the presence or absence of specific horizons and the possession of particular physical and chemical properties.

Another way of classifying soils is via 'texture' – as determined from the mix of sand, silt and clay that is present in samples from the B horizon (see Fig. 56.3).

Fig. 56.2 (a) Photograph of a soil profile, or pedon (a cutting of farmland soil, South Island, New Zealand), showing surface vegetation and various horizons (no scale available, but about 1 m deep). (b) diagrammatic representation of a model soil profile . Not all horizons will be present in all soil types, and they may vary in thickness. Typically, soil profiles as above might be 0.5–1.5 m deep. Further subdivisions of these coded soil horizons are made on the basis of physical and chemical composition (USDA-NRCS, 1999). A Leached soil (E horizon) is one in which minerals have been removed, generally by dissolving in water percolating from above.

Source: (a) peter vick167.123rf.com

Table 56.1 A classification of soils into 12 orders based on that published by USDA-NRCS (1999). For a diagram of horizon positions, see Fig. 56.1

Order	Outline of key properties
Alfisols	Frequently found under deciduous forest vegetation. Clay accumulation in B horizon; thick E horizon; moderately acidic; water often present.
Andisols	Contain material from volcanic sources. Dark A horizon; relatively high mineral and nutrient content.
Aridisols	Soils formed in dry climates. Horizons of lime (CaO), gypsum (CaSO$_4$) or salt (NaCl); dry, low in organic material.
Entisols	Recently formed soils such as river floodplains, often bare of vegetaton. Lack well-formed horizons, other than A.
Gelisols	Permafrost soils with a permanently frozen horizon. Other horizons affected by freeze–thaw cycles.
Histosols	Soil that is predominantly organic. Deep peaty A horizon.
Inceptisols	Recently formed soil of fine texture. Often found on sloping terrain. Lack well-formed horizons.
Mollisols	Typical of grassland and broadleaf forests. Deep, dark A horizon. May have significant B horizon and lime (CaO) accumulation.
Oxisols	Well-weathered, thick soils typical of tropical and subtropical climates. Low fertility, Fe and Al clays present and may be acidic.
Spodosols	Typical of temperate coniferous forests. O and B horizons present, the latter with Fe and Al oxide.
Ultisols	Established soil typical of tropical and subtropical climates. Thick A horizon; clay in B horizon; acidic.
Vertisols	Soil typical of region with pronounced wet and dry seasons. High clay content. Well-mixed upper horizons due to infill in cracking when dry.

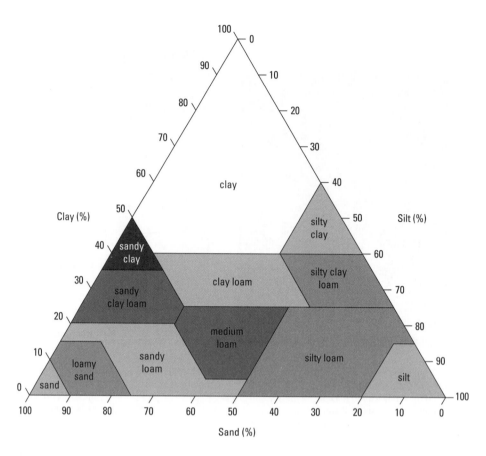

Fig. 56.3 A classification of soils based on the relative proportions of sand, silt and clay.

Sediment classification

As with soils, sediments can be classified according to:

- **Physical properties** – for example, the ratio of gravel, sand and mud. Examples of these types of classification, derived from the work of R.L. Folk, are provided by Long (2006). At their most simple, they involve just four categories: course sediments; mixed sediments; sands and muddy sands; muds and sandy muds.

- **Origin of the sediment** – for example, the presence, types and proportions of silts and clays originating from weathered rocks on land, biogenic material ($CaCO_3$ and SiO_2 muds and oozes composed of skeletal material from diatoms, radiolarians, nanoplankton, etc.), and material of volcanic origin.

Bulk properties of soils and sediments

> **KEY POINT**   Relatively straightforward measures of bulk soil properties can be used to calculate values for a range of variables that describe the nature and status of soils.

Bulk density

The weight of soil or sediment per unit volume (bulk density) is usually expressed as kg m^{-3}. This represents two things: the proportions of mineral and organic matter, and the relative amount of interstitial space. The latter indicates the degree of compaction of the material. Mineral matter is very dense (typically 2500–2600 kg m^{-3}) compared with organic

---

> **Box 56.1** How to measure the bulk density of a soil or sediment
>
> This method is most suitable for dryish coherent soils.
>
> 1. Collect a sample using a soil corer (a metal or plastic cylinder open at one end) by pushing it slowly and steadily (hammering it gently if absolutely necessary) into the soil/sediment to a measured depth. Trim the core as appropriate to the study. Replicate samples should be used to account for variability.
>
> 2. Measure the radius of the core.
>
> 3. The volume of the core in m³ is now calculated as volume = $\pi r^2 h$, where r = radius (m) and h = depth (m).
>
> 4. Transfer the material to an oven and dry to constant weight at 105 °C. Measure wet weight beforehand if required.
>
> 5. Find the bulk density from:
>
> Density = oven dry mass (kg) / sample volume (m³)
>
> [56.1]
>
> For substrates containing pebbles or roots, or for wet and sandy specimens which might flow, a different collection method should be used and the volume should be measured using e.g. a measuring cylinder.

matter (typically 500 kg m⁻³), so the density of the sample will indicate the relative proportions of mineral and organic constituents.

Box 56.1 describes one method of measuring the bulk density of a soil or sediment. The average bulk density of a mineral soil is about 1500 kg m⁻³ while that of (wholly organic) peat is about 500 kg m⁻³. As a general rule, therefore, soils and sediments decrease in average particle density by 1 kg m⁻³ for each 1% (by weight) increase in organic matter content.

Water content

Soil or sediment water content can be measured from the same sample by measuring the wet weight of the sample before Stage 4 (oven drying) of Box 56.1. Soil water content as percentage by weight then equals:

$$(W_1 - W_2) / W_1 \times 100\%$$

[56.2]

where W_1 = wet weight of sample and W_2 = dry weight of sample.

Also relevant in some studies are measurements of water status relative to 'field capacity', the relative volume of near-saturated soil (see Table 56.2). Measurement of water potential (Chapter 25), sometimes termed 'suction

> **Storing cores** – if the soil or sediment core is to be used for measuring water content, it must be placed in a watertight container (e.g. plastic bag) during transportation and storage.

Table 56.2 Some measurements of the water status of soils. The maximum water content and the relationship between water content and water potential depends on the soil type. Clay soils generally hold more water than sandy soils and have a lower (more negative) water potential for a given water content

Parameter	Description	Typical water content (% vol)	Typical water potential (kPa)
Saturated water content	Fully saturated	20–50	0
Field capacity	Water content 2–3 days after saturation (e.g. by rain)	10–35	−33
Permanent wilting point	Moisture content at which plants observed to wilt	1–25	−1500
Residual water content	Remaining water (unavailable)	0.1–10	−∞

Examples

1. If a sample of soil weighed 118.4 g on collection and 97.5 g after oven drying, then its water content would be $(118.4 - 97.5)/118.4 \times 100 = 17.6\%$

2. If a sample of soil had a volume of 252 mL in a measuring cylinder and when 252 mL of water was then added, the volume reading was now 423 mL, then the air space volume would be $(2 \times 252) - 423 = 81$ mL and the % air space volume $81 \times 100/252 = 32.1\%$

pressure', can be made with specialist equipment such as tensiometers, psychrometers and resistance sensors (Campbell, 1988; Nobel, 2009). The solute potential of soil water can be carried out using a psychrometer (Chapter 25) and hydrostatic pressure (i.e. the 'head of water') with various systems involving floats and pressure tensiometers (see IAEA, 2008).

Air space volume

This is a measure of interstitial space. The air space volume (a) is most simply estimated by measuring the volume of a sample soil in a measuring cylinder (v), then adding an equivalent volume of water. After any air bubbles have escaped, the total volume is remeasured. This value (x) is equivalent to $2v - a$, so the air space volume can be found from:

$$a = 2v - x \qquad [56.3]$$

The percentage air space volume is then simply calculated as $100a/v$.

Organic content

Soil organic content can be measured by the loss on ignition (LOI) technique (Box 56.2). This method is based on the loss in weight of a sample when organic matter is burnt off, converting it to carbon dioxide and water which are lost as gases. The percentage weight loss on ignition of soil and sediment samples is approximately equal to twice their organic carbon (C) content. There are, however, no standard conditions of temperature and time for combustion (550 °C for 4 h is typical): you must, therefore, always specify the conditions used.

For a quartz sand or silt, the LOI will provide a good approximation of the organic content. If clay minerals are present, water will be lost from their crystal lattices if combustion temperatures exceed about 550 °C. This could represent about 5–13% by weight of the clays according to their mineralogical type. Similarly, if $CaCO_3$ is present (e.g. in the form of shell fragments), it will break down to CaO at these temperatures with the loss of 44% by weight due to CO_2 release.

Carbon content can also be measured using chemical techniques or a carbon–nitrogen–sulphur (CNS) analyser. A further technique is

Sample heterogeneity – the measurement of organic content by LOI will include the living biota unless you have removed them (difficult). This makes it a very variable quantity, so replication is essential if meaningful results are to be obtained. One earthworm present in a sample will have a very large effect . . .

Box 56.2 How to measure the organic content of soil and sediment samples using the loss on ignition (LOI) technique

1. Take an aliquot of air-dried soil or sediment, typically 2 g in weight, and grind using a mortar and pestle.

2. Place the ground sample in a pre-weighed porcelain crucible and reweigh to give combined weight.

3. Heat the sample and crucible in a furnace for 4 h at 550 °C, for example. If you do not have access to a furnace that permits control of combustion temperature, you can determine LOI approximately by heating the sample in a crucible over a Bunsen flame for 30 min.

4. Allow to cool in a desiccator and reweigh when cold.

5. Calculate organic content as percentage weight loss by difference:

$$\% \text{ organic matter} = (W_2 - W_3) / (W_2 - W_1) \times 100 \qquad [56.4]$$

where W_1 = weight of crucible, W_2 = initial combined weight of sample plus crucible, and W_3 = final (after ignition) combined weight of sample plus crucible.

Table 56.3 The Udden–Wentworth particle size scale

Particle types	Particle size range (mm)
Boulders	>256
Large cobbles	128–256
Small cobbles	64–128
Very large pebbles	32–64
Large pebbles	16–32
Medium pebbles	8–16
Small pebbles	4–8
Granules	2–4
Very coarse sand	1–2
Coarse sand	0.5–1.0
Medium sand	0.25–0.5
Fine sand	0.125–0.25
Very fine sand	0.0625–0.125
Silt	0.004–0.0625
Clay	<0.004

oxidisation: a strong oxidising agent, such as hydrogen peroxide (H_2O_2), is used to digest the organic material which is then estimated from weight loss. However, it is important to note that H_2O_2 breaks down colloidal or humified organic matter but *not* fibrous (cellulosic) residues, and therefore leads to an underestimation of the total organic content of samples.

Mineral components

Particle size analysis

Since both soils and sediments are granular materials, their most fundamental property is particle size distribution. However, because component particle sizes may range through clay, silt, sand, granules, pebbles and cobbles to boulders (see Table 56.3), no single technique of particle size analysis is suitable for all soil types.

The sizes of individual coarse particles can be measured directly, but it is necessary to use indirect methods for samples of finer-grained materials. Typically, large pebbles and above are be measured using Vernier calipers (p. 151), while medium pebbles to very fine sand are measured using sieving techniques (Box 56.3). In the silt to clay ranges, sedimentation methods (see below) and electronic sizing techniques (e.g. a Coulter counter, p. 293) are used.

The particle size distribution can be characterised in terms of median grain size, mean size, standard deviation, skewness and kurtosis (Chapter 75). Particles can be assigned to appropriate size classes

---

Box 56.3   How to measure the grain size distribution of soils and sediments by dry sieving

1. **Prepare the sample.** Dry and disaggregate the sample thoroughly before sieving. If significant amounts of clay are present in your sample, disaggregation will be necessary so that the primary particles are collected by the appropriate sieves, rather than aggregates of particles. The simplest method of disaggregation is to grind the sample in a mortar with a rubber-tipped pestle, but take care not to grind down the primary particles. An alternative is to use a chemical dispersing agent in water. Marine sediments should be washed to remove salts prior to drying. About three washings in three volumes of fresh water, with thorough stirring, are needed to remove the salt from sediment. Take care not to lose fine particles when decanting.

2. **Set up the mechanical sieve shaker.** Use a set of stacked sieves with mesh sizes typically of 2 mm (1ϕ) to 62.5 µm (4ϕ) at intervals of either 0.5 or 0.25ϕ, together with a lid for the top of the stack and pan for the base (e.g. Endecott Test Sieve series). Note: grain size in ϕ (phi) = $\log_2$ grain size in mm.

3. **Weigh the initial sample or subsample.** This should be between 100 g and 500 g.

4. **Place the initial sample in the coarsest sieve at the top of the stack.** Note that it may be necessary to carry out sieving in several stages according to the number of sieve fractions you wish to measure and the number of sieves that the shaker will hold. In this case the material retained in the pan after the first sieving is transferred to the coarsest sieve of the second set, and so on.

5. **Mechanically agitate for 15 min.** Overloading of sieves must be avoided as this restricts the passage of particles through the meshes and may lead to distortion of the meshes. The load per sieve should be no more than 4–6 grains in thickness.

6. **Carefully remove all of the sediment retained by each sieve** using the appropriate brush, and weigh each fraction. Use a balance capable of weighing to an accuracy of 0.01 g. Care must be taken to ensure that all grains are removed from each sieve by careful brushing. Never use a brass wire brush on a fine mesh sieve or you will tear the wires of the delicate lattice. Brass wire brushes should only be used on sieves above 2 mm in aperture.

7. **Weigh the silt plus clay fraction retained in the pan.**

Aggregated structures (aggregates) in soil – these are formed when individual soil granules stick together in clumps. Biological, physical and chemical processes all contribute to aggregation, which is an important characteristic of soil for plant growth.

(e.g. Table 56.3); these can be counted or weighed and the fraction in each size class expressed as a percentage of the total sample weight or number as appropriate. Results can be plotted as a frequency histogram or cumulative percentage frequency curve.

> **KEY POINT** If aggregates are present, it is important to note that these may reflect the true structure of the soil in the natural environment. Destroying the aggregate structure to primary grains for the purposes of size analysis may lead to a false evaluation of the *functional* grain size distribution.

Soil texture

The standard method of characterising soil texture is by quantifying the relative mass ratios of sand, silt and clay once larger components have been removed by sieving (Fig. 56.3). Traditionally the particle size distribution of fine-grained materials was carried out by the 'pipette method'. This method is based on Stokes' law and involves measuring the weight of sediment retained in suspension (as sampled by pipette) after known periods of settling.

The relative masses of these three components can also be determined by a combination of sieving and hydrometer specific gravity readings taken at different standard times while an aqueous suspension of the soil is allowed to settle. The hydrometer readings, carefully temperature corrected, allow particle density to be calculated via Stokes' law. The protocols for such determinations are lengthy (typically over 7 h) and require careful sample preparation. An example of such a protocol is provided at http://www. environment.nsw.gov.au/resources/soils/testmethods/psa.pdf

Other techniques include a variety of rapid, automated, electronic methods. Some are based on the principle of Stokes' law (e.g. SediGraph®) while others measure particle size according to the degree of diffraction or scattering of a laser beam (e.g. the Coulter counter, p. 293).

Stokes' law – this states that the terminal fall velocity of a spherical particle (V) settling in a liquid (water) is directly proportional to the difference in density between the particle (ρ_s) and the liquid (ρ_w) and to the square of the particle diameter (d), and indirectly proportional to the viscosity of the liquid (μ):

$$V = [(\rho_s - \rho_w)\, gd^2]/18\mu \qquad [56.5]$$

where g = acceleration due to gravity (9.8 ms^{-2}).

Chemistry

The most commonly used method for extracting non-degradable analytes such as metals from soils and sediments is chemical digestion, also referred to as wet-ashing. The method typically uses a strong oxidising acid to break down the organic material and solubilise the analyte(s) of interest. The most commonly used acid is nitric acid, which may be added to the sample in a glass container, extracted, then evaporated to near dryness on a hotplate. The residue is then redissolved in an appropriate solvent such as dilute acid for further chemical analysis (Chapter 52). The mineralogy of the constituent particles is invariably complex, and is conventionally determined by X-ray spectroscopy, a specialist technique (see Dean *et al*, 2011).

Organic components

Soil and sediment organic matter consists of:

Definitions

Organic carbon (OC)– the carbon in a soil or sediment sample other than that in carbonates, usually expressed as percentage C by weight.

Organic matter (OM)– the total of organic molecules (i.e. containing C, H, O and N) in a soil or sediment sample, usually expressed as percentage LOI by weight.

- **humus,** stable decomposed material chiefly of plant origin (i.e. fragments of leaves, stems and roots);
- **partially decomposed material,** also chiefly of plant origin;

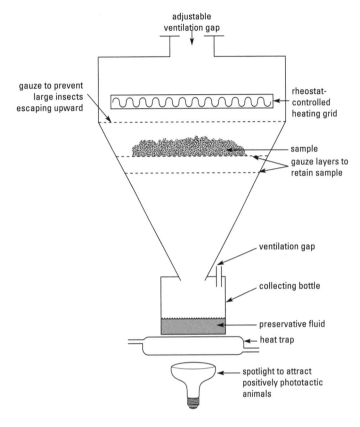

adjustable
ventilation gap

gauze to prevent
large insects
escaping upward

rheostat-
controlled
heating grid

sample
gauze layers to
retain sample

ventilation gap

collecting bottle

preservative fluid

heat trap

spotlight to attract
positively phototactic
animals

Fig. 56.4 A Berlese funnel for the extraction of invertebrates from soil or litter.

- **microbial biomass,** such as fungi, bacteria, etc;
- **fauna,** including small animals such as nematodes and amoebae, but also larger animals such as worms.

Methods of measuring the total organic content of soils is described on p. 358. There are a number of direct mechanical processes that can be used to separate animals from the soil and vegetable material. Variations of sieving, flotation and elutriation (washing out) are the most common, particularly when the fauna has been killed or preserved before analysis begins. Live mobile material is usually extracted from substrates using some form of gradient, such as the heat gradient in the Berlese funnel (Fig. 56.4) for soil or litter samples, or ice gradients for aquatic sediment samples. Methods for identifying extracted organisms are discussed in Chapter 40.

Interstitial components

The spaces between the particles may contain air, water (with dissolved solutes) and organisms belonging to the meiofauna or microbiota. The chemical composition of the interstitial water is influenced by the type of parent rock, the vegetation type, amounts of rainfall and average temperatures.

Collecting samples of pore waters and pore air from soils and sediments usually involves the use of a modified hypodermic syringe. This should be

Using a stain to help to identify organic materials – if samples are hand-sorted to identify larger organisms, staining of organic materials with Rose Bengal stain is useful, but it does stain both living and dead organic material.

Definitions

Meiofauna – small invertebrate animals that live in salt and fresh water.

Microbiota – the assemblage of microbial species living in a particular habitat.

of the large gauge, broad tip variety, with an opening of *c.* 4–5 mm in diameter. A piece of metal tubing, *c.* 15 cm in length, is attached to the tip. This should have a sealed end and be bored with several small holes around its perimeter at a specified sampling distance. Thus, if one or more holes become blocked, the others will still allow water or air to be drawn into the syringe. Samples of *c.* 5 mL in volume may be collected in this way. Centrifugation (Chapter 67) may also be used to obtain water from soil samples.

Chemical analysis of the gas and liquid contents may be carried out by a variety of techniques ranging from simple colorimetric tests through to chromatographic analysis (Chapter 69).

An important chemical characteristic of interstitial water is its pH (see Chapter 26) which can be measured *in situ* using a pH electrode system or in sampled soil water using colorimetric tests (inaccurate). Another important constituent is dissolved oxygen, which can be usefully measured using an electrode system (Chapter 65); alternative, less convenient chemical methods can also be used. Nutrients are commonly measured constituents. Again, various colorimetric tests are available in kit forms, but for quantitative studies, spectroscopic techniques (Chapter 68) are preferred.

Sampling for meiofauna and microbiota is relatively straightforward and usually consists of taking small subsamples of larger benthic samples or small cores where direct sampling is possible. The extraction and analysis of the organisms, however, are very specialised tasks and are frequently species or group-specific (see Chapters 38 and 39).

Text references

Campbell, G.S. (1988) Soil water potential measurement: an overview. *Irrigation Science,* **9**, 265–73.

Dean, J.R., Jones, A.M., Holmes, D., Reed, R.H., Jones, A.M. and Weyers, J.D.B. (2011) *Practical Skills in Chemistry,* 2nd edn. Prentice Hall, Harlow. [Chapter 50 covers X-ray spectroscopy.]

IAEA (2008) *Field Estimation of Soil Water Content: A Practical Guide to Methods, Instrumentation and Sensor Technology*. International Atomic Energy Agency, Vienna. Also available at: http://www-pub. iaea.org/mtcd/publications/pdf/tcs-30_web.pdf Last accessed 30/01/15.

Long, D. (2006) *Seabed Sediment Classification.* Available: http://www.emodnet.seabedhabitats.eu/pdf/ GmHm3_Detailed_explanation_of_seabed_sediment_classification.pdf Last accessed 30/01/15. [Report produced for Marine European Seabed Habitats, MESH.]

Nobel, P.S. (2009) *Physiochemical and Environmental Plant Physiology*, 4th edn. Academic Press, New York.

USDA-NRCS (1999) *Soil Taxonomy: A Basic System of Soil Classification for Making and Interpreting Soil Surveys,* 2nd edn. US Government Printing Office, Washington. Also available at: ftp://ftp-fc.sc.egov.usda.gov/NSSC/ Soil_Taxonomy/tax.pdf Last accessed 30/01/15. [Note that this highly technical reference document is some 871 pages long.]

Sources for further study

Eleftheriou, A. (2013) *Methods for the Study of Marine Benthos,* 4th edn. Wiley Blackwell, Oxford.

Manahan, S.E. (ed) (2013) *Fundamentals of Environmental and Toxicological Chemistry: Sustainable Science*, 4th edn. CRC Press, Boca Raton.

Smith, R.L. and Smith, T.M. (2001) Soils. In: *Ecology and Field Biology,* 6th edn. Benjamin Cummins, San Fransisco.

Tan, K.H. (2011) *Principles of Soil Chemistry,* 4th edn. CRC Press, Boca Raton.

STUDY EXERCISES

56.1 Identify the key differences between a soil and a sediment and consider the similarities and differences in their respective physical and chemical conditions.

56.2 Chose one of the soil types mentioned in Table 56.1. For this type, outline the following: geographical locations of examples of this soil type; prevailing vegetation (natural climax and, where relevant, cultivated) and nutritional status (both plentiful and rare nutrients).

56.3 A student takes a sample of soil and places it in a 1 L measuring cylinder, gently shakes it and obtains a reading of 564 mL. She measures out 564 mL of water, adds it slowly to the soil, taps the cylinder to remove air bubbles and remeasures. The reading is 980 mL. What percentage of the original soil sample was interstitial space?

Answers to these study exercises are available at www.pearsoned.co.uk/practicalskills

Methods of controlling a selection of atmospheric conditions are discussed in Chapter 24.

Definitions

Climate – average weather conditions over a long period of time (decades and longer).

Meteorology – the study of weather and its patterns on daily, seasonal and annual timescales.

Microclimate – atmospheric conditions prevailing within a small space, usually in the layer near the ground that is affected by topographical or vegetational features of the surface.

Topography – (in this context) the three-dimensional land surface features of a given location.

Weather – atmospheric conditions at a given point in time.

Examples The effects of atmospheric variables on biological processes and responses are myriad and occur at a wide range of spatial and temporal scales. For example:

- The seasonal rainfall in a given territory affects the nature of the climax vegetation
- Windy ecosystems have a relatively high proportion of wingless insects
- Changes in light quality affect plant leaf development
- The nest temperature during egg incubation determines the sex of alligator embryos
- Atmospheric pressure changes before storms are sensed by birds, resulting in behavioural changes.
- Atmospheric humidity affects the rate of fungal spore germination.

Measuring and controlling atmospheric variables – because of their great influence on living organisms, it is vital to provide measurements of relevant atmospheric variables to correlate with your ecological observations, and where experiments are concerned, to control important atmospheric conditions.

Atmospheric variables influence many biological processes: it is therefore important that they are measured and/or controlled when making observations or carrying out experiments. The study of atmospheric variables, or weather, is known as meteorology, and covers the following main conditions:

- **air pressure;**
- **temperature;**
- **humidity;**
- **precipitation;**
- **wind;**
- **solar radiation and cloud cover.**

The average results of weather over a period of time shape the climate of an area. This can influence the distribution of plants, microbes and animals, forming characteristic biomes or ecosystems.

Methods of controlling a selection of atmospheric conditions are discussed in Chapter 24. The focus in this chapter is on methods of measurement of atmospheric variables for environmental studies.

Sampling in space and time

The atmosphere and its subcomponents are subject to a high degree of spatial and temporal variation over a wide range of scales. For example, humidity varies at the sub-mm scale close to leaves but also at the km scale and beyond when considering bulk air conditions. It is therefore important that measurements and sampling protocols are meaningful (i.e. from an appropriate location) and representative (i.e. taking account of variation at the relevant scale).

KEY POINT Meteorological variables are subject to local effects caused by topographical features, trees and buildings, so it is important to consider carefully and report the location where measurements are carried out.

The appropriate frequency of measurement or sampling will be project-dependent. Continuous data records obtained from chart recorders or computer data loggers allow averages, maxima and minima to be computed on a daily or per-observation/per-experiment basis. Otherwise, manual readings should be taken on a timescale commensurate with the aims of the project.

Weather stations

A weather station generally provides continuous measurements for a specific location or a close set of positions. Modern stations are electronic and often feature wireless communication to a data logger or computer. Instruments may incorporate integral predefined sensors, or a range of plug-in sensors may be used to suit specific applications. For some studies,

(a)   (b)

Fig. 57.1 Typical weather station (a) setup for sensors in the field; (b) WMR200 base station, with user interface (note aerial for wireless communication with sensors).

Oregon Scientific Global Distribution Ltd. www.oregonscientific.com

Maintenance of equipment – instruments for measuring meteorological variables are extremely sensitive and may require maintenance including recalibration at frequent and regular intervals.

research-grade stations will be required, with greater accuracy and robustness, while for others it will be acceptable to use cheaper consumer-grade product, such as the range produced by Oregon Scientific Global Distribution Ltd. Figure 57.1(a) illustrates a typical mast-mounted weather station and Fig. 57.1(b) an example of a standard weather station interface.

Weather stations should be set up in open areas on a mast as close as possible to the area of interest. Box 57.1 provides guidance for site selection and the precautions required for setting up sensors.

Box 57.1 **How to site a weather station and specific meteorological sensors**

For accurate and meaningful results, the following principles should be borne in mind:

- A wind sensor should be placed at a suitable distance away from any object (e.g. a tree or building) that could interfere with local wind flow. A rule of thumb is to place the sensor at a horizontal distance from the potentially interfering object that is at least ten times the object's vertical height. Slopes and hollows in the terrain should be avoided. The standard measurements heights above ground are 2, 3 and 10 m.

- A temperature sensor should be placed in a ventilated radiation shield – this is usually provided as part of the setup. The area surrounding the sensor should be open with short grass or earth and at least 9 m in diameter with no paved area within 30 m.

- A rain gauge should incorporate a wind shield and be sited at a distance of four times the height of any nearby obstruction. It should be sited on short grass at a height that will avoid in-splashing from rain and above the likely height of snow accumulations.

- A solar radiation sensor should be sited to avoid shadows and reflections (including those from other weather station sensors) and sources of artificial light (e.g. street lamps). In the northern hemisphere, this sensor would normally be placed on the south side of a weather station.

The positioning of other sensors, such as those for air pressure, is not critical, although the recommendation for some instruments is to site them indoors. In practice, the sensors of a weather station would be connected on a single mast positioned to take account of the principles noted above.

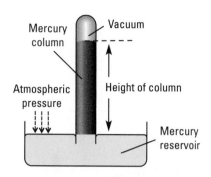

Fig. 57.2 Basic design of a mercury barometer. The height of the mercury column changes depending on the atmospheric pressure and instruments are calibrated with reference to published pressure data from a local meteorological office. This diagram is not to scale; the column height would be approximately 760 mm for an atmospheric pressure of 101.3 kPa.

Calibrating a barometer – this is most simply done during periods of high pressure using the prevailing reading published by the local meterological office.

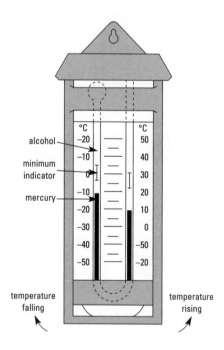

Fig. 57.3 Six's maximum–minimum thermometer. Readings are taken from the bottom of the indicators, which must be reset using a magnet after a reading has been taken.

Measurements of individual variables

Air pressure

For research purposes, atmospheric pressure is often measured in pascals (Pa). The pressure range at sea level is normally between 95 and 105 kPa with an average of 101.3 kPa. Extremes occur at the centre of hurricanes (low) and in continental high-pressure systems. Atmospheric pressure is important for monitoring and predicting weather and it influences other variables such as humidity. Its effects on living organisms tend not to be great within the normal range, although some behavioural responses have been observed (e.g. fish feeding).

Air pressure can be measured in three ways:

1. **Mercury barometer:** many designs, but essentially comprising a mercury column within a glass tube, with the upper end closed to form a vacuum, and the lower open to atmosphere pressure (Fig. 57.2). The column's position depends on the atmospheric pressure pushing up from below.

2. **Aneroid barometer:** this is composed of a series of hollow disks within which is a partial vacuum. As air pressure changes, the disks expand or contract, moving an arm which indicates pressure on a calibrated scale. A barograph consists of an aneroid barometer connected to an arm and inked pen which records changes in pressure continuously on a slowly revolving drum, giving a recording for a particular timeline.

3. **Electronic sensors:** there are a range of types, including silicon-capacitive models which use a micro-diaphragm and pressure cavity to create a variable capacitor. These are configured to produce an electrical signal proportional to absolute pressure, taking account of ambient temperature.

Temperature

Atmospheric temperature is measured in °C. It varies on a daily and seasonal basis and is affected by sunlight. A comfortable room temperature is about 22–25 °C. The rate of all chemical reactions, and hence all biochemical reactions, is temperature-dependent and hence this is an important variable to determine. The three most common ways of measuring temperature are:

1. **Thermometers:** simple thermometers may contain either mercury with alcohol and involve measurement of the thermal expansion of the liquid(s). Six's maximum–minimum thermometer (Fig. 57.3) uses metal indicators to allow recording of the maximum and minimum temperatures reached over a given period of time.

2. **Thermograph:** this is a temperature-recording device which uses a bimetallic strip connected to an arm with a pen to measure temperature change. The pen writes on chart paper on a continuously revolving drum, giving a recording over a particular timeline (Fig. 57.4).

3. **Electronic sensors (thermistors and thermocouples):** a thermistor measures temperature as changes in the resistance of a temperature-sensitive substance. A thermocouple measures the electrical current produced between two wires made of different metals. Thermistors and thermocouples are incorporated in a range of commercially available instruments.

Fig. 57.4 A thermograph.

Fig. 57.5 A non-recording rain gauge.

Humidity

Atmospheric humidity is generally reported as percentage water vapour saturation (relative humidity), but it is important to note that this varies very much with temperature for a particular air water vapour content. Humidity affects a wide range of biological processes, from leaf transpiration to animal activity. It can be measured in three ways:

1. **Wet and dry bulb hygrometer (psychrometer):** this method operates on the principle of evaporative cooling, the extent of which is related to atmospheric humidity. It requires two simple thermometers, one of which is wrapped in muslin which is dipped in a reservoir of water to keep it damp. This is the wet bulb and by comparing its temperature with that of a dry bulb, the humidity can be measured by reference to tables. A whirling (or sling) psychrometer comprises wet and dry bulb thermometers as above but is mounted on a frame which can be swung rapidly. Because evaporation rates are speeded up, the humidity can be measured more rapidly.

2. **Hydrograph:** this is an instrument which uses a specially treated hair, which changes in length with humidity change, to record changes in humidity using a pen and a slowly revolving recording drum (as per Fig. 57.4).

3. **Electronic sensors:** these are of four types. Dewpoint hydrometers operate by estimating the temperature at which air is saturated with water vapour and it starts condensing. Most incorporate a chilled mirror and an optical sensor. Capacitive and resistive sensors measure electrical properties of a material sensitive to humidity. Thermal conductivity sensors measure heat transfer in air and record absolute humidity rather than relative humidity.

Precipitation

The amount of rain, hail and snow – collectively, precipitation – is measured in mm (of water in the liquid state). As all life requires water, the amount and seasonality of precipitation affects the distribution of organisms and the availability of water also influences many biological processes at the whole organism and cellular levels. The main methods of recording precipitation are:

- **A non-recording gauge.** This is simply a cylinder which collects rain as it falls (Fig. 57.5). The diameter (hence area) of the funnel is standardised and the volume of water collected over a given period is measured manually using a calibrated glass container. Snow or hail falling in the gauge must be melted before taking any readings.

- **A recording gauge.** This records the amount of water collected on a rotating drum and provides detailed information on timing and intensity. All have some emptying mechanism, such as a tipping bucket. They allow precipitation intensity to be recorded on an hourly basis.

Wind

Air velocity (wind speed) is measured in m s⁻¹ and ranges from near zero in still air to 30 m s⁻¹ in a storm (Table 57.1). Air flow has effects at the micro scale, for example water and temperature loss, as well having larger-scale

Table 57.1 The Beaufort scale of wind force

Beaufort number	Descriptive title	Effect on land features	Speed (m s^{-1})
0	Calm	Smoke rises vertically	<0.3
1	Light air	Direction shown by smoke but not wind vanes	0.3–1.5
2	Light breeze	Wind felt on face; leaves rustle; vane moves	1.6–3.3
3	Gentle breeze	Leaves and twigs in constant motion	3.4–5.4
4	Moderate breeze	Raises dust and paper; small branches moved	5.5–7.9
5	Fresh breeze	Small trees begin to sway	8.0–10.7
6	Strong breeze	Large branches in motion; whistling in telephone wires	10.8–13.8
7	Near gale	Whole trees in motion	13.9–17.1
8	Gale	Breaks twigs off trees	17.2–20.7
9	Strong gale	Slight structural damage to roofs, etc.	20.8–24.4
10	Storm	Trees uprooted; considerable structural damage	24.5–28.4
11	Violent storm	Widespread damage	28.5–32.6
12	Hurricane	Widespread devastation	>32.6

disruptive effects on organisms and ecosystems. Wind speed and direction can be determined in three main ways:

1. **A revolving cup anemometer.** This consists of three revolving cups connected to a recording meter which counts the number of rotations in a given time period (see Fig. 57.1). The meter then converts this to wind speed. It may be fixed or hand-held.

2. **A ventimeter.** This is a calibrated tube over which wind passes, causing a reduction in pressure in the tube. This causes a pointer to rise in proportion to the wind speed (Fig. 57.6). These instruments are cheap and easy to use.

3. **By observation of the effects of the wind speed on objects.** This subjective method is based upon the Beaufort Scale (Table 57.1).

Wind direction is usually determined using a wind vane, which must be mounted away from obstacles that might locally affect the direction of the wind (see Box 57.1).

Solar radiation and cloud cover

Solar radiation is measured in various units, as outlined in Chapter 58, but those used in meteorology are usually W m^{-2}. As the ultimate source of energy for nearly all organisms, especially photosynthetic autotrophs, this is an important environmental variable. A radiometer (see pp. 373–4 for details) is the most commonly used instrument for light measurement. In some ecological studies, solar radiation may be measured using a glass ball (lens) mounted on a frame. The ball focuses the sun on a specially treated paper behind it, burning a line along the paper as it moves. The total length of the burnt line is a measure of the length of time for which the sun shone, typically expressed as 'hours of sunshine'.

The amount of cloud is usually recorded by estimating what proportion of the sky is covered by cloud. This is often expressed in eighths of the total visible area ('oktas').

Fig. 57.6 A ventimeter.

Primary pollutants – these are released directly into the atmosphere from the source, e.g. carbonaceous particles from a diesel engine or sulphur dioxide from a coal-fired power station.

Secondary pollutants – these are formed by later chemical reactions involving primary pollutants, e.g. sulphuric acid from sulphur dioxide emissions, or ground level ozone from reactions among combustion products.

Examples   Pollutants affect biological processes in a wide range of ways. For instance:

- Acid rain caused by industrial emissions results in leaf damage, lowered photosynthesis and plant death

- Eutrophication of aquatic habitats due to leaching of crop fertilizers effects ecosystem balance, causing species blooms and low oxygenation

- Pesticide chemicals become concentrated in the food chain, leading to physiological problems in apex species

- Oil pollution kills warm-blooded animals by coating feathers and fur, causing excess heat loss or loss of mobility

- Chemical pollution can result in abnormal growth patterns.

Definitions

Aerosol – a dispersion of solid or liquid particles in a gas medium, e.g. smoke.

Particulates – the solid particles suspended in a gas or liquid, e.g. soot.

Definition

PM_{10} – the concentration (mg m^{-3}) of particulate materials with a particle diameter of 10 μm or less. This is the component thought to be most damaging to human health.

Sampling for chemical analyses

Most of the atmospheric gas and particle analysis in environmental science is related to the problem of airborne pollutants. The sampling of gaseous and semi-volatile materials is termed ambient air sampling unless it derives specifically from a point source of gaseous emissions. Tests of ambient air masses are based on the assumption that the sample is representative, but this is usually a challenge because of the complexity of the factors influencing spatial and temporal variability.

Methods of sampling fall into three main categories:

1. **Sampling containers:** these must be chosen carefully so that gaseous analytes do not react irreversibly with filter or sample container surfaces or with collected aerosol particles; the contents are usually analysed in the gas phase by, for example, chromatographi techniques (Chapter 69).

2. **Filters:** including activated carbon filters, sometimes in combination with absorbent resins (e.g. Tenax® traps). These are suitable for volatile and semi-volatile compounds, and absorbed compounds can be recovered by eluting with ethanol or carbon disulphide.

3. **Cold traps:** these act to condense and collect volatile compounds on a surface or in solution.

Once collected, the gaseous sample is then subject to similar chemical analysis processes and measurement criteria as outlined in Chapter 52.

KEY POINT   Sorbent sampling via filters is fraught with problems of contamination, interferences, capture efficiency and recovery efficiency, but it is an important and valuable technique provided it is used with the appropriate controls and blanks.

Particulates

The atmosphere includes a range of types of suspended particulate material. This includes both organic and inorganic components with particle diameters from less than 10 nm to more than 100 mm. Owing to particle interactions, physical measurements of the components of suspended particulate matter (aerosols) can be complex and results variable.

A simple method for collecting particulates is to expose adhesive strips to the atmosphere, provided they can be protected from the direct effects of rain. Subsequent examination using a microscope allows an estimate of the size and number of particles to be made for qualitative and semi-quantitative assessments. Particulates can also be filtered from the air, but care is required in case filters are contaminated with materials that can cause errors as a result of chemical/physical transformation.

Quantitative measurement usually depends upon a gravimetric determination of particle mass after collection by filtration. The most important derived value is termed the PM_{10}. Because it is determined by weight, it includes both primary and secondary particles. Chemical analysis of the particulates usually requires solubilisation and other pretreatments of the filtered materials (Chapter 52).

> **KEY POINT** Larger particles, by virtue of their inertia, are not readily able to enter the inlet of some samplers. Their measurement is therefore dependent upon the orientation to the wind and the strength of the wind, both of which influence the efficiency of particle filtration.

Acid rain

Here, atmospheric oxidation of gaseous nitrogen dioxide (NO_2) and sulphur dioxide (SO_2) causes formation of the strong acids nitric acid (HNO_3) and sulphuric acid (H_2SO_4). These are removed from the atmosphere with precipitation, giving rise to acid rain. Hence, in this context, simple measurements such as the pH of rainwater can be informative. Pure water in equilibrium with atmospheric carbon dioxide has a pH of about 5.6 but pH values of 4 and below may be observed as a result of gaseous pollution. Methods of measuring pH in water samples are covered in Chapter 26.

Sources for further study

Barker, J.R. (1995) A brief introduction to atmospheric chemistry. In *Progress and Problems in Atmospheric Chemistry,* (Advanced Series in Physical Chemistry, Vol. 3), 1–33. World Scientific, Singapore.

Campbell Scientific Inc. (1997) *Weather Station Siting and Installation Tools.* Available: ftp://ftp. campbellsci.com/pub/outgoing/apnotes/siting.pdf Last accessed 10/02/15.

Greenwood, R., Mills, G. and Vrana, B. (2007) *Passive Sampling Techniques in Environmental Monitoring (Comprehensive Analytical Chemistry vol. 48),* 1–486. Elsevier, Amsterdam.

Jacob, D.J. (2000) *Introduction to Atmospheric Chemistry.* Princeton University Press, Princeton, NJ.

STUDY EXERCISES

57.1 What is the significance of microclimate in making measurements of wind speed and direction?

57.2 Obtain a set of temperature data from a local weather station and use *Excel* to plot mean daily temperatures and extreme high and low temperatures for the months of January and June in your chosen year. How well does the calculation of a mean daily temperature provide meaningful information for biological studies?

57.3 What would be the key elements of a protocol for collecting and analysing ozone in the lower atmosphere, e.g. in a city location?

Answers to these study exercises are available at www.pearsoned.co.uk/practicalskills

58 Measuring light

Table 58.1 Wavelength ranges for UV, visible and IR radiation, compiled from several sources

Type	Wavelength range (nm)
UVC	100–290*
UVB	290–320
UVA	320–400
Violet	390–450
Blue	450–500
Green	500–600
Yellow	560–600
Orange	600–620
Red	620–770
IR	>770

* Note the UVC is absorbed by the Earth's ozone layer.

Light measurement is directly relevant to several aspects of biology, including: photosynthesis, photomorphogenesis and photoperiodism in plants; perception and thermoregulation in animals; aquatic biology.

The nature of light

Light is most strictly defined as that part of the spectrum of electromagnetic radiation detected by the human eye. However, the term is also applied to radiation just outside that visible range (e.g. UV and infrared 'light'). Electromagnetic radiation is emitted by the sun and by other sources (e.g. an incandescent lamp) and the electromagnetic spectrum is a broad band of radiation, ranging from cosmic rays to radio waves (Fig. 58.1). Most biological experiments involve measurements within the UV, visible and infrared regions (generally, within the wavelength range 200–1000 nm, see Table 58.1).

Radiation has the characteristics of a particle and of a vibrating wave, travelling in discrete particulate units, or 'packets', termed photons. A quantum is the amount of energy contained within a single photon (it is important not to confuse these two terms, although they are sometimes used interchangeably in the literature). In some circumstances, it is appropriate to measure light in terms of the number of photons, usually expressed directly in moles (6.02×10^{23} photons = 1 mol); older textbooks may use the redundant term Einstein as the unit of measurement, where 1 Einstein = 1 mol photons. Alternatively, the energy content (power) may be measured (e.g. in $W\ m^{-2}$). Radiation also behaves as a vibrating electrical and magnetic field moving in a particular direction, with the magnetic and electrical components vibrating perpendicular to one another and perpendicular to the direction of travel.

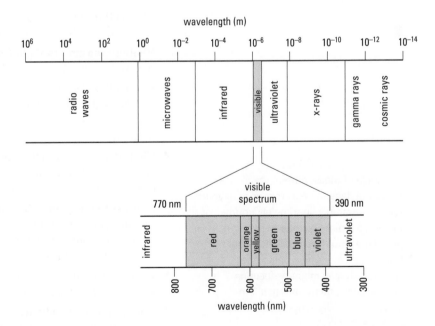

Fig. 58.1 The electromagnetic spectrum.

Methods for Physical and Chemical Analysis of Fresh Waters, 2nd edn, International Biological Programme, No. 8, Blackwell Scientific (Golterman, H. L., Clymo, R. S. and Ohnstad, M.A.M., 1978) reproduced by kind permission of Fisher Scientific UK Ltd.

Fig. 58.2 Relative efficiency of vision (solid line) and photosynthesis (dashed line) as a function of wavelength.

The wave nature of radiation gives rise to the concepts of wavelength (λ, usually measured in nm), frequency (v, measured in s^{-1}), speed (c, the speed of electromagnetic radiation, which is 3×10^8 m s^{-1} in a vacuum) and direction. In other words, radiation is a vector quantity, where

$$c = \lambda v \qquad\qquad [58.1]$$

Photometric and radiometric measurements

Photometric measurements

These are based on the energy perceived by a 'standard' human eye, with maximum sensitivity in the yellow–green region, around 555 nm (Fig. 58.2). The unit of measurement is the candela, a base unit in the SI system, defined in terms of the visual appearance of a specific quantity of platinum at its freezing point. Derived units are used for the luminous flow (lumen) and luminous flow per unit area (lux). These units were once used in photobiology and you may come across them in older literature. However, it is now recognised that such measurements are of little direct relevance to biologists, including even those who may wish to study visual responses, because they are not based on fundamental physical principles.

> **KEY POINT** The human eye rapidly compensates for changes in light climate by varying the size of the pupil and is a very poor source of information on light quantity. It is important to make light measurements using reliable instruments and to express these measurements in appropriate units.

Radiometric measurements

The radiometric system is based on physical properties of the electromagnetic radiation itself, expressed either as the number of photons, or as their energy content. The following terms are used (units of measurement in parentheses):

- **Photon flux** (mol photons s^{-1}) is the number of photons arriving at an object within the specified time interval.
- **Photon exposure** (mol photons m^{-2}) is the total number of photons received by an object, usually expressed per unit surface area.
- **Photon flux density** (mol photons m^{-2} s^{-1}) or PFD is the most commonly used term to describe the number of photons arriving at a particular surface, expressed per unit surface area and per unit time interval.
- **Photosynthetically active radiation** or PAR is radiation within the waveband 400–700 nm, since the photosynthetic pigments (chlorophylls, carotenoids, etc.) show maximum absorption within this band.
- **Photosynthetic photon flux density** (mol photons m^{-2} s^{-1}) or PPFD is the number of photons within the waveband 400–700 nm arriving at a particular surface, expressed per unit surface area and per unit time interval. Often this term is used interchangeably with PFD.
- **Irradiance** (J m^{-2} s^{-1} = W m^{-2}) is the amount of energy arriving at a surface, expressed per unit surface area and per unit time interval.
- **Photosynthetic irradiance** (W m^{-2}) or PI is the energy of radiation within the waveband 400–700 nm arriving at a surface, expressed per unit surface area and per unit time interval.

Relevance of measurement scale – your choice will depend on its relevance to your study. For example, in some environmental studies, you might be interested in total energy input, so irradiance would be a valid measurement; in another type of study, you might be focusing on plant productivity and therefore photosynthetically active radiation, PAR, would be more relevant.

Table 58.2 Approximate conversion factors for a photosynthetic irradiance (PI) of 1 Wm^{-2} to photosynthetic photon flux density (PPFD)

Source	PPFD (µmol photons m^{-2} s^{-1})
Sunlight	4.6
'Cool white' fluorescent tube	4.6
Osram 'daylight' fluorescent tube	4.6
Quartz-iodine lamp	5.0
Tungsten bulb	5.0

(*Source:* Lüning, 1981)

Fig. 58.3 Spectral distribution of energy output from various sources. (Adapted from Golterman *et al.*, 1978.)

Fig. 58.4 A typical radiometer (quantum) sensor. Photo courtesy of Skye Instruments Ltd.

Choice of measurement scale

Photon flux density

This is often the most appropriate unit of measurement for biological systems where individual photons are involved in the underlying process, e.g. in photosynthetic studies, where PPFD is measured, since each photochemical reaction involves the absorption of a single photon by a pigment molecule. Most modern light-measuring instruments (radiometers) can measure this quantity, giving a reading in µmol photons m^{-2} s^{-1}.

Irradiance

This is appropriate if you are interested in the energy content of the light, e.g. if you are studying energy balance, or thermal effects. Many radiometers measure photosynthetic irradiance within the waveband 400–700 nm, giving a reading in W m^{-2}. It is possible to make an approximate conversion between PPFD and PI measurements, providing the spectral properties of the light source are known (see Table 58.2).

Spectral distribution

This can be determined using a spectroradiometer, e.g. to compare different light sources (Fig. 58.3). A spectroradiometer measures irradiance or photon flux density in specific wavebands. This instrument consists of a monochromator (p. 449) to allow separate narrow wavebands (5–25 nm bandwidth) to be measured by a detector; some instruments provide a plot of the spectral characteristics of the source.

Measurements can be instantaneous or integrated over a time period (e.g. a day). Instantaneous measurements are used in physiological studies to record prevailing conditions while integrated measurements tend to be used in environmental research (Chapter 57).

Using a radiometer ('light meter')

The main components of a radiometer are:

- **Receiver (sensor):** either flat-plate, hemispherical or spherical, depending upon requirements. Most incorporate a protective diffuser, to reduce reflection.
- **Detector:** either thermoelectric or photoelectric. Some photoelectric detectors suffer from fatigue, with a decreasing response on prolonged exposure: check the manufacturer's handbook for exposure times.
- **Processor** and readout device to convert the output from the detector into a visible reading, in analogue or digital form.

Figure 58.4 shows a typical flat-plate radiation detector with diffusion plate. Modern instruments may incorporate two sets of receiver–detector assemblies to account for sources of radiation over 360°, and data loggers are capable of presenting instantaneous or integrated readings. Box 58.1 gives practical details of the steps involved in using a radiometer. In your write-up, give full details of how the measurement was made, e.g. the type of light source, instrument used, where the sensor was placed, whether an average was calculated, etc.

Box 58.1 Measuring photon flux density or irradiance using a battery-powered radiometer

1. Check the battery. Most instruments have a setting that gives a direct readout of battery voltage. Recharge if necessary before use.

2. Select the appropriate type of measurement (e.g. photon flux density or irradiance over the PAR waveband, or an alternative range): the simpler instruments have a selection dial for this purpose.

3. Place the sensor in the correct location and position for the measurement: it may be appropriate to make several measurements at different positions, and take an average.

4. Choose the most appropriate scale for the readout device: for needle-type meters, the choice of maximum reading is usually selected by a dial, within the range 0.3 to 30 000. Start at a high range and work down until the reading is on the scale. Your final scale should be chosen to provide the most accurate reading, e.g. a reading of 15 µmol photons $m^{-2} s^{-1}$ should be made using the 0–30 scale, rather than a higher range.

5. Read the value from the meter. For needle-type instruments there may be two scales, the upper one marked from 0 to 10 and the lower one from 0 to 3: make sure you use the correct one, e.g. a half-scale deflection on the 0–30 scale is 1.5.

6. Check that the answer is realistic, e.g. full sunlight has a PPFD of up to 2 000 µmol photons $m^{-2} s^{-1}$ (PI ≡ 400 W m^{-2}), though the value will be far lower on a dull or cloudy day, while the PPFD at a distance of 1 m from a mercury lamp is around 150 µmol photons $m^{-2} s^{-1}$, and 50 µmol photons $m^{-2} s^{-1}$ at the same distance from a fluorescent lamp.

Text references

Golterman, H.L., Clymo, R.S. and Ohnstad, M.A.M. (1978) *Methods for Physical and Chemical Analysis of Fresh Waters.* Blackwell Scientific, Oxford.

Lüning, K.J. (1981) 'Light', in C.S. Lobban and M.J. Wynne (eds), *The Biology of Seaweeds,* pp. 326–55. Blackwell Scientific, Oxford.

Sources for further study

Boyd, R.W. (1983) *Radiometry and the Detection of Optical Radiation.* Wiley, New York.

Ryer, A.D. (1997) *Light Measurement Handbook.* Available http://www.intl-light.com/handbook/

Last accessed 1/04/15.
[A commercial site, produced by International Light Technologies, Inc., Peabody, Massachusetts.]

STUDY EXERCISES

58.1 **Carry out a calculation involving interconversion of photosynthetic photon flux and photosynthetic irradiance.** For sunlight with a photosynthetic photon flux density (PPFD) of 1610 mol m^{-2}, what is the total amount of photosynthetic energy (in joules) falling on a leaf of area 45 cm^2 over a 30-min experimental period?

58.2 **Compare the photosynthetic photon flux densities (PPFD) of different locations.** The following values represent typical light levels for a range of different situations, expressed either as photosynthetic irradiance (PI) or PPFD:

(a) outside, on a sunny day: PI = 300 W m^{-2};

(b) in a room lit by 'cool white' fluorescent tubes: PPFD = 6.50 nmol photons $cm^{-2} s^{-1}$;

(c) under the leaf canopy in a forest: PPFD = 275 µmol photons $m^{-2} s^{-1}$;

(d) in a growth cabinet lit by a bank of 'daylight' fluorescent lights: PI = 35.2 W m^{-2};

(e) in a room lit by a tungsten bulb: PI = 1.15 mW cm^{-2}.

Convert all to PPFD in the same units (expressed to three significant figures) and then rank the locations in order of decreasing PPFD.

58.3 **Research the various types of commercial radiometers (light meters) available on the Web.** Make a list of the features that you might want to find out about and see how much information you can find.

Answers to these study exercises are available at www.pearsoned.co.uk/practicalskills

Advanced analytical techniques

59 Immunological methods

Definitions

Antibody – a protein produced in response to an antigen (an *anti*body-*gen*erating foreign macromolecule).

Epitope – a site on the antigen that determines its interaction with a particular antibody.

Hapten – a substance that contains at least one epitope, but is too small to induce antibody formation unless it is linked to a macromolecule.

Ligand – a molecule or chemical group that binds to a particular site on another molecule.

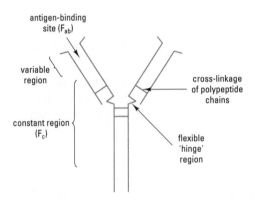

Fig. 59.1 Diagrammatic representation of IgG (antibody).

Producing polyclonal antibodies – in the UK, this is controlled by government regulations, since it involves vertebrate animals: personnel must be licensed by the Home Office and must operate in accordance with the Animals Scientific Procedures Act (1986) and with the Code of Practice for the Housing and Care of Animals (2005).

Antibodies are an important component of the immune system, which protects animals against certain diseases (see Delves *et al.*, 2011). They are produced by B lymphocytes in response to foreign macromolecules (antigens). A particular antibody will bind to a site on a specific antigen, forming an antigen–antibody complex (immune complex). Immunological bioassays use the specificity of this interaction for:

- **identifying macromolecules, cellular components or whole cells;**
- **quantifying a particular substance.**

Antibody structure

An antibody is a complex globular protein, or immunoglobulin (Ig). Although there are several types, IgG is the major soluble antibody in vertebrates and is used in most immunological assays. Its main features are:

- **Shape:** IgG is a Y-shaped molecule (Fig. 59.1), with two antigen-binding sites.
- **Specificity:** variation in amino acid composition at the antigen-binding sites explains the specificity of the antigen–antibody interaction.
- **Flexibility:** each IgG molecule can interact with epitopes that are different distances apart, including those on different antigen molecules.
- **Labelling:** regions other than the antigen-binding sites can be labelled, e.g. using a radioisotope or enzyme (pp. 380–2).

KEY POINT The presence of two antigen-binding sites on a single flexible antibody molecule is relevant to many immunological assays, especially the agglutination and precipitation reactions.

Antibody production

Polyclonal antibodies

These are commonly used at undergraduate level. They are produced by repeated injection of antigen into a laboratory animal. After a suitable period (3–4 weeks) blood is removed and allowed to clot, leaving a liquid phase (polyclonal antiserum) containing many different IgG antibodies, resulting in:

- **cross-reaction** with other antigens or haptens;
- **batch variation,** as individual animals produce slightly different anti-bodies in response to the same antigen;
- **non-specificity,** as the antiserum will contain many other antibodies.

Standardisation of polyclonal antisera therefore is difficult. You may need to assess the amount of cross-reaction, interbatch variation or non-specific binding using appropriate controls, assayed at the same time as the test samples.

Monoclonal antibodies

These are specific to a single epitope and are produced from individual clones of cells, grown using cell culture techniques (p. 300). Such cultures provide a stable source of antibodies of known, uniform specificity. Although monoclonal antibodies are likely to be used increasingly in future years, polyclonal antisera are currently employed for many routine immunological assays.

Agglutination tests

When antibodies interact with a suspension of a particulate antigen, e.g. cells or latex particles, the formation of immune complexes (Fig. 59.2) causes visible clumping, termed agglutination. A positive haemagglutination reaction gives an even 'carpet' of red cells over the base of the tube, while a negative reaction gives a tightly packed 'button' of red cells at the bottom of the tube. Agglutination tests are used in several ways:

- **Microbial identification:** at the species or subspecies level (serotyping), e.g. mixing an unknown bacterium with the appropriate antiserum will cause the cells to agglutinate.
- **Latex agglutination using bound antigens:** by coating soluble (non-particulate) antigens on to microscopic latex spheres, their reaction with a particular antibody can be visualised.
- **Latex agglutination using bound antibodies:** antibodies can be bound to latex microspheres, leaving their antigen-binding sites free to react with soluble antigen.
- **Haemagglutination:** red blood cells can be used as agglutinating particles. However, in some instances, such reactions do not involve antibody interactions (e.g. some animal viruses may haemagglutinate unmodified red blood cells).

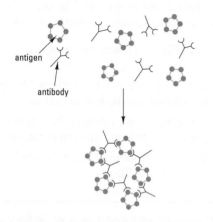

Fig. 59.2 Formation of an antigen-antibody complex.

Precipitin tests

Immune complexes of antibodies and soluble antigens (or haptens) usually settle out of solution as a visible precipitate: this is termed a precipitin test, or precipitation test. The formation of visible immune complexes in agglutination and precipitation reactions only occurs if antibody and antigen are present in an optimal ratio (Fig. 59.3). It is important to appreciate the shape of this curve: cross-linkage is maximal in the zone of equivalence, decreasing if either component is present in excess. The quantitative precipitin test can be used to measure the antibody content of a solution (Lefkovits, 1997, gives details). Visual assessment of precipitation reactions forms the basis of several other techniques, described below.

Immunodiffusion assays

These techniques are easier to perform and interpret than the quantitative precipitin test. Precipitation of antibody and antigen occurs within an agarose gel, giving a visible line corresponding to the zone of equivalence (Fig. 59.3). Details of the main techniques are given in Box 59.1.

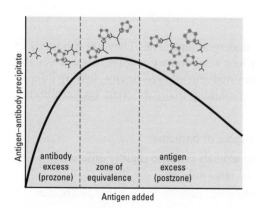

Fig. 59.3 Precipitation curve for an antigen titrated against a fixed amount of antibody.

Box 59.1 How to carry out immunodiffusion assays

The two most widespread approaches are (i) single radial immunodiffusion (Mancini technique) and (ii) double-diffusion immunoassay (Ouchterlony technique).

Single radial immunodiffusion (RID)

This is used to quantify the amount of antigen in a test solution, as follows:

1. Prepare an agarose gel (1.5% w/v), containing a fixed amount of antibody: allow to set on a glass slide or plate, on a level surface.

2. Cut several circular wells in the gel. These should be of a fixed size between 2 and 4 mm in diameter (see Fig. 59.4(a)). Cut your wells carefully. They should have straight sides and the agarose must not be torn or lifted from the glass plate. All wells should be filled to the top, with a flat meniscus, to ensure identical diffusion characteristics. Non-circular precipitin rings, resulting from poor technique, should not be included in your analysis.

3. Add a known amount of the antigen or test solution to each well.

4. Incubate on a level surface at room temperature in a moist chamber: diffusion of antigen into the gel produces a precipitin ring. This is usually measured after 2–7 days, depending on the molecular mass of the antigen.

5. Examine the plates against a black background (with side illumination), or stain using a protein dye (e.g. Coomassie blue).

6. Measure the diameter of the precipitin ring, e.g. using Vernier calipers (p. 151).

7. Prepare a calibration curve (Chapter 53) from the samples containing known amounts of antigen (Fig. 59.4(b)): the squared diameter of the precipitin ring is directly proportional to the amount of antigen in the well.

8. Quantify the amount of antigen in your test solutions, using a calibration curve prepared from standards assayed at the same time.

Double-diffusion immunoassay (Ouchterlony technique)

This technique is widely used to detect particular antigens in a test solution, or to look for cross-reaction between different antigens.

1. Prepare an agarose gel (1.5% w/v) on a level glass slide or plate: allow to set.

2. Cut several circular wells in the gel.

3. Add test solutions of antigen or polyclonal antiserum to adjacent wells. Both solutions diffuse outwards, forming visible precipitin lines where antigen and corresponding antibody are present in optimal ratio (Fig. 59.5).

The various reactions between antigen and antiserum are:

- Identity: two wells containing the same antigen, or antigens with identical epitopes, giving a fused precipitin line (identical interaction between the antiserum and the test antigens, Fig. 59.5(a)).

- Non-identity: where the antiserum contains antibodies to two different antigens, each with its own distinct epitopes, giving two precipitin lines that intersect without any interaction (no cross-reaction, Fig. 59.5(b)).

- Partial identity: where two antigens have at least one epitope in common, but where other epitopes are present, giving a fused precipitin line with a spur (cross-reaction, Fig. 59.5(c)).

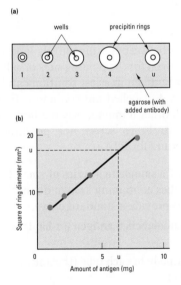

Fig. 59.4 Single radial immunodiffusion (RID). (a) Assay: four standards are shown (wells 1 to 4, each one double the strength of the previous standard), and an unknown (u), run at the same time, (b) Calibration curve. The unknown contains 6.25 μg of antigen. Note the non-zero intercept of the calibration curve, corresponding to the square of the well diameter: do not force such calibration lines through the origin.

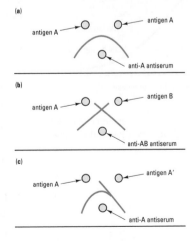

Fig. 59.5 Precipitin reactions in double-diffusion immunoassay: (a) identity; (b) non-identity; (c) partial identity.

Fig. 59.6 Laurell rocket immunoelectrophoresis. (a) Assay: precipitin rockets are formed by electrophoresis of five standards of increasing concentration (wells 1 to 5) and an unknown (u). (b) Calibration curve: the unknown sample contains 7.7 μg of antigen.

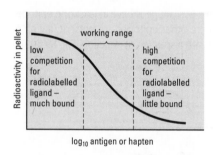

Fig. 59.7 Radioimmunoassay calibration curve. Note that the assay is insensitive at very low and very high antigen levels.

Immunoelectrophoretic assays

These methods combine the precipitin reaction with electrophoretic migration, providing sensitive, rapid assays with increased separation and resolution.

Crossover electrophoresis (countercurrent electrophoresis)

Similar to the Ouchterlony technique, since antigen and antibody are in separate wells. However, the movement of antigen and antibody towards each other is driven by a voltage gradient (p. 462): most antigens migrate towards the anode, while IgG migrates towards the cathode. This method is faster and more sensitive than double immunodiffusion, taking 15–20 min to reach completion.

Quantitative immunoelectrophoresis (Laurell rocket immunoelectrophoresis)

Similar to RID, as the antibody is incorporated into an agarose plate while the antigen is placed in a well. However, a voltage gradient moves the antigen into the gel, usually towards the anode, while the antibody moves towards the cathode, giving a sharply peaked, rocket-shaped precipitin line, once equivalence is reached (within 2–10 h). The height of each rocket shape at equivalence is directly proportional to the amount of antigen added to each well. A calibration curve for samples containing known amount of antigen can be used to quantify the amount of antigen present in test samples (Fig. 59.6).

Radioimmunological methods

These methods use radioisotopes to detect and quantify the antigen–antibody interaction, giving improved sensitivity over agglutination and precipitation methods. The principal techniques are radioimmunoassay and immunoradiometric assay.

Radioimmunoassay (RIA)

This is based on competition between a radioactively labelled antigen (or hapten) and an unlabelled antigen for the binding sites on a limited amount of antibody. The quantity of antigen in a test solution can be determined using a known amount of radiolabelled antigen and a fixed amount of antibody (Fig. 59.7). As with other immunoassay methods, it is important to perform appropriate controls, to screen for potentially interfering compounds. The basic procedure for RIA is:

1. **Add appropriate volumes of a sample to a series of small test tubes.** Prepare a further set of tubes containing known quantities of the substance to be assayed to provide a standard curve.

2. **Add a known amount of radiolabelled antigen (or hapten) to each tube** (sample and standard).

3. **Add a fixed amount of antibody to each tube** (the antibody must be present in limited quantity).

4. **Leave at constant temperature for a fixed time** (usually 24 h), to allow antigen–antibody complexes to form.

5. **Precipitate the antibody and bound antigen using saturated ammonium sulphate, followed by centrifugation.**

6. **Determine the radioactivity of the supernatant or the precipitate** (pp. 418–21).

7. **Prepare a calibration curve of radioactivity against $\log_{10}$ antigen** (Fig. 59.7). The curve is most accurate in the central region, so adjust the amount of antigen in your test sample to fall within this range.

Note the following:

- **Measure all volumes as accurately as possible** as the end result depends on the volumetric quantities of unlabelled (sample) antigen, radio-labelled antigen and antibody: an error in any of these reagents will invalidate the assay.

- **Incorporate replicates,** so that errors can be quantified.

- **Seek your supervisor's advice** about fitting a curve to your data: this can be a complex process (see Chapter 53).

Immunoradiometric assay (IRMA)

This technique uses radiolabelled antibody, rather than antigen, for direct measurement of the amount of antigen (or hapten) in a sample. Most immunoradiometric assays are similar to the double-antibody sandwich method described below, except that the second antibody is labelled using a radioisotope. The important advantages over RIA are:

- **linear relationship between amount of radioactivity and test antigen;**
- **wider working range for test substance;**
- **improved stability/longer shelf life.**

Enzyme immunoassays (EIA)

These techniques are also known as enzyme-linked immunosorbent assays (ELISA). They combine the specificity of the antibody–antigen interaction with the sensitivity of enzyme assays using either an antibody or an antigen conjugated (linked) to an enzyme at a site that does not affect the activity of either component. The enzyme is measured by adding an appropriate chromogenic substrate, which yields a coloured product. Enzymes offer the following advantages over radioisotopic labels:

Chromogenic substrates for enzyme immunoassay – these include tetra-methylbenzidine (Box 59.2), which is a highly sensitive substrate used to detect horseradish peroxidase activity, producing a blue-coloured product as a result of peroxidase activity. Chapter 60 provides further details of chromogenic and fluorogenic substrates used for enzyme assays.

- **Increased sensitivity:** a single enzyme molecule can produce many product molecules, amplifying the signal.

- **Simplified assay:** enzyme assays are usually easier than radioisotope assays (pp. 387–9).

- **Improved stability of reagents:** components are generally more stable than their radiolabelled counterparts, giving them a longer shelf life.

- **No radiological hazard:** no requirement for specialised containment/disposal facilities.

- **Automation is straightforward:** using disposable microtitre plates and an optical scanner.

The principal techniques are double-antibody-sandwich ELISA, indirect ELISA and competitive ELISA.

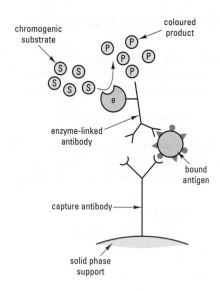

Fig. 59.8 Double-antibody sandwich ELISA.

Fig. 59.9 Indirect ELISA.

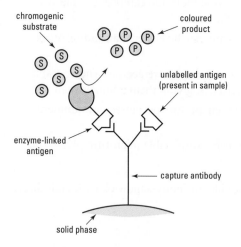

Fig. 59.10 Competitive ELISA.

Double-antibody-sandwich ELISA

This is used to detect specific antigens, involving a three-component complex between a capture antibody linked to a solid support, the antigen, and a second, enzyme-linked antibody (Fig. 59.8). This can be used to detect a particular antigen, e.g. a virus in a clinical sample, or to quantify the amount of antigen.

Indirect ELISA

This is used for antibody detection, with a specific antigen attached to a solid support. When the appropriate antibody is added, it binds to the antigen and will not be washed away during rinsing. Bound antibody is then detected using an enzyme-linked anti-immunoglobulin, e.g. a rabbit IgG antibody raised against human IgG (Fig. 59.9). One advantage of the indirect assay is that a single enzyme-linked anti-immunoglobulin can be used to detect several different antibodies, since the specificity is provided by the bound antigen.

Competitive ELISA

Here, any antigen present in a test sample competes with added enzyme-labelled antigen for a limited number of binding sites on the capture antibody (Fig. 59.10). Most commercial systems use 96-well microplates (12 columns by 8 rows), where each well is coated with the appropriate antibody. Following addition of known volumes of (i) sample and (ii) enzyme-labelled antigen, the plates are incubated (typically, up to 1 h), then washed thoroughly to remove all unbound material. Bound enzyme-labelled antigen is then detected using a suitable substrate. Quantitative results can be obtained by measuring the absorbance of each well using a spectrophotometric microplate reader: the absorbance at a particular wavelength is *inversely* proportional to the amount of antigen present in the test sample. Calibration standards (p. 326) are required to convert the readings to an amount or concentration. Alternatively, the test can be carried out in positive/negative format. Box 59.2 gives practical details for a sandwich-type ELISA.

Dipstick immunoassays

Improvements in technology have led to the development of a range of diagnostic kits that can be used either in a medical setting 'near-patient testing' or in the home, including immunoassays for cancer screening and drug testing. The first of these available to the general public was the home pregnancy test kit, which measures the hormone human chorionic gonadotrophin (hCG) in urine. The level of this hormone is raised substantially during pregnancy. The principle of this test, which is similar to other home test kits, is that coloured plastic microbeads are conjugated (attached) to an anti-hCG monoclonal antibody and then coated on to an absorbent plastic or cellulose strip. When this strip is dipped into urine, any hCG present will bind to the anti-hCG antibody, and both hormone–antibody complexes and free antibody will then move up the strip by capillary action until they reach a second hCG-specific antibody that is coupled chemically to the strip so that it cannot move (see Fig. 59.11). Any hormone–antibody complex will bind to this 'fixed' antibody (through the hCG component); due to the presence of the coloured plastic microbeads,

Box 59.2 How to perform an ELISA assay

While the following example is for a sandwich (capture) ELISA assay in microplate format, the same general principles apply to the other types of ELISA:

1. **Prepare the apparatus.** Switch on all equipment required:

   (a) The microplate reader – used to measure the absorbance of the solution in each well: set the reader to the required wavelength.

   (b) The microplate washer (where used) – each well must be washed at various stages during the procedure. When using an automated washer, first check that the wash bottle contains sufficient diluent and then test using an old microplate, to check that all wells are being washed correctly. Where required, use a wire needle to clean any blocked wash delivery tubes and repeat. For manual washing, use a wash bottle or multi-channel pipettor – make sure you fill each well and empty out all of the wash solution at the end of each wash stage.

   (c) The computer – this will contain the software required to label the wells, draw the calibration curve and calculate the results for test samples: fill out the ELISA template with details of the assays to be carried out.

2. **Prepare the various solutions to be analysed.** These include:

   (a) Test samples – make sure that each sample is identified with a code that enables you to record what each test well contains.

   (b) Calibrators/standards – including 'cutoff' calibrators and known positive standards.

   (c) Controls – positive and negative controls and blanks.

3. **Coat the wells with the capture antibody.** Typically 100 μL of a solution of the appropriate monoclonal or polyclonal antibody is added to each well and microplates are then incubated overnight at 4 °C to allow binding to the well.

4. **Wash the wells.** Transfer the microplate to the washer (or wash manually) – wash six times to remove excess coating (capture) antibody. The final rinse should be programmed so that the washer leaves the wells empty of diluent.

5. **Add blocking solution to each well.** Typically 100 μL of an inert protein solution (bovine serum albumin) is added to each well to block any free binding sites on the well. Microplates are incubated at room temperature for 30 min and then washed, as in step 4.

6. **Add test samples, calibrators and controls to wells.** Typically 100 μL of appropriately diluted test sample, control, etc. is added to each well. Microplates are incubated at room temperature for 90 min and then washed as in step 4.

7. **Add the detection antibody to each well.** Add 100 μL of monoclonal or polyclonal detection antibody labelled with a suitable enzyme (e.g. horseradish peroxidise, HRP) to each well. The microplate is then incubated at room temperature for 30–60 min and then washed, as in step 4.

8. **Add chromogenic substrate to each well.** For example, with HRP-labelled antigen add 100 μL of a standard solution of tetramethylbenzidine (TMB) and hydrogen peroxide to each well and re-incubate in darkness for 30 min, to allow colour development. The TMB is oxidised in the presence of hydrogen peroxide to produce a blue colour.

> **SAFETY NOTE** TMB and hydrogen peroxide are harmful by inhalation – use a fume hood.

9. **Stop the reactions.** For example, by adding 100 μL of 2 mol L^{-1} sulphuric acid to denature the enzyme. The colour of the oxidised TMB will change from blue to yellow as a result of the pH shift. While the human eye can readily distinguish different shades of blue, it is more difficult to visually assess different shades of yellow, once the reaction has been stopped.

10. **Measure the absorbance of each sample/calibrator/control well.** Transfer the microplate to the reader and assay at an appropriate wavelength: for TMB, use 450 nm.

11. **Interpret the results.** Check that the absorbance values of calibrators and control are within the required range. Then, for each sample, either record the absolute value (convert to concentration or amount, e.g. using a calibration curve, Chapter 53) or record as 'positive' or 'negative' (based on values for 'cutoff' calibrators), as appropriate.

zone 2: Any uncomplexed antibody binds to a further antibody, raised against it in a different animal (e.g. goat, anti-mouse IgG antibody)

zone 1: Any antigen–antibody complexes bind to a second antibody (double antibody sandwich)

Labelled antibody forms complexes with any antigen present in sample and travels up the dipstick by capillary action

sample

Fig. 59.11 Two-site dipstick immunoassay. For a positive test, as shown here, coloured bands will appear at zones 1 and 2 while a negative test will give colour in zone 2 only (any other colour options indicate a failed test).

a distinct line will then appear at this location on the strip, demonstrating the presence of the hormone. If no hormone–antibody complex is present, the first antibody will simply continue up the strip until it reaches an antibody raised against the anti-hCG antibody itself, coupled to the strip at a higher point. This will bind any free anti-hCG antibody in the sample and a coloured line will appear at this point. Since there will always be an excess of the original anti-hCG antibody conjugated to the plastic beads, the appearance of two coloured lines means that the hormone is present and the test result is positive, whereas if only the second coloured line appears, it means that the test is negative (this also serves as a control to confirm that the test is working – lack of any coloured lines indicates that the test has failed).

Text references

Delves, P.J., Martin, S.J., Burton, D.R. and Roitt, I.M. (2011) *Roitt's Essential Immunology,* 12th edn. Blackwell, Oxford.

Lefkovits, I. (1997) *Immunology Methods Manual: The Comprehensive Sourcebook of Techniques.* Academic Press, New York.

Sources for further study

Hay, F.C., Westwood, O.M.R. and Nelson, P.N. (2002) *Practical Immunology.* Blackwell, Oxford.

Van Emon, J.M. (2006) *Immunoassays and Other Bioanalytical Techniques.* CRC Press, Boca Raton.

Wild, D.G. (2013) *The Immunoassay Handbook,* 4th edn. Elsevier, Amsterdam.

STUDY EXERCISES

59.1 **Determine the amount of antigen in a test solution using Laurell rocket immunoelectrophoresis.** The figure below represents the results from a rocket immunoelectrophoresis assay for a series of standards. Determine the amount of antigen in a test solution giving a rocket (precipitin line) height of 18.5 mm.

59.2 **Determine the amount of antigen in a test solution using single radial immunodiffusion (RID).** Using the following data for a series of standards, determine the amount of antigen in a test solution giving a ring diameter of 6.5 mm.

Single radial immunodiffusion results for a series of standards containing 4–20 μg antigen

Amount of antigen (μg)	Ring diameter (mm)
4	3.5
8	5.5
12	7.0
16	8.0
20	9.0

59.3 **Interpret data from a haemagglutination test.** In some conditions, such as Hashimoto's disease, autoantibodies to the thyroid protein thyroglobulin are produced in large amounts. The level of these antibodies in serum can be measured by indirect haemagglutination: red blood cells coated with thyroglobulin are mixed with serial doubling dilutions of the test serum. The titre of antithyroglobulin autoantibodies is the lowest dilution at which haemagglutination occurs. The figure below represents the outcome of an antithyroglobulin haemagglutination test on a serum sample. The first well – numbered 1 – represents a ten-fold dilution of the serum sample, with serial doubling dilutions in subsequent wells.

(a) Write a brief explanation of the test results shown in the figure.

(b) Determine the autoantibody titre.

(c) Score the sample as 'positive' if the titre is more dilute than 1 in 5000 (i.e. 1: >5000) and 'negative' if it is less dilute than 1 in 5000 (i.e. 1: <5000).

```
 1   2   3   4   5   6   7   8   9  10  11  12
◉  ◉  ○  ○  ○  ○  ○  ○  ○  ○  ◉  ◉
```

Answers to these study exercises are available at www.pearsoned.co.uk/practicalskills

Enzymes are globular proteins that increase the rate of specific biochemical reactions. Each enzyme operates on a limited number of substrates of similar structure to generate products under well-defined conditions of concentration, pH, temperature, etc. In metabolism, groups of enzymes work together in sequential pathways to carry out a particular process, e.g. the multi-reaction conversion of glucose to lactate in glycolysis.

> **Example**   Enzyme EC 1.1.1.1 is usually known by its trivial name, alcohol dehydrogenase.

> **KEY POINT**   Enzymes are categorised according to the chemical reactions involved, leading to a four-figure Enzyme Commission code number and a systematic name for each enzyme. Most enzymes also have a recommended trivial name, often denoted by the suffix 'ase'.

Measuring enzyme reactions

Activity

This is measured in terms of the rate of enzyme reaction. Activity may be expressed directly as amount of substrate utilised per unit time (e.g. nmol min^{-1}, etc.), or in terms of the non-SI international unit (U, or sometimes IU), defined as the amount of enzyme that will convert 1 μmol of substrate to product(s) in 1 min under specified conditions. However, the recommended (SI) unit of enzyme activity is the katal (kat), which is the amount of enzyme that will convert 1 mol of substrate to product(s) in 1 s under optimal conditions, determined from the following equation:

> **Example**   Using eqn [60.1], the conversion of 4.5 μmol (4.5×10^{-6} mol) of substrate in 5 min (300 s) would give an enzyme activity of $4.5 \times 10^{-6} \div 300 = 1.5 \times 10^{-8}$ kat = 15 nkat.

$$\text{enzyme activity (kat)} = \frac{\text{substrate converted (mol)}}{\text{time (s)}} \quad [60.1]$$

This unit is relatively large (1 kat = 6×10^{7} U) so SI prefixes are often used, e.g. nkat or pkat (p. 177). Note that the units are *amount of substrate* (mol), not concentration (mol L^{-1} or mol m^{-3}).

For enzymes with macromolecular substrates of unknown molecular weight (e.g. deoxyribonuclease, amylase), activity can be expressed as the mass of substrate consumed (e.g. ng DNA min^{-1}), or amount of product formed (e.g. nmol glucose min^{-1}). You must ensure that your units clearly specify the substrate or product used, especially when the enzyme transformations involve different numbers of substrate or product molecules.

> **Example**   The hydrolysis of 1 molecule of maltose to give 2 glucose molecules by α-glucosidase means that enzyme activity specified in terms of substrate consumption (nmol maltose) would be half the value expressed with respect to product formation (nmol glucose).

The turnover number of an enzyme is the *amount of substrate* (mol) converted to product in 1 s by 1 mol of enzyme operating at maximum rate under optimum conditions. In practice, this requires information on the molecular weight of the enzyme, the amount of enzyme present and its maximum activity (pp. 390–1).

Purity

The specific activity of an enzyme preparation expresses the enzyme activity in terms of the quantity of protein present:

> **Example**   For an enzyme preparation with an activity of 1.5×10^{-8} kat in 24 mg (= 2.4×10^{-8} kg), eqn [60.2] gives a specific activity of $1.5 \times 10^{-8} \div 2.4 \times 10^{-8}$ = 0.625 kat kg^{-1}.

$$\text{specific activity} = \frac{\text{enzyme activity (kat)}}{\text{mass of protein (kg)}} \quad [60.2]$$

Alternative units may be used, e.g. U mg^{-1}, ng min^{-1} mg^{-1}. The specific activity is a useful way of comparing the purity of different enzyme

Preparing a calibration curve for protein (enzyme) assay – Box 53.1 gives practical details and Study exercise 53.2 (p. 331) gives typical data for a protein assay.

Fig. 60.1 The chromophore 3, 3′, 5, 5′-tetramethylbenzidine (TMB), used in peroxide assays, e.g. in ELISA (p. 381).

Example α-Glucosidase activity can be measured using the artificial substrate p-nitrophenol-α-D-glucose, which liberates p-nitrophenol (yellow) with an absorption maximum at 404 nm.

Example Phosphoenolpyruvate carboxylase can be assayed by coupling to malate dehydrogenase: the first enzyme converts phosphoenolpyruvate to oxaloacetate which is then oxidised to malate by the second enzyme (present in excess) with a stoichiometric (1 : 1) reduction of NAD^+. The coupled assay is monitored spectrophotometrically (pp. 449–50) as an increase in A_{340} against time.

Fig. 60.2 Structure of 4-methylumbelli-ferone (4-MU). The free hydroxyl group* is used to create fluorogenic substrates, e.g. coupled to galactose (4-MU-GAL) it can be used to assay β-galactosidase activity, since the galactoside derivative is non-fluorescent and hydrolysis can be detected by the fluorescence of liberated 4-MU.

preparations (purified enzyme preparations have high specific activities), comparing the stages in purification of an enzyme (specific activity increases as other proteins are eliminated), and assessing enzyme stability (an unstable enzyme will show a decrease in specific activity with time).

Enzyme activity and purity are often expressed in terms of the protein content of the enzyme preparation. Most assays for protein content do not give an absolute value, but require standard solutions, containing appropriate amounts of a particular protein, to be analysed at the same time, enabling a standard curve to be constructed. Bovine serum albumin (BSA) is commonly used as a general-purpose protein standard. However, you may need to use an alternative standard if your enzyme has an amino acid composition that is markedly different from BSA, depending on the specific method you use (Box 60.1).

Types of assay

The rate of substrate utilisation or product formation must be measured under controlled conditions, using some characteristic that changes in direct proportion to the concentration of the test substance.

Spectrophotometric assays

Many substrates and products absorb visible or UV light, and the change in absorbance at a particular wavelength provides a convenient assay method (Chapter 68). Artificial substrates are used where no suitable spectrophotometric assay is available for the natural substrates. In most cases, these are chromogenic analogues of the natural substrate, producing coloured products, e.g. tetramethylbenzidine (TMB: Fig. 60.1), used in enzyme immunoassay (pp. 381–2). In other cases, a product may be measured by a colorimetric chemical reaction.

Several assays are based on interconversion of the nicotinamide adenine dinucleotide coenzymes NAD^+ or $NADP^+$, which are reversibly reduced by many enzymes. The reduced form (either NADH or NADPH) can be detected at 340 nm, where the oxidised form has negligible absorbance (p. 449). An alternative approach is to use a coupled-enzyme assay, where the product of the test enzyme is used as a substrate for a second enzyme reaction that involves oxidation/reduction of nucleotide coenzymes. Such assays are particularly useful for continuous monitoring of enzyme activity (see below) and for reactions where the product from the test substance is too low to detect by other methods, since coupled assays are more sensitive.

Fluorimetric assays

Certain substrates liberate fluorescent products as a result of enzyme activity, providing a highly sensitive assay method. Fluorogenic substrates include fluorescein and methylumbelliferone derivatives (Fig. 60.2). Care is required, since impurities in the enzyme preparation may produce background fluorescence, or quenching (reduction) of the signal.

Radioisotopic assays

These are useful where the substrate and product can be easily separated, e.g. in decarboxylase assays using a ^{14}C-labelled substrate, where gaseous $^{14}CO_2$ is produced.

Box 60.1 Methods of determining the amount of protein in an aqueous solution

For all of the following methods, the amounts are appropriate for semi-micro cuvettes (1.5 mL volume, path length 1 cm). Appropriate controls (blanks) must be analysed, to assess possible interference (due to buffers, etc.).

Biuret method

This is based on the specific reaction between cupric ions (Cu^{++}) in alkaline solution and two adjacent peptide bonds, as found in proteins. As such, it is not significantly affected by differences in amino acid composition between proteins.

1. Prepare protein standards over an appropriate range (typically, between 1 and 20 mg mL^{-1}).

2. Add 1 mL of each standard solution to separate test tubes. Prepare a reagent blank, using 1 mL of distilled water, or an appropriate solution.

3. Add 1 mL of each unknown solution to separate test tubes.

4. Add 1 mL of Biuret reagent (1.5 g $CuSO_4 \cdot 5H_2O$, 6.0 g sodium potassium tartrate in 300 mL of 10 w/v NaOH) to all standard and unknown tubes and to the reagent blank.

5. Incubate at 37 °C for 15 min.

6. Read the absorbance of each solution at 520 nm against the reagent blank. The colour is stable for several hours.

The main limitation of the Biuret method is its lack of sensitivity – it is unsuitable for solutions with a protein content of less than 1 mg mL^{-1}.

Direct measurement of UV absorbance (Warburg-Christian method)

Proteins absorb EMR maximally at 280 nm (due to the presence of aromatic amino acids) and this forms the basis of the method. The principal advantages of this approach are its simplicity and the fact that the assay is non-destructive. The most common interfering substances are nucleic acids, which can be assessed by measuring the absorbance at 260 nm: a pure solution of protein will have a ratio of absorption (A_{280}/A_{260}) of approximately 1.8, decreasing with increasing nucleic acid contamination. Note also that any free aromatic amino acids in your solution will absorb at 280 nm, leading to an overestimation of protein content. The simplest procedure, which includes a correction for small amounts of nucleic acid, is as follows (use quartz cuvettes throughout):

1. Measure the absorbance of your solution at 280 nm (A_{280}): if A_{280} is greater than 1, dilute by an appropriate amount and remeasure.

2. Repeat at 260 nm (A_{260}).

3. Estimate the approximate Protein concentration using the following relationship:

$$[\text{protein}] \text{ mg mL}^{-1} = 1.45\, A_{280} - 0.74\, A_{260} \qquad [60.3]$$

This equation is based on the work of Warburg and Christian (1942), for enolase. For other proteins, it should not be used for quantitative work, since it gives only a rough approximation of the amount of protein present, due to variations in aromatic amino acid composition.

Lowry (Folin-Ciocalteau) method

This is a colorimetric assay, based on a combination of the Biuret method, described above, and the oxidation of tyrosine and tryptophan residues with Folin and Ciocalteau's reagent to give a blue-purple colour. The method is extremely sensitive (down to a protein content of 20 µg mL^{-1}), but is subject to interference from a wide range of non-protein substances, including many organic buffers (TRIS, HEPES, etc.), EDTA, urea and certain sugars. Choice of an appropriate standard is important, as the intensity of colour produced for a particular protein is dependent on the amount of aromatic amino acids.

1. Prepare protein standards within an appropriate range for your samples (the method can be used from 0.02–1.00 mg mL^{-1}).

2. Add 1 mL of each standard solution to separate test tubes. Prepare a reagent blank, using 1 mL of distilled water, or an appropriate solution.

3. Add 1 mL of each of your unknown solutions to separate test tubes.

4. Then, add 5 mL of 'alkaline solution' (prepared by mixing 2% w/v Na_2CO_3 in 0.1 mol L^{-1} NaOH, 1% w/v aqueous NaKtartrate in the ratio 100 : 1 : 1). Mix thoroughly and then allow to stand for 3 min.

5. Add 0.5 mL of Folin–Ciocalteau reagent (commercial reagent, diluted 1 : 1 with distilled water on the day of use). Mix rapidly and thoroughly and then allow to stand for 30 min.

6. Read the absorbance of each sample at 600 nm.

(continued)

Box 60.1 (continued)

Dye-binding (Bradford) method

Coomassie brilliant blue combines with proteins to give a dye-protein complex with an absorption maximum of 595 nm. This provides a simple and sensitive means of measuring protein content, with few interferences. However, the formation of dye-protein complex is affected by the number of basic amino acids within a protein, so the choice of an appropriate standard is important. The method is sensitive down to a protein content of approximately $5\ \mu g\ mL^{-1}$ but the relationship between absorbance and concentration is often nonlinear, particularly at high protein content.

1. Prepare protein standards over an appropriate range (between 5 and $100\ \mu g\ mL^{-1}$).

2. Add $100\ \mu L$ of each standard solution to separate test tubes. Prepare a reagent blank, using $100\ \mu L$ of distilled water, or an appropriate solution (note that these small volumes must be accurately dispensed, e.g. using a calibrated pipettor (pp. 140–1).

3. Add $100\ \mu L$ of your unknown solutions to separate test tubes.

4. Add 5.0 mL of Coomassie brilliant blue G250 solution $(0.1\ g\ L^{-1})$.

5. Mix and incubate for at least 5 min: read the absorbance of each solution at 595 nm.

Other methods are less widely used. They include determination of the total amount of nitrogen in solution (e.g. using the Kjeldahl technique) and calculating the protein content, assuming a nitrogen content of 16%. An alternative approach is to precipitate the protein (e.g. using trichloracetic acid, tannic acid, or salicylic acid) and then measure the turbidity of the resulting precipitate (using a nephelometer, or a spectrophotometer, p. 449).

Definition

pH stat – a system that measures the pH of a solution and automatically adds either acid or alkali to maintain a constant pH.

Electrochemical assays

Enzyme reactions involving acids and bases can be monitored using a pH electrode (p. 164), though the change in pH will affect the activity of the test enzyme. An alternative approach is to measure the amount of acid or alkali required to maintain a pre-selected pH in a pH-stat.

An oxygen electrode can be used if O_2 is a substrate or a product (p. 425). Other ion-specific electrodes can monitor ammonia, nitrate, etc.

Monitoring substrate utilisation/product formation

Continuous assays (kinetic assays)

The change in substrate or product is monitored as a function of time, to provide a progress curve for the reaction (Fig. 60.3). These curves start off in a near-linear manner, decreasing in slope as the reaction proceeds and substrate is used up. The initial velocity of the reaction (v_0) is obtained by drawing a tangent to the curve at zero time and measuring its slope. Continuous monitoring can be used when the test substance can be assayed rapidly and non-destructively, e.g. using a chromogenic substrate.

Discontinuous assays (fixed-time assays)

It is sometimes necessary to measure the amount of substrate consumed or product formed after a fixed time period, e.g. where the test substance is assayed by a (destructive) colorimetric chemical method. It is vital that the time period is kept as short as possible, with the change in substrate concentration limited to around 10%, so that the assay is within the linear part of the progress curve (Fig. 60.3). This may need to be established in a preliminary experiment, e.g. using a continuous assay.

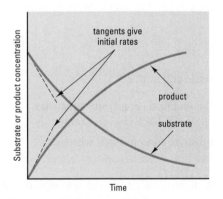

Fig. 60.3 Enzyme reaction progress curve: substrate utilisation/product formation as a function of time.

Disrupting tissues – for more details, see Reed *et al.*, 2016.

Enzyme purification

The purification of an enzyme usually involves several stages:

1. **Tissue/cell disruption,** often using mechanical/ultrasonic homogenisation.

2. **Differential centrifugation,** removing larger particulate components (p. 442).

3. **Ammonium sulphate fractionation:** selective precipitation by the step-wise addition of ammonium sulphate at low temperature (<10 °C). The precipitated protein is then redissolved in a fresh buffer solution.

4. **Chromatographic and/or electrophoretic separation,** including ion exchange and gel filtration chromatography and polyacrylamide gel electrophoresis or isoelectric focusing. This stage may involve several individual steps and can lead to a fully purified enzyme (purification to homogeneity), with maximum specific activity.

At each step in the procedure the percentage yield of enzyme in an individual fraction can be determined:

$$\text{yield} = \frac{\text{enzyme in fraction (kat or U)}}{\text{enzyme in original preparation (kat or U)}} \times 100 \qquad [60.4]$$

> **Example** If an original preparation of 50 mL had an activity of 100 pkat mL^{-1} (total 5 nkat) and a subsequent fraction of 5 mL volume had an activity of 200 pkat mL^{-1} (total 1 nkat), using eqn [60.4] this would represent a percentage yield of 20%.

Note that the yield equation does not use a concentration-based measure of enzyme activity, as this would be affected by the volumes of the solutions involved. The relative purification of an enzyme is usually expressed as 'n-fold purification', where:

$$n = \frac{\text{specific activity of fraction}}{\text{specific activity of original preparation}} \qquad [60.5]$$

> **Example** Using eqn [60.5] an increase in specific activity from 10 nkat (mg protein)$^{-1}$ to 120 nkat (mg protein)$^{-1}$ represents a 12-fold purification.

Enzyme kinetics

For most enzymes, when the initial velocity (v_0) using a fixed amount of an enzyme is plotted as a function of the concentration of a single substrate [S] with all other substrates present in excess, a rectangular hyperbola is obtained (Fig. 60.4). At low substrate concentrations v_0 is directly proportional to [S], with a decreasing proportional response as substrate concentration is increased until saturation is achieved. The shape of this plot can be described by a mathematical relationship, known as the Michaelis–Menten equation:

$$v_0 = \frac{V_{\text{max}}[\text{S}]}{K_{\text{m}} + [\text{S}]} \qquad [60.6]$$

This equation makes use of two kinetic constants:

1. V_{max}, the maximum velocity of the reaction (at infinite substrate concentration).

2. K_{m}, the Michaelis constant, is the substrate concentration where $v_0 = \frac{1}{2}V_{\text{max}}$.

V_{max} is a function of the amount of enzyme and is the appropriate rate to use when determining the specific activity of a purified enzyme (p. 386). The Michaelis constant can provide a measure of the substrate affinity of an enzyme and is an important characteristic of a particular enzyme. Thus, an enzyme with a large K_{m} usually has a low affinity for the substrate, whereas an enzyme with a small K_{m} usually has a high affinity.

Fig. 60.4 Effect of substrate concentration on enzyme activity.

(a)

(b)

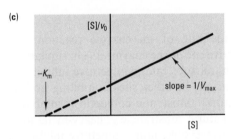

(c)

Fig. 60.5 Graphical transformations for determining the kinetic constants of an enzyme. (a) Lineweaver–Burk plot. (b) Eadie–Hofstee plot. (c) Hanes–Woolf plot.

Your first step in determining the kinetic constants for a particular enzyme is to measure the rate of reaction at several substrate concentrations, as in Fig. 60.5. There are various ways to obtain K_m and V_{max} from such data, mostly involving drawing a graph representing a linear transformation of eqn [60.6]:

- **The Lineweaver–Burk plot:** a graph of the reciprocal of the reaction rate $(1/v_0)$ against the reciprocal of the substrate concentration $(1/[S])$ gives $-1/K_m$ as the intercept of the x axis and $1/V_{max}$ as the intercept of the y axis (Fig. 60.5(a)). Note that the slope of the plot is most affected by the least accurate values, i.e. those measured at low substrate concentration.

- **The Eadie–Hofstee plot:** v_0 against $v_0/[S]$, where the intercept on the y axis gives V_{max} and the slope equals $-K_m$ (Fig. 60.5(b)).

- **The Hanes–Woolf plot:** $[S]/v_0$ against $[S]$, giving $-K_m$ as the intercept of the x axis and $1/V_{max}$ from the slope (Fig. 60.5(c)).

There are several computer packages that will plot the above relationships and calculate the kinetic constants from a given set of data using linear regression analysis (p. 525). While the Eadie–Hofstee and Hanes–Woolf plots distribute the data points more evenly than the Lineweaver–Burk plot, the best approach to such data is to use non-linear regression on untransformed data. This is usually outside the scope of the simpler computer programs, though tailor-made commercial packages can carry out such analyses. Note also that some enzymes do not show Michaelis–Menten kinetics, e.g. those with more than one active site per molecule (allosteric enzymes), which often give sigmoid curves of v_0 against [S]. Such enzymes are usually involved in the control of metabolism.

Factors affecting enzyme activity

If you want to measure the maximum rate of a particular enzyme reaction, you will need to optimise the following:

Substrate concentration

The substrates must be present in excess, to ensure maximum reaction velocity. Equation [60.6] shows that a substrate concentration equivalent to $10 \times K_m$ will give 91% of V_{max}, while a concentration of $10 \times K_m$ will give 99% of the maximum rate.

Cofactors

Many enzymes require appropriate concentrations of specific cofactors for maximum activity. These are subdivided into coenzymes (soluble, low molecular weight organic compounds that are actively involved in catalysis by accepting or donating specific chemical groups, i.e. they are co-substrates of the enzyme; examples include NAD^+ and ADP); and activators (inorganic metal ions, required for maximal activity, e.g. Mg^{2+}).

Using stored enzymes – when removing a bottle of freeze-dried enzyme from a freezer or fridge, do not open it until it has been warmed to room temperature or water may condense on the contents – this will make weighing inaccurate and may lead to loss of enzyme activity.

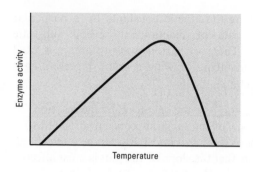

Fig. 60.6 Effect of temperature on enzyme activity.

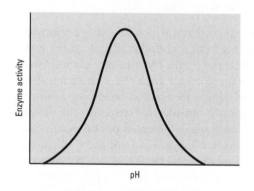

Fig. 60.7 Effect of pH on enzyme activity.

Temperature

Enzyme activity increases with temperature, until an optimum is reached. Above this point, activity decreases as a result of protein denaturation (Fig. 60.6). Note that the optimum temperature for enzyme *activity* may not be the same as that for maximum *stability* (enzymes are usually stored at temperatures near to or below 0 °C, to maximise stability).

pH

Enzymes work best at a particular pH, due to changes in ionisation of the substrates or of the amino acid residues within the enzyme (Fig. 60.7). Most enzyme assays are performed in buffer solutions (pp. 167–9), to prevent changes in pH during the assay.

> **KEY POINT** Note that temperature and pH optima are dependent upon reaction conditions – you should therefore specify the experimental conditions under which such optima are determined.

Inhibitors

Many compounds can reduce the rate of an enzyme reaction, e.g. substances that compete for the active site of the enzyme due to similarities in chemical structure with the natural substrate (competitive inhibitors), or substances that bind to the enzyme at other sites, inhibiting normal function (non-competitive inhibitors). Most non-competitive inhibitors are not chemically related to the natural substrate, e.g. heavy-metal ions, such as Hg^{2+} and Cd^+, acting as irreversible, non-competitive inhibitors. Competitive inhibitors reduce K_m but have no effect on V_{max}, whereas non-competitive inhibitors reduce V_{max} but have no effect on K_m.

Text references

Reed, R.H., Holmes, D., Weyers, J.D.B. and Jones, A.M. (2016) *Practical Skills in Biomolecular Sciences*, 5th edn. Pearson, Harlow.

Sources for further study

Bisswanger, H. (2008) *Enzyme Kinetics: Principles and Methods.* Wiley, New York.

Cook, P.F. and Cleland, W.W. (2006) *Enzyme Kinetics and Mechanisms.* Garland Science, Oxford.

Cornish-Bowden, A. (2012) *Fundamentals of Enzyme Kinetics*, 4th edn. Wiley, New York.

Warburg, O. and Christian, W. (1942) 'Isolierung und Kristallisation des Garungsferments Enolase', *Biochemische Zeitschrift,* **310**, 384–421.

[Covers basic principles, kinetics and experimental systems.]

Purich, D. (2010) *Enzyme Kinetics, Catalysis and Control: a Reference of Theory and Best-practice Methods.* Elsevier, Amsterdam.

STUDY EXERCISES

60.1 Determine enzyme activity and purity (specific activity). An extract of α-glucosidase containing 2.5 mg of protein was incubated with p-nitrophenol–α-D-glucose at 1.0 mmol L^{-1} and assayed by following the liberation of p-nitrophenol at 404 nm. This gave a progress curve with an initial velocity equal to an absorbance change of 0.25 min^{-1}. Given that the conversion of 10 nmol of substrate to product results in an absorbance change of 0.122:

(a) Determine the amount of product formed in nmol min^{-1}.

(b) Calculate the enzyme activity in kat.

(c) Determine the specific activity, in terms of (i) nmol min^{-1} (mg protein)$^{-1}$ and (ii) kat (mg protein)$^{-1}$.

Quote all answers to three significant figures.

60.2 Calculate yield and relative purification of an enzyme. The table below shows representative data for an enzyme purification involving five stages. Calculate the yield and n-fold purification at each stage (give all answers to three significant figures).

Volume, protein concentration and enzyme activity at several stages during a protein purification procedure

Stage	Total volume of extract (mL)	Enzyme activity (U mL^{-1})	Protein concentration (mg mL^{-1})
1	78	1.2	4.8
2	70	1.2	1.7
3	10	6.4	2.5
4	5	10.8	1.4
5	0.6	65.7	3.4

60.3 Estimate K_m and V_{max} for an enzyme. The table below shows representative data for enzyme activity as a function of substrate concentration. Use this to determine the kinetic constants K_m and V_{max} of the enzyme (give your answers to three significant figures).

Enzyme activity at various substrate concentrations

Substrate concentration (μmol L^{-1})	Enzyme activity (U)
0	0
5	16.2
7	20.5
10	28.1
15	32.7
25	39.2
50	56.7
100	68.2

Answers to these study exercises are available at www.pearsoned.co.uk/practicalskills

Gregor Mendel, an Austrian monk, made pioneering studies of the genetics of eukaryotic organisms in the middle of the nineteenth century. He made crosses between different forms of flowering plants. Through careful examination and numerical analysis of the observable characteristics, or phenotype, of the parents and their progeny, Mendel was able to deduce much about their genetic characteristics, or genotype. The principles derived from these experiments explain the basis of heredity, and hence underpin our understanding of sexual reproduction, bio-diversity and evolution. Mendelian genetics is concerned primarily with the transmission of genetic information, as opposed to molecular genetics, which deals with the molecular details of the genome and techniques for altering genes (see Chapter 62).

KEY POINT A common initial stumbling block in genetics is terminology. In many cases the definitions are interdependent, so your success in this subject depends on your grasp of all the definitions and underlying ideas explained below.

Important terms and concepts

Each character in the phenotype is controlled by the organism's genes, the basic units of inheritance. Each gene includes the 'genetic blueprint' (DNA) that usually defines the amino acid sequence for a specific polypeptide or protein – often an enzyme or a structural protein. The protein gives rise to the phenotype through its activity in metabolism or its contribution to the organism's structure. The full complement of genes in an individual is known as its genome. Individual genes can exist in different forms, each of which generally leads to a different form of the protein it codes for. These different gene forms are known as alleles.

In eukaryotes, the genes are located in a particular sequence on chromosomes within the nucleus. The number of chromosomes per cell is characteristic for each organism (its chromosome number, n). For example, the chromosome number for man is 23. In cells of most 'higher' organisms, there are two of each of the chromosomes (2n). This is known as the diploid state. As a result of the process of meiosis, which precedes reproduction, special haploid cells are formed (gametes) that contain only one of each chromosome (1n). In sexual reproduction, haploid gametes from two individuals fuse to form a zygote, a diploid cell with a new genome, which gives rise to a new individual through the process of mitosis. Cell numbers are increased by this process, producing genetically identical cells.

Organisms vary in the span of the diploid and haploid phases. In some 'lower' organisms, the haploid phase is the longer-lasting form; in most 'higher' organisms, the diploid phase is dominant. A life-cycle diagram can be used to show how the phases are organised and what life forms are involved in each case (see e.g. Fig. 30.6).

Since each diploid individual carries two of each chromosome, it has two copies of each gene in every cell. The number of alleles of each character present depends on whether the relevant genes are the same – the homozygous state – or whether they are different – the heterozygous state. Hence, while there may be many alleles for any given gene, an individual could have two at most, and might have only one if it were homozygous. This will depend on the alleles present in the parental gametes that fused when the zygote was formed. Offspring that inherit a different combination of alleles at the two loci compared with their parents are known as recombinants.

The basis of Mendel's experiments and of many exercises in genetics are *crosses*, where individuals showing particular phenotypes are mated and the phenotypes of the offspring, or F_1 generation, are studied (see Box 61.1). If homozygous individuals carrying alternative alleles for a character are crossed, one of the alleles may be *dominant*, and all the F_1 generation will show that character in their phenotype. The character not evident is said to be *recessive*.

In describing crosses, geneticists denote each character with a letter of the alphabet, a capital letter being used for the dominant allele and lower case for the recessive. Taking for example a gene with dominant and recessive forms A and a respectively, there are three possibilities for each individual: it can be (a) homozygous recessive aa; (b) homozygous dominant AA; or (c) heterozygous Aa.

The reasons for dominance might relate to the activity of an enzyme coded by the relevant gene; for example, Mendel's yellow pea allele is dominant because the gene involved codes for the breakdown of chlorophyll. In the homozygous recessive case, none of the functional enzyme will be present, chlorophyll breakdown cannot occur and the seeds remain green. Not all alleles exhibit dominance in this form. In some cases, the heterozygous state results in a third phenotype (incomplete or partial dominance); in others, the heterozygous individual expresses both genotypes (codominance). Another possible situation is epistasis, in which one gene affects the expression of another gene that is independently inherited. Note also that many genes, such as those coding for human blood groups, have multiple alleles.

Example In the garden pea, *Pisum sativum*, studied by Mendel, the yellow seed allele Y was found to be dominant over the green seed allele y. A cross of YY × yy genotypes would give rise to Yy in all of the F_1 generation, all of which would thus have the yellow seed phenotype. If the F_1 generation were interbred, this Yy × Yy cross would lead to progeny in the next, F_2, generation with the genotypes YY : Yy : yy in the expected ratio 1 : 2 : 1. The expected phenotype ratio would be 3 : 1 for yellow : green seed.

KEY POINT Genetics problems may well involve one of the 'standard' crosses shown in Box 61.1. Before tackling the problem, try to analyse the information provided to see if it fits one of these types of cross. Figure 61.1 is a flowchart detailing the steps you should take in answering genetics problems.

Unless otherwise stated or obvious from the evidence, you should assume that the genes being considered in any given case are on separate chromosomes. This is important because it means that they will assort independently during meiosis. Thus, the fact that an allele of gene A is present in any individual will not influence the possibility of an allele of gene B being present. This allows you to apply simple probability in predicting the genetic make-up of the offspring of any cross (see 398).

Where genes are present on the same chromosome, they are said to be linked genes, and thus it would appear that they would not be able to assort independently during metaphase 1 of meiosis, when homologous

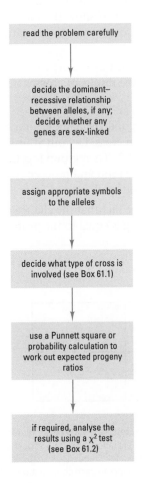

read the problem carefully

↓

decide the dominant–recessive relationship between alleles, if any; decide whether any genes are sex-linked

↓

assign appropriate symbols to the alleles

↓

decide what type of cross is involved (see Box 61.1)

↓

use a Punnett square or probability calculation to work out expected progeny ratios

↓

if required, analyse the results using a χ^2 test (see Box 61.2)

Fig. 61.1 Flowchart for tackling problems in Mendelian genetics.

Pedigree notation:

○   normal female

□   normal male

● ■   female or male displaying an inherited condition

○─□   mating

offspring

Example of simple family pedigree:

This diagram shows the offspring of a normal male and female. The two daughters are normal, but the son displays the inherited condition.

Fig. 61.2 Pedigree notation and family trees.

Denoting linked genes – these are often shown diagramatically, with a double line indicating the chromosome pair. For example, the two possible linkages for the genotype AaBb would be shown as:

$$\frac{A \quad b}{a \quad B} \quad \text{or} \quad \frac{A \quad B}{a \quad b}$$

chromosomes are independently orientated. However, although physically attached to each other, they may become separated when crossing-over occurs between the homologous chromosomes at an early stage of meiosis. Exchange of genetic information between homologous chromosomes is called recombination. Linkage can be detected from a cross between individuals heterozygous and homozygous recessive for the relevant genes, e.g. AaBb × aabb. If the genes A and B are on different chromosomes, we expect the ratio of AaBb, aabb, Aabb and aaBb to be 1 : 1 : 1 : 1 in the F_1. However, if the dominant alleles of both genes occur on the same chromosome, the last two combinations will occur, but rarely. Just how rarely depends on how far apart they lie on the relevant chromosome – the further apart, the more likely it is that crossing-over will occur. This is the basis of chromosome mapping (see Box 61.1).

Another complication you will come across is sex-linked genes. These occur on one of the X or Y chromosomes that control sex. Because one or other of the sexes – depending on the organism – is determined as XX and the other as XY (see Box 61.1), this means that rare recessive genes carried on the X chromosome may be expressed in XY individuals. Sex-linked genes are sometimes obvious from differences in the frequencies of phenotypes in male and female offspring. Pedigree charts (Fig. 61.2) are codified family trees that are often used to show the inheritance and expression of sex-linked characteristics through various generations.

Analysis of crosses

There are two basic ways of working out the results of crosses from known or assumed genotypes:

1. **The Punnett square method** provides a good visual indication of potential combinations of gametes for a given cross. Lay out your Punnett squares consistently as shown in Fig. 61.3. Then group together the like genotypes to work out the genotype ratio and proceed to work out the corresponding phenotype ratio if required.

2. **Probability calculations** are based on the fact that the chance of a number of independent events occurring is equal to the probabilities

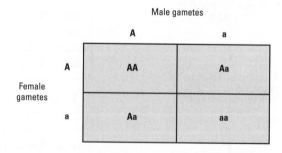

Fig. 61.3 Layout for a simple Punnett square for the cross Aa × Aa. The genotypic ratios for this cross are 1 : 2 : 1 for AA : Aa : aa, and the phenotypic ratio would be 3:1 for characteristic A to characteristic a. In this simple Punnett square, the allele frequencies are treated as equal ($f = 0.5$); if different from this, the probability of genotypes in each combination will be the relevant frequencies multiplied together.

Box 61.1 Types of cross and what you can (and cannot) learn from them

Monohybrid cross – the simplest form of cross, considering two alleles of a single gene.

Example: AA × aa

If only the parental *phenotypes* are known, you cannot always deduce the parental genotypes from the phenotype ratio in the F_1. An individual of dominant phenotype in the F_1 could arise from a homozygous dominant or heterozygous genotype. However, crossing the F_1 generation with themselves to give an F_2 generation may provide useful information from the phenotype ratios that are found.

Dihybrid cross – a cross involving two genes, each with two alleles.

Example: AaBB × AaBb

As with a monohybrid cross, you cannot always deduce the parental genotype from the phenotype ratio in the F_1 generation alone.

Test cross – a cross of an unknown genotype with a homozygous recessive.

Example: AABb × aabb

A test cross is one between an individual dominant for A and B with one recessive for both genes. The progeny will all be dominant for A, revealing the homozygous nature of the parent for this gene, but the progeny phenotypes will be split approximately 1 : 1 dominant to recessive for gene B, revealing the heterozygous nature of the parent for this gene. This type of cross reveals the unknown parental genotype in the proportions of phenotypes in the F_1.

Sex-linked cross – a cross involving a gene carried on the X chromosome; this can be designated as dominant or recessive using appropriate superscripts (e.g. X^A and X^a).

Example: $X^A X^a × X^A Y$

In sex-linked crosses you need to know the basis of sex determination in the species concerned – which of XX or XY is male (for example, the former in birds, butterflies and moths, the latter in mammals and *Drosophila*). The expected ratios of phenotypes in the offspring will depend on this. Note that the recessive genotype a will be expressed in the $X^a Y$ case.

Crosses with linked genes – genes are linked if they are on the same chromosome. This is revealed from a cross between individuals heterozygous and homozygous recessive for the relevant genes.

Example 1 (genes on separate chromosomes): expected offspring frequency from the cross AaBb × aabb is AaBb, aabb, Aabb and aaBb in the ratio 1 : 1 : 1 : 1.

Example 2 (linked genes): the frequencies of the last two combinations in example 1 might be skewed according to the direction of parental linkage.

Chromosome mapping uses the frequency of crossing-over of linked genes to estimate their distance apart on the chromosome on the basis that crossing-over is more likely the further apart are the genes. So-called 'map units' are calculated on the following basis:

$$\frac{\text{no. of recombiant progeny}}{\text{total no. of progeny}} \times 100 = \% \text{ crossing-over}$$

[61.1]

By convention, 1% crossing-over = 1 map unit (centimorgan, cM). The order of a number of genes can be worked out from their relative distances from each other. Thus, if genes A and B are 12 map units apart, while A and B are respectively 5 and 7 map units from C, the assumed order on the chromosome is ACB (Fig. 61.4).

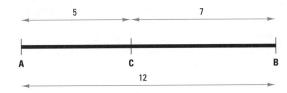

Fig. 61.4 Genetic map showing relative positions of genes A, B and C.

Using probability calculations – this can be simpler and faster than Punnett squares when two or more genes are considered.

Carrying out genetic crosses – typically these are performed using organisms that have a large number of offspring, to even out random variation. Other useful attributes include short generation time, ease of maintenance, and wide range of mutations readily observed in the phenotype.

Limitation of the Chi^2 test – note that the formula cited in Box 61.2 is valid only if expected numbers are greater than 5.

Table 61.1 Values of Chi^2 (χ^2) for which $P = 0.05$. The value for ($n - 1$) degrees of freedom (d.f.) should be used, where n = the number of categories (= phenotypes) considered (normally fewer than 4 in genetics problems). If χ^2 is less than or equal to this value, accept the null hypothesis that the observed values arose by chance; if χ^2 is greater than this value, reject the null hypothesis and conclude that the difference between the observed and expected values is statistically significant

Degrees of freedom	χ^2 value for which $P = 0.05$
1	3.84
2	5.99
3	7.82
4	9.49

of each event occurring, multiplied together. Thus if the probability P of a child being a boy is 0.5 and the probability of the child of particular parents being blue-eyed is 0.5, then the probability of that couple having a blue-eyed son is $0.5 \times 0.5 = 0.25$, and that of having two blue-eyed boys is:

$$P = (0.5 \times 0.5) \times (0.5 \times 0.5) = 0.0625$$

How do you decide whether the results of an experimental cross fit your expectation from theory as calculated above? This is not easy, because of the element of chance in fertilisation. Thus, while you might expect to see a 3 : 1 phenotype ratio of progeny for a given cross, in 500 offspring you might actually observe a ratio of 379 : 121, which is a ratio of just over 3.13 : 1. Can you conclude that this is significantly different from 3 : 1 in the context of random error? The answer to this problem comes from statistics. However, the answer is not certain, and your conclusion will be based on a balance of probabilities (see Box 61.2 and Table 61.1).

Box 61.2 Example of a Chi^2 (χ^2) test

This test allows you to assess the difference between observed (O) and expected (E) values and is extremely useful in biology. It is particularly valuable in determining whether progeny phenotype ratios fit your assumptions about their genotypes. The operation of the test is best illustrated by the use of an example. Assume that your null hypothesis (see p. 515) is that the phenotypic ratio is 3 : 1 and you observe that in 500 offspring the phenotype ratio is 379 : 121, whereas the expected ratio is 375 : 125.

Start the test by calculating the test statistic χ^2. The general formula for calculating χ^2 is:

$$\chi^2 = \sum \frac{(O - E)^2}{E} \qquad [61.2]$$

In this examle, this works out as:

$$\chi^2 = \frac{(379 - 375)^2}{375} + \frac{(121 - 125)^2}{125} = \frac{16}{375} + \frac{16}{125} = 0.171$$

The probability associated with this value can be obtained from χ^2 tables for ($n - 1$) degrees of freedom (d.f.), where n = the number of categories = number of phenotypes considered. Here the d.f. value is $2 - 1 = 1$. Since the χ^2 value of 0.171 is lower than the tabulated value for 1 d.f. (3.84, Table 61.1), we therefore accept the null hypothesis and conclude that the difference between observed and expected results is not significant (since P > 0.05). Had χ^2 been ≥ 3.84, then $P \leq 0.05$ and we would have rejected the null hypothesis and concluded that the difference was significant, i.e. that the progeny phenotype did not fit the expected ratio.

You can carry out Chi^2 calculations using the CHITEST function in Microsoft *Excel*. The *Help* function within that program contains useful guidance.

The Hardy–Weinberg Principle – this was named after its first, independent protagonists, Godfrey H. Hardy and Wilhelm Weinberg. It is sometimes referred to as the Hardy–Weinberg equilibrium, theorem, law or rule.

Population genetics

Population genetics is largely concerned with the frequencies of alleles in a population and how these may change in time. The Hardy–Weinberg Principle states that the frequency of alleles f remains the same between generations, unless influenced by some outside factor(s).

To understand why this is the case, consider alleles H and h for a particular gene, which exist in the breeding population at frequencies p and q respectively. If the individuals carrying these alleles interbreed randomly, then the expected phenotype and allele ratios in the F_1 generation can be calculated simply as:

$$f(HH) = p^2;$$
$$f(Hh) = 2pq; \text{ and}$$
$$f(hh) = q^2$$

If you wish to confirm this, lay out a Punnett square with appropriate frequencies for each allele. Now, by summation, the frequency of H in the $F_1 = p^2 + pq$ (a similar calculation can be made for allele h); and since in this example there are only two alleles, $p + q = 1$ and so $q = (1 - p)$ Substituting $(1 - p)$ for q, the frequency of H in the F_1 is thus:

$$p^2 - p(1 - p) = p^2 + p - p^2 = p \qquad [61.3]$$

i.e. the frequency of the allele is unchanged between generations. A similar relationship exists for the other alleles.

The Hardy–Weinberg Principle holds so long as the following criteria are satisfied:

1. **random mating** – so that no factors influence each individual's choice of a mate;

2. **large population size** – so that the laws of probability will apply;

3. **no mutation** – so that no new alleles are formed;

4. **no emigration, immigration or isolation** – so that there is no interchange of genes with other populations nor isolation of genes within the population;

5. **no natural selection** – so that no alleles have a reproductive advantage over others.

Population geneticists use the Hardy-Weinberg Principle and eqn [61.3] to gain an idea of the rate of evolution. By ensuring that criteria 1–4 hold, if there are any changes in allele frequency between generations, then the rate of change of allele frequencies indicates the rate of evolutionary change (natural selection).

Example Cystic fibrosis occurs in 0.04% of Caucasian babies. If this condition results from a double recessive allele aa, then following the Hardy-Weinberg Principle and eqn [61.3], $q^2 = 0.0004$ (0.04% expressed as a fraction of 1) and so $q = \sqrt{0.0004} = 0.02$, or 2%. Since $p = (1 - q)$, $p = 0.98$, or 98% The frequency of carriers of cystic fibrosis in the Caucasian population (people having the alleles Aa) is given by $2pq$. From the above, $2pq = 2(0.02 \times 0.98) = 0.0392$. Hence 3.92% of the Caucasian population are carriers (roughly one in 25).

Sources for further study

Blumberg, R.B. *MendelWeb.* Available: http://www.mendelweb.org/
Last accessed 15/07/15.
[A resource for students interested in the origins of classical genetics.]

Brooker, R. (2014) *Genetics: Analysis and Principles,* 5th edn. McGraw-Hill, Burr Hill.

Fletcher, H., Hickey, I. and Winter, P. (2012) *Instant Notes in Genetics,* 4th edn. BIOS, Oxford.

Hamilton, M. (2009) *Population Genetics.* Wiley, New York.

Klug, W.S. and Cummings, M.R. (2012) *Essentials of Genetics,* 8th edn. Prentice Hall, Upper Saddle River, NJ.

61.1 **Use a Punnett square to predict the outcome of a cross.** Lay out a Punnett square for a cross between genotypes RrOO × RrOo, where R is a semi-dominant gene for flower colour such that RR = red, Rr = pink and rr = white; and O is a dominant gene for corolla shape such that OO = closed corolla, Oo = closed corolla and oo = open corolla. From the Punnett square, derive both the genotypic and phenotypic ratios for the cross.

61.2 **Use probability to predict the outcome of a cross.** Two hazel-eyed parents are heterozygous for the eye-colour gene B. When expressed as bb, the individual is blue-eyed. Mum's hair is (genuine) blonde but Dad's is mousy-brown. In this case she is double recessive (mm) for the hair-colour gene M and he is heterozygous (Mm). What is the probability that they will have a blue-eyed, blonde daughter?

61.3 **Work out a likely genetic scenario for a given set of results.** Five tailless female mice were crossed with normal males (with tails). There were 31 normal mice and 28 tailless mice in the F_1 progeny. When pairs of tailless mice from the F_1 generation were crossed, their (F_2) progeny were as follows: normal, 27; tailless, 55; dead on birth, but tailless, 30. In each case the ratio of males to females was roughly 50 : 50. Provide a logical explanation for these results.

61.4 **Predict parental genotypes from the results of crosses involving sex-linked genes.** A red-eye gene is known to be sex-linked in *Drosophila*; that is, the alleles R (red-eyed) or r (white-eyed) are carried on the X chromosome, while the Y chromosome does not carry these eye-colour alleles. In the fruit-fly, XX = female and XY = male. Predict the possible parental genotypes from the following F_1 progeny ratios:

(a) 35 red-eyed female, 17 red-eyed male, 19 white-eyed male;

(b) 27 red–eyed male, 29 red-eyed female;

(c) 19 red-eyed female, 18 white-eyed female, 22 red-eyed male, 21 white-eyed male.

61.5 **Carry out a χ^2 (Chi^2) test.** A geneticist expects the results of a test cross to be in the phenotype ratio 1 : 2 : 1. He observes 548 progeny from his cross in the ratio 125 : 303 : 120. What should he conclude?

Answers to these study exercises are available at www.pearsoned.co.uk/practicalskills

Deoxyribonucleic acid (DNA) is the genetic material of all cellular organisms. The sequential arrangement of nucleotide subunits represents a genetic code for the synthesis of specific cellular proteins or functional RNA molecules. A portion of DNA that encodes the information for a single polypeptide, protein or polyribonucleotide is usually referred to as a gene while the entire genetic information of an organism is known as its genome.

Advances in the manipulation of nucleic acids in the laboratory have increased our understanding of the structure and function of genes at the molecular level. Additionally, these techniques can be used to alter the genome of an organism (genetic engineering or gene cloning), e.g. to create a bacterium capable of synthesising a foreign protein such as a potentially valuable hormone or vaccine, or a recombinant protein.

Conforming to regulations – in the UK, the *Genetically Modified Organisms (Contained Use) Regulations (2014)* provide the regulatory framework for all research procedures involving the genetic modification of organisms.

KEY POINT Although genetic manipulation must be carried out under strict containment, in accordance with appropriate legislation, the procedures involved in the isolation, amplification, recombination and cloning of DNA are often used at undergraduate level, to illustrate the general features of the techniques.

Basic principles

Gene cloning involves several steps:

1. **Isolation of the DNA sequence (gene) of interest** from the genome of an organism. This usually involves DNA purification followed by either enzymic digestion, mechanical fragmentation or PCR amplification, to liberate the target DNA sequence.

2. **Creation of an artificial recombinant DNA molecule** (sometimes referred to as rDNA), by inserting the gene into a DNA molecule capable of replicating in a host cell, i.e. a 'cloning vector'. Suitable cloning vectors for bacterial cells include plasmids, bacteriophages (bacterial viruses) and cosmids.

3. **Introduction of the recombinant DNA molecule into a suitable host,** e.g. *Escherichia coli* (transformation when a plasmid is used, or transfection for recombinant viral nucleic acid).

4. **Selection and growth of the transformed (or transfected) host cell,** using the techniques of cell culture (Chapter 48). Since a single transformed host cell can be grown to give a clone of genetically identical cells, each carrying the gene of interest, the technique is often referred to as 'gene cloning', or molecular cloning.

Definitions

Cosmids – hybrid plasmid vectors containing the cos (cohesive) sites from phage λ, enabling *in-vitro* packaging into phage capsids. Useful for cloning large segments of DNA, typically up to 50 kb.

Plasmids – circular molecules of DNA that are capable of autonomous replication. They can be isolated, manipulated and then reintroduced into bacterial cells.

Transfection – in bacteria: uptake of viral nucleic acid. In eukaryotes: uptake of any naked, foreign DNA.

Transformation – in bacteria: stable incorporation of external DNA, e.g. a plasmid. In eukaryotes: the conversion of a cell culture of finite life to a continuous (immortal) cell line (also occurs in cancer).

Extraction and purification of DNA

Specific details of the steps involved in the isolation of DNA vary, depending upon the source material. However, the following sequence shows the principal stages in the purification of plasmid DNA from bacterial cells.

Preparing glassware – all glassware for DNA purification must be siliconised, to prevent adsorption of DNA. All glass and plastic items must be sterilised before use.

1. **Cell wall digestion:** incubation of bacteria in a lysozyme solution will remove the peptidoglycan cell wall. This is often carried out under isotonic conditions, to stop the cells from bursting open and releasing

chromosomal DNA. Note that Gram-negative bacteria are relatively insensitive to lysozyme, requiring additional treatment to allow the enzyme to reach the cell wall layer, e.g. osmotic shock, or incubation with a chelating agent, e.g. ethylenediaminetetra acetic (EDTA). The latter treatment will also inactivate any bacterial deoxyribonucleases (DNases) in the solution, preventing enzymic degradation of plasmid DNA during extraction.

2. **Lysis using strong alkali (NaOH) and a detergent,** e.g. sodium dodecyl sulphate (SDS), to solubilise the cellular membranes and partially denature the proteins. Neutralisation of this solution (e.g. using potassium acetate) causes the chromosomal DNA to aggregate as an insoluble mass, leaving the plasmid DNA in solution.

3. **Removal of other macromolecules,** particularly RNA and proteins by, for example, enzymic digestion using ribonuclease and proteinase. Additional chemical purification steps give further increases in purity, e.g. proteins can be removed by mixing the extract with water-saturated phenol (50% v/v), or a phenol/chloroform mixture. On centrifugation, the DNA remains in the upper aqueous layer, while the proteins partition into the lower organic layer. Repeated cycles of phenol/chloroform extraction can be used to minimise the carry-over of these macromolecules. Additional purification can be obtained using isopycnic density gradient centrifugation in CsCl.

4. **Precipitation of DNA** using around 70% v/v ethanol: water (produced by adding two volumes of 95% v/v ethanol to one volume of aqueous extract), followed by centrifugation, to recover the DNA pellet. Further rinsing with 70% v/v ethanol: water will remove any salt contamination from the previous stages. The extracted DNA can then be redissolved in buffer solution and frozen for future use.

Separation of DNA using agarose gel electrophoresis

Electrophoresis is the term used to describe the movement of ions in an applied electrical field (as shown for DNA in Fig. 70.1). DNA molecules are negatively charged, migrating through an agarose gel towards the anode at a rate that is dependent upon molecular size – smaller, compact DNA molecules can pass through the sieve-like agarose matrix more easily than large, extended fragments. Electrophoresis of plasmid DNA is usually carried out using a submerged agarose gel. The amount of agarose is adjusted, depending on the size of the DNA molecules to be separated, e.g. 0.3% w/v agarose is used for large fragments (>20 000 bases) while 0.8% is used for smaller fragments. Note the following:

- **Individual samples are added to preformed wells using a pipettor.** The volume of sample added to each well is usually less than 25 μL, so a steady hand and careful dispensing are needed to pipette each sample.

- **The density of the samples is usually increased by adding a small amount of glycerol or sucrose,** so that each sample is retained within the appropriate well.

- **A water-soluble anionic tracking dye (e.g. bromophenol blue) is also added to each sample,** so that migration can be followed visually.

- **DNA fragment-size markers ('molecular weight' standards) are added to one or more wells.** After electrophoresis, the relative position of bands

Working with small volumes in molecular biology – use a pipettor of appropriate size (e.g. P2 or P20 Gilson Pipetman, p. 140) with a fine tip. For very small volumes, pre-wet the tip (draw up and dispense two or three times) before delivering the required volume.

Using DNA fragment-size markers – for accurate determination of fragment size (length) your standards must have the same conformation as the DNA in your sample, i.e. linear DNA standards for linear (restriction) fragments and closed circular standards for plasmid DNA.

SAFETY NOTE When using an electrophoresis tank, ensure that the top cover is in position, to prevent evaporation of buffer solution and to reduce the possibility of electric shock.

SAFETY NOTE Ethidium bromide is carcinogenic so always use gloves when handling stained gels and make sure you do not spill any staining solution, or use a safer alternative such as SYBR Safe.

SAFETY NOTE When working with any source of UV radiation, always wear suitable plastic safety glasses or goggles to protect your eyes.

Applications of PCR – these include:
- preparation of nuclei acid probes;
- DNA profiling in forensic science;
- clone screening, mapping and subcloning (Reed *et al.*, 2016).

of known size (length) can be used to prepare a calibration curve (usually, by plotting $\log_{10}$ of the molecular weight of each band against the distance travelled).

- **The gel should be run until the tracking dye has migrated across 80% of the gel** (see manufacturer's instructions for appropriate voltages/times).
- **After electrophoresis, the bands of DNA can be visualised** by soaking the gel for around 5 min in 0.1–0.5 mg L^{-1} ethidium bromide, which binds to DNA by intercalation between the paired nucleotides of the double helix.
- **Under UV light, bands of DNA are visible** due to the intense orangered fluorescence of the ethidium bromide. The limit of detection using this method is around 10 ng DNA per band. The migration of each band from the well can be measured using a ruler. Alternatively, a photograph can be taken, using a digital camera and adaptor or a dedicated system such as the BioRad GelDoc®.
- **If a particular band is required for further study** (e.g. a plasmid), the piece of gel containing that band is cut from the gel using a scalpel and the DNA extracted, e.g. using a silica membrane 'spin column'.

Identification of specific DNA molecules using Southern blotting

After electrophoretic separation in agarose gel, the fragments of DNA can be immobilised on a filter using a technique named after E.M. Southern, which has three main stages:

1. **Denaturation** – the double-stranded fragments are separated using concentrated alkali, giving single-stranded DNA (ssDNA); this is necessary to allow hybridisation with 'probe' DNA after blotting.
2. **Blotting** – a nitrocellulose filter is then placed directly on to the gel, followed by several layers of absorbent paper. The DNA is 'blotted' on to this filter as the buffer solution is pulled through.
3. **Probing** – specific sequences can be identified by incubation with labelled complementary 'probes' of single-stranded DNA, which will hybridise with a particular target sequence, followed by visualisation, often using an enzyme-based system.

DNA amplification using the polymerase chain reaction (PCR)

The polymerase chain reaction is particularly useful for amplifying small amounts of DNA, e.g. from a virus in a clinical specimen, or from a small tissue sample in forensic science. The three basic steps are:

1. **Denaturation** – double-stranded DNA (dsDNA) is separated into single strands by heating to 94–98 °C, to separate the individual DNA strands.
2. **Annealing** – oligonucleotide primers are added. These are short, synthetic DNA sequences that will hybridise at a specific position on the target DNA when the temperature is reduced to 37–65 °C.

Fig. 62.1 The polymerase chain reaction (PCR).

Assaying nucleic acids in solution – double-stranded DNA at 50 µg mL⁻¹ has an absorbance of 1 at 260 nm, and the same absorbance is obtained for singlestranded DNA at 33 µg mL⁻¹ and (singlestranded) RNA at 40 µg mL⁻¹. These values can be used to convert absorbance of a test solution to amount of nucleic acid; e.g. a solution of RNA giving A_{260} = 0.65 would contain 40×0.65 = 26 µg mL⁻¹ RNA.

3. **Primer extension** – this is the result of the action of thermostable DNA polymerase (e.g. *Taq* polymerase from *Thermus aquaticus*) at 72 °C.

Repeated cycling generates an exponentially increasing number of DNA fragments corresponding to the sequence between the two primers (see Fig. 62.1). Box 62.1 gives further practical details.

KEY POINT   PCR is so sensitive that one of the main problems associated with the technique is contamination with 'foreign' DNA. Great care is required to avoid sample contamination during *in vitro* amplification. It is good practice to carry out PCR amplification in a designated laboratory, to minimise the risk of contamination.

DNA assay

The relative amount of DNA in an agarose gel can be quantified by visual comparison of stained bands with 'marker' bands of known quantity. A more accurate method is to use band densitometry, e.g. using the GelDoc® system. A simple approach to quantifying the amount of nucleic acid in an aqueous solution is to measure the absorbance of the solution at 260 nm using a spectrophotometer, as detailed on p. 450. Note that the A_{260} value applies to purified DNA, whereas a plasmid DNA extract prepared using the protocol on p. 401 will contain a substantial amount of contaminating RNA, with similar absorption characteristics to DNA. Any contaminating protein would also invalidate the calculation; the protein can be detected by measuring the absorbance of the solution at 280 nm. Purified DNA has a value for A_{260}/A_{280} of around 1.8 and contaminating protein will give a lower ratio. If your solution gives a ratio substantially lower than this (<1.7), you should repeat the later steps of your extraction procedure (pp. 401–2).

Enzymic cleavage and ligation of DNA

Type II restriction endonucleases (commonly called restriction enzymes) recognise and cleave a particular sequence of double-stranded DNA (usually, four or six nucleotide pairs), known as the restriction site. Each enzyme is given a code name derived from the name of the bacterium from which it is isolated, e.g. *Hin* dIII was the third restriction enzyme isolated from *Haemophilus influenzae* strain Rd (Fig. 62.2). Most enzymes cleave

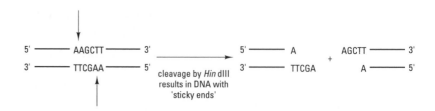

Fig. 62.2 Recognition site for the restriction enzyme *Hin* dIII. This is the conventional representation of double-stranded DNA, showing the individual bases, where A is adenine, C cytosine, G guanine and T thymine. The cleavage site on each strand is shown by an arrow.

Box 62.1 How to carry out the polymerase chain reaction (PCR)

The protocol below shows the main stages in a typical PCR: note that temperatures, incubation times and the number of cycles will vary with the particular application.

1. **Check you have the required apparatus and reagents to hand**, including (i) thermal cycler; (ii) template DNA (at least 50 ng µL^{-1}); (iii) stock solutions of deoxyribonucleoside triphosphates (dNTPs; at 5 mmol L^{-1} each); (iv) thermostable DNA polymerase (5 U µL^{-1}); (v) primers (e.g. at 30 pmol µL^{-1}); (vi) stock buffer solution, e.g. containing 100 mmol L^{-1} TRIS (pH 8.4), 500 mmol L^{-1} KCl, 15 mmol L^{-1} MgCl$_2$, 1% w/v gelatine, 1% v/v Triton X0100 (the stock buffer is often termed 10x PCR buffer).

2. **Prepare a reaction mixture for your template DNA:** for example, a mixture containing (i) 1.0 µL of template DNA; (ii) 2.5 µL of stock buffer solution; (iii)1.0 µL of each of the two primers; 1.0 µL of each of the four dNTPs; (iv) 0.1 µL of DNA polymerase; (v) 15.4 µL of sterile ultrapure water, to give a total reaction volume of 25.0 µL.

3. **Prepare appropriate positive and negative controls:** a positive control is a PCR template that is known to work under the reaction conditions used. The most commonly used negative control is the reaction mixture minus the template DNA, though negative controls can be set up lacking any one of the reaction components.

4. **Cycle in the thermal cycler:** for example, an initial period of 5 min at 94 °C, followed by 30 cycles of 94 °C for 1 min (denaturation), then 50 °C for 1 min (primer annealing), then 72 °C for 1 min (chain extension).

5. **Assess the effectiveness of the PCR:** for example, by agarose gel electrophoresis (Fig. 70.1) and SYBR Safe (or ethidium bromide) staining.

To avoid contamination in PCR:

- Use a laminar-flow cabinet (p. 285) as a dedicated workstation for PCR use, located in a separate lab from that used to carry out other DNA manipulation. For very sensitive work, use a short-wavelength UVC light inside the workstation for 20–30 min before starting work, to degrade any contaminating DNA. Make sure that you do not expose your skin or eyes to the short-wavelength UVC source (see Safety Note, p. 403).

- Keep exclusive supplies of pipettors, tips, tubes and reagents for sample preparation, reactions and product analysis.

- Autoclave all buffer solutions, pipette tips and tubes.

- Wear disposable gloves at all times and change them frequently.

Troubleshooting – common problems include:

- Lack of a PCR product (one or more components of the reaction mixture may have been left out; annealing temperature may be too high; denaturation temperature may be too low).

- Too many bands present (one or more components of the reaction mixture may be present in excess; primers may not be specific enough to the target sequence; annealing temperature may be too low).

- Bands corresponding to primer-dimers present (the primers may show partial complementarity – check sequences; primer concentration may be too high).

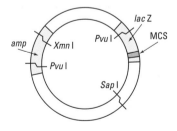

Fig. 62.3 Restriction map of the plasmid pUC19. The position of some individual restriction sites is shown together with the genes for ampicillin resistance (*bla*) and β-galactosidase (*lac Z*). MCS = multiple cloning site for 40 restriction enzymes within the *lac Z* gene.

each strand at a different position, producing short, single-stranded regions known as cohesive ends, or 'sticky ends'. A few enzymes cleave DNA to give blunt-ended fragments. Restriction enzymes that cleave DNA to give 'sticky ends' are widely used in genetic engineering, since two DNA molecules cut with the same restriction enzyme will have complementary single-stranded regions, allowing them to anneal (base pair), due to the formation of hydrogen bonds between individual bases within this region.

Restriction enzymes have two important applications in molecular genetics:

1. **Mapping.** A DNA molecule can be cleaved into several restriction fragments whose number, size and orientation relative to one another can be determined using agarose gel electrophoresis. The position of individual restriction sites can be used to create a restriction enzyme map for a particular molecule, e.g. a plasmid such as pUC19 (Fig. 62.3).

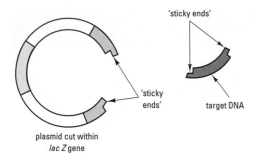

Fig. 62.4 Restriction of plasmid and foreign DNA e.g. with *Hin* dIII.

Fig. 62.5 Annealing and ligation of plasmid and target DNA to give a recombinant plasmid that confers resistance to ampicillin. As target DNA has been inserted within the *lac Z* gene, the gene is now discontinuous and inactive.

2. **Genetic engineering.** Two restriction fragments cut using the same enzyme (Fig. 62.4) and annealed by complementary base pairing can be permanently joined together using another bacterial enzyme (DNA ligase), which forms covalent bonds between the annealed strands, creating the recombinant molecule (Fig. 62.5). When the two molecules involved are the cloning vector and the target DNA, the size of the recombinant plasmid can be predicted (e.g. a plasmid of 4500 base pairs, plus a target DNA fragment of 2500 base pairs will give a recombinant molecule of 7000 base pairs), allowing separation by electrophoresis. Note that most of the plasmids used in genetic engineering code for two or more easily detectable marker genes (e.g. antibiotic resistance), with single restriction sites on each plasmid (Fig. 62.3).

Transformation/transfection of a suitable host cell

Once a recombinant plasmid has been produced *in vitro,* it must be introduced into a suitable host cell (e.g. *Escherichia coli*). Several procedures can be used:

- **Pretreatment with CaCl$_2$ at low temperature:** exponential phase cells (p. 291) are incubated with hypotonic CaCl$_2$ at 4 °C for around 30 min, followed by a brief heat shock (e.g. 42 °C for 2 min). The low-temperature incubation allows DNA to adhere to the cells, while the heat shock promotes DNA uptake.

- **Electroporation:** cells or protoplasts are subjected to electric shock treatment (typically, $> 1\,kV\ cm^{-1}$) for short periods (< 10 ms).

- **For animal and plant cells a range of techniques can be used,** including electroporation, various microinjection treatments, either using a microsyringe or microprojectiles ('biolistics'), or cationic lipids (Fig. 62.6), which form lipid–DNA complexes that fuse with cell membranes and allow DNA to enter the cell.

These treatments mostly cause a temporary increase in membrane permeability, leading to the uptake of plasmid DNA from the external medium. Such systems are often inefficient, with fewer than 0.1% of all cells showing stable transformation. However, this is not usually a significant problem, since a single viable transformant can be grown to give a large number of identical cells, using standard microbiological plating techniques (pp. 286–7).

To maximise the transformation efficiency of *E. coli* in the heat shock/ CaCl$_2$ procedure, use the minimum volume of solution in a thin-walled glass tube, so that the cells experience a rapid change in temperature.

Fig. 62.6 General structure of a synthetic cationic lipid, e.g. Tfx.

Selection and detection of transformants

As the efficiency of transformation is often very low, many of the plasmids used in genetic manipulation code for antibiotic resistance and other selectable traits, e.g. pUC19 (Fig. 62.3) carries the gene for the β-lactamase gene (*bla*) and also the gene for the enzyme β-galactosidase (*lac Z*). These genes act as markers in a single-step selection procedure. Expression of *bla* confers resistance to ampicillin, which can be demonstrated by including this antibiotic in the growth medium, thus selecting for transformants (non-transformed cells would be unable to grow on a medium containing ampicillin). The other gene is used as a marker to distinguish transformants that have acquired the recombinant plasmid from those that have taken up the original plasmid, as follows: an intact *lac Z* gene can be detected using a combination of (i) an inducer (e.g. isothiopropylgalactoside, IPTG) plus (ii) a suitable chromogenic substrate, e.g. 5-bromo-4-chloro-3-indolyl-β-D-galactoside (XGAL), which is broken down to release a blue-green indigo derivative. Thus a colony of cells containing the native plasmid would express *lac Z* and would appear blue, while the equivalent recombinant colony would have a discontinuous *lac Z* gene (Fig. 62.4) and would appear off-white, due to insertional inactivation of this gene. The multiple cloning site ('polylinker' region) within the *lac Z* gene enables use of any one of a number of different restriction enzymes, depending on the sites present in the target DNA sequence. Another useful feature of pUC19 and similar plasmids is that they are present in high copy number within a transformed cell, with several thousand identical copies per cell, giving improved yield of plasmid DNA on extraction (pp. 401–2).

Other plasmids make use of different markers; for example, the luciferase gene can be detected by bioluminescence in the presence of the substrate luciferin, or the green fluorescent protein (GFP, Fig 62.7), from the jellyfish *Aequorea victoria*. More specialised vectors are available for specialised functions, such as M13mp19, a single-stranded phage vector for DNA sequencing.

Box 62.2 provides an overview of Web-based resources for bioinformatics, while Reed *et al.* (2016) gives further practical information on additional procedures in molecular biology (including DNA sequencing, probes and hybridisation, and various forms of PCR including real-time PCR), bioinformatics (databases and tools) and proteomics (studying the expressed proteins in an organism).

Recognising transformants – after plating bacteria on to medium containing ampicillin, you may notice a few small 'feeder' or 'satellite' colonies surrounding a single larger (transformant) colony. The feeder colonies are derived from non-transformed cells that survive due to the breakdown of antibiotic in the medium around the transformant colony, and they should not be selected for subculture.

Fig. 62.7 The fluorophore of green fluorescent protein (GFP).

Box 62.2 Bioinformatics – Internet resources

Bioinformatics is a term used to describe the application of computers in biology and, in particular, to the analysis of sequence data for biopolymers such as proteins and nucleic acids. These complex molecules contain a large amount of information within their structures, and the only practical approach to understanding this is to use a computer – generally, by comparing sequence data for all or part of a specific biomolecule with that in a database. For ease of access, many bioinformatics databases and programs have been made freely accessible via the Internet.

The major databases holding primary sequence information for nucleic acids and proteins are operated by the European Bioinformatics Institute (http:// www.ebi.ac.uk/ Services/) and the National Center for Biotechnology Information (http://www.ncbi.nlm.nih.gov). These allow you to:

- find and retrieve a nucleotide or amino acid sequence from the database;

- translate a nucleotide sequence into an amino acid sequence and vice versa;

- search for similarity between a particular sequence or sequences within the database, e.g. by comparing and aligning the sequences for several proteins (or nucleic acids), to identify regions of sequence similarity (see Fig. 62.8); and

- carry out phylogenetic analysis, constructing 'ancestry trees' to show the most likely evolutionary relationships between sequences from various organisms.

Secondary databases have been created using primary sequence data, to provide information on the patterns identified within particular types of biomolecules, e.g. PROSITE (http://prosite.expasy.org/) and PRINTS (http://www.bioinf. manchester.ac.uk/dbbrowser/prints/index.php) databases, which identify protein families by their diagnostic 'signature' motifs, or the ExPaSy proteomics resource (http://.expasy. org/), for analysis of protein structure. Other websites have been constructed to bring together information for a particular organism, e.g. The *Saccharomyces* Genome Database (SGD, http://www.yeastgenome.org), The *Arabidopsis* Information Resource (https://www.arabidopsis.org/), The Caenorhabditis elegans Genome Project (http://www. sanger.ac.uk/research/ projects/caenorhabditisgenomics/) and the Human Genome Project (http://www.genome.gov/10001772). Many of these sites are rather complex and newcomers can find them a little difficult to navigate, especially if you are just browsing.

You are most likely to make use of bioinformatics databases in computer-based exercises: for example, where you are given a particular sequence and you then have to 'interrogate' a particular database, e.g. to identify similar sequences. Alternatively, you could try out the tutorials available at many of these sites.

```
H. sapiens    atgtatggcattgagaatgaagtcttcctgagccttccatgtatcctcaatgcccgggg
              ||||| ||||| ||||| |||||||||| || || || || ||||| ||||| |||||
R. norvegicus atgtacggcatcgagaacgaagtcttcctcagtctcccgtgcatccttaatgctcgggg
```

Fig. 62.8 Representative output from a BLAST (Basic Local Alignment Search Tool) alignment search for DNA sequences from part of the lactate dehydrogenase genes of a human (*Homo sapiens*) and a laboratory rat (*Rattus norvegicus*). Here, identical bases are shown in red, and non-identical bases are shown in black. This region shows 49 identical bases out of 59, i.e. 49 ÷ 59 × 100 = 83% identity (to the nearest integer).

Text reference

Reed, R.H., Holmes, D., Weyers, J.D.B. and Jones, A.M. (2016) *Practical Skills in Biomolecular Sciences*, 5th edn. Prentice Hall, Harlow.

Sources for further study

Brown, T.A. (2010) *Gene Cloning and DNA Analysis,* 6th edn. Blackwell, Oxford.

Bustin, S.A. (2009) *The PCR Revolution: Basic Technologies and Applications.* Cambridge University Press, Cambridge.

Howe, C. (2009) *Gene Cloning and Manipulation,* 2nd edn. Cambridge University Press, Cambridge.

Peterson, H. (2013) *Gene Cloning and DNA Analysis: An Introduction.* Koros Press, London.

Russell, P. (2009) *iGenetics: A Molecular Approach,* 3rd edn. Benjamin Cummings, San Francisco.

STUDY EXERCISES

62.1 **Explain the purpose of the various stages involved in the extraction and purification of DNA.** Having read through this chapter, briefly describe the main purpose of each of the following reagents within a DNA extraction protocol, and rearrange them into their order of use within the procedure:

(a) phenol : chloroform extraction;

(b) incubation in 70% v/v ethanol : water;

(c) ribonuclease and proteinase digestion;

(d) incubation with lysozyme/EDTA;

(e) sodium dodecyl sulphate incubation;

(f) NaOH treatment, followed by neutralization with potassium acetate.

62.2 **Construct a restriction map.** The figure below represents an electrophoretic separation of plasmid pARJ, digested with different restriction enzymes, both individually and in combination. All of the enzymes cut the plasmid only once. Use this information to construct a simple restriction map of the plasmid. Key to lanes:

1 = DNA standards (0.5–8 kbp);

2 = *Eco* RI;

3 = *Hin* dIII;

4 = *Pvu* II;

5 = *Eco* RI and *Hin* dIII;

6 = *Eco* RI and *Pvu* II;

7 = *Hin* dIII and *Pvu* II;

8 = *Eco* RI, *Hin* dIII and *Pvu* II;

9 = DNA standards (0.5–8 kbp).

62.3 **Calculate transformation efficiency.** A volume of 100 µL of *E. coli* cells was transformed using 20 µL of plasmid DNA (prepared at 0.1 ng µL^{-1}) and then made up to a final volume of 500 µL in buffer. A ten-fold dilution of this suspension was prepared and 200 µL of this dilution was surface-spread (p. 287) on to a suitable medium containing an antibiotic to select for transformants, giving 180 colonies on the spread plate after overnight incubation. What is the transformation efficiency, expressed as the number of transformants (colony-forming units, CFU) per µg DNA? (Give your answer in exponential notation, to three significant figures.)

Answers to these study exercises are available at www.pearsoned.co.uk/practicalskills

Fig. 63.1 Components of an electron impact mass spectrometer.

Fig. 63.2 Mass spectrum for methanol. m/z = charge ratio.

Stable isotopes of an element all have the same number of protons, but different numbers of neutrons. In contrast to radioisotopes (Chapter 64), they do not spontaneously decay to produce radiation. Stable isotopes have virtually identical gross chemical properties. When two or more stable isotopes of the same element exist in nature, they each contribute to the relative atomic mass (A_r) in direct proportion to their relative abundance.

Given their broad chemical equivalence, stable isotopes have a narrow range of practical applications in biology, and the techniques used to measure them are highly specialised and sensitive; they include mass spectrometry (MS) and nuclear magnetic resonance (NMR) spectrometry.

Measuring stable isotopes by mass spectrometry

Typically, stable isotopes are separated and quantified in terms of their mass-to-charge ratio using some form of mass spectrometry. This technique involves the vaporisation/ionisation of a sample in a gas phase. The ions are then accelerated to specific velocities using an electric field and then separated on the basis of their different masses and charges (Fig. 63.1). Each ion of a particular mass-to-charge ratio is detected sequentially with time. The most widely used method in MS is electron impact (EI) ionisation, where the electron source is a heated metal, such as a tungsten filament, subjected to an appropriate potential gradient. The stream of electrons emitted can then interact with elements and molecules in the sample, by either:

- **electron removal** – where an electron in a bond within the sample molecule is 'knocked out' by bombarding electrons, producing a positively changed ion (cation); or
- **electron capture** – where addition of an electron results in the production of a negatively charged ion (anion).

The ions are then subjected to a magnetic field (Fig. 63.1) which results in their detection in proportion to their abundance in the sample. Signal processing and analysis then give a mass spectrum, such as that shown in Fig. 63.2, which can be used to characterise and identify the compound.

For metabolic studies, the technique of isotope ratio mass spectrometry (IRMS) is very useful, since it removes the need to use radioactive isotopes. The technique exploits the ability of MS to distinguish between isotopes such as ^{13}C and ^{12}C. For example, a compound containing ^{13}C will have a greater overall mass than the same compound containing ^{12}C and can therefore be differentiated within the mass spectrum produced as a result of MS. By selectively labelling a key metabolite with a stable isotope that is naturally present at low concentration (e.g. ^{13}C), it is then possible to follow the fate of the metabolite by MS analysis of sequential samples. Typically, the metabolites would be combusted in an oxygen atmosphere, to give gaseous $^{13}CO_2$, followed by exposure of such gases to an EI source (Fig. 63.1); useful metabolic data can be obtained from the isotope ratio – in this instance $^{13}CO_2$ and $^{12}CO_2$.

Using stable carbon isotopes in photosynthetic studies

Measurements of stable carbon isotopes in organic material and in the air passing over photosynthetically active plant material can be used as a probe of photosynthetic physiology. 'Light' and 'heavy' isotopes of carbon (^{12}C and ^{13}C respectively) have different diffusion rates within the leaf and the key photosynthetic enzyme Rubisco (ribulose bisphosphate carboxylase oxygenase) also discriminates against ^{13}C, leading to changes in the isotope composition of photo-assimilated carbon compared to that of air. By measuring this discrimination, the physiological processes associated with carbon fixation and metabolism can be investigated in a time-integrated manner. Analysis of stable isotope composition makes use of mass spectrometric measurements of the ratio of heavy-to-light isotopes in the sample under investigation and in a predefined standard.

The mass spectrometer compares the mass-to-charge ratio *(m/z)* of the 44 and 45 masses of $^{12}CO_2$ and $^{13}CO_2$ respectively, to give the isotopic composition of (i) a sample and (ii) a standard, in terms of the ^{13}C molar abundance ratio (R), where:

$$R = \frac{[^{13}CO_2]}{[^{12}CO_2]} \qquad [63.1]$$

These values are then used to calculate the isotopic composition of the sample relative to the standard as a $\delta^{13}C$ value, usually expressed in parts per thousand (‰ – note that this is a proportion, and *not* a unit – spoken as 'per mil'). The original standard was CO_2 derived from a fossil <u>Be</u>lemnite from the <u>Pee</u> <u>Dee</u> deposits in South Carolina (PDB standard: $R = 0.01124$). The $\delta^{13}C$ value of a sample in parts per thousand is then calculated as:

$$\delta^{13}C = \frac{R_{\text{sample}} - R_{\text{standard}}}{R_{\text{standard}}} \times 1000 \qquad [63.2]$$

Isotopic composition values may also be used to calculate discrimination (Δ), which takes into account the isotopic composition of the standard, source and sample, according to the following equation:

$$\Delta = \frac{\delta_a - \delta_p}{1 + \delta_p} \qquad [63.3]$$

where δ_a and δ_p are the carbon isotope discrimination values for air (source) and plant (sample) respectively, expressed in fractional terms (see worked examples in margin). In contrast to $\delta^{13}C$ values which are negative when PDB is used as the standard, values for Δ are positive for plant material, since ^{12}C is favoured over ^{13}C. The advantage in using Δ notation is that it is more straightforward to make comparisons, avoiding the possible confusion involved in discussing differences in negative $\delta^{13}C$ values. Like $\delta^{13}C$ values, Δ values are usually expressed in parts per thousand (‰). Values for Δ values in C_3 plants, where Rubisco is the main carboxylating enyme, usually fall within the range 16–25‰, since Rubisco shows substantial discrimination against ^{13}C. In contrast, Δ values for C_4 plants, where PEP carboxylase is the main carboxylating enzyme, are lower – typically in the range 3–7‰ – since PEP carboxylase shows minimal discrimination against ^{13}C.

Advances in the study of hydrogen, oxygen and nitrogen isotope ratios have further extended the range of tools available to physiologists to investigate areas including plant water relations and nitrogen economy at

Example For air, with $[^{13}CO_2]$ at 0.144 µmol L^{-1} and $[^{12}CO_2]$ at 12.915 µmol L^{-1}, using eqn [63.1] gives a molar abundance ratio, $R = 0.144 \div 12.915 = 0.01115$ (to four significant figures).

Example For air, where $R = 0.01115$, $\delta^{13}C = (0.01115 - 0.01124) \div 0.01124 \times 1000 = -8.0‰$ (to one decimal place). $\delta^{13}C$ for plant tissue where Rubisco is the main carboxylating enzyme is *more negative* than $\delta^{13}C$ for C_4 plant tissue, since Rubisco discriminates strongly against ^{13}C (e.g. $R = 0.01095$, $\delta^{13}C = -25.8‰$). $\delta^{13}C$ for C_4 plant tissue where PEP carboxylase is the main carboxylating enzyme is *less negative* than $\delta^{13}C$ for C_3 plant tissue, since this enzyme shows little discrimination against ^{13}C, and the main effect is due to diffusion within the leaf (e.g. $R = 0.01111$, $\delta^{13}C = -11.6‰$).

Example Using eqn [63.3], Δ for C_3 plant tissue with a $\delta^{13}C$ of $-25.8‰$ is $[-0.008 - (-0.0258)] \div [1 + (-0.0258)] = 0.0182714 = 18.2‰$.
Using eqn [63.3], Δ for C_4 plant tissue with a $\delta^{13}C$ of $-11.5‰$ is $[-0.008 - (-0.0115)] \div [1 + (-0.0115)] = 0.0035407 = 3.5‰$.

(a)

(b)

Fig. 63.3 Effect of an applied magnetic field, B_0, on magnetic nuclei. (a) Nuclei in magnetic field have one of two orientations – either with the field or against the field (in the absence of an applied field, the nuclei would have random orientation). (b) Energy diagram for magnetic nuclei in applied magnetic field.

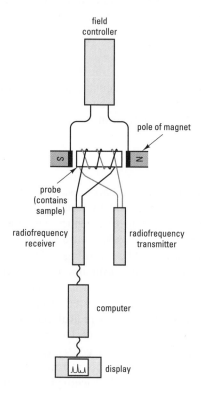

Fig. 63.4 Components of an NMR spectrometer.

the individual plant and ecosystem levels. Since the stable carbon isotope ratio of a plant is a broad measure of the overall balance between carbon uptake (photosynthesis) and water loss (transpiration), Δ is found to vary with changes in transpiration rate, with a lower value corresponding to a higher water use efficiency. It has been proposed that concurrent analysis of $^{18}O/^{16}O$ ratios can enable researchers to interpret variations in Δ in terms of water-use efficiency: this can be useful in plant breeding studies, where improved water-use efficiency results in closely coupled ^{13}C and ^{18}O signals.

NMR analysis of stable isotopes

Electromagnetic radiation at radiofrequencies of 1–500 MHz can be used to identify and monitor biological compounds and metabolic processes. This is possible because of differences in the magnetic states of atomic nuclei of different stable isotopes, involving very small transitions in energy level. The atomic nuclei of isotopes of many elements are magnetic, because they are charged and have spin. Typical stable isotopes that are magnetic include ^{1}H, ^{13}C ^{14}N, ^{15}N, ^{19}F and ^{31}P. When these nuclei interact with a uniform external magnetic field, they behave like tiny compass needles and align themselves in a direction either (i) parallel or (ii) antiparallel to the external magnetic field. The two orientations have different energies, with the parallel direction having a lower energy than the antiparallel direction (Fig. 63.3). The energy difference between these two levels corresponds to a precise electromagnetic frequency.

When a sample containing an isotope with a magnetic nucleus is placed in a magnetic field and exposed to an appropriate radiofrequency in an NMR spectrometer (Fig. 63.4), transitions between the energy levels of magnetic nuclei will occur when the energy gap and the applied frequency are in *resonance* (i.e. when they are matched exactly). This frequency is a characteristic of the particular isotope and can be used to study the isotope in biological systems. Typically, this is described in terms of the 'chemical shift' of the nuclei involved, which is a field-independent dimensionless value measured against the resonance frequency of a reference compound (for example, tetramethylsilane for ^{1}H and ^{13}C NMR), as:

$$\frac{\text{resonance frequency of sample} - \text{resonance frequency of reference}}{\text{resonance frequency of the reference}} \qquad [63.4]$$

Because the difference in frequency (numerator in [63.4]) is very small in relation to the resonance frequency of the reference (denominator in [63.4]), chemical shifts are usually expressed in parts per million (p.p.m.).

The resonance frequency of magnetic nuclei is affected by the presence of shielding due to small secondary fields generated by electrons in their vicinity, giving rise to different chemical shifts for the same isotope in different parts of a molecule. By convention, the chemical shift is positive if the nucleus is less shielded than the reference compound and negative if it is more shielded. The end result is an NMR spectrum where individual magnetic nuclei produce a 'signal' with peaks representing the different locations within a molecule.

Table 63.1 The relative merits and disadvantages of various magnetic nuclei in biomolecular studies

Nucleus	Relative sensitivity	Natural abundance	Comments
^1H	100	99.98%	Multiple, but specific, spectral lines are obtained for individual biomolecules. For mixtures, the ubiquitous occurrence of ^1H gives complex, overlapping signals that are often difficult to interpret. Gives a large solvent peak with aqueous samples (can be avoided by using D_2O as solvent). Mainly used for structural studies of pure macromolecules. Essential for MRI.
^{31}P	6.6	100%	Very useful for studies on living systems, with narrow resonance peaks and a wide range of chemical shifts for different molecules. Spectra are simpler and easier to interpret than for ^1H, but are not as distinctive: different compounds may give similar ^{31}P spectra. Several important P–containing compounds (including ATP, ADP and inorganic phosphate) can be detected in intact cells – useful in bioenergetic studies.
^{13}C	0.016	1.1%	Gives narrow signals and a wide range of chemical shifts. Resolution is better than for ^1H, and a wide range of organic biomolecules can be detected. Low natural abundance gives low sensitivity, extending the time required to accumulate spectra. However, low natural abundance also means that specific metabolites can be selected for ^{13}C isotope enrichment, allowing particular metabolic pathways to be investigated, e.g. carbon assimilation.

Biomolecular applications of NMR

The sensitivity of NMR has improved dramatically with the development of more powerful magnets. Details of the major uses of various magnetic nuclei are given in Table 63.1. The principal applications include:

- **studies of the structure and function of macromolecules and biological systems**, such as membranes;
- **metabolic investigations on living organisms**, including humans, since NMR can be used to obtain a 'fingerprint' of a particular molecule, and changes in the intensity of spectra can be used for kinetic studies (e.g. Fig. 63.5). This involves use of 'surface coils' as sources of radiation, or placing the organism within the core of an electromagnet;

Using NMR – in contrast to most conventional metabolic studies, NMR is non-invasive and the time course of metabolic reactions can be followed using a single experimental subject or preparation, eliminating variation between samples.

Fig. 63.5 Typical ^{31}P NMR spectrum from intact skeletal muscle (a) at rest and (b) after a fatiguing series of contractions. In (b) ATP levels are preserved at the expense of phosphocreatinine (PCr). Also, hydrolysis of ATP during contraction results in a large phosphate (P$_i$) peak which is shifted to the right compared with (a), reflecting the decrease in intracellular pH that accompanies glycolysis.

Fig. 63.6 A ¹³C NMR spectrum of *Himanthalia elongata*, with asterisks showing the peaks corresponding to individual nuclei within the carbohydrate altritol. m = mannitol
(courtesy J.A. Chudek).

- **measurement of intracellular pH** by determination of the chemical shift of the phosphate peak, as this changes with pH in a predictable manner;

- **magnetic resonance imaging (MRI)**, which is a form of proton NMR that uses a field gradient (as opposed to a uniform field) to produce signals that are translated by computers into anatomical images (Chary and Govil, 2008). Body tissues have different water contents (typically 60–90% w/w). Signals arising from protons in water in different tissues can be used to differentiate between tissues (e.g. 'grey' and 'white' matter in the brain), and between normal tissue and tumours. Fat deposits can be detected, due to the difference between the ¹H signal from fatty acids and that of the water in the surrounding tissue.

¹³C NMR analysis is particularly useful in photosynthetic studies, for example, in identifying the most abundant organic low molecular weight solutes in plants, and for studying how the concentrations of these solutes are adjusted in response to salt and water stress. The relative insensitivity of this approach, with ¹³C representing only 1.1% of the naturally occurring carbon atoms in biomolecules, becomes an advantage in such studies. Only those soluble organic molecules that are present in large quantities within a cell will be detected, giving a spectrum that typically shows signals from only one or two solutes. As an example, Fig. 63.6 shows a natural abundance ¹³C NMR spectrum of the seaweed *Himanthalia elongata*, which was the first report of the carbohydrate altritol in any living organism (Chudek *et al.*, 1984) – previous chromatographic studies had failed to separate this compound from other isomers, such as mannitol. The relatively low natural abundance of ¹³C also means that 'labelling' studies can be carried out, e.g. using ¹³CO₂, to follow the metabolism of photosynthetically fixed C within the plant.

Text references

Chary, K.V.R. and Govil, G. (2008) *NMR in Biological Systems: from Molecules to Humans.* Springer, Dordrecht.

Chudek, J.A., Footer, R., Davison, I.R. and Reed, R.H. (1984) Altritol in the brown alga *Himanthalia elongata. Phytochemistry* 23, 1081–1082.

Sources for further study

Keeler, J. (2010) *Understanding NMR Spectroscopy*, 2nd edn. Wiley, New York.

Smith, R.M. and Butsch, K.L. (2004) *Understanding Mass Spectra: A Basic Approach.* Wiley, New York.

Westbrook, C. and Roth, C. (2011) *MRI in Practice,* 4th edn. Blackwell, Oxford.
[Covers the basics of magnetic resonance imaging in medicine.]

STUDY EXERCISES

63.1 **Test your understanding of the terminology of stable isotopes.** What do the following terms/acronyms mean? (a) NMR; (b) MRI; (c) EI; (d) chemical shift; (e) electron capture.

63.2 **Explain the operating principles of a nuclear magnetic resonance (NMR) spectrometer.** Make sure that your explanation covers the major components of an NMR spectrometer.

63.3 **Calculate the molar abundance ratio ($R = [^{13}CO_2]/[^{12}CO_2]$ and $\delta^{13}C$ [$R_{sample} - R_{standard}$]/$R_{standard} \times 1000$) of a sample.** Determine R and $\delta^{13}C$ for an air sample with $[^{13}CO_2] = 0.152$ μmol L^{-1} and $[^{12}CO_2]$ at 12.825 μmol L^{-1}, and PDB standard of $R = 0.011\ 24$.

Answers to these study exercises are available at www.pearsoned.co.uk/practicalskills

The isotopes of a particular element have the same number of protons in the nucleus but a different number of neutrons, giving them the same proton number (atomic number) but a different nucleon number (mass number, i.e. number of protons + number of neutrons). Isotopes may be stable (Chapter 63) or radioactive. Radioactive isotopes (radioisotopes) disintegrate spontaneously at random to yield radiation and a decay product.

Radioactive decay

There are three forms of radioactivity (Table 64.1) arising from three main types of nuclear decay:

- **Alpha decay.** This involves the loss of a particle equivalent to a helium nucleus. Alpha (α) particles, being large and positively charged, do not penetrate far in living tissue, but they do cause ionisation damage and this makes them generally unsuitable for tracer studies.

- **Beta decay.** This involves the loss or gain of an electron or its positive counterpart, the positron. There are three subtypes:

  (a) Negatron (β^-) emission: loss of an electron from the nucleus when a neutron transforms into a proton. This is the most important form of decay for radioactive tracers used in biology. Negatron-emitting isotopes of biological importance include ^3H, ^{14}C, ^{32}P and ^{35}S.

  (b) Positron (β^+) emission: loss of a positron when a proton transforms into a neutron. This only occurs when sufficient energy is available from the transition and may involve the production of gamma rays when the positron is later annihilated by collision with an electron.

  (c) Electron capture (EC): when a proton 'captures' an electron and transforms into a neutron. This may involve the production of X-rays as electrons 'shuffle' about in the atom (as with ^{125}I) and it frequently involves electron emission.

- **Gamma emission.** Internal transition involves the emission of electromagnetic radiation in the form of gamma (γ) rays from a nucleus in a metastable state and always follows initial alpha or beta decay. Emission of gamma radiation leads to no further change in atomic number or mass.

Note from the above that more than one type of radiation may be emitted when a radioisotope decays. The main radioisotopes used in biology and their properties are listed in Table 64.2.

Examples $^{12}_{6}C$, $^{13}_{6}C$ and $^{14}_{6}C$ are three of the isotopes of carbon. About 98.9% of naturally occurring carbon is in the stable $^{12}_{6}C$ form. $^{13}_{6}C$ is also a stable isotope but it only occurs at 1.1% natural abundance. Trace amounts of radioactive $^{14}_{6}C$ are found naturally; this is a negatron-emitting radioisotope (see Table 64.2).

Examples ^{226}Ra decays to ^{222}Rn by loss of an alpha particle, as follows:

$$^{226}_{88}Ra \rightarrow \, ^{222}_{86}Rn + \, ^{4}_{2}He^{2+}$$

^{14}C shows beta decay, as follows:

$$^{14}_{6}C \rightarrow \, ^{14}_{7}N + \beta^-$$

^{22}Na decays by positron emission, as follows:

$$^{22}_{11}Na \rightarrow \, ^{22}_{10}Ne + \beta^+$$

^{55}Fe decays by electron capture and the production of an X-ray, as follows:

$$^{55}_{26}Fe \rightarrow \, ^{55}_{25}Mn + X$$

The decay of ^{22}Na by positron emission (β^+) leads to the production of a γ ray when the positron is annihilated on collision with an electron.

Table 64.1 Types of radioactivity and their properties

Radiation	Range of maximum energies (MeV*)	Penetration range in air (m)	Suitable shielding material
Alpha (α)	4–8	0.025–0.080	Unnecessary
Beta (β)	0.01–3	0.150–16	Plastic (e.g. Perspex)
Gamma (γ)	0.03–3	1.3–13†	Lead

* Note that $1\,MeV = 1.6 \times 10^{-13}\,J$.
† Distance at which radiation intensity is reduced to half.

Table 64.2 Properties of some isotopes used commonly in biology. Physical data obtained from Haynes (2014)

Isotope	Emission(s)	Maximum energy (MeV)	Half-life	Main uses	Advantages	Disadvantages
^{3}H	β^-	0.01861	12.3 years	Suitable for labelling organic molecules in a wide range of positions at high specific activity	Relatively safe	Low efficiency of detection, high isotope effect, high rate of exchange with environment
^{14}C	β^-	0.15648	5715 years	Suitable for labelling organic molecules in a wide range of positions	Relatively safe, low rate of exchange with environment	Low specific activity
^{22}Na	β^+ (90%) + γ, EC	2.842 (β^+)	2.60 years	Transport studies	High specific activity	Short half-life, hazardous
^{32}P	β^-	1.710	14.3 days	Labelling proteins and nucleotides (e.g. DNA)	High specific activity, ease of detection	Short half-life, hazardous
^{35}S	β^-	0.167	87.2 days	Labelling proteins and nucleotides	Low isotope effect	Low specific activity
^{36}Cl	β^-, β^+, EC	0.709 (β^-) 1.142 (β^+, EC)	300 000 years	Transport studies	Low isotope effect	Low specific activity, hazardous
^{125}I	EC + γ	0.178 (EC)	59.9 days	Labelling proteins and nucleotides	High specific activity	Hazardous
^{131}I	β^- + γ	0.971 (β^-)	8.04 days	Labelling proteins and nucleotides	High specific activity	Hazardous

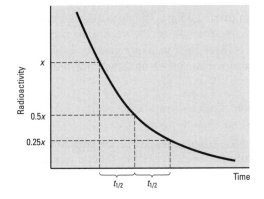

Fig. 64.1 Decay of a radioactive isotope with time. The time taken for the radioactivity to decline from x to $0.5x$ is the same as the time taken for the radioactivity to decline from $0.5x$ to $0.25x$, and so on. This time is the half-life ($t_{1/2}$) of the isotope.

Example For ^{35}S, with a half-life of 87.2 days (Table 64.2), the fraction remaining after 28 days would be worked out as follows:

$x = (-0.693 \times 28) \div 87.2 = -0.222522936$,

then using eqn [64.1],

$f = e^{-0222522936} = 0.800496646$

(approximately 80.0% of original activity).

Each radioactive particle or ray carries energy, usually measured in electron volts (eV). The particles or rays emitted by a particular radioisotope exhibit a range of energies, termed an energy spectrum, characterised by the maximum energy of the radiation produced, E_{max} (Table 64.2). The energy spectrum of a particular radioisotope is relevant to the following:

- **Safety:** isotopes with the highest maximum energies will have the greatest penetrating power, requiring appropriate shielding (Table 64.1).
- **Detection:** different instruments vary in their ability to detect isotopes with different energies.
- **Discrimination:** some instruments can distinguish between isotopes, based on the energy spectrum of the radiation produced (p. 420).

The decay of an individual atom (a 'disintegration') occurs at random, but that of a population of atoms occurs in a predictable manner. The radioactivity decays exponentially, having a characteristic half-life ($t_{1/2}$). This is the time taken for the radioactivity to fall from a given value to half that value (Fig. 64.1). The $t_{1/2}$ values of different radioisotopes range from fractions of a second to more than 10^{19} years (see also Table 64.2). If the $t_{1/2}$ is very short, as with ^{15}O ($t_{1/2} \approx 2$ min), then it is generally impractical to use the isotope in experiments because you would need to account for the decay during the experiment and counting period.

To calculate the fraction (f) of the original radioactivity left after a particular time (t), use the following relationship:

$$f = e^x, \text{ where } \quad x = -0.693t/t_{1/2} \qquad [64.1]$$

Note that the same units must be used for t and $t_{1/2}$ in the above equation.

Table 64.3 Relationships between units of radioactivity. For abbreviations, see text

1 Bq = 1 d.p.s.	1 Sv = 100 rem
1 Bq = 60 d.p.m.	1 Gy = 100 rad
1 Bq = 27 pCi	1 Gy ≈ 100 roentgen
1 d.p.s. = 1 Bq	1 rem = 0.01 Sv
1 d.p.m. = 0.0167 Bq	1 rad = 0.01 Gy
1 Ci = 37 GBq	1 roentgen ≈ 0.01 Gy
1 mCi = 37 MBq	
1 µCi = 37 kBq	

Example If 0.4 mL of a ^{32}P-labelled DNA solution at a concentration of 50 µmol L^{-1} (amount = 0.4 × 50 ÷ 1000 = 0.02 µmol L^{-1}) gave a count of 2490 d.p.m. (= 41.5 Bq), using eqn [64.2] this would correspond to a specific activity of 2490 ÷ 0.02 = 124 500 d.p.m. µmol L^{-1} (or 2075 Bq µmol L^{-1}).

Fig. 64.2 Components of a Geiger–Müller (G–M) tube.

Fig. 64.3 Components of a scintillation counter. Note that in most modern instruments, all components are enclosed within a single cabinet.

Measuring radioactivity

The SI unit of radioactivity is the becquerel (Bq), equivalent to one disintegration per second (d.p.s.), but disintegrations per minute (d.p.m.) are also used. The curie (Ci) is a non-SI unit equivalent to the number of disintegrations produced by 1 g of radium (37 GBq). Table 64.3 shows the relationships between these units. In practice, most instruments are not able to detect all of the disintegrations from a particular sample, i.e. their efficiency is less than 100% and the rate of decay may be presented as counts min^{-1} (c.p.m.) or counts s^{-1} (c.p.s.). Most modern instruments correct for background radiation and inefficiencies in counting, converting count data to d.p.m. Alternatively, the results may be presented as the measured count rate, although this is only valid where the efficiency of counting does not vary greatly among samples.

> **KEY POINT** The specific activity is a measure of the quantity of radioactivity present in a known amount of the substance.
>
> $$\text{specific activity} = \frac{\text{radioactivity (Bq, Ci, d.p.m., etc.)}}{\text{amount of substance (mol, g, etc.)}} \qquad [64.2]$$

This is an important concept in practical work involving radioisotopes, since it allows interconversion of disintegrations (activity) and amount of substance (see Box 64.1).

Two SI units refer to doses of radioactivity and these are used when calculating exposure levels for a particular source. The sievert (Sv) is the amount of radioactivity giving a dose in man equivalent to 1 gray (Gy) of X-rays: 1 Gy = an energy absorption of 1 J kg^{-1}. The dose received in most biological experiments is a negligible fraction of the maximum permitted exposure limit. Conversion factors from older units are given in Table 64.3.

The most important methods of measuring radioactivity for biological purposes are described below.

The Geiger–Müller (G–M) tube

This operates by detecting radiation when it ionises gas between a pair of electrodes across which a voltage has been applied (Fig. 64.2). You should use a hand-held Geiger–Müller tube for routine checking for contamination (although it will not pick up ^{3}H activity).

The scintillation counter

This operates by detecting the scintillations (fluorescence 'flashes') produced when radiation interacts with certain chemicals called fluors (Fig. 64.3). In solid (or external) scintillation counters (often referred to as 'gamma counters') the radioactivity causes scintillations in a crystal of fluorescent material held close to the sample. This method is only suitable for radioisotopes producing penetrating radiation.

Liquid scintillation counters are mainly used for detecting beta decay and they are especially useful in biology. The sample is dispersed or dissolved in a suitable solvent containing the fluor(s) – the 'scintillation cocktail'. The radiation first interacts with the solvent, and the energy from this interaction is passed to the fluors, which produce detectable light. The scintillations are measured by photomultiplier tubes (Fig. 64.3) that turn the light pulses into electronic pulses, the magnitude of which is directly related to the energy of the original radioactive event. The spectrum of electronic pulses is thus related to the energy spectrum of the radioisotope.

Box 64.1 How to determine the specific activity of an experimental solution

Suppose you need to make up a certain volume of an experimental solution, to contain a particular amount of radioactivity. For example, 50 mL of a mannitol solution at a concentration of 25 mmol L^{-1}, to contain 5 Bq μL^{-1} – using a manufacturer's stock solution of ^{14}C-labelled mannitol (specific activity = 0.1 Ci/mmol^{-1}).

1. Calculate the total amount of radioactivity in the experimental solution, in this example 5×1000 (to convert μL to mL $\times 50$ (50 mL required) = 2.5×10^5 Bq (i.e. 250 kBq).

2. Establish the volume of stock radioisotope solution required: for example, a manufacturer's stock solution of ^{14}C-labelled mannitol contains 50 μCi of radioisotope in 1 mL of 90% v/v ethanol: water. Using Table 64.3, this is equivalent to an activity of $50 \times 37 = 1850$ kBq. So, the volume of solution required is 250/1850 of the stock volume, i.e. 0.1351 mL (135 μL).

3. Calculate the amount of non-radioactive substance required as for any calculation involving concentration (see pp. 147, 156), e.g. 50 mL (0.05 L) of a 25 mmol L^{-1} (0.025 mol L^{-1}) mannitol (relative molecular mass 182.17) will contain $0.05 \times 0.025 \times 182.17 = 0.2277 g$.

4. Check the amount of radioactive isotope to be added. In most cases, this represents a negligible amount of substance, e.g. in this instance, 250 kBq of stock solution at a specific activity of 14.8×10^6 kBq/mmol^{-1} (converted from 0.4 Ci mmol^{-1} using Table 64.3) is equal to $250/14\,800\,000 = 16.89$ nmol, equivalent to approximately 3 μg mannitol. This can be ignored in calculating the mannitol concentration of the experimental solution.

5. Make up the experimental solution by adding the appropriate amount of non-radioactive substance and the correct volume of stock solution.

6. Measure the radioactivity in a known volume of the experimental solution. If you are using an instrument with automatic correction to Bq, your sample should contain the predicted amount of radioactivity, e.g. an accurately dispensed volume of 100 μL of the mannitol solution should give a corrected count of $100 \times 5 = 500$ Bq (or $500 \times 60 = 30000$ d.p.m.).

7. Note the specific activity of the experimental solution: in this case, 100 μL (1×10^{-4} L) of the mannitol solution at a concentration of 0.025 mol L^{-1} will contain 25×10^{-7} mol (2.5 μmol) mannitol. Dividing the radioactivity in this volume (30000 d.p.m.) by the amount of substance (eqn [64.2]) gives a specific activity of $30000/2.5 = 12000$ d.p.m. μmol^{-1}, or 12 d.p.m. nmol^{-1}. This value can be used:

   (a) To assess the accuracy of your protocol for preparing the experimental solution: if the measured activity is substantially different from the predicted value, you may have made an error in making up the solution.

   (b) To determine the counting efficiency of an instrument: by comparing the measured count rate with the value predicted by your calculations.

   (c) To interconvert activity and amount of substance: the most important practical application of specific activity is the conversion of experimental data from counts (activity) into amounts of substance. This is only possible where the substance has not been metabolised or otherwise converted into another form; e.g. a tissue sample incubated in the experimental solution described above with a measured activity of 245 d.p.m. can be converted to nmol mannitol by dividing by the specific activity, expressed in the correct form. Thus $245/12 = 20.417$ nmol mannitol.

Correcting for quenching – find out how your instrument corrects for quenching and check the quench indication parameter (QIP) on the printout, which measures the extent of quenching of each sample. Large differences in the QIP would indicate that quenching is variable among samples and might give you cause for concern.

Modern liquid scintillation counters use a series of electronic 'windows' to split the pulse spectrum into two or three components. This may allow more than one isotope to be detected in a single sample, provided their energy spectra are sufficiently different (Fig. 64.4). A complication of this approach is that the energy spectrum can be altered by pigments and chemicals in the sample, which absorb scintillations or interfere with the transfer of energy to the fluor; this is known as quenching (Fig. 64.4). Most instruments have computer-operated quench-correction facilities (based on measurements of standards of known activity and energy spectrum) that correct for such changes in counting efficiency.

Many liquid scintillation counters treat the first sample as a 'background', subtracting whatever value is obtained from the subsequent measurements as part of the procedure for converting to d.p.m. If not, you will need to

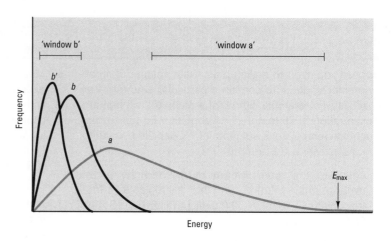

Fig. 64.4 Energy spectra for three radioactive samples, detected using a scintillation counter. Sample *a* is a high-energy β-emitter while *b* contains a low-energy β-emitter, giving a lower spectral range. Sample *b'* contains the same amount of low-energy β-emitter, but with quenching, shifting the spectral distribution to a lower energy band. The counter can be set up to record disintegrations within a selected range (a 'window'). Here, 'window a' could be used to count isotope *a* while 'window b' could give a value of isotope *b*, by applying a correction for the counts due to isotope *a*, based on the results from 'window a'. Dual counting allows experiments to be carried out using two isotopes (double labelling).

Liquid scintillation counting of high-energy β-emitters – β-particles with energies greater than 1 MeV can be counted in water (Čerencov radiation), with no requirement for additional fluors (e.g. ^{32}P).

subtract the background count from all other samples. Make sure that you use an appropriate background sample, identical in all respects to your radioactive sample but with no added radioisotope, in the correct position within the machine. Check that the background reading is reasonable (15–30 c.p.m.). Tips for preparing samples for liquid scintillation counting are given in Box 64.2.

Gamma-ray (γ-ray) spectrometry

This is a method by which a mixture of γ-ray-emitting radionuclides can be resolved quantitatively by pulse-height analysis. It is based on the fact that pulse heights (voltages) produced by a photomultiplier tube are proportional to the amounts of γ-ray energy arriving at the scintillant or a lithium-drifted germanium detector. The lithium-drifted germanium detector, which is abbreviated to Ge(Li) – pronounced 'jelly' – provides high resolution (narrow peaks), essential in the analysis of complex mixtures.

Autoradiography

This is a method where photographic film is exposed to the isotope. It is used mainly to locate radioactive tracers in thin sections of an organism or on chromatography papers and gels, but quantitative work is possible. The radiation interacts with the film in a similar way to light, silver grains being formed in the developed film where the particles or rays have passed through. The radiation must have enough energy to penetrate into the film, but if it has too much energy the grain formation may be too distant from the point where the isotope was located to identify precisely the point of origin (e.g. high-energy β-emitters). Autoradiography (Fig. 64.5) is a relatively specialised method and individual lab protocols should be followed for particular isotopes/applications.

Fig. 64.5 Autoradiograph showing the localisation of a growth factor in epithelial cells of rat kidney tubules (scale bar is 10 μm). The black-silver grains are due to radioactivity from ^{125}I-labelled growth factor. From Alexander *et al.* (1998), with permission.

Box 64.2 Tips for preparing samples for liquid scintillation counting

Modern scintillation counters are very simple to operate; problems are more likely to be due to inadequate sample preparation than to incorrect operation of the machine. Common pitfalls are the following:

- Incomplete dispersal of the radioactive compound in the scintillation cocktail. This may lead to underestimation of the true amount of radioactivity present:

  (a) Water-based samples may not mix with the scintillation cocktail – change to an emulsifier-based cocktail. Take care to observe the recommended limits, upper and lower, for amounts of water to be added or the cocktail may not emulsify properly.

  (b) Solid specimens may absorb disintegrations or scintillations: extract radiochemicals using an intermediate solvent like ethanol (ideally within the scintillation vial) and then add the cocktail. Tissue-solubilising compounds such as Soluene are effective, particularly for animal material, but extremely toxic, so the manufacturer's instructions must be followed closely. Radioactive compounds on slices of agarose or polyacrylamide gels may be extracted using a product such as Protosol. Agarose gels can be dissolved in a small volume of boiling water.

  (c) Particulate samples may sediment to the bottom of the scintillation vial – suspend them by forming a gel. This can be done with certain emulsifier-based cocktails by adding a specific amount of water.

- Chemiluminescence. This is where a chemical reacts with the fluors in the scintillation cocktail causing spurious scintillations, a particular risk with solutions containing strong bases or oxidising agents. Symptoms include very high initial counts that decrease through time. Possible remedies are:

  (a) Leave the vials at room temperature for a time before counting. Check with a suitable blank that counts have dropped to an acceptable level.

  (b) Neutralise basic samples with acid (e.g. acetic acid or HCl).

  (c) Use a scintillation cocktail that resists chemiluminescence such as Hiconicfluor.

  (d) Raise the energy of the lower counts detected to about 8 keV – most chemiluminescence pulses are weak (0–7 keV). This approach is not suitable for weak emitters, e.g. ^{3}H.

Biological applications for radioactive isotopes

The main advantages of using radioactive isotopes in biological experiments are:

- **Radioactivity is readily detected.** Methods of detection are sufficiently sensitive to measure extremely small amounts of radioactive substances.

- **Studies can be carried out on intact, living organisms.** If care is taken, minimal disruption of normal conditions will occur when radiolabelled compounds are introduced.

- **Protocols are simple** compared with equivalent methods for chemical analysis.

The main disadvantages are:

- **The 'isotope effect'.** Molecules containing different isotopes of the same atom may react at slightly different rates and behave in slightly different ways from the natural isotope. The isotope effect is more extreme the smaller the atom and is most important for ^{3}H-labelled compounds of low molecular mass.

- **The possibility of mistaken identity.** The presence of radioactivity does not tell you anything about the compound in which the radioactivity is present: it could be different from the one in which it was applied, due to metabolism or spontaneous breakdown of a ^{14}C-containing organic compound.

Metabolism of radiolabelled compounds – you may need to separate individual metabolites before counting, e.g. using chromatography (Chapter 69), or electrophoresis (Chaper 70).

The main types of experiments are:

- **Investigations of metabolic pathways:** a radioactively labelled substrate is added (often to an *in vitro* experimental 'system' rather than a whole organism) and samples taken at different time intervals. By identifying the labelled compounds and plotting their appearance through time, an indication of the pathway of metabolism can be obtained.

- **Translocation studies:** radioisotopes are used to follow the fate of molecules within an organism. Uptake and translocation rates can be determined with relative ease.

- **Ecological studies:** radioisotopic tracers provide a convenient method for determining food web interrelationships and for investigating behaviour patterns, while environmental monitoring may involve following the 'spectral signature' of isotopes deliberately or accidentally released.

- **Radio-dating:** the age of plant or mineral samples can be determined by measuring the amount of a radioisotope in the sample. The age of the specimen can be found using the $t_{1/2}$ by assuming how much was originally incorporated.

- **Mutagenesis and sterilisation:** radioactive sources can be used to induce mutations, particularly in micro-organisms. Gamma emitters of high energy will kill microbes and are used to sterilise equipment such as disposable Petri dishes.

- **Bioassays:** radioisotopes are used in several quantitative detection methods of value to biologists. Radioimmunoassay is described on p. 380. Isotope dilution analysis works on the assumption that introduced radiolabelled molecules will equilibrate with unlabelled molecules present in the specimen. The amount of substance initially present can be worked out from the change in specific activity of the radioisotope when it is diluted by the 'cold' material. A method is required whereby the substance can be purified from the sample and sufficient substance must be present for its mass to be measured accurately.

Lowenthal and Airey (2005) give further details and applications.

Working practices when using radioactive isotopes

By law, undergraduate work with radioactive isotopes must be very closely supervised. In practical classes, the protocols will be clearly outlined, but in project work you may have the opportunity to plan and carry out your own experiments, albeit under supervision. Some of the factors that you should take into account, based on the assumption that your department and laboratory are registered for radioisotope use, are discussed below:

- **Must you use radioactivity?** If not, it may be a legal requirement that you use an alternative method.

- **Have you registered for radioactive work?** Usually, all users must register with a local Radiation Protection Supervisor. Details of the project may have to be approved by the appropriate administrator(s). You may have to have a short medical examination before you can start work.

- **What labelled compound will you use?** Radioactive isotopes must be ordered well in advance through your department's Radiation

---

Example Carbon dating – living organisms have essentially the same ratio of ^{14}C to ^{12}C as the atmosphere; however, when an organism dies, its $^{14}C/^{12}C$ falls because the radioactive ^{14}C isotope decays. Since we know the half-life of ^{14}C (5715 years), a sample's $^{14}C/^{12}C$ ratio will allow us to estimate its age; e.g. if the ratio were exactly 1/8 of that in the atmosphere, the sample is three half-lives old and was formed 17 145 years before present. Such estimates carry an error of the order of 10% and are unreliable for samples older than 50 000 years, for which longer-lived isotopes can be used.

Registering for radioisotope work – in he UK, institutions must be registered for work with specific radioisotopes under the *Radioactive Substances Act* (1993).

Supervision of work with radioisotopes – in the UK, the *Ionising Radiations Act* (1985) provides details of local arrangements for the supervision of radioisotope work.

Protection Supervisor. Aspects that need to be considered include:

(a) **The radionuclide.** With many organic compounds this will be confined to ^3H and ^{14}C (but see Table 64.2). The risk of a significant 'isotope effect' may influence this decision (see p. 421).

(b) **The labelling position.** This may be a crucial part of a metabolic study. Specifically labelled compounds are normally more expensive than those that are uniformly ('generally') labelled.

(c) **The specific activity.** The upper limit for this is defined by the isotope's half-life but, below this, the higher the specific activity, the more expensive the compound.

- **Are suitable facilities available?** You will need a suitable work area, preferably out of the way of general lab traffic and within a fume cupboard for those cases where volatile radioactive substances are used or may be produced.

In conjunction with your supervisor, decide whether your method of application will introduce enough radioactivity into the system, how you will account for any loss of radioactivity during recovery of the isotope and whether there will be enough activity to count at the end. You should be able to predict approximately the amount of radioactivity in your samples, based on the specific activity of the isotope used, the expected rate of uptake/exchange and the amount of sample to be counted. Use the isotope's specific activity to estimate whether the non-radioactive ('cold') compound introduced with the radiolabelled ('hot') compound may lead to excessive concentrations being administered. Advice for handling data is given in Box 64.1.

Safety and procedural aspects

Make sure the bench surface is one that can be easily decontaminated by washing (e.g. Formica) and always use a disposable surfacing material such as Benchkote. It is good practice to carry out as many operations as possible within a Benchkote-lined plastic tray so that any spillages are contained. You will need a lab coat to be used exclusively for work with radioactivity, safety spectacles and a supply of thin latex or vinyl disposable gloves. Suitable vessels for liquid waste disposal will be required and special plastic bags for solids – make sure you know beforehand the disposal procedures for liquid and solid wastes. Wash your hands after handling a vessel containing a radioactive solution and again before removing your gloves. Gloves should be placed in the appropriate disposal bag as soon as your experimental procedures are complete.

It is important to comply with the following guidelines:

- **Read and obey the local rules** for safe usage of radiochemicals.
- **Maximise the distance between you and the source** as much as possible.
- **Minimise the duration of exposure.**
- **Wear protective clothing** (properly fastened lab coat, safety glasses, gloves) at all times.
- **Use appropriate shielding at all times** (Table 64.1).
- **Monitor your working area** for contamination frequently.
- **Mark all glassware, trays, bench work areas,** etc., with tape incorporating the international symbol for radioactivity (Fig. 64.6).

SAFETY NOTE Each new experiment should be planned carefully and experimental protocols laid down in advance so you work as safely as possible and do not waste expensive radioactively labelled compounds.

Carrying out a 'dry run' – consider doing this before working with radioactive compounds, perhaps using a dye to show the movement or dilution of introduced liquids, as this will lessen the risks of accident and improve your technique.

SAFETY NOTE The correct way to use Benchkote and similar products is with the waxed surface down (to protect the bench or tray surface) and the absorbent surface up (to absorb any spillage). Write the date in the corner when you put down a new piece. Monitor using a G–M tube and replace regularly under normal circumstances. If you are aware of spillage, replace immediately and dispose of correctly.

Fig. 64.6 Tape showing the international symbol for radioactivity.

- **Keep adequate records** of what you have done with a radioisotope – the amount remaining and that disposed of in waste form must agree.
- **Store radiolabelled compounds appropriately** and return them to storage immediately after use.
- **Dispose of waste promptly** and with due regard for local rules.
- **Make the necessary reports about waste disposal**, etc., to your departmental Radiation Protection Supervisor.
- **Clear up** after you have finished each experiment.
- **Wash thoroughly** after using radioactivity.
- **Monitor the work area and your body** when finished.

Text references

Alexander, S., Boes, M., Dake, B.L., Stokes, J.B. and Bar, R.S. (1998) Infused IGF-I/IGFBP-3 complex causes glomerular localization of IGF-I in the rat kidney. *American Journal of Physiology*, **275**, 32–37.

Haynes, W.M. (ed.) (2014) *CRC Handbook of Chemistry and Physics,* 95th edn. CRC Press, Boca Raton.

Lowenthal, G. and Airey, P. (2005) *Practical Applications of Radioisotopes and Nuclear Radiations.* Cambridge University Press, Cambridge.

Sources for further study

L'Annunziata, M.F. (ed.) (2012) *Handbook of Radioactivity Analysis*, 3rd edn. Academic Press, San Diego.

Slater, J. (ed). (2002) *Radioisotopes in Biology: A Practical Approach,* 2nd edn. Oxford University Press, Oxford.

STUDY EXERCISES

64.1 Carry out a half-life calculation. A rat dropping found in a pyramid in Egypt had a $^{14}C : ^{12}C$ ratio that was 57.25% of a modern-day standard. Use this value to estimate the approximate date when the rat visited the pyramid, to the nearest century.

64.2 Practise radioactivity interconversions. Express the following values in the alternative units indicated, with appropriate prefixes as necessary. Answer to three significant figures.

(a) 72 000 d.p.m. as Bq;

(b) 20 µCi as d.p.m.;

(c) 44 400 Bq as µCi;

(d) 6.3×10^5 d.p.m. mol^{-1} as Bq g^{-1}, for a compound with a relative molecular mass of 350;

(e) 3108 d.p.m. as pmol, for a sample of a standard where the specific activity is stated as 50 Ci mol^{-1}.

64.3 Use the concept of specific activity in calculations. A researcher wishes to estimate the rate of uptake of the sugar galactose by carrot cells in a suspension culture. She prepares 250 mL of the cell culture medium containing 10^7 cells per mL and unlabelled galactose at a concentration of 5 mmol L^{-1}. She then 'spikes' this with 5 µL (regard this as an insignificant volume) of radioactive standard containing 55 MBq of ^{14}C- labelled galactose (regard as an insignificant concentration). Answer to two significant figures.

(a) Calculate the specific activity of the galactose in the culture solution in Bq mol^{-1}.

(b) If the total cell sample takes up 79.2×10^5 Bq in a 2-h period, calculate the galactose uptake rate in mol s^{-1} $cell^{-1}$.

Answers to these study exercises are available at www.pearsoned.co.uk/practicalskills

There are two main approaches to the measurement of dissolved oxygen in aqueous solution:

1. **Electrometric methods,** e.g. oxygen electrodes. This is the most versatile and straightforward approach, if a suitable probe/electrode and calibrated meter are available.

2. **Chemical assays,** e.g. the Winkler method. This approach provides a highly reliable way of quantifying the amount of dissolved oxygen in water, but requires access to a laboratory and analytical equipment – it is often used with chemically 'fixed' samples, obtained from field sites (e.g. p. 352). This assay is also used to calibrate electrometric systems.

Oxygen electrodes and probes

These are used to measure oxygen in solution, by monitoring the current flow between two electrodes when a voltage is applied: this current is dependent on the oxygen level in solution. The most widespread electrode is the Clark type (Fig. 65.1), manufactured in the UK by Rank Bros, Cambridge, which is suitable for measuring O_2 concentrations in cell, organelle and enzyme suspensions. Platinum and silver electrodes are in contact with a solution of electrolyte (normally saturated KCl). The electrodes are separated from the medium by a Teflon (polytetrafluoroethylene, PTFE) membrane, permeable to O_2. When an electrical potential is applied to the electrodes, this generates a current proportional to the O_2 concentration (Wilson and Walker, 2010). The reactions can be summed up as:

$$4Ag \rightarrow 4Ag^+ + 4e^-$$ (at silver anode)

$$O_2 + 2e^- + 2H^+ \rightleftharpoons H_2O_2$$ (in electrolyte solution; O_2 replenished by diffusion from test solution)

$$H_2O_2 + 2e^- + 2H^+ \rightleftharpoons 2H_2O$$ (at platinum cathode)

Setting up and using a Clark (Rank) oxygen electrode

Box 65.1 describes the steps involved in setting up this type of electrode, and Box 65.2 explains how to analyse the results you obtain. If you are setting up from scratch, perform steps 1–13 in Box 65.1. If a satisfactory membrane is already in place, start at step 7.

The temperature of the incubation vessel can be controlled by passing water (e.g. from a water bath) through the outer chamber. Cells or organelles may be present in the solution added to the incubation chamber or can be added via the hole in the stopper using a syringe, as can any chemicals you wish to introduce, such as metabolic substrates or inhibitors. Take care not to introduce air bubbles and remove any that appear by gently raising and lowering the stopper. The electrode can be used repeatedly, providing the membrane is satisfactory: remove solutions carefully (e.g. using a pipette, or vacuum line and trap). Keep water in the chamber when not in use. Replace the membrane if:

- **the reading becomes noisy;**
- **the electrode will not zero** after adding sodium dithionite;

Fig. 65.1 Transverse section through a Clark (Rank) oxygen electrode.

Labels:
stopper (volume adjuster)
locking nut
rubber 'O' ring
incubation chamber
magnetic stirrer bar
rubber 'O' ring
+ − to recorder
central platinum cathode
Teflon membrane
circular silver anode
saturated KCl solution
magnetic stirrer base

Using polytetrafluoroethylene (PTFE, Teflon) membranes – these are readily permeable to many small molecules and are used for monitoring of dissolved gases in aqueous fluids.

Box 65.1 How to set up a Clark (Rank) oxygen electrode

1. Detach the base of the incubation vessel (see Fig. 65.1) by unscrewing the locking ring.

2. Add enough saturated KCl to cover the electrodes.

3. Cut a 1 mm square hole in the centre of a 10×10 mm square of lens tissue and place on to the KCl solution with hole over the central cathode.

4. Cut a 10×10 mm square of Teflon (PTFE) membrane and place over the lens tissue; seal by gently lowering the incubation vessel and tightening the locking ring, making sure that the rubber O-ring is correctly positioned over the membrane.

   (a) Do not overtighten the locking ring.

   (b) Take care not to trap air bubbles beneath the membrane.

   (c) Make sure that the membrane does not become twisted.

5. Clamp the electrode over the magnetic stirrer base using the clamping screw.

6. Connect the electrode leads to the polarising unit/recording device (silver anode to positive, platinum cathode to negative). You can use either a digital readout system on the control unit or the output can be passed to a chart recorder, giving a readout of oxygen status as a function of time (see Box 65.2). Check that the polarising voltage is set to 0.60 V and adjust, if necessary, using the 'polarising voltage' control (typically, this requires a small screwdriver to adjust).

7. Add air-saturated experimental solution and a small Teflon-coated magnetic stirrer bar to the chamber. The volume of the incubation chamber can be adjusted by moving the locking nut on the stopper. To adjust, add the appropriate amount of liquid to the chamber using a pipette, insert the stopper and screw the locking nut until the solution just fills the incubation chamber.

8. Gently push the stopper (volume adjuster) into position, making sure that no air bubbles are trapped in the chamber, and switch on the stirrer. Adjust the rate of stirring using the 'stirrer' control, if required.

9. Set the zero – remove the stopper and add a few crystals of sodium dithionite – the dithionite ions ($S_2O_4^{2-}$) will be oxidised to sulphite (SO_3^{2-}) and sulphate ions (SO_4^{2-}), thereby consuming all of the O_2 in the solution within 3–5 min (an alternative is to bubble N_2 through the solution for > 10 min). Once the reading has stabilised, if this is not at zero, then adjust using the 'set zero' control (in practice, the zero is usually quite stable and often needs no adjustment).

10. Adjust the sensitivity – rinse out the dithionite solution thoroughly/repeatedly, then replace with air-equilibrated water and allow the reading to stabilise (this may take 5–10 min). Then adjust the 'sensitivity' control of the electrode system to set the oxygen-saturated value at an appropriate point (for a controller with a digital readout, this is best set to 100, and then the values represent percentage oxgen saturation). To check that the instrument is working correctly, switch off the stirrer for a few moments – the reading should drop quickly as oxygen is consumed at the electrode surface and should then rise rapidly when the stirrer is switched back on.

11. Rinse the incubation chamber thoroughly and add fresh experimental solution.

12. Carry out your experiment.

13. Remove the solution and check the calibration. If the reading for air-equilibrated water is different, the electode's sensitivity or the temperature may have changed and you may need to recalibrate and repeat the measurement.

- **the response becomes too slow** (check by switching off the stirrer – the reading should drop rapidly as the available O_2 is consumed).

> **KEY POINT** Successful operation of a Clark (Rank) oxygen electrode requires an intact Teflon membrane and clean electrodes. When replacing the Teflon membrane, you should always check that the electrode surfaces are clean (shiny): if necessary, use a mild abrasive paste to remove any oxidised material.

Oxygen probes

Clark-type oxygen electrodes are also available in probe form for immersion in the test solution (Fig. 65.2). Such probes are particularly useful for field studies since they allow direct measurement of oxygen status *in situ*, in contrast to chemical assays (e.g. the Winkler method, p. 428). Wetzel and Likens (2010) discuss field sampling protocols for measuring dissolved oxygen in fresh waters.

Box 65.2 How to convert a chart recorder trace to a rate of O_2 consumption or production

Having set up your oxygen electrode (Box 65.1) or probe, the most common experiments are those in which you measure the rate of oxygen consumption (respiration) or production (photosynthesis) in a fixed volume of solution, e.g. the chamber of a Rank (Clark) oxygen electrode. Typically, this involves attaching the electrode to a chart recorder and following the change in oxygen concentration with time. The principal steps are as follows:

1. Calculate the amount of oxygen in the electrode chamber – multiply the appropriate oxygen-saturation concentration, from Table 65.1, by the volume of the chamber. For example, if you are working in mmol L^{-1} ($=$ μmol mL^{-1}) then at 20 °C, air-equilibrated (oxygen-saturated) distilled water contains 0.285 μmol mL^{-1}. For an electrode chamber of volume 5 mL, there will be $0.285 \times 5 = 1.425$ μmol oxygen in the water within the chamber at saturation.

2. Set the zero on the recorder – following the addition of a small amount of sodium dithionite to the chamber, to remove oxygen (Box 65.1), the trace is monitored until it has stabilised. Then adjust the 'zero' or 'back off' control of the recorder until the chart trace is set to zero on the chart paper: this is marked as 'A' on Fig. 65.3.

3. Set the oxygen-saturated reading on the recorder – rinse out the dithionite solution thoroughly/repeatedly, replace with air-equilibrated water and allow the reading to stabilise (this may take 3–5 min). Then adjust the 'sensitivity' control of the chart recorder to set the oxygen-saturated value at an appropriate point on the chart – for respiration measurements, this can be close to the full width of the chart paper, whereas for photosynthesis measurements it is more usual to set it closer to the mid-point of the chart paper, to allow measurements over 100% to be recorded. In the example shown in Fig. 65.3, where the chart paper is calibrated in mm divisions, the oxygen saturated reading is set at 80 divisions (80 mm), marked as 'B'. Note that the recorder pen can be lifted between readings, and during rinsing of the chamber, to reduce the amount of 'noise' on the chart.

4. Calculate the amount of oxygen equivalent to a single division of the chart paper – divide the amount of oxygen in the chamber by the number of divisions (mm) to give the amount per division, e.g. $1.425 \div 80 = 0.0178125$ μmol ($=17.8125$ nmol per division).

5. Carry out your experiment – the example shown in Fig. 65.3 represents the consumption of oxygen during respiration of a suspension of yeast cells, marked as 'C'. Continue the recording until you are satisfied that the rate has remained stable for at least 5 min. In many experiments, you will be adding substances that change the rate (e.g. metabolic inhibitors, or substrates); this can be done by injecting a solution containing the substance through the small hole in the stopper of the electrode (keep the volume of this solution to $<1\%$ of the total volume of the chamber, to minimise the effect of the added solution).

6. Convert the gradient of the trace into a rate of oxygen consumed or produced – using a transparent ruler, draw the line of best fit through that part of the trace showing a stable relationship (see Fig. 65.3). Use this line of best fit to work out the gradient of the trace, by converting from linear dimensions to an amount per unit time: e.g. the gradient shown in Fig. 65.3 is equivalent to a decrease of 24 chart divisions (mm) in 3 cm. Using the value calculated in step 4, the amount of oxygen consumed is thus $17.8125 \times 24 = 427.5$ nmol. For a chart speed of 0.2 cm min^{-1}, this represents the change occurring over a time interval of $3 \div 0.2 = 15$ min. Thus the rate of oxygen consumed is $427.5 \div 15 = 28.5$ nmol min^{-1}.

7. Express the rate on an appropriate basis – in most instances, this will be either per cell (e.g. determined using a haemocytometer, p. 294) or per unit mass of protein (pp. 388–9), chlorophyll (p. 450), or an equivalent marker. For the example shown in Fig. 65.3, if there were 2.4×10^7 yeast cells per mL of suspension, then in the 5 mL chamber of the electrode there would be a total of $2.4 \times 10^7 \times 5 = 1.2 \times 10^8$ yeast cells, giving a rate of $28.5 \div (1.2 \times 10^8) = 2.375 \times 10^{-7}$ nmol min^{-1} $cell^{-1}$, probably better expressed as 237.5 amol min^{-1} $cell^{-1}$. Note that when inhibitors/substrates are used, you should always wait until a new stable rate has been achieved (for 5 min) before using the trace to determine the new rate of oxygen consumption/production.

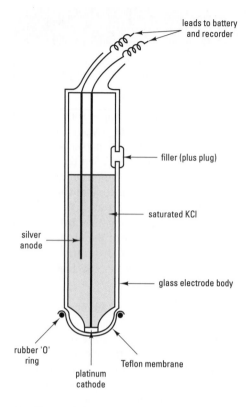

leads to battery and recorder

filler (plus plug)

saturated KCl

silver anode

glass electrode body

rubber 'O' ring

Teflon membrane

platinum cathode

Fig. 65.2 A Clark-type oxygen probe.

KEY POINT Note that the solution must be stirred continuously during measurement, to replenish the oxygen consumed by the electrode ('boundary layer' effect): gentle movement of the probe during measurement provides sufficient agitation.

Chemical methods

The Winkler (iodometric) method is the standard volumetric assay for measuring dissolved oxygen in water. The basis of the test involves a specific sequence of chemical reactions:

1. Initial 'fixation' of dissolved oxygen, by reaction with manganese ions (Mn^{2+}) in alkaline solution to form manganous dioxide.
2. Oxidation of iodide ($2I^-$) to liberate iodine (I_2) by manganous dioxide, under acidic conditions.
3. Titration of I_2 against thiosulfate ($S_2O_3^{2-}$). Typically, the reagents are prepared in concentrations that enable the direct conversion of the volume of thiosulfate solution titrated in mL to the amount of dissolved oxygen in solution, in mg L^{-1}.

The reactions can be summed up as:

$$Mn^{2+} + 2(OH)^- + \tfrac{1}{2}O_2 \rightarrow MnO_2 + H_2O \qquad \text{(fixation of oxygen)}$$
$$MnO_2 + 2I^- + 4H + \rightarrow Mn^{2+} + I_2 + 2H_2O \quad \text{(liberation of iodine)}$$
$$2S_2O_3^{2-} + I_2 \rightarrow S_4O_6^{2-} + 2I^- \qquad \text{(thiosulfate titration of iodine)}.$$

Box 65.3 gives practical details of the method.

Table 65.1 Oxygen-saturation values for distilled water and sea water at standard atmospheric pressure and a range of temperatures (derived from Green and Carritt, 1967)

| Temperature °C | Distilled water | | | Sea water | | |
	mol m^{-3} (mmol L^{-1})	mg L^{-1} (p.p.m.)	mL L^{-1}	mol m^{-3} (mmol L^{-1})	mg L^{-1} (p.p.m.)	mL L^{-1}
0	0.460	14.7	10.3	0.359	11.5	8.04
2	0.435	13.9	9.75	0.342	10.9	7.65
4	0.413	13.2	9.24	0.326	10.4	7.30
6	0.392	12.5	8.78	0.311	9.95	6.97
8	0.373	11.9	8.35	0.298	9.54	6.66
10	0.355	11.4	7.95	0.285	9.12	6.37
12	0.339	10.8	7.59	0.273	8.74	6.11
14	0.324	10.4	7.25	0.261	8.35	5.85
16	0.310	9.92	6.94	0.251	8.03	5.62
18	0.297	9.50	6.65	0.241	7.71	5.40
20	0.285	9.12	6.38	0.232	7.42	5.19
22	0.274	8.77	6.12	0.224	7.17	5.00
24	0.263	8.42	5.89	0.215	6.88	4.82
26	0.253	8.1	5.67	0.208	6.66	4.64
28	0.244	7.81	5.46	0.200	6.40	4.48
30	0.235	7.52	5.27	0.193	6.18	4.32
37	0.211	6.75	4.71	0.174	5.57	3.88

Note: tabulated values assume atmospheric pressure = 101.3 kPa (≈ 760 mm Hg); for more accurate work, a correction for any deviation can be made by multiplying the appropriate figures from the table by the ratio of the real pressure to the assumed pressure.

Box 65.3 How to carry out the Winkler method for determination of dissolved oxygen in water

1. Collect the water sample – typically, this will be in a 'Winkler bottle' with a ground glass stopper, 300 mL in volume. The water should fill the bottle, with no trapped air bubbles.

2. 'Fix' the dissolved oxygen – remove the stopper and add 1 mL of manganous sulfate solution (500 g $MnSO_4.5H_2O$ in 1 L of solution), then 1 mL of alkaline iodide solution (500 g NaOH + 135 g NaI in 1 L of solution: include sodium azide, NaN_3 at 10 g L^{-1} if nitrite may be present in the test water, e.g. for wastewater treatment plant effluent). Additions should be made with the tip of the delivery pipette held at the water meniscus. Replace the ground glass stopper, again avoiding any trapped air bubbles.

3. Mix thoroughly – invert the bottle several times over a period of at least 20 s, taking care not to dislodge the stopper (keep your thumb over the stopper to prevent it from falling out). A brown floc will form during mixing: this is hydrated MnO_2.

4. Allow the floc to settle – a clear supernatant should form above the floc as it settles. Leave this until the floc has settled into the lower half of the bottle, then repeat the mixing process at step 3, before leaving the floc to settle into the lower third of the bottle, when it is ready for titration.

5. Liberate iodine – remove the ground glass stopper and add 1 mL of concentrated sulfuric acid, H_2SO_4 (safety note – take care to rinse any spilled drops of acid with plenty of water). The sulfuric acid should be added dropwise from the pipette, without splashing. Replace the ground glass stopper, again avoiding any trapped air bubbles.

6. Mix thoroughly – invert the bottle several times over a period of at least 20 s, taking care to prevent the stopper from falling out. The floc should disappear and a brown colour should appear in the solution. For samples with no dissolved oxygen, the floc formed at step 3 would have been white and the solution will now be colourless. If necessary, the first 6 steps can be carried out under field conditions and the 'fixed' samples are then returned to the laboratory for subsequent assay within 6 h of initial fixation.

7. Titrate against thiosulfate – transfer exactly 202 mL of the solution from step 6 to a flask (this volume is required to provide exactly 200 mL of original water sample, since it takes into account the dilution provided by the 3 mL of reagents added at steps 2 and 5). Then, titrate this against 0.025 mL L^{-1} $Na_2S_2O_3$ solution, stirring continuously until the brown colour almost disappears. To find the exact end point of the titration, add a couple of drops of 5 g L^{-1} starch solution at this stage, which will cause the solution to turn blue: titrate dropwise until the blue colour just disappears. Record the volume of thiosulfate solution titrated.

8. Determine the dissolved oxygen content of the water sample. Simply convert the amount of thiosulfate titrated in mL to amount of dissolved oxygen in mg L^{-1}. Convert to mmol L^{-1}, or mL L^{-1}, if required, as shown in Table 65.1.

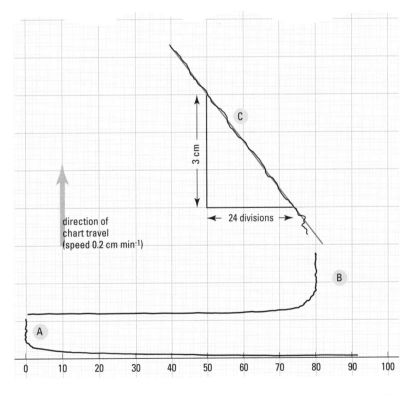

Fig. 65.3 Chart recorder output for a Clark (Rank) oxygen electrode. A = zero oxygen; B = oxygen saturation; C = respiration of yeast.

direction of chart travel (speed 0.2 cm min^{-1})

3 cm

24 divisions

Text references

Green, E.J. and Carritt, D.E. (1967) 'New tables for oxygen saturation of seawater', *Journal of Marine Research,* **25**, 140–7.

Wetzel, R.G. and Likens, G.E. (2010) *Limnological Analysis,* 3rd edn. Springer, New York.

Source for further study

Eaton, A.D., Clesceri, L.S., Rice, E.W. and Greenberg, A.E. (2012) *Standard Methods for the Examination of Water and Wastewater,* 22nd edn. American Public Health Association, Washington.

Wilson, K. and Walker, J. (eds.) (2010) *Principles and Techniques of Biochemistry and Molecular Biology,* 7th edn. Cambridge University Press, Cambridge.

STUDY EXERCISES

65.1 Calculate the oxygen content of specified volumes of water at defined temperatures. Using the information in Table 65.1 and assuming air equilibration at 101.3 kPa, work out the amount of oxygen in each of the following (give all answers to three significant figures):

(a) 4 mL of distilled water at 20 °C (express your answer in µmol);

(b) 20 mL of sea water at 12 °C (express your answer in µmol);

(c) 10 mL of distilled water at 15 °C (express your answer in µmol);

(d) 250 mL of distilled water at 37 °C (express your answer in mg);

(e) 200 mL of sea water at 25 °C (express your answer in mL).

65.2 Calculate respiration using data from an oxygen electrode readout. An oxygen electrode was set up to contain 5.0 mL of air-equilibrated distilled water at 30 °C, giving a full-scale reading on a chart recorder (200 divisions on the chartrecorder paper), while a zero reading gave no deflection (0 divisions on the chart recorder paper). A yeast cell suspension (containing 4.1×10^9 cells in total) was added to the same electrode system along with a suitable growth medium, giving an approximately linear trace on the chart recorder with a slope of −25 units per cm, at a chart speed of 0.5 cm min^{-1}. What is the respiratory oxygen consumption, expressed to three significant figures in nmol min^{-1} (10^9 cells)$^{-1}$?

65.3 Determine net photosynthetic rate from oxygen-electrode readings. The following data represent oxygen–electrode readings taken every minute for an illuminated suspension of a cyanobacterium in a 5.0 mL chamber of an oxygen electrode at 20 °C with readings from 0 min to 8 min of: 70.4; 73.6; 75.3; 78.5; 83.6; 85.9; 88.2; 91.4; 94.1. The system was first calibrated to read 100.0 for air-equilibrated distilled water and 0.0 for anaerobic water, in effect making the scale read in terms of percentage oxygen saturation. The chlorophyll a content of an equivalent amount of cyanobacterial suspension to that used in the electrode chamber was measured at 2.13 µg mL^{-1}. What is the net photosynthetic rate of the cyanobacterial suspension, in µmol min^{-1} (mg chlorophyll a)$^{-1}$?

Answers to these study exercises are available at www.pearsoned.co.uk/practicalskills

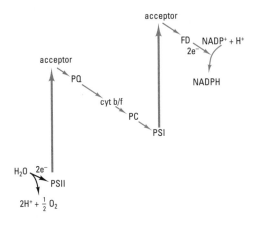

Fig. 66.1 Non-cyclic electron transport from H_2O to $NADP^+$ in plant photosynthesis (Z scheme) showing the principal components: PS II, photosystem II; PQ, plastoquinone; cyt, cytochrome; PC, plastocyanin; PS I, photosystem I; FD, ferredoxin; $NADP^+$, nicotinamide adenine dinucleotide phosphate.

Techniques for investigating photosynthesis and respiration are considered together, since:

- **Gas exchange processes in these two processes may be studied using similar techniques,** and interpreted in similar stoichiometric terms, e.g. by relating changes in O_2 or CO_2 to those of other reactants or products.

- **Membrane-bound electron transport systems are involved in the production of a transmembrane H^+ gradient** that drives ATP synthesis via an ATPase embedded in the same membrane. Measurement of electron transport in photosynthesis and respiration often makes use of redox dyes as artificial electron donors or acceptors, with spectrophotometric analysis of the reactions.

- **Some electron transport inhibitors can affect both photosynthesis and respiration,** while others are specific to one or the other process.

> **KEY POINT** Photosynthesis and respiration are complex, multistage metabolic pathways involving a large number of cellular components: they may be studied as a whole, or as isolated components and individual metabolic reactions, to gain a deeper insight into the underlying processes.

Photosynthesis

This is the conversion of light energy to chemical energy by photoautotrophic plants and bacteria. The overall process is often summarised in terms of the synthesis of carbohydrate, $(CH_2O)_n$, driven by the energy within photons of light (hv), as:

$$nCO_2 + 2nH_2O^* + 4nhv \rightarrow (CH_2O)_n + nO_2^* + nH_2O \qquad [66.1]$$

where the asterisks show that all of the O_2 is derived from the photolysis of water. Carbohydrate synthesis occurs as a result of two distinct processes, namely (i) light reactions and (ii) dark reactions. In the former, light energy is absorbed by pigments within membrane-bound reaction centres (photosystems). The water-splitting light reactions generate electrons, protons and molecular oxygen. Electron and proton movement via a series of membrane-bound redox carriers leads to the reduction of $NADP^+$ and the generation of ATP (photophosphorylation), driven by a transmembrane H^+ gradient, or proton-motive force (p. 438). The light reactions of plant-type photosynthesis are often represented by the 'Z scheme' (Fig. 66.1).

The dark reactions use the products of the light reactions to 'fix' CO_2, via a series of soluble enzymes known as the Calvin cycle, or reductive pentose phosphate cycle. The primary carboxylating enzyme in temperate green plants (C_3 plants) and cyanobacteria is ribulose bisphosphate carboxylase ('Rubisco'). This enzyme catalyses the addition of CO_2 to ribulose bisphosphate (C_5), producing two molecules of phosphoglyceric acid (C_3) which are then reduced to triose phosphate (C_3) by enzymes that utilise the NADPH and ATP generated by the light reactions. Thus both light and dark reactions operate in a coupled manner when photosynthetically active cells are illuminated with visible light.

Measuring photosynthetic pigments – equations for spectrophotometric analysis of chlorophylls are given on p. 450.

Definitions

C_3 **plants** – those in which phosphoglyceric acid is the first stable product.

C_4 **plants** – those in which malate or aspartate are the principal products, via oxaloacetic acid, in a reaction catalysed by phosphoenolpyruvate (PEP) carboxylase.

CAM plants – those in which malic acid is produced, via oxaloacetic acid, by PEP carboxylase activity at night, with little or no net assimilation of CO_2 during the day (crassulacean acid metabolism).

Measuring photosynthetic activity

Equation [66.1] shows that the rate of photosynthesis can be measured in terms of the amount of O_2 evolved or CO_2 fixed. However, several other metabolic processes may lead to concurrent changes in O_2 and/or CO_2 status, particularly:

- **Photorespiration** – light-dependent O_2 uptake, due to the oxygenase activity of Rubisco, leading to the production of glycollic acid (C_2) and the subsequent release of CO_2 due to the operation of a 'scavenging' pathway where two molecules of glycollate are converted to one of phosphoglyceric acid (C_3). The relative rates of photosynthesis and photorespiration depend upon the concentration of O_2 and CO_2 at the active site of Rubisco – in C_3 plants, photorespiration is highest under conditions of high temperature, high light and low water availability, while C_4 plants show low levels of photorespiration.

- **Respiration** – generation of ATP via carbohydrate oxidation, as in non-photosynthetic cells (p. 436). In most photoautotrophs, respiratory activity is likely to be lower during the day than at night. However, plant tissues often respond to wounding or environmental stress by *increasing* their rate of respiration – this may be significant if you are measuring photosynthetic activity in tissue fragments, isolated cells or protoplasts. If you wish to calculate gross rates of photosynthesis, you must make appropriate allowances for both respiration and photorespiration.

Measurement of oxygen production

This is most often used with aquatic systems, including algae and photosynthetic bacteria, higher plant cells, protoplasts, chloroplast suspensions or isolated thylakoids. Oxygen in solution can be determined by end-point chemical analysis (e.g. the Winkler method, p. 429) or, more conveniently, by continuous monitoring using an oxygen electrode and chart recorder, to give the rate of net O_2 production (Chapter 65) under various conditions, e.g. light, temperature, etc. For photosynthesis–irradiance (*P–I*) curves using an O_2 electrode, you should note that the electrode assembly can act as a lens, so the light within the chamber can be higher than that measured at the outside surface. The rate of gross photosynthesis is obtained by correcting for O_2 uptake when the electrode assembly is transferred to darkness (e.g. using a thick black cloth).

Measurement of carbon dioxide uptake

For higher plant studies, this is most easily achieved using an infrared gas analyser, or IRGA (Fig. 66.2). Most modern instruments are portable and generally incorporate: (i) a cuvette, with a transparent window that attaches to a whole leaf or a known area of leaf providing a gas-tight seal, and within which the air is stirred by a fan; (ii) a gas supply system that allows control over input gases; (iii) miniaturised infrared analysers to detect differences in CO_2 and H_2O content of input and output gas streams; (iv) systems for measuring (and possibly controlling) other environmental variables, e.g. light and temperature; (v) an on-board microprocessor for calculating results as rates of photosynthesis and transpiration and estimating the leaf's internal pCO_2 (C_i) and for exporting these and other data (e.g. time, leaf temperature, photosynthetic photon flux density) to PCs. Portable IRGAs are particularly

---

Definitions

Gross photosynthesis – the rate of CO_2 uptake or O_2 production, allowing for gas exchange due to respiratory and photorespiratory activity.

Net photosynthesis – the net rate of CO_2 uptake or O_2 production, including respiratory and photorespiratory gas exchange.

Correcting for photorespiration and respiration – measuring the rate of O_2 consumption or CO_2 production in the first few minutes after the plant material has been transferred from the light to darkness will give you the best estimate of photorespiratory and respiratory gas exchange. Use this rate to convert net photosynthesis to gross photosynthesis.

Preventing oxygen supersaturation in oxygen electrode studies – lower the O_2 content of your experimental solutions by bubbling them with N_2 before use, so that the O_2 evolved during the experiment remains in solution.

Fig. 66.2 A portable infrared gas analyser, used to measure photosynthetic CO_2 uptake (leaf cuvette/probe shown in left hand, with analysis unit and user interface below).

Fig. 66.3 Example of a photosynthesis–irradiance (*P–I*) curve for a typical plant leaf. Such a curve might be obtained using an IRGA, manipulating *I* in steps down from high values and measuring *P* rates when stable (see Parsons *et al.*, 1997). Typical values for estimated parameters might be: dark respiration rate, − 4 μmol CO_2 m^{-2} s^{-1}; light compensation point, 30 μmol photons m^{-2} s^{-1}; maximum photosynthetic efficiency, 8 photons per CO_2 molecule fixed (the theoretical maximum is about 20); maximum photosynthetic rate, 20 μmol CO_2 m^{-2} s^{-1}, which might be achieved at *I* = 1000 μmol photons m^{-2} s^{-1} (full sunlight is about 1800 μmol photons m^{-2} s^{-1}). Various methods are used to fit lines to data points and thereby estimate the above parameters – there remains controversy as to which is the best mathematical function to use for this.

Maximum photosynthetic rate: phase where *P* reaches an asymptote and systems are effectively saturated with light (**photosynthetic capacity**)

Photo-inhibition: phase in high light where *P* may be reduced in some cases

Maximum photosynthetic efficiency: where slope of curve is steepest, measured in mol CO_2 assimilated per photon received

Light compensation point: where *P* is 0 and photosynthesis balances respiration

Net rate of photosynthesis, *P* (μmol CO_2 m^{-2} s^{-1})

Irradiance, *I* (μmol photons m^{-2} s^{-1})

Low light phase where photosynthesis exceeds respiration and *P* is negative: (at *I* = 0, an estimate of the **dark respiration rate**)

Definitions

Compensation point – PFD or pCO_2 at which net photosynthesis = 0, i.e. where gross photosynthesis, respiration and photorespiration are balanced: from the intercept of a *P–I* or *P–C_i* curve (x-axis).

Photosynthetic efficiency – the rate of photosynthesis when limited either (i) by supply of photons or (ii) by supply of CO_2, determined from the initial gradient of a *P–I* or *P–C_i* curve respectively.

Photosynthetic capacity – the rate of photosynthesis under saturating photon flux density (PFD) or CO_2 supply, determined from the upper asymptote of a *P–I* or *P–C_i* curve respectively.

Interpreting ^{14}C fixation data – short-term studies of a few minutes may estimate gross photosynthesis, while longer-term studies will give a rate closer to net photosynthesis, since some of the fixed ^{14}C will be respired.

Studying the release of photo assimilated ^{14}C in aquatic photoautotrophs – the loss of glycollate and other metabolites can be quantified by acidifying a known amount of medium, to drive off unfixed ^{14}C, then counting directly, or after an appropriate separation procedure.

useful for the rapid construction of photosynthesis–irradiance (*P–I*) and photosynthesis–pCO_2 (*P–Ci*) curves used to estimate photosynthetic efficiency, photosynthetic capacity and compensation points (Fig. 66.3).

Using carbon isotopes

Radioisotopic ^{14}C and stable ^{13}C are used to study photosynthesis. Radiocarbon (^{14}C) can be supplied as $^{14}CO_2$ for gas exchange studies or, more readily, as $H^{14}CO_3^-$ for studies in aqueous systems, since there will be an interconversion of soluble CO_2, HCO_3^- and $H_2CO_3^{2-}$. In aqueous systems, the plant material is incubated in medium containing ^{14}C–labelled bicarbonate for a known time period, then removed and prepared for liquid scintillation counting (pp. 418–20). Microalgal cells and photosynthetic bacteria can be separated from the experimental medium by filtration, or silicone oil microcentrifugation (p. 443) – mild acid treatment (e.g. 50 mmol L^{-1} HCl) will ensure that any unfixed ^{14}C is released as $^{14}CO_2$. To express the results in terms of the amount of C assimilated, you will need to calculate (i) the total inorganic C content of the medium, i.e. CO_2 + HCO_3^- + $H_2CO_3^{2-}$, obtained from pH and alkalinity measurements or by IR spectroscopy of the CO_2 produced when a known amount of the medium is acidified and (ii) the relationship between the total inorganic C content and the amount of ^{14}C tracer added (i.e. the specific activity of the experimental solution, p. 419).

One of the advantages of studying the photoassimilation of ^{14}C is that the radiotracer can be used to follow the fate of fixed carbon, by separating and fractionating the various cellular components prior to counting, e.g. by sequential solvent extraction of the plant material in 80% ethanol (low molecular weight solutes) then boiling water (polysaccharides), leaving a residual fraction (structural polysaccharides, proteins and nucleic acids). More sophisticated separation techniques must be used to quantify the amount of radioactivity within individual biomolecules, e.g. using column chromatography (p. 455) and autoradiography (p. 420), or 2-D thin layer chromatography (p. 454) and autoradiography. High specific activity $H^{14}CO_3^-$ and short incubation times of a few seconds are required to study the early stages in carbon photoassimilation.

The use of the stable carbon isotopes ^{12}C and ^{13}C in photosynthetic studies is covered in detail within Chapter 63, including example calculations of the relative abundance of each stable isotope, and $\delta^{13}C$ (p. 411).

Chlorophyll fluorescence

This can be used to provide a non-destructive indication of photosynthetic function. When light energy is absorbed by the light-harvesting apparatus of green plants, several reactions compete for the deactivation of excited chlorophyll molecules. Principally, there are three possible outcomes: (i) the light energy may be trapped and used to reduce photosystem II (PS II) and drive photochemical reactions (Fig. 66.1); (ii) the energy may be lost as heat; or (iii) the energy may be re-emitted as fluorescence, measured at wavelengths around 685 nm using a purpose-built chlorophyll fluorometer. Since these three processes are mutually exclusive, measurements of chlorophyll fluorescence can provide information about changes in photochemistry and heat dissipation, since the maximum potential fluorescence yield is 'quenched' (reduced) by (i) photochemical and (ii) non-photochemical processes. Most measurements are made using pulse-amplified modulated (PAM) fluorometers (Fig. 66.4), where the light source used to generate the chlorophyll fluorescence can be turned on and off at high frequency (pulse modulated), enabling the chlorophyll fluorescence signal to be separated from all other light, and therefore allowing measurements to be made in light-adapted leaves under 'natural' conditions.

To interpret fluorescence changes, the relative contribution of the two quenching processes must be established. The starting point is to consider what happens when a strong light is shone on a dark-adapted leaf: there is a rapid rise in chlorophyll fluorescence to a maximum value, as photosynthetic electron transport is rapidly saturated and no further electrons can be accepted by PS II. If the light remains on, the chlorophyll fluorescence falls progressively, due to photochemical processes (e.g. activation of carbon-fixing enzymes) and non-photochemical processes (dissipation of chlorophyll energy as heat). By measuring changes in chlorophyll fluorescence, the relative contributions of these two processes can be quantified. At a practical level, this is achieved by first shining a single flash of high-intensity light (the saturating pulse) on to a leaf in the dark-adapted state and the resulting maximum fluorescence signal, F_m (see Fig. 66.5 for detail) is then used to determine the potential maximum quantum yield of PS II as:

$$\text{Maximum quantum yield PS II} = \frac{F_m - F_0}{F_m} \qquad [66.2]$$

The leaf is then allowed to adapt to ambient light, and the saturation pulse is then reapplied: the reduction in maximum fluorescence (to F_m') is the result of non-photochemical quenching (NPQ), i.e. dissipation of energy as heat (Fig. 66.5). This can be determined from the relationship:

$$\text{NPQ} = \frac{F_m - F_{m'}}{F_{m'}} \qquad [66.3]$$

NPQ processes protect the protein components of the photosynthetic apparatus from oxidative damage due to excess light. The remaining

Fig. 66.4 A pulse-amplified modulated fluorometer (PAM), used to measure chlorophyll fluorescence (probe shown in left hand, attached to leaf, with analysis/recording unit on shoulder strap).
Courtesy of Heinz Walz GmBH, Effeltrich, Germany (http://www.walz.com).

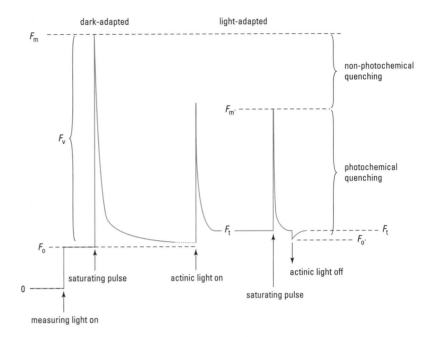

Fig. 66.5 Representation of chlorophyll fluorescence measurement using the saturation pulse method (for details, see text).

Measuring photosynthetic quotients – in studies where both O_2 and CO_2 are measured, the photosynthetic quotient (PQ) can be determined from the relationship: PQ = O_2 evolved ÷ CO_2 consumed. In the simplest case, where fixed carbon accumulates as carbohydrate, a value of 1 should be obtained. However, the PQ may vary, depending on the amount of carbon incorporated into fats, proteins, etc. and the utilisation of photosynthetic energy for other metabolic processes and growth, e.g. NO_3^- assimilation, carbon storage.

fraction is equivalent to photochemical quenching (qP), a measure of the proportion of PS II centres that are 'open', calculated from the relationship:

$$qP = \frac{F_{m'} - F_t}{F_{m'} - F_{0'}} \qquad [66.4]$$

The quantum yield of PS II ($\Phi_{PS\,II}$), which is a measure of the efficiency of PS II photochemistry, can be determined as:

$$\Phi_{PS\,II} = \frac{F_{m'} - F_t}{F_{m'}} \qquad [66.5]$$

The light and dark reactions of photosynthesis are coordinated, since electron transport generates the ATP and NADPH required for CO_2 fixation. For example, a reduction in the rate of CO_2 assimilation due to a metabolic inhibitor will be mirrored by a decrease in $\Phi_{PS\,II}$ and, in many cases, by an increase in NPQ as the light intensity is increased to levels in excess of those required for photosynthesis.

Studying photosynthetic electron transport

Robert Hill first showed that isolated chloroplasts can generate O_2 in the absence of CO_2 fixation, as long as they are provided with an artificial electron acceptor that intercepts electrons from the photosynthetic electron transport chain. The net result of this 'Hill reaction' is the photolysis of water and the reduction of the Hill acceptor (A), as follows:

$$2nH_2O + 4nA \rightarrow 4nAH + nO_2 \qquad [66.6]$$

Measuring the Hill reaction using DCPIP – since the dye will revert to its oxidised (blue) form as soon as the chloroplasts are removed from the light, you must measure the absorbance as quickly as possible.

The practical significance of the Hill reaction is that it allows the photochemical reactions of the electron transport system to be studied independently of the dark reactions of photosynthesis. Many of the Hill acceptors are redox dyes that show changes in absorbance as they are reduced, e.g. ferricyanide ($Fe(CN)_6^{3-}$), which accepts electrons from the PS I complex, or 2,6-dichlorophenolindophenol (DCPIP), which intercepts

Table 66.1 Some inhibitors of photosynthetic electron transport

Inhibitor – target site
Hydroxylamine (NH_2OH) – photolysis of H_2O
DCMU (3(3,4–dichlorophenyl)–1,1–dimethylurea) – electron transport from PS II to plastoquinone
DBMIB (2,5–dibromo–3–methyl–6–isopropyl–*p*–benzoquinone) – electron flow from plastoquinone to cytochrome *f*
Methyl viologen (Paraquat) – electron flow from PS I to $NADP^+$
Atrazine (2–chloro–4–(2–propylamino)–6–ethylamine–5–triazine) – electron flow to plastoquinone
HOQNO (2–heptyl–4–hydroxyquinoline–*N*–oxide) – electron flow between quinones and cytochromes
DSPD (disalicylidinepropanediamine) – electron flow via ferredoxin

electrons from the electron transport chain between PS II and PS I (Fig. 66.1). The reduction of these artificial electron acceptors can be followed spectrophotometrically, allowing the Hill reaction to be quantified. Alternatively, an O_2 electrode can be used to follow O_2 evolution from PS II in the presence of various artificial electron acceptors. Redox mediators can be used to investigate the following aspects of photosynthetic electron flow:

- **the activity of PS II and PS I operating in series** can be studied in whole chloroplasts using a suitable terminal acceptor, e.g. ferricyanide;

- **the activity of PS II can be studied in whole chloroplasts** using DCPIP (membrane-permeable), or in fragmented chloroplasts using ferricyanide;

- **the activity of PS I can be measured if PS II is blocked by the herbicide diuron (DCMU)** and if electron flow is maintained by the addition of an artificial electron donor, e.g. ascorbate plus DCPIP. Note that this cannot be measured by O_2 evolution, since PS II is inoperative – methyl viologen can be used as an electron acceptor from PS I and is rapidly reoxidised, consuming O_2 in a reaction which can be measured using an O_2 electrode (as O_2 *uptake*, rather than O_2 *production*);

- **the sites of action of inhibitors (Table 66.1) can be determined** by measuring their effects on the various components of the electron transport system, measured using redox dyes or the methyl viologen/O_2 electrode system.

Respiration

At the cellular level, respiration can be considered as the oxidation of organic compounds coupled to the production of so-called high-energy intermediates, such as ATP. The overall process is often represented as:

$$C_6H_{12}O_6 + 12O_2^* + 6H_2O + nADP + nP_i \rightarrow$$
$$6CO_2 + 12H_2O^* + nATP + nH_2O \qquad [66.7]$$

the asterisks showing that O_2 is converted to water, and where n has a *maximum* value of 38. The process can be divided into three principal stages, namely (i) glycolysis, (ii) the tricarboxylic acid (TCA) cycle and (iii) oxidative phosphorylation. The glycolytic pathway operates in the cytosol and results in the partial breakdown of glucose (C_6) to two molecules of pyruvate (C_3) plus two molecules of NADH and two molecules of ATP. The mitochondrial TCA cycle involves the sequential dismantling of pyruvate to CO_2 (via an intermediate step involving decarboxylation to acetyl-coA), producing one molecule of $FADH_2$, four molecules of NADH and one of GTP for each pyruvate. The final stage occurs at the inner mitochondrial membrane – the movement of electrons and protons along the respiratory transport chain from reductant to O_2 as the terminal electron acceptor creates a transmembrane H^+ gradient that leads to the net synthesis of ATP, as summarised in Fig. 66.6. Thus, as in photosynthesis, respiration involves

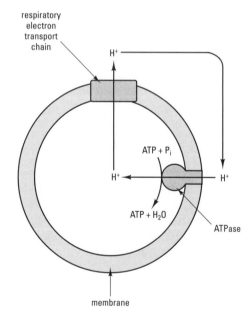

Fig. 66.6 Proton flow in mitochondrial oxidative phosphorylation: the creation of a trans-membrane H^+ gradient due to the respiratory electron transport chain drives ATP synthesis via a membrane-bound ATPase.

Fig. 66.7 Representative data for O_2 uptake by a suspension of mitochondria in response to additions of substrate and ADP.

> **Interpreting respiratory quotients** – the complete oxidation of carbohydrates should give values close to 1.0, in agreement with eqn [66.8], while the oxidation of fats will give values close to 0.7 and protein oxidation will produce values of about 0.8.

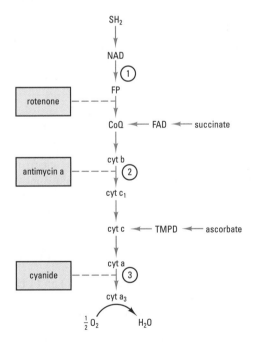

Fig. 66.8 The mitochondrial electron transport system, showing the sites of action of various inhibitors (dotted lines) and the three sites of ATP synthesis (numbered 1–3). S, substrate; FP, flavoprotein; CoQ, coenzyme Q; cyt, cytochrome; TMPD, tetramethylphenylenediamine.

enzyme-catalysed interconversion of organic compounds plus membrane-bound electron transport reactions that are coupled to ATP synthesis.

The enzymic reactions of glycolysis and the TCA cycle can be studied using ^{14}C-labelled intermediates and pulse-chase experiments, using techniques similar to those used to investigate the dark reactions of photosynthesis (p. 431), or by purification and characterisation of the individual enzymes. Oxidative phosphorylation can be studied using techniques appropriate for electron transport reactions.

Measuring the respiratory activity of cells and tissues

The overall rate of cellular respiration is usually measured in terms of the amount of O_2 consumed by a known amount of material in a given time, e.g. as μmol O_2 mg protein^{-1} min^{-1}.

The principal techniques include:

- **Manometry** – this traditional approach involves measuring either the pressure change (Warburg manometer) or volume change (Gilson manometer) as gases are produced or consumed during respiration. Manometry is relatively insensitive and may be subject to large measurement errors: the principal application is in studying the relationship between the amounts of O_2 consumed and CO_2 evolved in terms of the respiratory quotient (RQ) for a particular substrate, from the relationship:

$$RQ = \frac{CO_2 \text{ evolved}}{O_2 \text{ consumed}} \qquad [66.8]$$

The calculations involved in determining CO_2 production or O_2 consumption are complex and are specific to individual instruments – manufacturer's guidelines should be followed carefully.

- **Oxygen electrode studies** – this has largely replaced manometry, as the apparatus is more versatile and less complex to set up and interpret, giving a continuous readout of O_2 status via a chart recorder. Net O_2 evolution can be measured using whole cells, mitochondrial suspensions or submitochondrial preparations.

Using an oxygen electrode to study respiratory electron transport

Intact mitochondria suspended in an isotonic medium will show little respiratory activity unless supplied with (i) a suitable electron donor or substrate (e.g. NADH), (ii) ADP and (iii) P_i: this is termed respiratory control and the mitochondria are said to be tightly coupled, since there is a close link between electron transport, O_2 consumption and phosphorylation. The extent of this coupling can be determined using an O_2 electrode (p. 425) to measure O_2 consumption in the presence of substrate, P_i and ADP, with that in the absence of ADP (Fig. 66.7), as:

$$\text{respiratory control ratio} = \frac{\text{rate of } O_2 \text{ consumption with ADP}}{\text{rate of } O_2 \text{ consumption without ADP}} \qquad [66.9]$$

Freshly prepared, tightly coupled mitochondria should have a respiratory control ratio of ≥ 4.

To investigate the relationship between the number of ATP molecules produced per substrate molecule, substrate and P_i are supplied in excess and the O_2 uptake produced by a known amount of ADP is measured

(a)

(b)

Fig. 66.9 Representative data for O_2 uptake by mitochondria in response to the addition of (a) electron transport inhibitors and substrates, or (b) inhibitors of ATP synthesis and uncouplers. Steeper slopes indicate faster rates of O_2 uptake.

using an O_2 electrode (Fig. 66.8), allowing the P/O ($\equiv$ADP/O) ratio for a particular substrate to be calculated as:

$$P/O \text{ ratio} = \frac{\mu mol \text{ ADP added}}{2 \times \mu mol \text{ } O_2 \text{ consumed}} \qquad [66.10]$$

Note that the amount of O_2 is multiplied by 2, since O_2 is a diatomic molecule. The P/O ratio is determined by the site at which electrons are transferred to the respiratory chain (Fig. 66.9). Thus NADH should give a P/O ratio of 3, while succinate and $FADH_2$ should have values of 2. Artificial electron donors to cytochrome c have P/O ratios of 1, e.g. ascorbate/tetramethylphenylenediamine (TPMD). However, measured P/O ratios are always less than these values, due to partial uncoupling and the action of ATPases.

The effects of inhibitors of respiratory electron transport on the rate of oxygen consumption of mitochondria can be used to determine their site of action (Fig. 66.8). Thus, an inhibitor of electron flow will prevent O_2 uptake while an artificial substrate that donates electrons at a location beyond the site of inhibition will restore O_2 uptake, as shown in Fig. 66.9.

'Uncouplers' act by increasing the permeability of membranes to protons, causing the dissipation of the transmembrane H^+ gradient. Addition of an uncoupler increases electron flow along the respiratory chain in the absence of phosphorylation, i.e. there is loss of respiratory control. Uncouplers can reverse the effects of inhibitors such as oligomycin, whose target is the mitochondrial ATPase. Oligomycin prevents the return of H^+ to the interior of the mitochondrion, thereby inhibiting respiratory electron transport and O_2 uptake – the addition of an uncoupler (e.g. CCCP) reverses this inhibition.

> **KEY POINT** When working with inhibitors, make sure you treat all of these as potentially toxic, observing appropriate safety precautions (Chapter 22), and that you know the procedure to be followed in case of accident or spillage.

Measuring the components of the proton-motive force in chloroplasts and mitochondria

According to chemiosmotic principles, the transmembrane electrochemical potential gradient of protons drives ATP synthesis in photophosphorylation and oxidative phosphorylation via a membrane-bound ATPase (Fig. 66.6). This gradient is often expressed as the proton-motive force, or PMF (Δp), expressed in mV as:

$$\Delta p = E_m - 59(pH_i - pH_0) \qquad [66.11]$$

at 25 °C, where E_m is the transmembrane electrical potential (mV) and pH_i and pH_o are internal and external pH values. The individual components of the proton-motive force can be determined by a variety of methods:

- **Transmembrane electrical potential**, by the distribution of a suitable radiolabelled lipophilic cation, e.g. tetraphenylphosphonium (TPP$^+$),

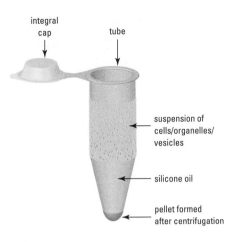

integral
cap

tube

suspension of
cells/organelles/
vesicles

silicone oil

pellet formed
after centrifugation

Fig. 66.10 Silicone oil microcentrifugation system: cells, organelles or membrane vesicles are pelleted below the silicone oil layer as a result of their higher density, while the less dense suspension medium remains about the silicone oil layer on centrifugation.

in accordance with the Nernst equation; by the measurement of K^+ or Rb^+ in the presence of the ionophore valinomycin; or by the quenching of a fluorescent dye (e.g. cyanine or oxanol dyes).

- **Transmembrane pH gradient** from the equilibrium distribution of a radiolabelled weak permeant acid, e.g. 5,5-dimethyl-2,4-oxazolidinedione (DMO); by quenching of a fluorescent pH probe, e.g. 9-aminoacridine; or by ^{31}P-NMR.

For radiolabelled probes, organelles or membrane vesicles can be separated from the bathing medium by silicone oil microcentrifugation (Fig. 66.10), with suitable correction for carry-over of external medium, e.g. using a membrane-impermeant solute, labelled with a second radioisotope. Care is required in such double-labelled experiments, as carry-over can represent a significant component, leading to substantial measurement errors if uncorrected.

Typical measurements for Δp across energy-transducing membranes are 180–200 mV (inside-negative): in mitochondria, E_m is the principal component while ΔpH represents the largest component of Δp in thylakoid membranes.

Text references

Maxwell, K. and Johnson, G.N. (2000) Chlorophyll fluorescence – a practical guide. *Journal of Experimental Botany,* **51**, 659–68.

Parsons, R., Weyers, J.D.B., Lawson, T. and Godber, I. (1997) Rapid and straightforward estimates of photosynthetic characteristics using a portable gas exchange system. *Photosynthetica*, **34**, 265–279.

Sources for further study

Lambers, H., Chapin, F.S. and Pons, T.L. (2008) *Plant Physiological Ecology,* 2nd edn. Springer, New York.

Nobel, P.S. (2009) *Physicochemical and Environmental Plant Physiology*, 4th edn. Academic Press, Oxford.

Ridge, I. (ed.) (2002) *Plants.* Oxford University Press, Oxford.
[Chapter 2 is a good general introduction to the physiology of photosynthesis.]

Taiz, L. and Zeiger, E. (2014) *Plant Physiology,* 6th edn. Sinauer, Sunderland, USA.
[Provides a detailed treatment of photosynthesis.]

West, J.B. (2011) *Respiratory Physiology: The Essentials,* 9th edn. Lippincott, Williams and Wilkins, Philadelphia.
[Provides a comprehensive guide to the human respiratory system.]

Wilson, K. and Walker, J. (eds) (2010) *Principles and Techniques of Practical Biochemistry and Molecular Biology,* 7th edn. Cambridge University Press, Cambridge.
[Also covers other topics, including spectroscopy and centrifugation.]

STUDY EXERCISES

(See also Study exercises 65.2 and 65.3 for calculations of respiratory oxygen consumption and photosynthetic oxygen evolution based on oxygen electrode measurements.)

66.1 Check your understanding of photosynthesis and respiration. Having read through this chapter, distinguish between each of the following pairs of terms:

(a) net photosynthesis and gross photosynthesis;

(b) C_3 photosynthesis and C_4 photosynthesis;

(c) the oxygenase and carboxylase functions of Rubisco;

(d) coupling and uncoupling;

(e) respiratory quotient and respiratory control ratio.

66.2 Calculate rates of photosynthetic carbon fixation from ^{14}C data. A sample of the photosynthetic cyanobacterium *Spirulina platensis* containing 0.29 mg chlorophyll *a* (chl a) was incubated in the light for 3 min in 10 mL of an aqueous solution containing inorganic carbon (HCO_3^-) at a concentration of 12 mmol L^{-1} and at a specific activity of 12 Bq pmol^{-1} ($\equiv$ 0.72 d.p.m. nmol^{-1}) At the end of this period, the cyanobacterial cells were separated from the medium and assayed by liquid scintillation counting, giving a quench-corrected value of 2808 d.p.m. What is the photosynthetic rate, expressed in terms of μmol C fixed per mg chl a per minute? (Express your answer to three significant figures.)

66.3 Interpret results from an experiment using DCPIP to study photosynthetic electron transport. The table below shows data for the change in A_{600} of a chloroplast suspension incubated with the Hill acceptor DCPIP, plus different levels of the herbicide simazine.

A_{600} measurements for a chloroplast suspension incubated with DCPIP and various levels of simazine

Time (s)	Control (no simazine)	Simazine at:		
		1.0 nmol L^{-1}	10.0 nmol L^{-1}	100.0 nmol L^{-1}
0	0.462	0.461	0.464	0.462
30	0.372	0.394	0.41	0.436
60	0.320	0.345	0.371	0.413
90	0.284	0.315	0.343	0.397
120	0.253	0.284	0.317	0.385

(a) What is the reason for the rapid decrease in A_{600} in the control sample, with no added simazine?

(b) How do you account for the fact that increasing amounts of simazine lead to a reduction in the extent of change in A_{600} with time?

(c) Plot a graph of A_{600} against time to show the composite data from the table. Use this to determine the initial rate of change in A_{600}. Then, calculate the percentage inhibition at each simazine concentration. Finally, plot percentage inhibition against simazine concentration, to determine the concentration causing 50% inhibition (I_{50}); give your answer to three significant figures.

66.4 Calculate the P/O ratio for a respiring mitochondrial suspension. An oxygen electrode chamber containing a mitochondrial suspension (0.5 mL), NADH (0.5 mL) and buffer (4 mL) showed a fall in oxygen concentration from 218 to 174 μ mol L^{-1} on addition of 0.05 mL of a 25 mmol L^{-1} ADP solution. What was the P/O ratio over this period?

66.5 Calculate mitochondrial proton-motive force. A suspension of mitochondria in buffer at an external pH of 7.2 had a measured internal (matrix) pH of 8.6 and a measured transmembrane potential (inner membrane) of −85.6 mV. Calculate the proton-motive force across the inner membrane (express your answer to three significant figures). What is your interpretation of the calculated value?

Answers to these study exercises are available at www.pearsoned.co.uk/practicalskills

67 Centrifugation

Table 67.1 Relationship between speed (r.p.m.) and acceleration (relative centrifugal field, RCF) for a typical bench centrifuge with an average radius of rotation, $r_{av} = 115$ mm

r.p.m.	RCF*
500	30
1000	130
1500	290
2000	510
2500	800
3000	1160
3500	1570
4000	2060
4500	2600
5000	3210
5500	3890
6000	4630

*RCF values rounded to nearest 10.

Particles suspended in a liquid will move at a rate that depends on:

- **the applied force** – particles in a liquid within a gravitational field, e.g. a stationary test tube, will move in response to the earth's gravity;
- **the density difference between the particles and the liquid** – particles less dense than the liquid will float upwards while particles denser than the liquid will sink;
- **the size and shape of the particles;**
- **the viscosity of the medium.**

For most biological particles (cells, organelles or molecules) the rate of flotation or sedimentation in response to the earth's gravity is too slow to be of practical use in separation.

> **KEY POINT**   A centrifuge is an instrument designed to produce a centrifugal force far greater than the earth's gravity, by spinning the sample about a central axis (Fig. 67.1 and Table 67.1). Particles of different size, shape or density will thereby sediment at different rates, depending on the speed of rotation and their distance from the central axis.

Fig. 67.1 Principal components of a low-speed bench centrifuge.
From diagram of low-speed centrifuge *model MSE Centaur 2*, supplied by Fisher Scientific UK Ltd, reproduced by kind permission of Fisher Scientific UK Ltd.

Working in SI units – to convert RCF to acceleration in SI units, multiply by 9.80 ms^{-2}: e.g. an RCF of 290 is equivalent to an acceleration of $290 \times 9.80 = 2842$ ms^{-1}.

Examples Suppose you wanted to calculate the RCF of a bench centrifuge with a rotor of $r_{av} = 95$ mm running at a speed of 3000 r.p.m. Using eqn [67.1] the RCF would be: $1.118 \times 95 \times (3)^2 = 956\,g$.

You might wish to calculate the speed (r.p.m.) required to produce a relative centrifugal field of 2000 g using a rotor of $r_{av} = 85$ mm. Using eqn [67.2] the speed would be: $945.7 \sqrt{(2000 \div 85)} = 4587$ r.p.m.

How to calculate centrifugal acceleration

The acceleration of a centrifuge is usually expressed as a multiple of the acceleration due to gravity ($g = 9.80$ ms^{-2}), termed the relative centrifugal field (RCF, or 'g value'). The RCF depends on the speed of the rotor (n, in revolutions per minute, r.p.m.) and the radius of rotation (r, in mm) where:

$$\text{RCF} = 1.118r\left(\frac{n}{1000}\right)^2 \qquad [67.1]$$

This relationship can be rearranged, to calculate the speed (r.p.m.) for specific values of r and RCF:

$$n = 945.7\sqrt{\left(\frac{\text{RCF}}{r}\right)} \qquad [67.2]$$

However, you should note that RCF is not uniform within a centrifuge tube: it is highest near the outside of the rotor (r_{max}) and lowest near the central axis (r_{min}). In practice, it is customary to report the RCF calculated from the average radius of rotation (r_{av}), as shown in Fig. 67.5 below. It is also worth noting that RCF varies relative to the *square* of the speed: thus the RCF will be doubled by an increase in speed of approximately 41% (Table 67.1).

Centrifugal separation methods

Differential sedimentation (pelleting)

By centrifuging a mixed suspension of particles at a specific RCF for a particular time, the mixture will be separated into a pellet and a supernatant (Fig. 67.2). The successive pelleting of a suspension by spinning for a fixed time at increasing RCF is widely used to separate organelles from cell homogenates. The same principle applies when cells are harvested from a liquid medium.

Density-gradient centrifugation

The following techniques use a density gradient, a solution that increases in density from the top to the bottom of a centrifuge tube (Fig. 67.3).

- **Rate-zonal centrifugation.** By layering a sample on to a shallow preformed density gradient, followed by centrifugation, the larger particles will move faster through the gradient than the smaller ones, forming several distinct zones (bands). This method is time-dependent, and centrifugation *must* be stopped before any band reaches the bottom of the tube (Fig. 67.4).

- **Isopycnic centrifugation.** This technique separates particles on the basis of their buoyant density. Several substances form density gradients during centrifugation (e.g. sucrose, CsCl, Ficoll, Percoll, Nycodenz). The sample is mixed with the appropriate substance and then centrifuged – particles form bands where their density corresponds to that of the medium (Fig. 67.4). This method requires a steep gradient and sufficient time to allow gradient formation and particle redistribution, but is unaffected by further centrifugation.

Bands within a density gradient can be sampled using a fine Pasteur pipette, or a syringe with a long, fine needle. Alternatively, the tube may be punctured and the contents (fractions) collected dropwise in several tubes. For accurate work, an upward displacement technique can be used (see Ford and Graham, 1991).

Fig. 67.2 Differential sedimentation. (a) Before centrifugation, the tube contains a mixed suspension of large, medium and small particles of similar density. (b) After low-speed centrifugation, the pellet is predominantly composed of the largest particles. (c) Further high-speed centrifugation of the supernatant will give a second pellet, predominantly composed of medium-sized particles. (d) A final ultracentrifugation step pellets the remaining small particles. Note that all of the pellets apart from the final one will have some degree of cross-contamination.

Fig. 67.3 Density gradients. (a) A continuous (linear) density gradient. (b) A discontinuous (stepwise) density gradient, formed by layering solutions of decreasing density on top of each other. (c) A single-step density barrier, designed to allow selective sedimentation of one type of particle.

Density-barrier centrifugation

A single-step density barrier (Fig. 67.3(c)) can be used to separate cells from their surrounding fluid, e.g. using a layer of silicone oil adjusted to the correct density using dinonyl phthalate. Blood cell types can be separated using a density barrier of e.g. Ficoll.

Types of centrifuge and their uses

Low-speed centrifuges

These are bench-top instruments for routine use, with a maximum speed of 3000–6000 r.p.m. and RCF up to 6000 g (Fig. 67.1). They are used to harvest cells, larger organelles (e.g. nuclei, chloroplasts) and coarse precipitates (e.g. antibody–antigen complexes, p. 378). Most modern machines also have a sensor that detects any imbalance when the rotor is spinning and cuts off the power supply (Fig. 67.1). However, some of the older models do not, and must be switched off as soon as any vibration is noticed, to prevent damage to the rotor or harm to the operator. Box 67.1 gives details of operation for a low-speed centrifuge.

Microcentrifuges (microfuges)

These are bench-top machines, capable of rapid acceleration up to 12 000 r.p.m. and 10 000 g. They are used to sediment small sample volumes (up to 1.5 mL) of larger particles (e.g. cells, precipitates) over short timescales (typically, 0.5–15 min). They are particularly useful for the rapid separation of cells from a liquid medium, e.g. silicone oil microcentrifugation.

Continuous-flow centrifuges

Useful for harvesting large volumes of cells from their growth medium. During centrifugation, the particles are sedimented as the liquid flows through the rotor.

High-speed centrifuges

These are usually larger, free-standing instruments with a maximum speed of up to 25 000 r.p.m. and RCF up to 60 000 g. They are used for microbial cells, many organelles (e.g. mitochondria, lysosomes) and protein precipitates. They often have a refrigeration system to keep the rotor cool at high speed. You would normally use such instruments only under direct supervision.

Ultracentrifuges

These are the most powerful machines, having maximum speeds in excess of 30 000 r.p.m. and RCF up to 600 000 g, with sophisticated refrigeration and vacuum systems. They are used for smaller organelles (e.g. ribosomes, membrane vesicles) and biological macromolecules. You would not normally use an ultracentrifuge, though your samples may be run by a member of staff.

Rotors

Many centrifuges can be used with tubes of different size and capacity, either by changing the rotor, or by using a single rotor with different buckets/adaptors.

- **Swing-out rotors:** sample tubes are placed in buckets that pivot as the rotor accelerates (Fig. 67.5(a)). Swing-out rotors are used on many low-speed centrifuges: their major drawback is their extended path length and the resuspension of pellets due to currents created during deceleration.

Working with silicone oil – the density of silicone oil is temperature-sensitive, so work in a location with a known, stable temperature or the technique may fail.

Fig. 67.4 Density-gradient centrifugation. The central tube shows the position of the sample prior to centrifugation, as a layer on top of the density gradient medium. Note that particles sediment on the basis of size during rate-zonal centrifugation (a), but form bands in order of their densities during isopycnic centrifugation (b). ●, large particles, intermediate density; ▲, medium-sized particles, low density; ◀━, small particles, high density.

Recording usage of high-speed centrifuges and ultracentrifuges – most departments have a log book (for samples/speeds/times): make sure you record these details, as the information is important for servicing and replacement of rotors.

SAFETY NOTE When changing a rotor, make sure that you carry it properly (do not knock/drop it), that you fit it correctly (do not cross-thread it, and tighten to the correct setting using a torque wrench) and that you store it correctly (clean it after use and do not leave it lying around).

(a)

(b)

(c)

Fig. 67.5 Rotors: (a) swing-out rotor; (b) fixed-angle rotor; (c) vertical-tube rotor.

- **Fixed-angle rotors:** used in many high-speed centrifuges and microcentrifuges (Fig. 67.5(b)). With their shorter path length, fixed rotors are more effective at pelleting particles than swing-out rotors.
- **Vertical-tube rotors:** used for isopycnic density-gradient centrifugation in high-speed centrifuges and ultracentrifuges (Fig. 67.5(c)). They cannot be used to harvest particles in suspension as a pellet is not formed.

Box 67.1 **How to use a low-speed bench centrifuge**

1. Choose the appropriate tube size and material for your application, with caps where necessary. Most low-speed machines have four-place or six-place rotors – use the correct number of samples to *fill* the rotor assembly whenever possible.

2. Fill the containers to the appropriate level: do not overfill, or the sample may spill during centrifugation.

3. It is vital that the rotor is balanced during use. Therefore, *identical* tubes must be prepared, to be placed opposite each other in the rotor assembly. This is particularly important for density-gradient samples, or for samples containing materials of widely differing densities, e.g. soil samples, since the density profile of the tube will change during a run. However, for low-speed work using small amounts of particulate matter in aqueous solution, it is sufficient to counterbalance a sample with a second tube filled with water, or a saline solution of similar density to the sample.

4. Balance each pair of sample tubes (plus the corresponding caps, where necessary) to within 0.1 g using a top-pan balance; add liquid dropwise to the lighter tube, until the desired weight is reached. Alternatively, use a set of scales. For small sample volumes (up to 10 mL) added to disposable, lightweight plastic tubes, accurate pipetting of your solution may be sufficient for low-speed use.

5. For centrifuges with swing-out rotors, check that each holder/bucket is correctly positioned in its locating slots on the rotor and that it is able to swing freely. All buckets must be in position on a swing-out rotor, even if they do not contain sample tubes – buckets are an integral part of the rotor assembly.

6. Load the sample tubes into the centrifuge. Make sure that the outside of the centrifuge tubes, the sample holders and sample chambers are dry: any liquid present will cause an imbalance during centrifugation, in addition to the corrosive damage it may cause to the rotor. For sample holders where rubber cushions are provided, make sure that these are correctly located. Balanced tubes must be placed opposite each other – use a simple code if necessary, to prevent mix-ups.

7. Bring the centrifuge up to operating speed by gentle acceleration. Do not exceed the maximum speed for the rotor and tubes used.

8. If the centrifuge vibrates at any time during use, switch off and find the source of the problem.

9. Once the rotor has stopped spinning, release the lid and remove all tubes. If any sample has spilled, make sure you clean it up thoroughly using a noncorrosive disinfectant, e.g. Virkon, so that it is ready for the next user.

10. Close the lid (to prevent the entry of dust) and return all controls to zero.

Centrifuge tubes

These are manufactured in a range of sizes (from 1.5 mL up to 1000 mL) and materials. The following aspects may influence your choice:

- **Capacity.** This is obviously governed by the volume of your sample. Note that centrifuge tubes must be completely full for certain applications, e.g. for high-speed work.

- **Shape.** Conical-bottomed centrifuge tubes retain pellets more effectively than round-bottomed tubes, while the latter may be more useful for density-gradient work.

- **Maximum centrifugal force.** Detailed information is supplied by the manufacturers. Standard Pyrex glass tubes can only be used at low centrifugal force (up to 2000 g).

- **Caps.** Most fixed-angle and vertical-tube rotors require tubes to be capped, to prevent leakage during use and to provide support to the tube during centrifugation. For low-speed centrifugation, caps must be used for any hazardous samples. Make sure you use the correct caps for your tubes.

- **Solvent resistance.** Glass tubes are inert, polycarbonate tubes are particularly sensitive to organic solvents (e.g. ethanol, acetone), while polypropylene tubes are more resistant. See manufacturer's guidelines for detailed information.

- **Sterility.** Disposable plastic centrifuge tubes are often supplied in sterile form. Glass and polypropylene tubes can be repeatedly sterilised. Cellulose ester tubes should *not* be autoclaved. Repeated autoclaving of polycarbonate tubes may lead to cracking/stress damage.

- **Opacity.** Glass and polycarbonate tubes are clear, while polypropylene tubes are more opaque.

- **Ability to be pierced.** If you intend to harvest your sample by puncturing the tube wall, cellulose acetate and polypropylene tubes are readily punctured using a syringe needle.

> **SAFETY NOTE** When working with centrifuge tubes never *be tempted to use a tube or bottle that was not designed to fit the machine you are using (e.g. a general-purpose glass test tube, or a screw-capped bottle), or you may damage the centrifuge and cause an accident.*

> Using microcentrifuge tubes – the integral push-on caps of microcentrifuge tubes must be correctly pushed home before use or they may come off during centrifugation.

Balancing the rotor

For the safe use of centrifuges, the rotor must be balanced during use (Box 67.1), or the spindle and rotor assembly may be damaged permanently; in severe cases, the rotor may fail and cause a serious accident.

Safe practice

Given their speed of rotation and the extremely high forces generated, centrifuges have the potential to be extremely dangerous, if used incorrectly.

> **SAFETY NOTE** *Never* balance centrifuge tubes 'by eye' – use a balance. Note that a 35 mL tube full of liquid at an RCF of 3000 g has an effective weight greater than a large adult man.

> **KEY POINT** It is vital that you balance your loaded centrifuge tubes before use. As a general rule, balance all sample tubes to within 1% or better, using a top-pan balance or scales. Place balanced tubes opposite each other.

You should only use a particular centrifuge if you fully understand the operating principles: if unsure, check with a member of staff. For safety reasons, all centrifuges are manufactured with an armoured casing that should contain any fragments in cases of rotor failure. Machines usually have a safety lock to prevent the motor from being switched on unless the lid is closed and to stop the lid from being opened while the rotor is moving. Do not be tempted to use older machines without a safety lock, or centrifuges where the locking mechanism is damaged/inoperative. Be particularly careful to make sure that hair and clothing are kept well away from moving parts.

Text reference

Ford, T.C. and Graham, J.M. (1991) *An Introduction to Centrifugation.* Bios Scientific Publishers, Oxford.

Sources for further study

Bonner, P. and Hargreaves, A. (2011) *Basic Bioscience Laboratory Techniques: A Pocket Guide.* Wiley, New York.

Boyer, R.F. (2011) *Biochemistry Laboratory: Modern Theory and Techniques*, 2nd edn. Benjamin Cummings, San Francisco. [Also covers other topics, including chromatography and spectroscopy.]

STUDY EXERCISES

67.1 **Decide on the type of centrifuge required for a particular application.** What centrifuge would you use for each of the following?

   (a) Separating mitochondria from a cell homogenate.

   (b) Separating yeast cells from their surrounding medium in an experiment to study the uptake of a radiolabelled amino acid as a function of time.

   (c) Harvesting cells from a bioreactor containing 25 L of growth medium.

67.2 **Determine centrifugal acceleration for a specific centrifuge.** Calculate the centrifugal acceleration (RCF, i.e. the *g* value) of the following (express your answers to three significant figures):

   (a) A bench centrifuge with an average radius of rotation of 125 mm operating at 4000 r.p.m.

   (b) A bench-top microcentrifuge with an average radius of rotation of 60 mm, operating at 12 000 r.p.m.

   (c) An ultracentrifuge with an average radius of rotation of 186 mm, operating at 30 000 r.p.m.

67.3 **Determine the speed required to give a particular centrifugal acceleration (*g* value).** Calculate the speed required (r.p.m. value) for each of the following (express your answers to three significant figures):

   (a) RCF of 1500 *g*, using a bench centrifuge with an average radius of rotation of 95 mm.

   (b) RCF of 50 000 *g*, using a high-speed centrifuge with an average radius of rotation of 135 mm.

   (c) RCF of 13 000 *g*, using a bench-top microcentrifuge with an average radius of rotation of 5.5 cm.

67.4 **Calculate the difference in centrifugal acceleration (*g* value) between the top, middle and bottom of a centrifuge tube.** Assuming that the minimum, average and maximum radial distances of a centrifuge tube in a swing-out rotor of a bench centrifuge operating at 5000 r.p.m. are 45 mm, 70 mm and 95 mm respectively, what are the corresponding *g* values at the top, middle and bottom of the tube when the centrifuge is operating? (Express your answer to three significant figures.)

Answers to these study exercises are available at www.pearsoned.co.uk/practicalskills

The absorption and emission of electromagnetic radiation of specific energy (wavelength) are characteristic features of many molecules, involving the movement of electrons between different energy states, in accordance with the laws of quantum mechanics. Spectroscopic techniques are used to measure and interpret such interactions between molecules and radiation.

UV/visible spectrophotometry

This is a widely used technique for measuring the absorption of radiation in the visible and UV regions of the spectrum. A spectrophotometer is an instrument designed to allow precise measurement at a particular wavelength, while a colorimeter is a simpler instrument, using filters to measure broader wavebands (e.g. light in the green, red or blue regions of the visible spectrum).

Principles of light absorption

Two fundamental principles govern the absorption of light by a solution:

1. **The absorption of light passing through a solution is exponentially related to the number of molecules of the absorbing solute, i.e. the solute concentration [C].**

2. **The absorption of light passing through a solution is exponentially related to the length of the absorbing solution, l.**

These principles are combined in the Beer–Lambert relationship (sometimes referred to simply as 'Beer's Law'), usually expressed in terms of absorbance (A)–the logarithm of the ratio of incident light (I_0) to emergent light (I):

$$A = \varepsilon l[C] \qquad [68.1]$$

where A is absorbance, ε is a constant for the absorbing substance at a specific wavelength, termed the absorptivity, or absorption coefficient, and $[C]$ is expressed either as mol L^{-1} or g L^{-1} (see p. 155) and l is given in cm. This relationship is extremely useful, since most spectrophotometers are constructed to give a direct measurement of absorbance, sometimes also termed extinction (E), of a solution (older texts may use the outdated term optical density, OD).

> **KEY POINT** The Beer–Lambert relationship states that there is a direct linear relationship between the concentration of a substance in a solution, [C], and the absorbance of that solution, A.

Absorbance at a particular wavelength is often shown as a subscript, e.g. A_{550} represents the absorbance at 550 nm. The proportion of light passing through the solution is known as the transmittance (T), and is calculated as the ratio of the emergent and incident light intensities.

Some instruments have two scales:

1. **an exponential scale** from zero to infinity, measuring absorbance;

2. **a linear scale** from 0 to 100, measuring (per cent) transmittance.

For most practical purposes, the Beer–Lambert relationship (eqn [68.1]) applies and you should use the absorbance scale.

Using spectroscopy – this technique is valuable for:

- tentatively identifying compounds, by determining their absorption or emission spectra;
- quantifying substances, either singly or in the presence of other compounds, by measuring the signal strength at an appropriate wavelength;
- determining molecular structure;
- following reactions, by measuring the disappearance of a substance, or the appearance of a product as a function of time.

Definitions

Absorbance (A) – this is given by:

$$A = \log_{10}(I_0/I)$$

usually shown as A_x where 'x' is the wavelength in nanometres.

Transmittance (T) – this is usually expressed as a percentage often at a particular wavelength, T_x, where

$$T_X = (I/I_0) \times 100(\%).$$

Example For incident light (I_0) = 1.00 and emergent light (I) = 0.16 (expressed in relative terms) then $A = \log_{10}(1.00 \div 0.16) = \log_{10} 6.25 = 0.796$ (to three significant figures). The corresponding transmittance, $T = (0.16 \div 1.00) \times 100 = 16\%$.

light condenser photocell amplifier/
source lens   readout
        sample coloured
        tube   filter

Fig. 68.1 Components of a colorimeter.

Colorimeter

This can be used with solutions where the test substance is highly coloured and present as the major constituent, e.g. haemoglobin in blood, or where a substance is assayed by adding a reagent that gives a coloured product (a chromophore), e.g. amino acid assay using ninhydrin reagent. Quantification of a particular substance requires a calibration curve, constructed using known amounts of the compound measured at the same time as the test samples, rather than using the Beer–Lambert relationship.

The light source is usually a tungsten filament bulb, focused by a condenser lens to give a parallel beam of light that passes through a glass sample tube or cuvette containing the solution, then through a coloured filter to a photocell detector, which develops an electrical potential in direct proportion to the intensity of the light falling on it (Fig. 68.1). The signal from the photocell is then amplified and passed to a galvanometer, or digital readout, calibrated on a logarithmic scale (see Box 68.1).

The broad bandwidth of most filters means that colorimetry cannot be used to identify a particular compound, nor to distinguish between two compounds with closely related absorption characteristics, e.g. in mixed solution. The photocells used in colorimeters have coefficients of variation (p. 509) of around 0.5%, so they are not suitable for work requiring a high degree of precision. In the simplest instruments, the logarithmic measurement scale has arbitrary units, adjusted via sensitivity/scale zero controls, and values obtained on one instrument will not be directly comparable with other instruments, or with the same instrument on different settings. A colorimeter is not suitable for quantitative work at a particular wavelength.

Box 68.1 How to use a colorimeter

1. **Switch on and stabilise**, allowing at least 5 min for the lamp to warm up before use.

2. **Choose a filter that is complementary** to the colour of the substance to be measured. Thus haemoglobin is red because it absorbs light within the blue/green regions of the spectrum and should be measured using a blue filter. Similarly, a blue substance should be measured using a red filter.

3. **Set the scale to zero** using an appropriate solution blank.

4. **Adjust the sensitivity** to give a reasonable deflection on the galvanometer/readout device, e.g. for a standard solution containing a known amount of the test substance. You would normally adjust the sensitivity so that your highest standard gave a reading close to the maximum on the readout device.

5. **Analyse your samples and standard solutions**, making sure you take all the measurements you need in a single 'run', otherwise they will not be directly comparable.

6. **Use the same tube in the same orientation** in the sample holder to improve precision, since individual tubes may differ in their light-absorbing characteristics, wall thickness, etc.

7. **Rinse the tube between samples**, making sure that the rinsing solution does not dilute the next sample and that the outside of the tube is dry before it is loaded into the instrument.

8. **Make frequent checks on reproducibility** by repeat measurements of the same solution (e.g. after every six to eight samples). You should also prepare test and standard solutions in duplicate.

9. **Plot a calibration curve for your standard solutions** and draw the line of best fit through the points. Do not worry if it is a curve, rather than a straight line, or if it does not pass through the origin. Do not extrapolate the calibration curve. If a sample has a reading greater than the highest standard, it should be diluted and reassayed. Chapter 53 gives further advice on preparing and using calibration curves.

Fig. 68.2 Components of a UV/visible spectrophotometer.

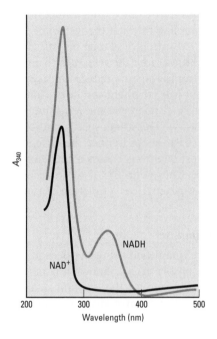

Fig. 68.3 Absorption spectra of nicotinamide adenine dinucleotide in oxidised (NAD⁺) and reduced (NADH) form. Note the 340 nm absorption peak, used for quantitative work (p. 387).

Using plastic disposable cuvettes – these are adequate for work in the near-UV region, e.g. for enzyme studies using nicotinamide coenzymes, at 340 nm, as well as the visible range. Table 23.2 gives details of the spectral cut-off wavelengths of different materials.

Measuring low absorbances – for accurate readings with dilute solutions, use a cuvette with a longer optical path length (e.g. 5 cm, rather than 1 cm).

UV/visible spectrophotometer

The principal components of a UV/visible spectrophotometer are shown in Fig. 68.2. High-intensity tungsten bulbs are used as the light source in basic instruments, capable of operating in the visible region (i.e. 400–700 nm). Deuterium lamps are used for UV spectrophotometry (200–400 nm); these lamps are fitted with quartz envelopes, since glass does not transmit UV radiation.

A major improvement over the simple colorimeter is the use of a diffraction grating to produce a parallel beam of monochromatic light from the (polychromatic) light source. In practice the light emerging from such a monochromator does not have a single wavelength, but is a narrow band of wavelengths. This bandwidth is an important characteristic, since it determines the wavelengths used in absorption measurements – the bandwidth of basic spectrophotometers is around 5–10 nm, while research instruments have bandwidths of less than 1 nm.

Bandwidth is affected by the width of the exit slit (the slit width), since the bandwidth will be reduced by decreasing the slit width. To obtain accurate data at a particular wavelength setting, the narrowest possible slit width should be used. However, decreasing the slit width also reduces the amount of light reaching the detector, decreasing the signal-to-noise ratio. The extent to which the slit width can be reduced depends upon the sensitivity and stability of the detection/amplification system and the presence of stray light.

Most UV/visible spectrophotometers are designed to take cuvettes with an optical path length of 1 cm. Disposable plastic cuvettes are suitable for routine work in the visible range using aqueous and alcohol-based solvents, and glass cuvettes are useful for other organic solvents. Glass cuvettes are manufactured to more exacting standards, so use optically matched glass cuvettes for accurate work, especially at low absorbances (<0.1), where any differences in the optical properties of cuvettes for reference and test samples will be pronounced. Glass and plastic absorb UV light, so quartz cuvettes must be used at wavelengths below 300 nm.

KEY POINT Before taking a measurement, make sure that cuvettes are clean, unscratched, dry on the outside, filled to the correct level and located in the correct position in their sample holders.

Proteins and nucleic acids in biological samples can accumulate on the inside faces of glass/quartz cuvettes, so remove any deposits using acetone on a cotton bud, or soak overnight in 1 mol L⁻¹ nitric acid. Corrosive and hazardous solutions must be used in cuvettes with tightly fitting lids, to prevent damage to the instrument and to reduce the risk of accidental spillage.

Basic instruments use photocells similar to those used in colorimeters or photodiode detectors. In many cases, a different photocell must be used at wavelengths above and below 550–600 nm, due to differences in the sensitivity of such detectors over the visible waveband. The detectors used in more sophisticated instruments give increased sensitivity and stability when compared with photocells.

Digital displays are increasingly used in preference to needle-type meters, as they are not prone to parallax errors and misreading of the scale. Some digital instruments can be calibrated to give a direct readout of the concentration of the test substance.

Measuring high absorbances in colorimetric analysis – if any final solution has an absorbance that is too high to be read with accuracy on your spectrophotometer (i.e. A > 1), it is not appropriate to dilute the solution so that it can be measured. This dilutes both the sample molecules and the colour reagents to an equal extent. Instead, you should dilute the original sample and re-assay.

KEY POINT Basic spectrophotometers are most accurate within the absorbance range from 0.00 to 1.00 and your standard and test solutions should be prepared to give readings within this range.

Types of UV/visible spectrophotometer

Basic instruments are single-beam spectrophotometers in which there is only one light path. The instrument is set to zero absorbance using a blank solution, which is then replaced by the test solution, to obtain an absorbance reading. An alternative approach is used in double-beam spectrophotometers, where the light beam from the monochromator is split into two separate beams, one beam passing through the test solution and the other through a reference blank. Absorbance is then measured by an electronic circuit that compares the output from the reference (blank) and sample cuvettes. Double-beam spectrophotometry reduces measurement errors caused by fluctuations in output from the light source or changes in the sensitivity of the detection system, since reference and test solutions are measured at the same time (Box 68.2). Recording spectrophotometers are double-beam instruments, designed to record either the difference in absorbance between reference and test solutions across a predetermined waveband to give an absorption spectrum (Fig. 68.3), or to record the change in absorbance at a particular wavelength as a function of time (e.g. in an enzyme assay, see Chapter 60).

Examples in quantitative analysis
The molar absorptivity of NADH is 6.22×10^3 L mol^{-1} cm^{-1} at 340 nm. For a test solution giving an absorbance of 0.21 in a cuvette with a light path of 5 mm, using eqn [68.1] this is equal to a concentration of:

$0.21 = 6.22 \times 10^3 \times 0.5 \times [C]$
$[C] = 0.0000675$ mol L^{-1}
(or 67.5 μmol L^{-1}).

The specific absorptivity (10 g L^{-1}) of double-stranded DNA is 200 at 260 nm, therefore a solution containing 1 g L^{-1} will have an absorbance of 200/10 = 20. For a DNA solution, giving an absorbance of 0.35 in a cuvette with a light path of 1.0 cm, using eqn [68.1] this is equal to a concentration of:

$0.35 = 20 \times 1.0$ [C]
$[C] = 0.0175$ g L^{-1}
(equivalent to 17.5 μg mL^{-1}).

Chlorophylls a and b in vascular plants and green algae can be assayed in 90% v/v acetone/water by measuring the absorbance of the mixed solution at two wavelengths, according to the formulae:

Chlorophyll a (mg L^{-1}) = $11.93 A_{664} - 1 93 A_{647}$
Chlorophyll b (mg L^{-1}) = $20.36 A_{647} - 5.5 A_{664}$

Note: different equations are required for other solvents.

Quantitative spectrophotometric analysis

A single substance in solution can be quantified using the Beer–Lambert relationship (eqn [68.1]), provided its absorptivity is known at a particular wavelength (usually the absorption maximum for the substance, since this will give the greatest sensitivity). The molar absorptivity is the absorbance given by a solution with a concentration of 1 mol L^{-1} (= 1 kmol m^{-3}) of the compound in a light path of 1 cm. The appropriate value may be available from tabulated spectral data (e.g. Anon, 1963), or it can be determined experimentally by measuring the absorbance of known concentrations of the substance (Box 68.2) and plotting a standard curve (Chapter 53). This should confirm that the relationship is linear over the desired concentration range and the slope of the line will give the molar absorptivity.

The specific absorptivity is the absorbance given by a solution containing 10 g L^{-1} (i.e. 1 % w/v) of the compound in a light path of 1 cm. This is useful for substances of unknown molecular weight, e.g. proteins or nucleic acids, where the amount of substance in solution is expressed in terms of its mass, rather than as a molar concentration. For use in eqn [68.1], the specific absorptivity should be divided by 10 to give the solute concentration in g L^{-1}.

This simple approach cannot be used for mixed samples. In such cases, it may be possible to estimate the amount of each substance by measuring the absorbance at several wavelengths, e.g. protein estimation in the presence of nucleic acids (p. 387). Further details on spectrophotometric methods of determining the amount of protein in an aqueous sample are given in Chapter 60.

Atomic spectroscopy

Atoms of certain metals will absorb and emit radiation of specific wavelengths when heated in a flame, in direct proportion to the number of atoms present. Atomic spectrophotometric techniques measure the

SAFETY NOTE In atomic spectroscopy, the use of high-pressure gas cylinders can be particularly hazardous. Always consult a member of staff before using such apparatus.

Box 68.2 How to use a spectrophotometer

1. Switch on and select the correct lamp for your measurements (e.g. deuterium for UV, tungsten for visible light).

2. Allow at least 15 min for the lamp to warm up and for the instrument to stabilise before use.

3. Select the appropriate wavelength: on older instruments a dial is used to adjust the monochromator, whereas newer machines have microprocessor-controlled wavelength selection.

4. Select the appropriate detector: some instruments choose the correct detector automatically (on the basis of the specified wavelength), and others have manual selection.

5. Choose the correct slit width (if available): this may be specified in the protocol you are following, or may be chosen on the manufacturer's recommendations.

6. Insert appropriate reference blank(s): single-beam instruments use a single cuvette, whereas doublebeam instruments use two cuvettes (a matched pair for accurate work). The reference blanks should match the test solution in all respects apart from the substance under test, i.e. they should contain all reagents apart from this substance. *Make sure that the cuvettes are positioned correctly, with their polished (transparent) faces in the light path, and check that they are accurately located in the cuvette holder(s).*

7. Check/adjust the 0% transmittance: most instruments have a control that allows you to zero the detector output in the absence of any light (so-called dark current correction). Some microprocessor-controlled instruments carry out this step automatically.

8. Set the absorbance reading to zero: usually via a dial, or digital readout.

9. Analyse your samples: replace the appropriate reference blank with a test sample, allow the absorbance reading to stabilise (5–10 s) and read the absorbance value from the meter/readout device. For absorbance readings greater than 1 (i.e. < 10% transmittance), the signal-to-noise ratio is too low for accurate results. Your analysis may require a calibration curve (Chapter 53), or you may be able to use the Beer–Lambert relationship (eqn [68.1]) to determine the concentration of test substance in your samples.

10. Check the scale zero at regular intervals using *a* reference blank, e.g. after every 10 samples.

11. Check the reproducibility of the instrument: measure the absorbance of a single solution several times during your analysis. It should give the same value.

Problems (and solutions): inaccurate/unstable readings are most often due to incorrect use of cuvettes, e.g. dirt, fingerprints or test solution on outside of cuvette (wipe the polished faces using a soft tissue before insertion into the cuvette holder), condensation (if cold solutions are not allowed to reach room temperature before use), air bubbles (which scatter light and increase the absorbance; tap gently to remove), insufficient solution (causing refraction of light at the meniscus), particulate material in the solution (centrifuge before use, where necessary) or incorrect positioning in light path (locate in correct position).

Fig. 68.4 Components of a flame photometer.

SAFETY NOTE When carrying out acid digestion, *always* work within a fume hood and wear gloves and safety glasses throughout the procedure. Rinse any spillages with a large volume of water.

---

absorption or emission of particular wavelengths of UV and visible light, to identify and quantify such metals.

Flame atomic-emission spectrophotometry (or flame photometry)

The principal components of a flame photometer are shown in Fig. 68.4. A liquid sample is converted into an aerosol in a nebuliser (atomiser) before being introduced into the flame, where a small proportion (typically less than 1 in 10 000) of the atoms will be raised to a higher energy level, releasing this energy as light of a particular wavelength, which is passed through a filter to a photocell detector. Flame photometry is used to measure the alkali metal ions K^+, Na^+ and Ca^{++} in biological fluids. Box 68.3 gives details on how to use a flame photometer.

Atomic-absorption spectrophotometry (or flame absorption spectrophotometry)

This technique is applicable to a broad range of metal ions, including those of Pb, Cu, Zn, etc. It relies on the absorption of light of a specific

---

Box 68.3 How to use a flame photometer

1. **Switch on the instrument and allow it to stabilise.** Light the flame and then wait for at least 5 min before analysing your solutions.

2. **Check for impurities in your reagents.** For example, if you are measuring K^+ in an acid digest of some biological material, e.g. plant tissue, you should check the K^+ content of a reagent blank (a solution containing everything except the plant tissue, prepared in exactly the same way as the samples). Once converted to concentration (see below) you should then subtract this value from the concentration determined for each of your sample solutions, to obtain the true K^+ content.

3. **Quantify your samples using a calibration curve** (Chapter 53). Calibration standards should cover the expected concentration range for the test solutions – your calibration curve may be non-linear, especially at higher concentrations (e.g. above 1 mmol L^{-1}, i.e. 1 mol m^{-3} in SI units).

4. **Assay all solutions in duplicate,** so that repeatability can be assessed (if duplicate readings do not agree, repeat until you are satisfied that the measurements are reliable).

5. **Check your calibration.** Make repeated measurements of a standard solution of known concentration after every six or seven samples, to confirm that the instrument calibration is still valid.

6. **Consider the possibility of interference.** Other metal atoms may emit light that is detected by the photocell, since the filters cover a wider waveband than the emission wavelength of a particular element. This can be a serious problem if you are trying to measure low concentrations of a particular metal in the presence of high concentrations of other metals (e.g. Na^+ in sea water), or if there are other substances present that form complexes with the test metal, suppressing the signal (e.g. phosphate).

Plotting calibration curves in quantitative analysis – do not force your calibration line to pass through zero if it clearly does not. There is *no* reason to assume that the zero value is any more accurate than any other reading you have made (see also Chapter 53).

wavelength by atoms dispersed in a flame. The appropriate wavelength is provided by a cathode lamp, coated with the element to be analysed, focused through the flame and on to the detector. When the sample is introduced into the flame, it will decrease the light detected in direct proportion to the amount of metal present. Practical advantages over flame photometry include improved sensitivity, increased precision and decreased interference. Newer variants of this method include flameless atomic absorption spectrophotometry and atomic fluorescence spectrophotometry, both of which are more sensitive than the flame-absorption technique.

Other spectroscopic methods

The range of advanced techniques used in biological research is detailed in Table 68.1 (see also Reed *et al.* (2016) for e.g. fluorescence spectrophotometry).

Table 68.1 Advanced spectroscopic techniques

Technique	Biological applications
Spectrofluorimetry	Quantification of fluorescent molecules, e.g. chlorophyll, porphyrins Enzyme assay, using fluorogenic substrates
Infrared (IR) spectrometry	Gas analysis Identification of molecular structure
Electron spin resonance (ESR) spectrometry	Metalloprotein analysis Free radical behaviour
Mass spectrometry	Biochemical structure/identification Used with gas chromatography (pp. 456–7)
Nuclear magnetic resonance (NMR) spectrometry	Macromolecule structure (^1H and ^{13}C NMR) Metabolic studies (^{13}C and ^{31}P NMR) Intracellular pH (^{31}P NMR)
Circular dichroism spectroscopy	Macromolecule structure, e.g. proteins and nucleic acids

Text references

Anon. (1963) *Tables of Spectrophotometric Absorption Data for Compounds used for the Colorimetric Detection of Elements (International Union of Pure and Applied Chemistry)*. Butterworth-Heinemann, London.

Sources for further study

Boyer, R.F. (2011) *Biochemistry Laboratory: Modern Theory and Techniques,* 2nd edn. Prentice Hall, Harlow.

Gore, M.G. (ed.) (2000) *Spectrophotometry and Spectrofluorimetry: A Practical Approach,* 2nd edn. Oxford University Press, Oxford.

Ninfa, A.J., Ballou, D.P. and Benore, M. (2009) *Fundamental Laboratory Approaches for*

Reed, R.H., Holmes, D.H., Weyers, J.D.B. and Jones, A.M. (2016) *Practical Skills in Biomolecular Sciences,* 5th edn. Pearson, Harlow.

Biochemistry and Biotechnology, 2nd edn. Wiley, New York.

Wilson, K. and Walker, J. (eds.) (2010) *Principles and Techniques of Biochemistry and Molecular Biology,* 7th edn. Cambridge University Press, Cambridge. [Also covers a range of other topics, including chromatography and centrifugation.]

STUDY EXERCISES

68.1 Write a protocol for using a spectrophotometer. After reading this chapter, prepare a detailed stepwise protocol explaining how to use one of the spectrophotometers in your department. Ask another student or a tutor to evaluate your protocol and provide you with feedback.

68.2 Use the Beer–Lambert relationship in quantitative spectrophotometric analysis. Calculate the following (express your answer to three significant figures):

(a) The concentration of NADH (μmol L^{-1}) in a test solution giving an absorbance at 340 nm (A_{340}) of 0.53 in a cuvette with a path length of 1 cm, based on a molar absorptivity for NADH of 6220 L mol^{-1} cm^{-1} at this wavelength.

(b) The amount of NADH (nmol) in 20 mL of a test solution where A_{340} = 0.62 in a cuvette with a path length of 5 cm, based on a molar absorptivity for NADH of 6220 L mol^{-1} cm^{-1} at this wavelength.

(c) The mass concentration (μgmL) of double-stranded DNA in a test solution giving an absorbance at 260 nm (A_{260}) of 0.57 in a cuvette of path length 5 mm, based on an absorptivity of 20 L g^{-1} cm^{-1}.

(d) The amount (ng) of double-stranded DNA in a 50 μL subsample from of a test solution where A_{260} = 0.31 in a cuvette of path length 1 cm, based on an absorptivity of 20 L g^{-1} cm^{-1}.

(e) The chlorophyll *a* content, in μg (g fresh weight)$^{-1}$, of a plant leaf of fresh weight 1.56 g extracted in 50 mL of 90% acetone, where A_{664} = 0.182 at 664 nm and A_{647} = 0.035, based on the equation: chlorophyll *a* (mg L^{-1}) = 11.93 A_{664} − 1.93 A_{647}.

68.3 Determine the molar absorptivity of a substance in aqueous solution. A solution of *p*-nitrophenol containing 8.8 μg mL^{-1} gave an absorbance of 0.535 at 404 nm in a cuvette of path length 1 cm. What is the molar absorptivity of *p*-nitrophenol at 404 nm, expressed to three significant figures? (Note: M_r of *p*-nitrophenol is 291.27.)

68.4 Determine the concentration of metal ions based on atomic spectroscopy of test and standard solutions. The following data represent a set of calibration standards for K^+ in aqueous solution, measured by flame photometry:

Absorbance of standard solutions containing K^+ at up to 0.5 mmol L^{-1}

K^+ concentration (mmol L^{-1})	Absorbance
0	0.000
0.1	0.155
0.2	0.279
0.3	0.391
0.4	0.537
0.5	0.683

Draw a calibration curve using the above data and use this to estimate the amount of K^+ in a test sample prepared by digestion of 0.482 g of tissue in a final volume of 25 mL of solution, giving an absorbance of 0.429 when measured at the same time as the standards shown above. Express your answer in μmol K^+ (g tissue)$^{-1}$, to three significant figures. See also Study exercise 53.1 for a similar exercise, based on atomic absorption spectrophotometry of Zn.

Answers to these study exercises are available at www.pearsoned.co.uk/practicalskills

Chromatography is used to separate the individual constituents within a sample on the basis of differences in their physical characteristics, e.g. molecular size, shape, charge, volatility, solubility and/or adsorptivity. The essential components of a chromatographic system are:

- **A stationary phase,** either a solid, a gel or an immobilised liquid, held by a support matrix.
- **A chromatographic bed:** the stationary phase may be packed into a glass or metal column, spread as a thin layer on a sheet of glass or plastic, or adsorbed on cellulose fibres (paper).
- **A mobile phase,** either a liquid or a gas that acts as a solvent, carrying the sample through the stationary phase and eluting from the chromatographic bed.
- **A delivery system** to pass the mobile phase through the chromatographic bed.
- **A detection system** to monitor the test substances.

Individual substances interact with the stationary phase to different extents as they are carried through the system, enabling separation to be achieved.

> **KEY POINT** In a chromatographic system, those substances that interact strongly with the stationary phase will be retarded to the greatest extent while those that show little interaction will pass through with minimal delay, leading to differences in distances travelled or elution times.

Types of chromatographic system

Chromatographic systems can be categorised according to the form of the chromatographic bed, the nature of the mobile and stationary phases and the method of separation.

Thin-layer chromatography (TLC) and paper chromatography

Here, you apply the sample as a single spot near one end of the sheet or plate, by microsyringe or microcapillary. This sheet is allowed to dry fully, then it is transferred to a glass tank containing a shallow layer of solvent (Fig. 69.1). Remove the sheet when the solvent front has travelled across 80–90% of its length.

You can express movement of an individual substance in terms of its relative frontal mobility, or R_F value, where:

$$R_F = \frac{\text{distance moved by substance}}{\text{distance moved by solvent}} \qquad [69.1]$$

Alternatively, you may express movement with respect to a standard of known mobility, as R_X, where:

$$R_X = \frac{\text{distance moved by test substance}}{\text{distance moved by standard}} \qquad [69.2]$$

The R_F (or R_X) value is a constant for a particular substance and solvent system (under standard conditions) and closely reflects the partitioning of

---

> Making compromises in chromatography – the process is often a three-way compromise between:
>
> 1. separation of analytes;
> 2. time taken for analysis;
> 3. volume of eluent.
>
> Thus, if you have a large sample volume and want to achieve a good separation of a mixture of analytes, the time taken for the chromatography will be lengthy.

> **SAFETY NOTE** The solvents used as the mobile phases of chromatographic systems are often toxic and may produce noxious fumes – where necessary, work in a fume hood.

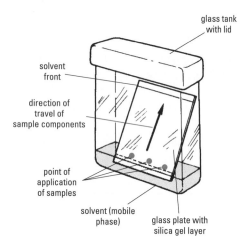

Fig. 69.1 Components of a TLC system.

> Using a TLC system – it is essential that you allow the solvent to pre-equilibrate in the chromatography tank for at least 2 h before use, to saturate the atmosphere with vapour. Deliver drops of sample with a blunt–ended microsyringe. Make sure you know exactly where each sample is applied, so that R_F values can be calculated.

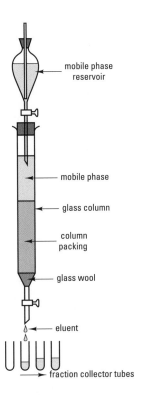

Fig. 69.2 Equipment for column chromatography (gravity-feed system).

Fig. 69.3 Peak characteristics in a chromatographic separation, i.e. a chromatogram.

Using HPLC – HPLC is a versatile form of chromatography, used with a wide variety of stationary and mobile phases, to separate individual compounds of a particular class of molecules on the basis of size, polarity, solubility or adsorption characteristics.

the substance between the stationary and mobile phases. Tabulated values are available for a range of biological molecules and solvents (e.g. Stahl, 1969). However, you should analyse one or more reference compounds on the same sheet/plate as your unknown sample, to check their R_F values.

Column chromatography

Here, you pack a glass column with the appropriate stationary phase and equilibrate the mobile phase by passage through the column, either by gravity (Fig. 69.2), or using a low-pressure peristaltic pump. You can then introduce the sample to the top of the column, to form a discrete band of material. This is then flushed through the column by the mobile phase. If the individual substances have different rates of migration, they will separate within the column, eluting at different times as the mobile phase travels through the column.

You can detect eluted substances by collecting the mobile phase as it elutes from the column in a series of tubes (discontinuous monitoring), either manually or with an automatic fraction collector. Fractions of 2–5% of the column bed volume are usually collected and analysed, e.g. by chemical assay. You can now construct an elution profile (or chromatogram) by plotting the amount of substance against either time, elution volume or fraction number, which should give a symmetrical peak for each substance (Fig. 69.3).

You can express the migration of a particular substance at a given flow rate in terms of its retention time (t), or elution volume (V_e). The separation efficiency of a column is measured by its ability to distinguish between two similar substances, assessed in terms of:

- **Selectivity** (α), measured using the following equation, which takes into account the retention times of the two peaks (i.e. t_a and t_b), plus the column dead time (t_0):

$$\alpha = \frac{t_b - t_0}{t_a - t_0} \qquad [69.3]$$

The column dead time is the time it takes for an unretained compound to pass through the column without interacting with the stationary phase.

- **Resolution** (R), quantified in terms of the retention time and the base width (W) of each peak:

$$R = \frac{2(t_a - t_b)}{W_a + W_b} \qquad [69.4]$$

The subscripts a and b in eqns [69.3] and [69.4] refer to substances a and b respectively (Fig. 69.3). For most practical purposes, R values of 1 or more are satisfactory, corresponding to 98% peak separation for symmetrical peaks.

High-performance liquid chromatography (HPLC)

Column chromatography originally used large 'soft' stationary phases that required low-pressure flow of the mobile phase to avoid compression; separations were usually time-consuming and of low resolution ('low performance'). Subsequently, the production of small, incompressible, homogeneous particulate support materials and high-pressure pumps with reliable, steady flow rates have enabled high-performance systems to be developed. These systems operate at pressures up to 10 MPa, forcing the mobile phase through the column at a high flow rate to give rapid separation with reduced band broadening, due to smaller particle size.

HPLC columns are usually made of stainless steel, and all components, valves, etc., are manufactured from materials that can withstand the high pressures involved. The two main solvent-delivery systems are:

1. **Isocratic separation:** a single solvent (or solvent mixture) is used throughout the analysis.
2. **Gradient elution separation:** the composition of the mobile phase is altered using a microprocessor-controlled gradient programmer, which mixes appropriate amounts of two different substances to produce the required gradient.

Most HPLC systems are linked to a continuous monitoring detector of high sensitivity, e.g. proteins may be detected spectrophotometrically by monitoring the absorbance of the eluent at 280 nm as it passes through a flow cell (cuvette). Other detectors can be used to measure changes in fluorescence, refractive index, ionisation, radioactivity, etc. The detector is linked to a recorder or microcomputer to produce an elution profile. Most detection systems are non-destructive, which means that you can collect eluent with an automatic fraction collector for further study (Fig. 69.4).

Separation of macromolecules (especially proteins and nucleic acids) usually requires 'biocompatible' systems in which stainless steel components are replaced by titanium, glass or fluoroplastics, using lower pressures to avoid denaturation, e.g. the Pharmacia FPLC system. Such separations are carried out using ion-exchange, gel-permeation and/or hydrophobic-interaction chromatography (p. 459).

Fig. 69.4 Components of an HPLC system.

Gas chromatography (GC)

Modern GC uses capillary chromatography columns (internal diameter 0.1–0.5 mm) up to 50 m in length (Fig. 69.5). The stationary phase is generally a cross-linked silicone polymer, coated as a thin film on the inner wall of the capillary: at normal operating temperatures, this behaves in a similar manner to a liquid film, but is far more robust. The mobile phase ('carrier gas') is usually nitrogen or helium. Selective separation is achieved as a result of the differential partitioning of individual compounds between the carrier gas and silicone polymer phases. The separation of most biomolecules is influenced by the temperature of the column, which may be constant during the analysis ('isothermal' – usually 50–250 °C) or, more commonly, may increase in a pre-programmed manner (e.g. from 50 °C to

---

Learning from experience – if you are unable to separate your test substance(s) using a particular method, do not regard this as a failure, but instead, think about what this tells you about either the substance(s) or your sample.

Applications of high-performance liquid chromatography – the speed and sensitivity of HPLC makes this the method of choice for the separation of many small molecules of biological interest, normally using reverse-phase partition chromatography (p. 457).

Applications of gas chromatography – GC is used to separate volatile, nonpolar compounds: substances with polar groups must be converted to less polar derivatives prior to analysis, to prevent adsorption on the column, leading to poor resolution and peak tailing.

Fig. 69.5 Components of a GC system.

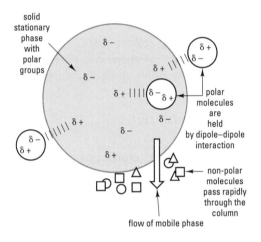

Fig. 69.6 Adsorption chromatography (polar stationary phase).

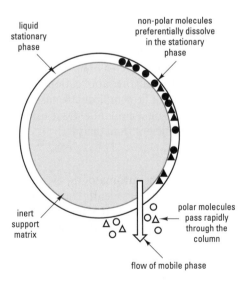

Fig. 69.7 Liquid–liquid partition chromatography, e.g. reverse-phase HPLC.

250 °C at 10 °C per min). Samples are injected on to the 'top' of the column, through an injection port containing a gas-tight septum. The output from the column can be monitored by:

- **Flame ionisation:** the outflow gas is passed through a flame where any organic compounds will be ionised and subsequently detected by an electrode mounted near the flame tip.

- **Electron capture:** using a beta-emitting radioisotope as the means of ionisation. This is capable of detecting extremely small amounts (pmol) of electrophilic compounds.

- **Spectrometry:** including mass spectrometry (GC-MS) and infrared spectrometry (GC-IR).

- **Thermal conductivity:** changes in the composition of the gas at the outflow alter the resistance of a platinum wire.

GC can only be used directly with samples capable of volatilisation at the operating temperature of the column, e.g. short-chain fatty acids. Other substances need to be chemically modified to produce more volatile compounds, e.g. long-chain fatty acids are usually analysed as methyl esters while monosaccharides are converted to trimethylsilyl derivatives.

Separation methods

Adsorption chromatography

This is a form of solid–liquid chromatography. The stationary phase is a porous, finely divided solid that adsorbs molecules of the test substance on its surface due to dipole–dipole interactions, hydrogen bonding and/or van der Waals interactions (Fig. 69.6). The range of adsorbents is limited, e.g. polystyrene-based resins (for non-polar molecules), silica, aluminium oxide and calcium phosphate (for polar molecules). Most adsorbents must be activated by heating to 110–120 °C before use, since their adsorptive capacity is decreased by bound water. Adsorption chromatography can be carried out in column or thin-layer form, using a range of organic solvents.

Partition chromatography

This is based on the partitioning of a substance between two liquid phases, in this instance the stationary and mobile phases. Substances that are more soluble in the mobile phase will pass rapidly through the system, whereas those that favour the stationary phase will be retarded (Fig. 69.7). In normal phase partition chromatography the stationary phase is a polar solvent, usually water, supported by a solid matrix (e.g. cellulose fibres in paper chromatography) and the mobile phase is an immiscible, non-polar organic solvent. For reverse-phase partition chromatography the stationary phase is a non-polar solvent (e.g. a C_{18} hydrocarbon, such as octadecylsilane) that is chemically bonded to a porous support matrix (e.g. silica), while the mobile phase can be chosen from a wide range of polar solvents, usually water or an aqueous buffered solution containing one or more organic solvents, e.g. acetonitrile. Solutes interact with the stationary phase through non-polar interactions and so the *least* polar solutes elute last from the column. Solute retention and separation are controlled by changing the composition of the mobile phase (e.g. % v/v acetonitrile). Reverse-phase high-performance liquid chromatography (RP-HPLC) can separate a range of non-polar, polar and ionic biomolecules, including peptides, proteins, oligosaccharides and vitamins.

Selecting a separation method – it is often best to select a technique that involves direct interaction between the substance(s) and the stationary phase (e.g. ion-exchange, affinity or hydrophobic-interaction chromatography), due to the increased capacity and resolution compared with other methods (e.g. partition or gel-permeation chromatography).

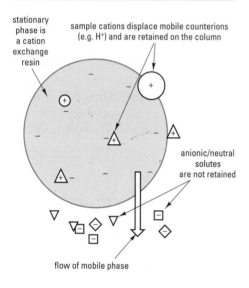

Fig. 69.8 Ion-exchange chromatography (cation exchanger).

Using a gel-permeation system – keep your sample volume as small as possible, as low resolution means that the sample will be diluted during passage through the column.

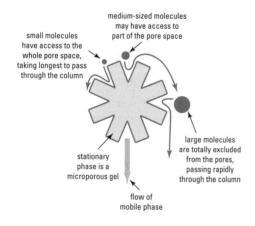

Fig. 69.9 Gel-permeation chromatography.

Ion-exchange chromatography

Here, separations are carried out using a column packed with a porous matrix that has a large number of ionised groups on its surfaces, i.e. an ion-exchange resin. The groups may be cation or anion exchangers, depending upon their affinity for positive or negative ions. The net charge of a resin depends on the pK_a of the ionisable groups and the solution pH, in accordance with the Henderson–Hasselbalch equation (p. 167).

For most practical applications, you should select the ion-exchange resin and buffer pH so that the test substances are strongly bound by electrostatic attraction to the ion-exchange resin on passage through the system, while the other components of the sample are rapidly eluted (Fig. 69.8). You can then elute the bound ions by changing the pH (which will alter the affinity of the test substance for the resin) or by raising the salt concentration of the mobile phase (to displace the bound ions); for instance, if you pass a solution of increasing concentration through the system, the weakly bound ions will elute first while the strongly bound ions will be elute at a higher concentration.

Ion-exchange chromatography can be used to separate mixtures of a wide range of ions, including amino acids, peptides, proteins and nucleotides. Electrophoresis is an alternative means of separating charged molecules, as described in Chapter 70.

Gel-permeation chromatography (molecular-exclusion chromatography)

Here, the stationary phase is a cross-linked gel containing a network of minute pores and channels of various sizes. Large molecules may be completely excluded from these pores, passing through the interstitial spaces and eluting rapidly from the column in the liquid mobile phase. Smaller molecules will penetrate the gel matrix to an extent that will depend upon their size and shape, retarding their progress through the column (Fig. 69.9). Substances will therefore be eluted from a gel-permeation column in decreasing order of their molecular size.

Cross-linked dextrans (e.g. Sephadex), agarose (e.g. Sepharose) and polyacrylamide (e.g. Bio-Gel) are used to separate mixtures of macromolecules, particularly enzymes, antibodies and other globular proteins. Calibration of a gel-filtration column using molecules of similar shape and known molecular mass enables the molecular mass of other components to be estimated, since a plot of elution volume (V_e) against $\log_{10}$ molecular mass is approximately linear. A further application of gel-permeation chromatography is for separating low-molecular-mass and high-molecular-mass components, e.g. desalting a protein extract using a Sephadex G–25 column is faster and more efficient than dialysis.

Affinity chromatography

This is a highly specific form of adsorption chromatography, where a particular binding molecule (or ligand) is covalently attached to a solid matrix in such a way that the affinity of the ligand for its complementary molecule is unchanged. The immobilised ligand is then packed into a chromatography column where it will selectively adsorb the complementary molecule, preventing its passage through the system (Fig. 69.10). Once the sample has been applied to the column and the contaminating substances washed through with buffer, the purified complementary molecule can be eluted, e.g. by changing the pH or salt

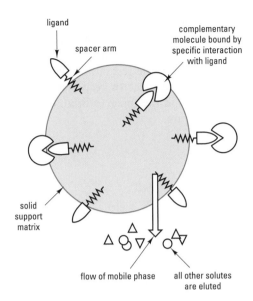

Fig. 69.10 Affinity chromatography.

Elution of substances from an affinity system – make sure that your elution conditions do not affect the interaction between the ligand and the stationary phase, or you may elute the ligand from the column.

Quantifying biomolecules – note that quantitative analysis often requires assumptions about the *identity* of separated compounds (p. 321) and that further techniques may be needed to provide information about the nature of the compounds present.

Problems with peaks – non-symmetrical peaks may result from column overloading, co-elution of solutes, poor packing of the stationary phase, or interactions between the substance and the support material.

Using external standardisation – samples and standards should be analysed more than once, to confirm the reproducibility of the technique.

concentration, or by adding other substances with a greater affinity for the ligand. Providing a suitable ligand is available, affinity chromatography can be used for single-step purification of small quantities of a particular molecule in the presence of large amounts of contaminating substances. Examples of ligands include:

- **triazine dyes,** for protein purification;
- **enzyme substrates and co-factors,** for certain enzymes;
- **antibodies,** for specific antigens (see p. 377);
- **protein A,** for IgG antibody purification;
- **single-stranded oligonucleotides,** for complementary nucleic acids, e.g. mRNA, or particular single-stranded DNA sequences;
- **lectins,** for specific monosaccharide subunits.

Hydrophobic-interaction chromatography

This technique is used to separate proteins. The stationary phase consists of a non-polar group (e.g. an octyl or phenyl group) bound to a support matrix. This binds proteins differentially, according to the number and position of hydrophobic surface groups. The principle is similar to reverse-phase HPLC (Fig. 69.7), but samples are analysed under non-denaturing conditions, to retain biological activity.

Quantitative analysis

Most detectors and chemical assay systems give a linear response with increasing amounts of the test substance over a given 'working range' of concentration. Alternative ways of converting the measured response to an amount of substance are:

- **External standardisation:** this is applicable where the sample volume is sufficiently precise to give reproducible results (e.g. HPLC, column chromatography). Peak areas (or heights) of known amounts of the substance give a calibration factor or calibration curve, used to calculate the amount of test substance in the sample (Chapter 53).
- **Internal standardisation:** where you add a known amount of a reference substance (not originally present in the sample) to the sample, to give an additional peak in the elution profile. You determine the response of the detector to the test and reference substances by analysing a standard containing known amounts of both substances, to provide a response factor (*r*), where:

$$r = \frac{\text{peak area (or height) of test substance}}{\text{peak area (or height) of reference substance}} \qquad [69.5]$$

Use this response factor to quantify the amount of test substance (Q_t) in a sample containing a known amount of the reference substance (Q_r), from:

$$Q_t = \frac{\text{peak area (or height) of test substance}}{\text{peak area (or height) of reference substance}} \times (Q_r \div r) \qquad [69.6]$$

Internal standardisation should be the method of choice wherever possible, since it is unaffected by small variations in sample volume

Using an internal standard – you should add an internal standard to the sample at the first stage in the extraction procedure, so that any loss or degradation of test substance during purification is accompanied by an equivalent change in the internal standard.

(e.g. for GC microsyringe injection). The internal standard should be chemically similar to the test substance(s) and must give a peak that is distinct from all others in the sample. An additional advantage of an internal standard that is chemically related to the test substance is that it may show up problems due to changes in detector response, etc. A disadvantage is that it may be difficult to find the space to fit an internal standard peak into a complex chromatogram.

Text reference

Stahl, E. (1969) *Thin Layer Chromatography – A Laboratory Handbook,* 2nd edn. Springer-Verlag, Berlin.

Sources for further study

Cazes, J. (2005) *Encyclopedia of Chromatography.* CRC Press, Boca Raton.

McMaster, M. (2009) *HPLC: A Practical User's Guide.* Wiley, New York.

McNair, H.M. and Miller, J.M. (2009) *Basic Gas Chromatography,* 2nd edn. Wiley, New York.

Wilson, K. and Walker, J. (eds) (2010) *Principles and Techniques of Biochemistry and Molecular Biology,* 7th edn. Cambridge University Press, Cambridge.

STUDY EXERCISES

69.1 Calculate R_F and R_X values from a chromatogram. The figure below represents the separation of three pigments by thin-layer chromatography.

(a) What is the R_F value of each pigment?
(b) What is the mobility of B and C, relative to A (R_A)?

Express both answers to three significant figures.

Thin-layer chromatographic separation of a mixture of pigments (A, B and C)

69.2 Calculate the resolution and selectivity of two components from a chromatogram. Two compounds were separated by column chromatography, giving retention times of 4 min 30 s for A and 6 min 12 s for B, while a compound that was completely excluded from the stationary phase was eluted in 1 min 35 s. The base width of peak A was 40 s and the base width of peak B was 44 s. Calculate (a) the selectivity and (b) the resolution for these two compounds (express all answers to three significant figures).

69.3 Calculate the amount of substance in a chromatographic separation using an internal standard. The chromatogram below represents the separation of three carbohydrates by GC.

Peak A corresponds to mannitol, peak B to sucrose and peak C to the internal standard, trehalose, at 1.50 mg. The table below gives retention times and areas of the peaks:

Gas–liquid chromatographic separation of three carbohydrates. Note that the plotter shows peaks A and B to be off-scale: however, the instrument still gives a valid area measurement (see table below).

Retention times and peak areas for carbohydrates A, B and C

Peak	Retention time (min)	Area (relative)
A	3.92	2060
B	5.82	1898
C	6.03	604

Given a response factor (r) of 1.26 for mannitol and 0.92 for sucrose (relative to the internal standard), determine the amount of A and B in the sample (express your answer to three significant figures).

Answers to these study exercises are available at www.pearsoned.co.uk/practicalskills

70 Electrophoresis

Fig. 70.1 Components of an electrophoresis system.

Definitions

Anode – positive electrode (negatively charged ions which move to this electrode are termed anions).

Cathode – negative electrode (positively charged ions which move to this electrode are termed cations).

Handling cellulose acetate electrophoresis strips – these are fragile and must be handled carefully – avoid touching the flat surfaces with your fingers.

Preparing agarose gels – agarose 'sets' at around 36–38°C, so gels must be allowed to cool fully after pouring, or they will be distorted/damaged.

Electrophoresis is used to separate individual charged molecules (ionic substances) in a sample on the basis of their differential movement in an electric field. The essential components of an electrophoretic system (Fig. 70.1) are:

- **Two electrodes,** the anode and the cathode, between which a voltage can be applied using a microprocessor-controlled 'power pack' that provides control over voltage (V) and current (A) at constant power (W) for a specified time.
- **A buffer (salt) solution,** through which the current flows and in which the ionic substances move at a controlled, constant pH.
- **A supporting medium,** which limits diffusion and thermal convection, leading to enhanced and consistent separation of the charged molecules.
- **A visualisation/detection system,** to view the results following electrophoresis.

KEY POINT Biomolecules move at different rates within an electrophoretic system, leading to their separation on the basis of their size, shape and/or charge, depending on the conditions used.

Different buffer systems are used in electrophoresis, depending on the pH requirement of the procedure and the molecules being separated. Buffers are usually prepared as standard mixtures, including a chelating agent such as ethylenediaminetetraacetic acid (EDTA), and are often referred to by their acronyms, for example TBE (for Tris/boric acid/EDTA) and TAE (for Tris/acetic acid/EDTA).

Types of electrophoretic system

These can be characterised according to the supporting medium used. The main types that you are likely to come across are detailed below.

Cellulose acetate

Here, the support medium has a large and uniform pore structure that enables unrestricted movement of large biomolecules. Strips of cellulose acetate are often used for the rapid separation of plasma proteins in biochemistry practicals, allowing the procedure to take less than an hour.

Agarose

This is a purified carbohydrate polymer obtained from seaweed. It is prepared in gel format by mixing powder with electrophoresis buffer at concentrations of 0.5–3.0%, boiled until the mixture becomes clear, then poured on to a glass plate or a casting tray and allowed to cool until it forms a gel. The size of pores within an agarose gel depends on the concentration of agarose used: high concentrations of agarose produce gels with small pores, which means that large biomolecules such as proteins and DNA fragments are slowed in their movement through such gels, leading to 'molecular sieving'. This is the basis of the electrophoretic separation of DNA and RNA fragments based on their size (p. 402).

> **SAFETY NOTE** Preparing gels for PAGE – acrylamide and bisacrylamide are extremely potent neurotoxins. You must wear plastic gloves when handling solutions containing these reagents, and when handling polyacrylamide gels.

Polyacrylamide

Polyacrylamide gel electrophoresis (PAGE) has a major role in the analysis of proteins. The gel is formed by polymerising acrylamide monomers into long cross-linked chains using bisacrylamide under standard conditions to give a gel with a pore size that can be 'tailored' to suit the size of biomolecule to be separated.

PAGE is especially useful for protein separation, where it is used in either of two main ways:

1. **Under non-dissociating conditions** – where proteins will retain their native 3-D configuration.

2. **Under dissociating conditions** – where the addition of the detergent sodium dodecyl sulphate (SDS) to the buffer causes proteins to denature, producing linear polypeptides with a uniform negative charge per unit length of polypeptide. Since SDS-dissociated polypeptides will have identical charge densities, when they are subjected to PAGE they will then migrate in direct relationship to their size. This not only gives effective separation, but additionally enables the molecular mass of each polypeptide to be determined, by comparing its mobility with polypeptide standards of known mass, run under the same conditions (Fig. 70.2).

Applications of electrophoresis

Agarose gel electrophoresis of DNA and RNA

These molecules carry a net negative charge, due to the phosphate groups in the polymer. As a result, they move towards the anode in electrophoretic systems. Typically, the support medium is agarose, used at low concentration (e.g. 0.3%) for large DNA fragments and at higher concentration (e.g. 0.8%) for smaller fragments, while very small fragments are best separated using a polyacrylamide gel. Box 70.1 gives advice on running agarose gel electrophoresis of DNA.

Polyacrylamide gel electrophoresis of proteins

PAGE can be used for high-resolution one-dimensional and two-dimensional separation of proteins. In addition to the choice of using SDS (dissociating conditions) or not, there are a number of additional choices available for PAGE, including:

- **Whether to use rod gels or slab gels.** The former are useful for individual large samples, where separation of a protein is required in large amounts, while the latter can be used for multiple samples. Slab gels enable 20 or more individual samples to be separated under identical conditions.

- **Whether to use a continuous or discontinuous buffer system.** Discontinuous systems have different buffers in the gel compared with the reservoirs – they are more time-consuming to prepare but they give better separation and resolution with dilute samples of large volume.

- **What pH to use.** The pH is not critical for SDS-PAGE, since SDS-treated polypeptides are negatively charged over a wide pH range. However, for non-dissociating conditions, the pH can be critical, particular when the biological activity of proteins is to be retained, e.g. in enzyme separations.

(a)

(b)

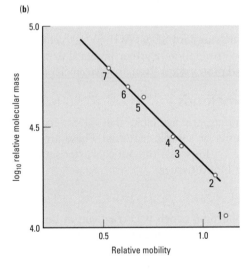

Fig. 70.2 Determination of relative molecular mass (M_r) of proteins by SDS-PAGE. (a) Gel samples: 1 cytochrome *c*, 2 myoglobin, 3 γ-globulin, 4 carbonic anhydrase, 5 ovalbumin, 6 albumin, 7 transferrin, 8 mixture of samples 1–7. Photo courtesy of Pharmacia Biotech. (b) Plot of log M_r against distance travelled through gel.

Box 70.1 How to carry out agarose gel electrophoresis of DNA

1. **Prepare the gel.** Typically, a small volume (10–20 mL) of buffer plus agarose is heated gently until the powder dissolves – take care not to overheat, or it will boil over. Nowadays, gels are often cast with a small amount of a visualising dye, such as SYBR®Safe.

2. **Prepare the samples.** A small amount of sucrose or glycerol is usually added, to increase the density of the sample. A water-soluble anionic 'tracking' dye (e.g. bromophenol blue or xylene cyanol) is also added to each sample, so that migration can be followed visually.

3. **Load the samples on to the gel.** Individual samples are added to the pre-formed wells using a pipettor (the sample should be retained within the well due to its higher density, compared with the buffer solution). The volume of sample added to each well is small – typically less than 25 µL, so a very steady hand and careful dispensing are needed to pipette each sample accurately.

4. **Load the DNA markers on to the gel.** Typically, these standards of known size are added to the first and last wells of the gel: after electrophoresis, the relative positions of the bands of known size can be used to prepare a calibration curve (Chapter 53), usually by plotting $\log_{10}$ size (length) against distance travelled.

5. **Carry out ('run') the electrophoresis.** DNA separation is usually carried out at 100–150 V for 30–60 min (see manufacturer's instructions for specific details, according to which 'power pack' you are using): the gel should be run until the 'tracking' dye has migrated across 80% of the gel.

6. **Examine the result.** If you have used a visualising dye within the agarose, then you simply transfer the gel to a UV transilluminator and look for 'bands' of fluorescence corresponding to each DNA fragment.

> **SAFETY NOTE** Always wear suitable UV-filtering safety glasses when working with UV radiation, to protect your eyes. Use a digital camera to photograph your gel. Alternatively, a dedicated image capture system can be used, e.g. Bio-Rad GelDoc®.

7. **Extract any DNA bands of interest.** If a particular band is required for further study, the piece of gel containing that band can be cut from the gel using either a clean scalpel or a specialised gel band cutter.

- **What gel concentration to use.** For separation of unknown mixtures, a gradient gel can be used.

Box 70.2 gives advice on running standard SDS-PAGE in slab gel format for protein separation and size determination. Further practical details for electrophoresis of DNA, RNA and proteins are given in Westermeier (2005).

Advanced electrophoretic methods

A number of advanced techniques are available that give very high resolution with very small amounts of sample material. These are more likely to be used in project work, e.g. for isolation and characterisation of individual protein constituents in blood/tissues.

Isoelectric focusing (IEF)

> **Definition**
>
> **Isoelectric point** – the pH at which a biomolecule (e.g. a protein) carries no net charge.

In contrast to standard electrophoresis, which is carried out at constant pH using a buffer system, IEF is carried out using a pH gradient across the gel. When a sample containing a mixture of proteins is applied to this gradient, the individual proteins will migrate to their isoelectric point on the pH gradient. This enables the proteins to be separated according to their isoelectric points, which is, in turn, a function of the relative proportions of different amino acids present.

Box 70.2 How to carry out SDS-PAGE for protein separation

1. **Prepare the gel.** Nowadays, many laboratories use pre-cast gels, bought from a manufacturer (e.g. Bio-Rad). If you are preparing your own gel, you will need to follow the protocol very carefully. Typically the correct proportions of acrylamide, bisacrylamide and SDS are mixed together and degassed, under vacuum. Then ammonium persulphate and tetramethylethylenediamine (TEMED) are added to trigger the polymerisation. Once the latter two constituents are added, the gel should be poured immediately into the casting tray (including the well former 'comb').

2. **Prepare the samples.** The protein sample is mixed with a buffer solution containing SDS (to bind to the dissociated proteins), plus dithiothreitol (DTT) (to cleave disulphide bonds in the proteins) and a 'tracker' dye (e.g. bromophenol blue), then heated for 5 min at 95 °C, to disrupt the tertiary structure and 'linearise' the polypeptide chains.

3. **Load the samples on to the gel.** Individual samples are added to the wells using a pipettor (p. 141). The volume of sample added to each well is typically less than 100 μL, so a very steady hand and careful dispensing are needed to accurately pipette each sample (steady the pipettor using both hands if this helps). To optimise the separation of proteins, the volume added should be kept as small as possible.

4. **Load the molecular mass standards.** Nowadays, many labs use 'rainbow' markers, with a wide range of proteins of known molecular mass, each of which is stained a different colour, to enable estimation of the molecular masses of unknown proteins, by visual comparison.

5. **Run the electrophoresis.** The gel is positioned with the well/samples closest to the cathode (negative electrode), since they will move towards the anode during electrophoresis as a result of their negative charge. Protein separation is typically carried out at 80–100 V for 1–2 h (see manufacturer's instructions for specific details, according to which 'power pack' you are using): the gel should be run until the 'tracking' dye has migrated across 80% of the gel). Higher voltages give faster separation, but poorer separation of protein 'bands', and may cause denaturation of proteins due to heating.

6. **Fix and stain the gel.** One of the most widely used approaches uses Coomassie Brilliant Blue stain. Typically, this involves immersing the gel in 0.25% w/v Coomassie Blue for 1 h then destaining overnight in a methanol/acetic acid solution. A safer, water-based alternative can be bought from a manufacturer (e.g. Bio-Rad Bio-Safe Coomassie Stain) which allows the gel to be destained in water overnight. After destaining, separated proteins are visible as blue bands against an unstained background. For higher resolution, a silver stain can be used instead of Coomassie Brilliant Blue.

7. **Examine the results.** After destaining, separated proteins are visible as blue bands against an unstained background (Fig. 70.2(a)). The position of these bands can be compared with the molecular mass standards to determine the size of each band in the test sample. For greater accuracy, the distance moved by the molecular mass standards can be plotted against $\log_{10}$ relative molecular mass and the resulting relationship (Fig. 70.2(b)) used to determine the size of unknown bands.

Two-dimensional electrophoresis

The most commonly used version of this technique involves separating a mixture of proteins in one dimension using IEF, followed by separation by molecular mass using denaturing SDS-PAGE in a second dimension, at right angles to the first. This technique allows up to 1000 proteins to be separated from a single sample. Analysis of the complex patterns that result from 2-D electrophoresis requires computer-aided gel scanners to acquire and process data from the gel image. These systems can compare, match and quantify protein 'spots' from several gels; slight variation in the patterns found in different 'runs' are accounted for using internal reference 'landmark' proteins, which are either added as standard proteins, or are proteins known to be present in all samples.

Avoiding streaking in two-dimensional electrophoresis – Check carefully to ensure that the sample contains no particulate material (e.g. from protein aggregation); filter or centrifuge before use.

high-voltage
supply

capillary
coil

detector

⊖

outlet vial

⊕

inlet vial

data acquisition
and analysis

Fig. 70.3 Components of a capillary electrophoresis system.

Capillary electrophoresis (CE)

This technique, which is also known as capillary zone electrophoresis (CZE) combines the high resolving power of electrophoresis with the speed and versatility of high-performance liquid chromatography (pp. 455–6). The technique largely overcomes the poor resolution problems encountered when carrying out electrophoresis without a supporting medium by using a capillary tube with a high surface area to volume ratio. A further advantage is that extremely small sample volumes can be analysed (e.g. 5–10 nL). The versatility of CE is demonstrated by its use in the separation of a range of biomolecules, including amino acids, proteins, DNA/RNA, vitamins and drugs.

The components of a typical CE setup are shown in Fig. 70.3. The capillary (internal diameter 25–50 μm) is made of fused silica and externally coated with a polymer for added mechanical strength – a gap in the polymer provides a 'window' for detection purposes. Samples are injected into the capillary and the components then move through the capillary at different rates, depending on their interaction with the negative-charged surface of the silica capillary, which means that they leave the other end of the capillary as discrete 'peaks', often detected using UV/visible absorption spectrophotometry. Several advanced variants are used for more specialised work; these include micellar electrokinetic capillary chromatography (MECC), capillary gel electrophoresis (CGE) and capillary isoelectric focusing (CIEF).

Text reference

Westermeier, R. (2005) *Electrophoresis in Practice: A Guide to Methods and Applications of DNA and Protein Separations,* 4th edn. Wiley-VCH, Weinheim.

Sources for further study

American Electrophoresis Society *Homepage.* Available: http://www.aesociety.org/ Last accessed 16/07/15. [Includes details on electrophoretic techniques including IEF 2D electrophoresis and CE.]

British Society for Proteome Research. *Homepage.* Available: http://www.bspr.org/ Last accessed 16/07/15. [Includes links to a wide range of web resources on electrophoresis.]

Garcia, C.D., Chumbimuni-Torres, K.V. and Carrilho, E. (2013) *Capillary Electrophoresis and Microchip Capillary Electrophoreses; Principles, Applications and Limitations.* Wiley, New York.

Guttmann, A. (2009) *Capillary Gel Electrophoresis and Related Microseparation Techniques.* Elsevier, Amsterdam.

Walker, J.M (2009) *The Protein Protocols Handbook,* 3rd edn. Humana Press, New Jersey.

STUDY EXERCISES

70.1 Find out why the net charge on a protein molecule varies with pH. Identify the amino acids primarily responsible for determining the net charge on a protein molecule and draw simple diagrams to represent the ionisation of their side chains, indicating how you would expect these side chains to be charged at acid, neutral and alkaline pH values.

70.2 Test your understanding of acronyms used in electrophoresis. What do each of the following stand for? (a) SDS; (b) PAGE; (c) IEF; (d) CE.

70.3 Explain the function of the various reagents used in SDS-PAGE analysis of proteins. What is the function of each of the following SDS-PGAE reagents? (a) acrylamide; (b) *N,N'*-methylene bisacrylamide; (c) ammonium persulfate; (d) *N,N,N',N'*-tetramethylethylenediamine; (e) sodium dodecyl sulphate; (f) 2-mercaptoethanol; (g) buffer; (h) Coomassie blue R-250; (i) bromophenol blue.

70.4 Determine the M_r of subunits of a protein by SDS-PAGE. A number of proteins of known molecular weight were analysed by SDS-PAGE and the results are shown in the table. In the same experiment, a sample of purified lactate dehydrogenase (LDH) was treated in identical fashion and run, producing a band with a relative mobility of 0.77. (a) What is the M_r of the polypeptide produced from the LDH sample?

Relative mobilities of molecular weight standards by SDS-PAGE

Protein	M_r	Relative mobility
Myoglobin	17 200	1.00
Carbonic anhydrase	29 000	0.84
Ovalbumin	43 000	0.67
Human albumin	68 000	0.51
Transferrin	77 000	0.46

(b) If the M_r of native LDH is 136000, what do you conclude about the subunit structure of this enzyme?

70.5 Consider the requirements for sample application in PAGE and IEF. Explain why in PAGE the sample is applied in a discrete narrow band, usually at the cathodic end of the gel, while in IEF the sample can be applied at any point along the length of the gel without concern about location or narrowness of the sample zone.

Answers to these study exercises are available at www.pearsoned.co.uk/practicalskills

Analysis and presentation of data

The process of discovering the meaning within your results can be fascinating. There are two main elements to this process:

1. **Exploratory data manipulation** – this is used to investigate the nature of your results and suggest possible patterns and relationships within the data set. The aim is to generate hypotheses for further investigation. Exploratory techniques allow you to visualise the form of your data. They are ideal for examining results from pilot 'studies', but should be used throughout your investigations.

2. **Confirmatory analysis** – this is used to test the hypotheses generated during the exploratory phase. The techniques required are generally statistical in nature and are dealt with in Chapter 76.

> **KEY POINT** Spreadsheets (Chapter 13) are invaluable tools for data manipulation and transformation: complex mathematical procedures can be carried out rapidly and the results visualised almost immediately using the inbuilt graphing functions. Spreadsheets also facilitate the statistical analysis of data (see Chapters 75 and 76).

Summarising your results – original, unsummarised data belong only in your primary record, either in laboratory books or as computer records. You should produce summary tables to organise and condense original data, as shown above.

Organising numbers

In order to organise, manipulate and summarise data, you should:

- **Simplify the numbers,** e.g. by rounding or taking means. This avoids the detail becoming overwhelming.
- **Rearrange your data** in as many ways as possible for comparison.
- **Display in graphical form;** this provides an immediate visual summary that is relatively easy to interpret.
- **Look for an overall pattern** in the data – avoid getting lost in the details at this stage.
- **Look for any striking exceptions** to that pattern (outliers) – they often point to special cases of particular interest or to errors in the data produced through mistakes during the acquisition, recording or copying of data.
- **Move from graphical interpretations to well-chosen numerical summaries and/or verbal descriptions,** including where applicable an explanatory hypothesis.

After collecting data, the first step is often to count how frequently each value occurs and to produce a frequency table. The frequency is simply the number of times a value occurs in the data set, and is, therefore, a count. The raw data could be acquired using a tally chart system to provide a simple frequency table. To construct a tally chart (e.g. Fig. 71.1):

Colour	Tally	Total
Green	III	3
Blue	IIII III	8
Red	IIII	4
White	IIII IIII II	12
Black	I	1
Maroon	III	3
Yellow	II	2
		33

Fig. 71.1 An example of a tally chart.

Producing a histogram – a *neatly* constructed tally chart will double as a rough histogram, as illustrated above.

- **enter only one tally at a time;**
- **if working from a data list, cross out each item on the list as you enter it on to the tally chart,** to prevent double entries;
- **check that all values are crossed out at the end** and that the totals agree.

Table 71.1 An example of a frequency table

Size class	Frequency	Relative frequency (%)
0–4.9	7	2.6
5–9.9	23	8.6
10–14.9	56	20.9
15–19.9	98	36.7
20–24.9	50	18.7
25–29.9	30	11.2
30–34.9	3	1.1
Total	267	99.8*

*≠ 100 due to rounding error.

stem	leaves
7	23
7	55
7	6
7	9
8	000
8	233
8	45555
8	77
8	888899
9	0000111111
9	2333333
9	44555555555
9	66677777
9	88888999
10	00

Fig 71.2 A simple 'stem and leaf' plot of a data set. The 'stem' shows the common component of each number, while the 'leaves' show the individual components, e.g. the top line in this example represents the numbers 72 and 73.

Example Allometry involves a logarithmic transformation of data that reveals aspects of the relationship between the dimensions of organisms (see p. 314).

Convert the data to a formal table when complete (e.g. Table 71.1). Because proportions are easier to compare than class totals, the table may contain a column to show the relative frequency of each class. Relative frequency can be expressed in decimal form (as a proportion of 1) or as a percentage (as a proportion of 100).

Graphing data

Graphs are an effective way to investigate trends in data and can reveal features that are difficult to detect from a table, e.g. skewness of a frequency distribution. The construction and use of graphs is described in detail in Chapter 72. When investigating the nature of your data, the main points are as follows:

- **Make the values stand out clearly:** attention should focus on the actual data, not the labels, scale markings, etc. (contrast with the requirements for constructing a graph for data presentation, see p. 475).
- **Avoid clutter in the graph:** leave out grid lines and try to use the simplest graph possible for your purpose.

> **KEY POINT** Use a computer spreadsheet with graphics options whenever possible: the speed and flexibility of these powerful tools should allow you to explore every aspect of your data rapidly and with relatively little effort (see Chapters 13 and 72).

Displaying distributions

A visual display of a distribution of values is often useful for variables measured on an interval or ratio scale (p. 174). The distribution of a variable can be displayed by a frequency table for each value or, if the possible values are numerous, groups (classes) of values of the variable. Graphically, there are two main ways of viewing such data:

1. **histograms** (see pp. 477, 483), generally used for large samples;
2. **stem and leaf plots** (e.g. Fig. 71.2), often used for samples of less than 100: these retain the actual values and are faster to draw by hand. The main drawback is the limitation imposed by the choice of stem values since these class boundaries may obscure some features of the distribution.

These displays allow you to look at the overall shape of a distribution and to observe any significant deviations from the idealised theoretical ones. Where necessary, you can use data transformations to investigate any departures from standard distribution patterns such as the normal distribution or the Poisson distribution.

Transforming data

Transformations are mathematical functions applied to data values. They are particularly valuable where your results are related to areas and volumes (e.g. leaf area, body mass).

The most common use of transformations is to prepare data sets so that specific statistical tests may be applied. For instance, if you find that your data distribution is unimodal but not symmetrical (p. 504), it is often useful to apply a transformation that will redistribute the data values to form a symmetrical distribution. The object of this exercise is often to find the function that most closely changes the data into a standard normal distribution, allowing you to apply a wide range of parametric hypothesis-testing statistics (see Chapter 76). A frequently used transformation is to take logarithms of one or more sets of values: if the data then approximate to a normal distribution, the relationship is termed 'log-normal'.

Some general points about transformations are:

- **They should be made on the raw data, not on derived data values:** this is simpler, mathematically valid, and more easily interpreted.

- **The transformed data can be analysed like any other numbers.**

- **Transformed data can be examined for outliers,** which may be more important if they remain after transformation.

Figure 71.3 presents a ladder of transformations that will help you decide which transformations to try (see also Table 76.1). Note that percentage and proportion data are usually arc-sine transformed, which is a more complex procedure; consult Sokal and Rohlf (2011) for details.

Figure 71.4 illustrates the following 'quick-and-easy' way to choose a transformation:

1. **Calculate the 'five-number summary'** for the untransformed data (p. 508).
2. **Present the summary graphically** as a 'box-and-whisker' plot (p. 508).
3. **Decide whether you need to correct for positive or negative skew** (p. 504).
4. **Apply one of the 'mild' transformations in Fig. 71.3** *on the five-number summary values only.*
5. **Draw a new box-and-whisker plot** and see whether the skewness has been corrected.
6. **If the skewness has been undercorrected, try again** with a stronger transformation. If it has been overcorrected, try a milder one.

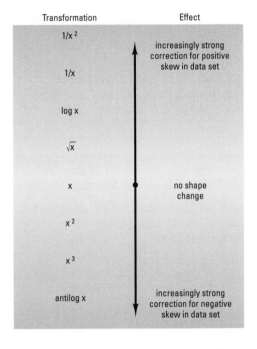

Fig. 71.3 Ladder of transformations (after J.W. Tukey).

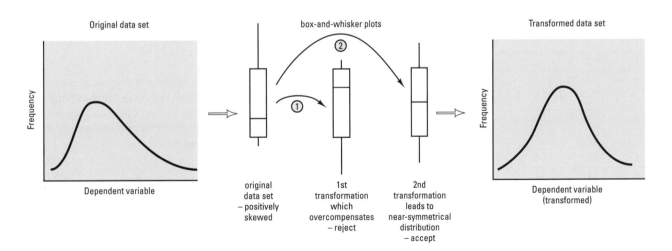

Fig. 71.4 Illustration of the processes of transforming a data set.

7. **When the distribution appears to be acceptable, transform the full data set** and recalculate the summary statistics. If necessary use a statistical test to confirm that the transformed data are normally distributed (p. 519).

8. **If no simple transformation works well, you may need to use nonparametric statistics when comparing data sets.**

Text reference

Sokal, R.R. and Rohlf, F.J. (2011) *Biometry,* 4th edn. W.H. Freeman, San Francisco.

Sources for further study

Heath, D. (1995) *An Introduction to Experimental Design and Statistics for Biology.* UCL Press, London.

Quinn, G.P. and Keough, M.J. (2002) *Experimental Design and Data Analysis for Biologists.* Cambridge University Press, Cambridge.

STUDY EXERCISES

The following study exercises can be most easily carried out using a spreadsheet such as Microsoft *Excel*.

71.1 Compare frequency distributions for samples with different sample sizes. Compute relative frequencies for each class in the table below. Graph the data as a frequency polygon.

Frequency distribution data

Hyphal length (μm)	Treatment B	Treatment A	Hyphal length (μm)	Treatment B	Treatment A
0	0	0	13	6	30
1	1	0	14	3	26
2	2	0	15	1	22
3	4	1	16	1	18
4	8	1	17	0	14
5	13	2	18	0	12
6	22	5	19	0	8
7	36	9	20	0	5
8	48	14	21	0	4
9	46	21	22	0	2
10	32	26	23	0	2
11	18	31	24	0	1
12	11	33	25	0	1

71.2 Use a stem and leaf plot. Work out the mean and standard deviation of the data contained in the following stem and leaf plot:

6	1334
6	56788
7	013344
7	556667899
8	11224
8	569
9	03
9	6

Stem and leaf plot

71.3 Use transformations to correct skew in a data set. Find a transformation that will make the data of treatment B in Study exercise 71.1 approximately symmetrical about the mean. Demonstrate graphically that you have accomplished this.

Answers to these study exercises are available at www.pearsoned.co.uk/practicalskills

Fig. 72.1 Effect of antibiotic on yield of two bacterial isolates: ○, sensitive isolate; □, resistant isolate. Vertical bars show standard errors ($n = 6$).

Graphs can be used to show detailed results in an abbreviated form, displaying the maximum amount of information in the minimum space. Graphs and tables present findings in different ways. A graph (figure) gives a visual impression of the content and meaning of your results, while a table provides an accurate numerical record of data values. You must decide whether a graph should be used, e.g. to illustrate a pronounced trend or relationship, or whether a table (Chapter 73) is more appropriate.

A well-constructed graph will combine simplicity, accuracy and clarity. Planning of graphs is needed at the earliest stage in any write-up as your accompanying text will need to be structured so that each graph delivers the appropriate message. Therefore, it is best to decide on the final form for each of your graphs before you write your text. The text, diagrams, graphs and tables in a laboratory write-up or project report should be complementary, each contributing to the overall message. In a formal scientific communication it is rarely necessary to repeat the same data in more than one place (e.g. as a table and as a graph). However, graphical representation of data collected earlier in tabular format may be applicable in laboratory practical reports.

Practical aspects of graph drawing

The following comments apply to graphs drawn for laboratory reports. Figures for publication, or similar formal presentation, are usually prepared according to specific guidelines provided by the publisher/organiser.

> **KEY POINT** Graphs should be self-contained – they should include all material necessary to convey the appropriate message without reference to the text. Every graph must have a concise explanatory title to establish the content. If several graphs are used, they should be numbered, so they can be quoted in the text.

- **Consider the layout and scale of the axes carefully.** Most graphs are used to illustrate the relationship between two variables (x and y) and have two axes at right angles (e.g. Fig. 72.1). The horizontal axis is known as the abscissa (x axis) and the vertical axis as the ordinate (y axis).

- **The axis assigned to each variable must be chosen carefully.** Usually the x axis is used for the independent variable (e.g. treatment) while the dependent variable (e.g. biological response) is plotted on the y axis (p. 202). When neither variable is determined by the other, or where the variables are interdependent, the axes may be plotted either way round.

- **Each axis must have a descriptive label showing what is represented,** together with the appropriate units of measurement, separated from the descriptive label by a solidus or 'slash' (/), as in Fig. 72.1, or by brackets, as in Fig. 72.2.

- **Each axis must have a scale with reference marks** ('tics') on the axis to show clearly the location of all numbers used.

- **A figure legend should be used to provide explanatory detail,** including a key to the symbols used for each data set.

Selecting a title – it is a common fault to use titles that are grammatically incorrect: a widely applicable format is to state the relationship between the dependent and independent variables within the title, e.g. 'The relationship between enzyme activity and external pH'.

Remembering which axis is which – a way of remembering the orientation of the x axis is that x is a 'cross', and it runs 'across' the page (horizontal axis) while y is the first letter of yacht, with a large vertical mast (vertical axis).

Fig. 72.2 Frequency distribution of masses for a sample of animals (sample size 24 085); the size class interval is 2 g.

Example For a data set where the smallest number on the log axis is 12 and the largest number is 9000, three-cycle log–linear paper would be used, covering the range 10–10000 (Fig. 72.3).

Handling very large or very small numbers

To simplify presentation when your experimental data consist of either very large or very small numbers, the plotted values may be the measured numbers multiplied by a power of 10: this multiplying power should be written immediately before the descriptive label on the appropriate axis (as in Fig. 72.2). However, it is often better to modify the primary unit with an appropriate prefix (p. 177) to avoid any confusion regarding negative powers of 10.

Size

Remember that the purpose of your graph is to communicate information. It must not be too small, so use at least half an A4 page and design your axes and labels to fill the available space without overcrowding any adjacent text. If using graph paper, remember that the white space around the grid is usually too small for effective labelling. The shape of a graph is determined by your choice of scale for the x and y axes which, in turn, is governed by your experimental data. It may be inappropriate to start the axes at zero (e.g. Fig. 72.1). In such instances, it is particularly important to show the scale clearly, with scale breaks where necessary, so the graph does not mislead. Note that Fig. 72.1 is drawn with 'floating axes' (i.e. the x and y axes do not meet in the lower left-hand corner), while Fig. 72.2 has clear scale breaks on both x and y axes.

Graph paper

In addition to conventional linear (squared) graph paper, you may be asked to use the following:

- **Probability graph paper.** This is useful when one axis is a probability scale (e.g. Fig. 76.5).
- **Log–linear graph paper.** This is appropriate when one of the scales shows a logarithmic progression, e.g. the exponential growth of cells in liquid culture (Fig. 48.3). Log–linear paper is defined by the number of logarithmic divisions (usually termed 'cycles') covered (e.g. Fig. 72.3), so make sure you use a paper with the appropriate number of cycles for your data. An alternative approach is to plot the log-transformed values on 'normal' graph paper.

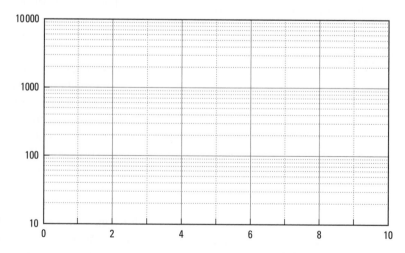

Fig. 72.3 Representation of three-cycle log–linear graph paper, marked up to show a y axis (log) scale from 10 to 10000 and an x axis (linear) scale from 0 to 10.

- **Log–log graph paper.** This is appropriate when both scales show a logarithmic progression, e.g. in allometry (p. 314).

Types of graph

Different graphical forms may be used for different purposes, including:

- **Plotted curves** are used for data where the relationship between two variables can be represented as a continuum (e.g. Fig. 72.4).
- **Scatter diagrams** are used to visualise the relationship between individual data values for two interdependent variables (e.g. Fig. 72.5) often as a preliminary part of a correlation analysis (p. 523).
- **Three-dimensional graphs** show the interrelationships of three variables, often one dependent and two independent (e.g. Fig. 72.6). A contour diagram is an alternative method of representing such data.
- **Histograms** represent frequency distributions of continuous variables (e.g. Fig. 72.7). An alternative is the tally chart (p. 471).
- **Frequency polygons** emphasise the form of a frequency distribution by joining the coordinates with straight lines, in contrast to a histogram. This is particularly useful when plotting two or more frequency distributions on the same graph (e.g. Fig. 72.8).
- **Bar charts** represent frequency distributions of a discrete qualitative or quantitative variable (e.g. Fig. 72.9). An alternative representation is the line chart (Fig. 76.3).
- **Pie charts** illustrate portions of a whole (e.g. Fig. 72.10).
- **Pictographs** give a pictorial representation of data (e.g. Fig. 72.11).

The plotted curve

This is the most common form of graphical representation used in biology. The key features are outlined below and given in checklist form in Box 72.1, while Box 72.2 gives advice on using Microsoft *Excel*.

> **Choosing between a histogram and a bar chart** – use a histogram for *continuous quantitative* variables and a bar chart for *discrete* variables (see Chapter 27 for details of these types of measurement scale).

> **Using computers to produce graphs** – never allow a computer program to dictate size, shape and other aspects of a graph: find out how to alter scales, labels, axes, etc., and make appropriate selections (see Box 72.2 for an example involving Microsoft *Excel*). Draw curves freehand if the program only has the capacity to join the individual points by straight lines.

Box 72.1 Checklist for the stages in drawing a graph

The following sequence can be used whenever you need to construct a plotted curve: it will need to be modified for other types of graph.

1. Collect all of the data values and statistical values (in tabular form, where appropriate).
2. Decide on the most suitable form of presentation: this may include transformation (p. 273) to convert data to linear form.
3. Choose a concise descriptive title, together with a reference (figure) number and date, where necessary.
4. Determine which variable is to be plotted on the *x* axis and which on the *y* axis.
5. Select appropriate scales for both axes and make sure that the numbers and their location (scale marks) are clearly shown, together with any scale breaks.
6. Decide on appropriate descriptive labels for both axes, with SI units of measurement, where appropriate.
7. Choose the symbols for each set of data points and decide on the best means of representation for statistical values.
8. Plot the points to show the coordinates of each value with appropriate symbols.
9. Draw a trend line for each set of points. Use a see-through ruler, so you can draw the line to have an equal number of points on either side of it.
10. Write a figure legend, to include a key that identifies all symbols and statistical values and any descriptive footnotes.

Fig. 72.4 Plotted curve: the rate of photosynthetic O_2 evolution as a function of photon flux density.

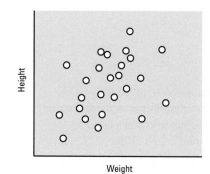

Fig. 72.5 Scatter diagram: height and weight of individual animals in a sample.

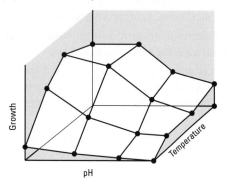

Fig. 72.6 Three-dimensional graph: growth of an organism as a function of temperature and pH.

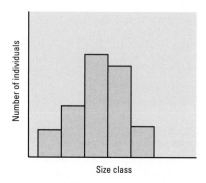

Fig. 72.7 Histogram: the number of plants within different size classes.

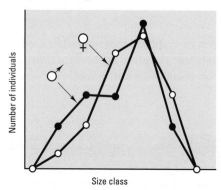

Fig. 72.8 Frequency polygon: frequency distributions of male and female animals according to size.

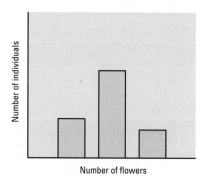

Fig. 72.9 Bar chart: number of flowers per plant.

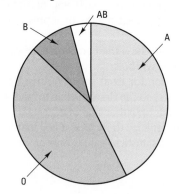

Fig. 72.10 Pie chart: relative abundance of human blood groups.

Fig. 72.11 Pictograph: distribution of plants on a rocky shore.

Choosing graphical symbols – plotted curves are usually drawn using a standard set of symbols: ●, ○, ■, □, ▲, △, ◆, ◇. By convention, paired symbols ('closed' and 'open') are often used to represent 'plus' (treatment) and 'minus' (control) treatments.

Adding error bars to Microsoft *Excel* graphs – you can do this as follows:

1. On your graph, right-click on one of the data points of the series to which you want to add error bars.

2. From *Layout* > *Analysis* > *Error Bars* menu select the appropriate option. Typically, you will wish to select *More Error Bars Options* > *Custom* and chose appropriate options from that menu. You may need to select a data array into which you have entered calculate values for the upper and lower error limits for your chosen statistic over the range of the plotted values. You can adjust the format of the error bars from the *More Error Bars Options* menu.

Note that the commands illustrated above may vary among software versions.

Conveying the correct message – the golden rule is: 'always draw the simplest line that fits the data reasonably well and is biologically reasonable'.

Extrapolating plotted curves – try to avoid the need to extrapolate by better experimental design.

Data points

Each data point must be shown accurately, so that any reader can determine the exact values of x and y. In addition, the results of each treatment must be readily identifiable. A useful technique is to use a dot for each data point, surrounded by a hollow symbol for each treatment (see Fig. 72.1). An alternative is to use symbols only (Fig. 72.4), though the coordinates of each point are defined less accurately. Use the same symbol for the same entity if it occurs in several graphs and provide a key to all symbols.

Statistical measures

If you are plotting average values for several replicates and if you have the necessary statistical knowledge, you can calculate the standard error (p. 509), or the 95% confidence limits (p. 523) for each mean value and show these on your graph as a series of vertical bars (see Fig. 72.1). Make it clear in the legend whether the bars refer to standard errors or 95% confidence limits and quote the value of n (the number of replicates per data point). Another approach is to add a least significant difference bar (p. 520) to the graph.

Interpolation

Once you have plotted each point, you must decide whether to link them by straight lines or a smoothed curve. Each of these techniques conveys a different message to your reader. Joining the points by straight lines may seem the simplest option, but may give the impression that errors are very low or non-existent and that the relationship between the variables is complex. Joining points by straight lines is appropriate in certain graphs involving time sequences (e.g. the number of animals at a particular site each year), or for repeat measurements where measurement error can be assumed to be minimal (e.g. recording a patient's temperature in a hospital, to emphasise any variation from one time point to the next). However, in most plotted curves the best straight line or curved line should be drawn (according to appropriate mathematical or statistical models, or by eye), to highlight the relationship between the variables – after all, your choice of a plotted curve implies that such a relationship exists. Do not worry if some of your points do not lie on the line: this is caused by errors of measurement and by biological variation (p. 175). Most curves drawn by eye should have an equal number of points lying on either side of the line. You may be guided by 95% confidence limits, in which case your curve should pass within these limits wherever possible.

Curved lines can be drawn using a flexible curve, a set of French curves, or freehand. In the latter case, turn your paper so that you can draw the curve in a single, sweeping stroke by a pivoting movement at the elbow (for larger curves) or wrist (for smaller ones). Do not try to force your hand to make complex, unnatural movements, as the resulting line will not be smooth.

Extrapolation

Be wary of extrapolation beyond the upper or lower limit of your measured values. This is rarely justifiable and may lead to serious errors. Whenever extrapolation is used, a dotted line ensures that the reader is aware of the uncertainty involved. Any assumptions behind an extrapolated curve should also be stated clearly in your text.

Box 72.2 How to create and amend graphs within a spreadsheet for use in coursework reports and dissertations

Spreedsheets can be used to create graphs of reasonable quality, as long as you know how to amend the default settings so that your graph meets the formal standards required for practical and project reports. As with a hand-drawn graph, the basic stages in graph drawing (Box 72.1) still apply. The following instructions explain how to produce an X–Y graph (plotted curve, p. 477), bar graph (p. 477), pie graph and histogram using Microsoft *Excel*, where all types of graphs are termed 'charts'. Note that the commands illustrated above may vary among software versions.

Producing an X–Y graph (*Scatter* chart in *Excel*)

1. Create the appropriate type of graph for your data. Enter the numeric values for your X variable data in the cells of a single column and the equivalent values for the Y variable in the adjacent cells of the next column to the right. Then select the whole data array (highlight the appropriate cells by clicking and holding down the left mouse button and dragging the cursor across the cells so that all values are included). Then select the *Insert* tab at the top of the sheet, and select (left-click) the *Scatter* chart from the options provided in the upper ribbon. Note that you should never use the *Line* chart option, as it is based on an x axis that is not quantitative, so all X values will appear as equally spaced categories, rather than having a true scale). Select the first option from the *Scatter* menu (*Scatter with only Markers*). Once selected, this will produce an embedded scatter chart of the type shown in Fig. 72.12(a). The line is then added later, as described below.

2. Change the default settings to improve the appearance of your graph. Consider each element of the image in turn, including the overall size, height and width of the graph (resize by clicking and dragging one of the 'sizing handles' around the edge of the chart). The graph shown in Fig. 72.12(b) was produced by altering the default settings, typically by moving the cursor over the feature and then clicking the right mouse button to reveal an additional menu of editing and formatting options. (Note that the example given below is for illustrative purposes only, and should not necessarily be regarded as prescriptive.)

Example for an X–Y graph (compare Fig. 72.12(a) with Fig. 72.12(b)):

- Unnecessary legend box on right-hand side can be removed using the (right-click) *Delete* option.

- Border to chart can be removed using *Format Chart Area* function (available by right-clicking within chart area).

- Gridlines can either be removed, using the *Delete* function or, if desired (as in Fig. 72.12(b)), changed by clicking on each axis and using the *Add Minor Gridlines* and *Format Gridlines* options to alter the *Colour* and *Style* of the gridlines to make them more like those of conventional graph paper.

- x and y axes can be reformatted by selecting each in turn, and using the *Format Axis* menu options to select appropriate scales for major and minor units, style of tick marks, etc. Remember that it is better to use a figure legend in *Word* (Chapter 13), rather than the *Chart Title* option within *Excel*.

- x and y axis labels can be added by selecting the *Layout* tab, then *Axis Titles*, then *Primary Horizontal Axis Title* and *Primary Vertical Axis Title* options, which will produce a text box beside each axis into which can be

(a)

(b)

Fig. 72.12 Examples of a plotted curve produced in Microsoft *Excel* using (a) default settings and (b) modified (improved) settings.

(continued)

Box 72.2 (continued)

typed the axis label and any corresponding units. This can be then changed from the default font using the *Home* tab options.

- Data point style can be changed by selecting (right-clicking) any data point and following the *Format Data Series* options to choose appropriate styles, colours and fill of the data markers.

- A straight line of best fit can be added by selecting any data point and using the *Add Trendline* option to choose a *Linear* line type with appropriate colour and style (explore other options within the *Format, Layout* and *Design* tabs at the top of the worksheet).

Producing a bar graph (*Column* chart in *Excel*)

1. Create the appropriate type of graph for your data. Enter the category names (for X axis) in one column and the numeric values (for Y axis) in the next column. Select (highlight) the data array, then select the *Insert* tab, and choose the *Column* chart from the options provided. For a standard bar graph, select the first option from the *Column* menu (*Clustered Column*). Once selected, this will produce an embedded bar graph of the type shown in Fig. 72.13(a).

2. Change the default settings to improve the appearance of your graph. The bar graph shown in Fig. 72.13(b) was produced by selecting each feature and altering the default settings, as detailed below (illustrative example).

Example for a bar graph (compare Fig. 72.13(a) with Fig. 72.13(b)):

- Unnecessary legend box on right-hand side can be removed using the *Delete* option.

- Border to chart can be removed using *Format Chart Area* function available by right-clicking within chart area).

- Gridlines can either be removed, using the *Delete* function or changed by selecting the gridlines and using the *Format Gridlines* option to alter the *Colour* and *Style*.

- The *y* axis can be reformatted by selecting the axis, then using the *Format Axis* menu options to select appropriate scales, tick marks, line colour, etc. Note that the *x* axis should already contain category labels from the spreadsheet cells (modify original cells to update spreadsheet, if necessary).

- *x* and *y* axis labels can be added by selecting the *Layout* tab, then *Axis Titles*, then *Primary Horizontal Axis Title* and *Primary Vertical Axis Title* options, as detailed for the plotted curve (Fig. 72.12).

- Bar colour can be modified using the *Format Data Series* menu, selecting appropriate *Fill* and *Border* colours, e.g. white and black respectively in Fig. 72.13(b).

- Individual *Y* data values can be shown using the *Add Data Labels* option (other options and adjustments can be made using the *Format, Layout* and *Design* tabs at the top of the worksheet).

Note that for all types of graph, it is better not to use the *Chart Title* option within *Excel*, which places the title at the top of the chart (as in Fig. 72.13(a)), but to copy and paste your untitled graph into a word-processed document, such as a Microsoft *Word* file (p. 483 gives details of the procedure), and then type a formal figure legend below the graph, as in Fig. 72.12(b) and Fig. 72.13(b). However, once your graph is embedded into a *Word* file, it is generally best not to make further amendments – you should go back to the original *Excel* file, make the required changes and then repeat the copy-paste procedure to reinsert the graph into the *Word* file.

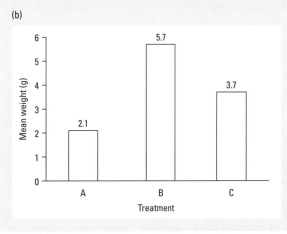

Figure 72.13 Examples of a bar graph produced in Microsoft *Excel* using (a) default settings and (b) modified (improved) settings.

(continued)

Box 72.2 (continued)

Producing a pie graph (*Pie* chart in *Excel*)

1. **Create the appropriate type of graph for your data.** Enter the category names for each part of the pie chart in one column, and the corresponding numbers (counts, fractions or percentages) in the next column. Select (highlight) the data array, then select (left-click) the *Insert* tab, and choose the *Pie* chart from the options provided. For a standard pie graph, select the first option from the menu (*Pie*). Once selected, this will produce an embedded pie graph.

2. **Change the default settings to improve the appearance of your graph.** For example, you can show the data values (*Data labels*), adjust colours and shading (e.g. switch from multi-colour to shades of grey), remove chart border, etc., as required (an illustrative example is shown as Fig. 72.14).

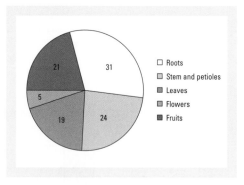

Fig. 72.14 Example of a pie graph produced in Microsoft *Excel* using format modified from the default settings.

Producing a histogram

The histogram function in *Excel* requires a little more effort to master, compared with other chart types. Essentially, a histogram is a graphical display of frequencies (counts) for a continuous quantitative variable (pp. 477, 483–6), where the data values are grouped into classes. It is possible to select the upper limit for each class into which you want the data to be grouped (these are termed 'bin range values' in *Excel*). An alternative approach is to let the software select the class intervals (bin range values) for you: *Excel* selects evenly distributed bins between the minimum and maximum values. However, this is often less effective than selecting your own class intervals.

To use the *Excel* histogram function, you will need to make sure that the *Analysis ToolPak* is loaded (check under the *Data* tab and, if not loaded, select via the *Office Button > Excel Options > Add-ins > Manage Excel Add-ins > Analysis*

ToolPak. The *Histogram* function should then be available in the *Analysis ToolPak* option on the *Data* tab/ribbon.

The following steps outline the procedure used to create the histograms shown in Fig. 72.15 for the table of data below (length, in mm, of 24 leaf petioles from a single plant).

7.2	6.5	7.1	8.5	6.6	7.2
7.0	7.3	8.6	9.1	7.5	8.3
7.1	5.7	7.3	7.6	6.9	7.1
8.3	7.6	5.4	8.6	7.9	8.0

1. **Enter the raw data values,** e.g. as a single column of numbers, or as an array, as above.

2. **Decide on the class intervals to be used.** Base your choice on the number of data points and the maximum and minimum values (for a small data set such as that shown above, you can do this by visual examination, whereas for a large data set, use the *Excel* functions COUNT, MAX and MIN (find these under the Σ symbol (*More Functions*) on the *Editing* section on the *Home* tab/ribbon, or use the *Descriptive Statistics > Summary statistics* option of the *Data Analysis* component on the *Data* tab/ribbon). A typical histogram would have 4–10 classes, depending on the level of discrimination required. Enter the upper limit for each class (*bin range values*) in ascending order in a separate array of cells, e.g. in a column close to the data values (in the above example, 6, 7, 8 and 9 were chosen – the few data values above the final bin value will be shown on the histogram as a group labelled 'more').

3. **Select the histogram function, then input your data and class interval values.** From the *Data* tab/ribbon, select *Data Analysis > Histogram*. A new window will open: input your data into the *Input Range* box (highlight the appropriate cells by clicking on the first data value and dragging to the final data value while holding down the left mouse button). Next, input the *Bin Range* values into the appropriate box (if this is left empty, *Excel* will select default bin range values). Most of the remaining boxes can be left empty, though you must click the last box to get a *Chart Output*, otherwise the software will give the numerical counts for each group, without drawing a histogram. Click *OK* and entries will be created within a new worksheet, showing the upper limits of each group (in a column labelled *Frequency*), plus a poorly constructed chart based on *Excel 2007* default settings, as shown in Fig. 72.15(a) (note that the default output is a bar chart, rather than a histogram, since there are gaps between the groups).

(continued)

Box 72.2 (continued)

Example for a histogram (compare Fig. 72.15(a) with Fig. 72.15(b):

- Chart can be resized to increase height, using the 'sizing handles' at the edges of the chart and border line around graph can be removed using *Format Chart Area* > *Border Color* > *No line* (menu available by right-clicking within chart area).

- Title and unnecessary legend box can be removed using the *Delete* option.

- Axis scales can be reformatted using the *Format Axis* menu options (e.g. scales, tick marks, line colour).

- *X* axis labels (class intervals) can be amended by typing directly into the cells containing the bin range values.

- Axis titles can be changed by typing directly into the axis title box (double-click to access).

- Bar colour can be changed (e.g. to grey, with black outline) using the *Format Data Series* > *Fill* and *Border Colour* options.

- Bar chart converted to correct histogram format (no gaps between bars) using the *Format Data Series* > *Series Options*, setting *Gap Width* to 0%.

- Figure legend can be added below figure in Microsoft *Word*, following copying and pasting of the Excel histogram into a *Word* file.

Importing an *Excel* chart into a *Word* document

One problem encountered with certain Microsoft *Office* products is that the standard *Cut* > *Paste* procedure gives a poor quality figure, with grainy appearance and fuzzy lines/text; similar problems occur using the *Insert* tab/ribbon in some *Word* versions. The simplest approach is to follow the step-wise procedure below:

1. Select your *Excel* chart: right-click outside the chart itself, near to the edge, then choose the *Copy* option from the drop-down menu.

2. Open your *Word* file, go to the *Home* tab/ribbon and select *Paste special* > *Microsoft Office Word Document Object* from the *Clipboard* options (left-hand side of ribbon).

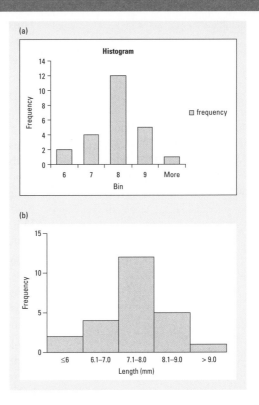

Figure 72.15 Examples of histogram output from Microsoft *Excel* using (a) default settings and (b) modified (improved) settings.

3. This should give a graph with the same crisp axis/line/text formatting as the original chart in *Excel*.

The alternative approach is to use *Excel* to print the entire graph (chart) as a single sheet, and then add this to the print-out from your word-processed document. However, the disadvantage with this approach is that you cannot produce a professional looking figure legend below your graph.

Spreadsheet commands illustrated here may vary among the various versions of Microsoft *Office* programs and for different spreadsheet products. Check the precise functions and syntax using the insert function menu option (f_x button in some versions of *Excel*).

Drawing a histogram - each datum is represented by a column with an area proportional to the magnitude of *y*: in most cases, you should use columns of equal width, so that the height of each column is then directly proportional to *y*. Shading or stippling may be used to identify individual columns, according to your needs.

The histogram

Whereas a plotted curve assumes a continuous relationship between the variables by interpolating between individual data points, a histogram involves no such assumptions. Histograms are also used to represent frequency distributions (p. 471), where the *y* axis shows the number of times a particular value of *x* was obtained (e.g. Fig. 72.2). As in a plotted curve, the *x* axis represents a continuous quantitative variable that can take any value within a given range (e.g. plant height), so the scale must be broken down into discrete classes and the scale marks on the x axis should show either the mid-points (mid-values) of each class, or the boundaries between the classes.

The columns are contiguous (adjacent to each other) in a histogram, in contrast to a bar chart, where the columns are separate because the x axis of a bar chart represents discrete values.

Interpreting graphs

The process of analysing a graph can be split into five phases:

1. **Consider the context.** Look at the graph in relation to the aims of the study in which it was reported. Why were the observations made? What hypothesis was the experiment set up to test? This information can usually be found in the Introduction or Results section of a report. Also relevant are the general methods used to obtain the results. This might be obvious from the figure title and legend, or from the Materials and Methods section.

2. **Recognise the graph form and examine the axes.** First, what kind of graph is presented (e.g. histogram, plotted curve)? You should be able to recognise the main types summarised on pp. 477–8 and their uses. Next, what do the axes measure? You should check what quantity has been measured in each case and what units are used.

3. **Look closely at the scale of each axis.** What is the starting point and what is the highest value measured? For the x-axis, this will let you know the scope of the treatments or observations (e.g. whether they lasted for 5 min or 20 years; whether a concentration span was two-fold or fifty-fold). For each axis, it is especially important to note whether the values start at zero; if not, then the differences between any treatments shown may be magnified by the scale chosen (see Box 72.3).

4. **Examine the symbols and curves.** Information will be provided in the key or legend to allow you to determine what these refer to. If you have made your own photocopy of the figure, it may be appropriate to note this directly on it. You can now assess what appears to have happened. If, say, two conditions have been observed while a variable is altered, when exactly do they differ from each other; by how much; and for how long?

5. **Evaluate errors and statistics.** It is important to take account of variability in the data. For example, if mean values are presented, the underlying errors may be large, meaning that any difference between two treatments or observations at a given x-value could simply have arisen by chance. Thinking about the descriptive statistics used (Chapter 75) will allow you to determine whether apparent differences could be significant in both statistical and biological senses.

Sometimes graphs are used to mislead. This may be unwitting, as in an unconscious favouring of a 'pet' hypothesis of the author. Graphs may be used to 'sell' a product in the field of advertising or to favouring of a viewpoint as, perhaps, in politics. Experience in drawing and interpreting graphs will help you spot these flawed presentations, and understanding how graphs can be erroneously presented (Box 72.3) will help you avoid the same pitfalls.

Examining graphs – do not be tempted to look at the data displayed within a graph before you have considered the graph's context, read the legend and decided the scale of each axis.

Understanding graphs within scientific papers – the legend should be a succinct summary of the key information required to interpret the figure without further reference to the main text. This is a useful approach when 'skimming' a paper for relevant information (pp. 21–22).

Box 72.3 How graphs can misrepresent and mislead

1. The 'volume' or 'area' deception – this is mainly found in histogram or bar chart presentations where the size of a symbol is used to represent the measured variable. For example, the amount of hazardous waste produced in different years might be represented on a chart by different sizes of a chemical drum, with the y axis (height of drum) representing the amount of waste. However, if the symbol retains its *shape* for all heights as in Fig. 72.16(a), its apparent *volume* will increase as a cubic function of the height, rather than in direct proportion. To the casual observer, a two-fold increase may look like an eight-fold one, and so on. Strictly, the *height* of the symbol should be the measure used to represent the variable, with no change in symbol width, as in Fig. 72.16(b).

2. Effects of a non-zero axis – A non-zero axis acts to emphasise the differences between measures by reducing the range of values covered by the axis. For example, in Fig. 72.17(a), it looks as if there are large differences in mass between males and females; however, if the scale is adjusted to run from zero, then it can be seen that the differences are not large as a proportion of the overall mass. Always scrutinise the scale values carefully when interpreting any graph.

3. Use of a relative rather than absolute scale – this is similar to the above, in that data compared using relative scales (e.g. percentage or ratio) can give the wrong impression if the denominator is not the same in all cases. In Fig. 72.18(a), two treatments are shown as equal in *relative* effect, both resulting in 50 relative response compared (say) to the respective controls. However, if treatment A is 50% of a control value of 200 and treatment B is 50% of a control value of 500, then the actual difference in *absolute* response would have been masked, as shown by Fig. 72.18(b).

4. Effects of a non-linear scale – when interpreting graphs with non-linear (e.g. logarithmic) scales, you may interpret any changes on an imagined linear scale. For example, the pH scale is logarithmic, and linear changes on this scale mean less in terms of absolute H^+ concentration at high (alkaline) pH than they do at low (acidic) pH. In Fig. 72.19(a), the cell density in two media is compared on a logarithmic scale, while in Fig. 72.19(b), the same data are graphed on a linear scale. Note, also, that the log y axis scale in Fig. 72.19(a) cannot be shown to zero, because there is no logarithm for 0.

Fig. 72.16 Increase in pesticide use over a 30-year period.

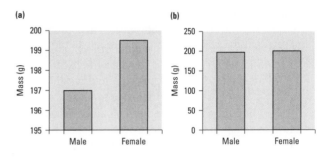

Fig. 72.17 Average mass of males and females in test group.

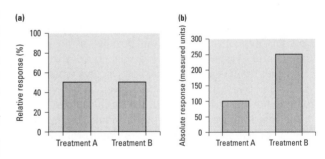

Fig. 72.18 Responses to treatments A and B.

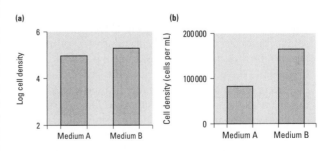

Fig. 72.19 Effect of different media on cell density.

(continued)

Box 72.3 (continued)

5. Unwarranted extrapolation – a graph may be extrapolated to indicate what would happen if a trend continued, as in Fig. 72.20(a). However, this can be done only under certain assumptions (e.g. that certain factors will remain constant or that relationships will hold under new conditions). There may be no guarantee that this will actually be the case. Figure 72.20(b) illustrates other possible outcomes if the experiment were to be repeated with higher values for the *x* axis.

6. Failure to account for data point error – this misrepresentation involves curves that are overly complex in relation to the scatter in the underlying data. When interpreting graphs with complex curves, consider the errors involved in the data values. It is probably unlikely that the curve would pass through all the data points unless the errors were very small. Figure 72.21(a) illustrates a curve that appears to assume zero error and is thus overly complex, while Fig. 72.21(b) shows a curve that takes possible errors of the points into account.

7. Failure to reject outlying points – this is a special case of the previous example. There may be many reasons for outlying data, from genuine mistakes to statistical 'freaks'. If a curve is drawn through such points on a graph, it indicates that the point carries equal weight with the other points when, in fact, it should probably be ignored. To assess this, consider the accuracy of the measurement, the number and position of adjacent points, and any special factors that might be involved on a one-off basis. Figure 72.22(a) shows a curve where an outlier (arrowed) has perhaps been given undue weight when showing the presumed relationship. If there is good reason to think that the point should be ignored, then the curve shown in Fig. 72.22(b) would probably be more valid.

8. Inappropriate fitted line – here, the mathematical function chosen to represent a trend in the data might be inappropriate. A straight line might be fitted to the data, when a curve would be more correct, or vice versa. These cases can be difficult to assess. You need to consider the theoretical validity of the model used to generate the curve (this is not always stated clearly). For example, if a straight line is fitted to the points, the implicit underlying model states that one factor varies in direct relation to another, when the true situation may be more complex. In Fig. 72.23(a), the relationship has been shown as a linear relationship, whereas an exponential relationship, as shown in Fig. 72.23(b), could be more correct.

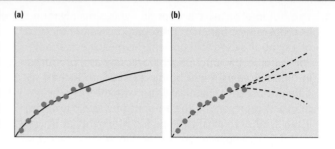

(a) **(b)**

Fig. 72.20 Extrapolation of data under different assumptions.

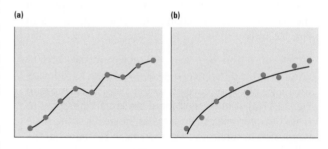

(a) **(b)**

Fig. 72.21 Fitted curves under different assumptions of data error.

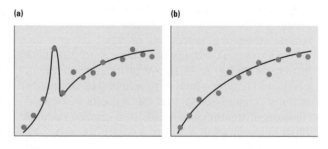

(a) **(b)**

Fig. 72.22 Curves with and without outlier taken into account.

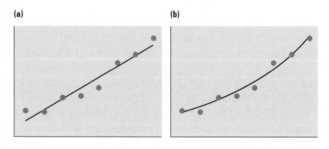

(a) **(b)**

Fig. 72.23 Different mathematical model used to represent trends in data.

Sources for further study

Briscoe, M.H. (1996) *Preparing Scientific Illustrations: A Guide to Better Posters, Presentations and Publications.* Springer-Verlag, New York.

Carter, M. *et al. Graphing Resources.* Available: https://www.ncsu.edu/labwrite/res/res-homepage.htm Last accessed 19/7/15.
[Collection of useful resources including a tutorial on using *Excel* to produce graphs.]

Institute of Biology (2000) *Biological Nomenclature: Recommendations on Terms, Units and Symbols.* Institute of Biology, London.
[Includes a section on presentation of data.]

Robbins, N.B. (2005) *Creating More Effective Graphs.* Wiley, New York.

Yuk, M. and Diamond, S. (2014) *Data Visualization for Dummies.* Wiley, New York.

STUDY EXERCISES

72.1 Select appropriate graphical presentations. Choose an appropriate graphical form for each of the following examples:

(a) Interaction between pH and cation concentration on enzyme activity.

(b) Proportion of different petal shapes in a countywide survey of *Primula vulgaris* flowers.

(c) Relationship between pulse rate and age in humans.

(d) Number of bacteria per field of view.

(e) Effect of copper concentration on the activity of an enzyme.

72.2 Create a pie chart. Display the following information in the form of a pie chart. Do *not* use a spreadsheet for this exercise.

Behaviour of male brown bear under zoo conditions

Behaviour category	Occasions observed
Eating	37
Scratching	8
Sexual display	13
Mating	2
Sleeping	25

72.3 Create a frequency distribution histogram. The following table gives data for the haemoglobin levels of 100 people. Plot a histogram showing the frequency distribution of the data. Write a brief description of the important features of the distribution.

Haemoglobin content (gL⁻¹) in blood

11.1	14.2	13.5	9.8	12.0	13.9	14.1	14.6	11.0	12.3
13.4	12.9	12.9	10.0	13.1	11.8	12.6	10.7	8.1	11.2
13.8	12.4	12.9	11.3	12.7	12.4	14.6	15.1	11.2	9.7
11.3	14.7	10.8	13.3	11.9	11.4	12.5	13.0	11.6	13.1
9.3	13.5	14.6	11.2	11.7	10.9	12.4	12.0	12.1	12.6
10.9	12.1	13.4	9.5	12.5	11.6	12.2	8.8	10.7	11.1
10.2	11.7	10.4	14.0	14.9	11.5	12.0	13.2	12.1	13.3
12.4	9.4	13.2	12.5	10.8	11.7	12.7	14.1	10.4	10.5
13.3	10.6	10.5	13.7	11.8	14.1	10.3	13.6	10.4	13.9
11.7	12.8	10.4	11.9	11.4	10.6	12.7	11.4	12.9	12.1

72.4 Find examples of misleading graphs. Create a portfolio of examples of misleading graphs taken from newspapers. For each graph, state what aspect is misleading (see Box 72.3) and, where possible, attempt to show the data correctly in a new graph.

Answers to these study exercises are available at www.pearsoned.co.uk/practicalskills

A table is often the most appropriate way to present numerical data in a concise, accurate and structured form. Assignments and project reports should contain tables that have been designed to condense and display results in a meaningful way and to aid numerical comparison. The preparation of tables for recording primary data is discussed on p. 209.

Decide whether you need a table, or whether a graph is more appropriate. Histograms and plotted curves can be used to give a visual impression of the relationships within your data (p. 477). On the other hand, a table gives you the opportunity to make detailed numerical comparisons.

> **KEY POINT** Always remember that the primary purpose of your table is to communicate information and allow appropriate comparison, not simply to put down the results on paper.

Preparation of tables

Title

Every table must have a brief descriptive title. If several tables are used, number them consecutively so they can be quoted in your text. The titles within a report should be compared with one another, making sure they are logical and consistent and that they describe accurately the numerical data contained within them.

Constructing titles – take care over titles as it is a common mistake in student practical reports to present tables without titles, or to misconstruct the title.

Structure

Display the components of each table in a way that will help the reader understand your data and grasp the significance of your results. Organise the columns so that each category of like numbers or attributes is listed vertically, while each horizontal row shows a different experimental treatment, organism, sampling site, etc. (as in Table 73.1). Where appropriate, put control values near the beginning of the table. Columns that need to be compared should be set out alongside each other. Use rulings to subdivide your table appropriately, but avoid cluttering it up with too many lines.

Using rulings in tables – modern style is to use horizontal rulings only, so if your word processor or spreadsheet includes vertical lines as default, you may wish to remove these.

Table 73.1 Characteristics of selected photoautotrophic microbes

Division	Species	Optimum [NaCl]* (mol m^{-3})	Intracellular carbohydrate	
			Identity	Quantity† (nmol (g dry wt)$^{-1}$)
Chlorophyta	*Scenedesmus quadruplicatum*	340	Sucrose	49.7
	Chlorella emersonii	780	Sucrose	102.3
	Dunaliella salina	4700	Glycerol	910.7
Cyanobacteria	*Microcystis aeruginosa*	< 20‡	None	0.0
	Anabaena variabilis	320	Sucrose	64.2
	Rivularia atra	380	Trehalose	ND

*Determined after 28-day incubation in modified Von Stosch medium.
†Individual samples, analysed by gas–liquid chromatography.
‡Poor growth in all media with added NaCl (minimum NaCl concentration 5 mol m^{-3}).
ND Sample lost: no quantitative data.

Headings and subheadings

These should identify each set of data and show the units of measurement, where necessary. Make sure that each column is wide enough for the headings and for the longest data value.

Numerical data

Within the table, do not quote values to more significant figures than necessary, as this will imply spurious accuracy (pp. 175, 495–7). By careful choice of appropriate units for each column you should aim to present numerical data within the range 0 to 1000. As with graphs, it is less ambiguous to use derived SI units, with the appropriate prefixes, in the headings of columns and rows, rather than quoting multiplying factors as powers of 10. Alternatively, include exponents in the main body of the table (see Table 26.1), to avoid any possible confusion regarding the use of negative powers of 10.

Examples If you measured the width of a fungal hypha to the nearest one-tenth of a micrometre, quote the value in the form '52.6 μm'.

Quote the width of a fungal hypha as 52.6 μm, rather than 0.0000526 m or 52.6×10^{-6} m.

Other notations

Avoid using dashes in numerical tables, as their meaning is unclear; enter a zero reading as '0' and use 'NT' not tested or 'ND' if no data value was obtained, with a footnote to explain each abbreviation. Other footnotes, identified by asterisks, superscripts or other symbols in the table, may be used to provide relevant experimental detail (if not given in the text) and an explanation of column headings and individual results, where appropriate. Footnotes should be as condensed as possible. Table 73.1 provides examples.

Saving further space in tables – in some instances a footnote can be used to replace a whole column of repetitive data.

Statistics

In tables where the dispersion of each data set is shown by an appropriate statistical parameter, you must state whether this is the (sample) standard deviation, the standard error (of the mean) or the 95% confidence limits and you must give the value of n (the number of replicates). Other descriptive statistics should be quoted with similar detail, and hypothesis-testing statistics should be quoted along with the value of P (the probability). Details of any test used should be given in the legend, or in a footnote.

Text

Sometimes a table can be a useful way of presenting textual information in a condensed form (see examples on pp. 417 and 452).

When you have finished compiling your tabulated data, carefully double-check each numerical entry against the original information, to ensure that the final version of your table is free from transcriptional errors. Box 73.1 gives a checklist for the major elements of constructing a table.

Using spreadsheets and wordprocessing packages – these can be used to prepare high-quality versions of tables for project work (Box 73.2).

Box 73.1 **Checklist for preparing a table**

Every table should have the following components:

1. **A title,** plus a reference number and date where necessary.
2. **Headings for each column and row,** with appropriate units of measurement.
3. **Data values,** quoted to the nearest significant figure and with statistical parameters, according to your requirements.
4. **Footnotes** to explain abbreviations, modifications and individual details.
5. **Rulings to emphasise groupings** and distinguish items from each other.

Box 73.2 How to use a word processor or a spreadsheet to create a table for use in coursework reports and dissertations

Creating tables with Microsoft *Word* – Word-processed tables are suitable for text-intensive or number-intensive tables, although in the second case entering data can be laborious. When working in this way, the natural way to proceed is to create the 'shell' of the table, add the data, and then carry out final formatting on the table.

1. **Move the cursor to the desired position in your document** – this is where you expect the top left corner of your table to appear. Click the *Insert* tab, then choose *Table*.

2. **Select the appropriate number of columns and rows** – do not forget to add rows and columns for headings. As default, a full-width table will appear, with single rulings for all cell boundaries, with all columns of equal width and all rows of equal height.

   Example of a 4 × 3 table:

3. **Customise the columns** – by placing the cursor over the vertical rulings then 'dragging', you can adjust their width to suit your heading text entries, which should now be added.

Heading 1	Heading 2	Heading 3	Heading 4

4. **Work through the table adding the data** – entries can be numbers or text.

Heading 1	Heading 2	Heading 3	Heading 4
xx	xx	xx	xx
xx	xx	xx	

5. **Make further adjustments to column and row widths to suit** – for example, if text fills several rows within a cell, consider increasing the column width, and if a column contains only single or double digit numbers, consider shrinking its width. To combine cells, first highlight them, then right-click the mouse button (or equivalent) and click on *Merge Cells*. You may wish to reposition text within a cell by right-clicking the mouse button and choosing an appropriate position in the cell alignment menu.

Heading 1	Heading 2	Heading 3	Heading 4
xx	xx	xx	xx
	xx	xx	

6. **Finally, remove selected borders to cells** – one way is to highlight the table, then click on the *Design* tab and choose an appropriate style from the template options. If you do not want any shading, click on *Design > Shading* and choose *No color*. Another way is to highlight the cells in the table, then right-click the mouse button and choose *Borders and Shading*. You can choose the style you wish from this submenu so that your table looks like the examples shown in this chapter.

7. **Add a table title** – this should be positioned above the table (cf. a figure title and legend, p. 478), legend and footnotes.

   Final version of the table:

   Table xx. A table of some data

Heading 1	Heading 2[a]	Heading 3	Heading 4
xx	xx	xx	xx
	xx	xx	xx

   [a]An example of a footnote.

Creating tables with Microsoft *Excel*:
Tables derived from spreadsheets are effective when you have lots of numerical data, especially when these are stored or created using the spreadsheet itself. When working in this way, you can design the table as part of an output or summary section of the spreadsheet, add explanatory headings, format, then possibly export to a word processor when complete.

(continued)

Box 73.2 (continued)

1. **Design the output or summary section.** Plan this as if it were a table, including adding text headings within cells.

16				
17				
18	Heading 1	Heading 2	Heading 3	Heading 4
19				
20				
21				
22				

2. **Insert appropriate formulae within cells to produce data.** If necessary, formulae should draw on the other parts of the spreadsheet.

16				
17				
18	Heading 1	Heading 2	Heading 3	Heading 4
19	Aaa	=A1	=C3*5	=SDEV(A1:A12)
20	Bbb	=A2	=F45/G12	=SDEV(B1:B12)
21				
22				

3. **Format the cells.** This is important to control the number of decimal places presented (*Home > Number > Format > Cells*).

4. **Adjust column width to suit.** You can do this via the column headings, by placing the cursor over the rulings between columns then 'dragging'.

5. **Add rulings as appropriate.** One way is to use the *Home > Font > Borders* menu, having selected the relevant cells.

16				
17				
18	Heading 1	Heading 2	Heading 3	Heading 4
19	Aaa	=A1	=C3*5	=SDEV(A1:A12)
20	Bbb	=A2	=F45/G12	=SDEV(B1:B12)
21				

6. **Add 'real' data values to the spreadsheet.** This should result in the summary values within the table being filled. Check that these are presented with the appropriate number of significant figures (pp. 495–7).

7. **The table can now be copied and pasted to a *Word* document.** For best results, use the *Paste Special...* > Microsoft Office Excel Worksheet Object option.

Note: instructions and commands illustrated here may vary among the various versions of Microsoft *Office* programs and for different products.

Sources for further study

Kirkup, L. (1994) *Experimental Methods: An Introduction to the Analysis and Presentation of Data.* Wiley, New York.

New York State University. *Labwrite Resources: Graphing Resources.* Available: https://www.ncsu.edu/labwrite/res/gh/gh-tables.html
Last accessed 19/7/15.
[Useful resources on both tables and their links to graphics.]

Simmonds, D. and Reynolds, L. (1994) *Data Presentation and Visual Literacy in Medicine and Science.* Butterworth-Heinemann, London.

Willis, J. (2004) *Data Analysis and Presentation Skills: An Introduction for the Life and Medical Sciences.* Wiley, Chichester.

73.1 Redesign a table of data. Using the following example, redraft the table to improve layout and correct inconsistencies.

Concentrations of low molecular weight solutes in bacteria

Bacterium	Constituent	Concentration
Escherichia coli	Proline	21.0 mmol L^{-1}
Escherichia coli	Trehalose	1.547×10^2 kmol m^{-3}
Bacillus subtilis	Proline	39.7 mmol L^{-1}
Bacillus subtilis	Glutamate	0.0521 mmol cm^{-3}
*Staphylococcus aureus**	Glutamate	15 260 mmol m^{-3}
Escherichia coli	Glutamate	0.50% w/v
Bacillus subtilis	Trehalose	<0.001% w/v

*Proline and trehalose were not measured.

73.2 Devise a text-based table. After reading through this chapter and working from memory, draw up a table listing the principal components of a typical table in the first column, and brief comments on the major features of each component in the second column.

73.3 Interpret data from a table. The table below shows representative data for energy, protein and niacin (vitamin B3) as a function of age for human males and females. The typical average mass ('weight') of each class is also shown.

a) In what age class does the mass of females exceed that of males?

b) In what age class is the energy requirement of males greatest in proportion to the energy requirement of females?

c) Over what age range do males and females have the same protein requirement?

d) In what age class do males have their highest niacin requirement, and how does this compare with females?

Selected age-related characteristics of human males and females

Age (years)	Average mass (kg)		Energy requirement (MJ)		Protein requirement (g)		Niacin requirement (mg)	
	Male	Female	Male	Female	Male	Female	Male	Female
<1	8	8	3.4	3.2	14	14	6	6
1–3	13	13	5.2	4.9	16	16	9	9
4–6	20	20	7.2	6.5	24	24	12	12
7–10	28	28	8.2	7.3	28	28	13	13
11–14	45	46	9.3	7.7	45	46	17	15
15–18	66	55	11.5	8.8	59	44	20	15
19–50	79	63	10.6	8.1	63	50	19	15
>50	77	65	9.5	7.7	63	50	15	13

Answers to these study exercises are available at www.pearsoned.co.uk/practicalskills

The following boxes give advice on dealing with numerical procedures:

- Box 24.2: preparing solutions (p. 147)
- Box 25.1: molar concentrations (p. 156)
- Box 48.1: cell-counting chambers (p. 294)
- Box 48.2: plate (colony) counts (p. 295)
- Box 53.1: calibration curves (p. 327)
- Box 64.1: radioactivity (p. 419)

Biology often requires a numerical or statistical approach. Not only is mathematical modelling an important aid to understanding, but computations are also often needed to turn raw data into meaningful information or to compare them with other data sets. Moreover, calculations are part of laboratory routine, perhaps required for making up solutions of known concentration (see p. 147 and text below) or for the calibration of a microscope (see p. 277). In research, 'trial' calculations can reveal what input data are required and where errors in their measurement might be amplified in the final result (see p. 204).

> **KEY POINT** If you find numerical work difficult, practice at problem-solving is especially important.

Practising at problem-solving:

- **demystifies the procedures involved**, which are normally just the elementary mathematical operations of addition, subtraction, multiplication and division (Table 74.1);
- **allows you to gain confidence** so that you do not become confused when confronted with an unfamiliar or apparently complex form of problem;
- **helps you recognise the various forms a problem can take** as, for instance, in crossing experiments in classical genetics (Box 61.1).

Table 74.1 Sets of numbers and operations

Sets of numbers	
Whole numbers:	0, 1, 2, 3, . . .
Natural numbers:	1, 2, 3, . . .
Integers:	. . . −3, −2, −1, 0, 1, 2, 3, . . .
Real numbers:	integers and anything between (e.g. −5, 4.376, 3/16, π, $\sqrt{5}$)
Prime numbers:	subset of natural numbers divisible by 1 and themselves only (i.e. 2, 3, 5, 7, 11, 13, ...)
Rational numbers:	p/q where p (integer) and q (natural) have no common factor (e.g. 3/4)
Fractions:	p/q where p is an integer and q is natural (e.g. −6/8)
Irrational numbers:	real numbers with no exact value (e.g. π)
Infinity:	(symbol ∞) is larger than any number (technically not a number as it does not obey the laws of algebra)
Operations and symbols	
Basic operators:	+, −, × and ÷ will not need explanation; however, / may substitute for ÷, * may substitute for × or this operator may be omitted
Powers:	a^n, i.e. 'a to the power n', means a multiplied by itself n times (e.g. $a^2 = a \times a = $ 'a squared', $a^3 = a \times a \times a = $ 'a cubed'). n is said to be the index or exponent. Note $a^0 = 1$ and $a^1 = a$
Logarithms:	the common logarithm (log) of any number x is the power to which 10 would have to be raised to give x (i.e. the log of 100 is 2; $10^2 = 100$); the antilog of x is 10^x. Note that there is no log for 0, so take this into account when drawing log axes by breaking the axis. Natural or Napierian logarithms (ln) use the base e (= 2.71828 . . .) instead of 10
Reciprocals:	the reciprocal of a real number a is $1/a$ ($a \neq 0$)
Relational operators:	$a > b$ means 'a is greater (more positive) than b', $<$ means less than, $\leq$ means less-than-or-equal-to and $\geq$ means greater-than-or-equal-to
Proportionality:	$a \propto b$ means 'a is proportional to b' (i.e. $a = kb$, where k is a constant). If $a \propto 1/b$, a is inversely proportional to b ($a = k/b$)
Sums:	$\sum x_i$ is shorthand for the sum of all x values from $i = 0$ to $i = n$ (more correctly the range of the sum is specified under the symbol)
Moduli:	$|x|$ signifies modulus of x, i.e. its absolute value (e.g. $|4| = |-4| = 4$)
Factorials:	$x!$ signifies factorial x, the product of all integers from 1 to x (e.g. 3! = 6). Note 0! = 1! = 1

Tracing errors in mathematical problems
– this is always easier when all the stages in
a calculation are laid out clearly.

Using a spreadsheet for numerical
problems – this may be very useful in
repetitive work or for 'what if?' case
studies (see Chapter 13). When using
a spreadsheet, much of the guidance
contained in this chapter remains relevant.

Table 74.2 Simple algebra – rules for manipulating

If $a = b + c$, then $b = a - c$ and $c = a - b$
If $a = b \times c$, then $b = a \div c$ and $c = a \div b$
If $a = b^c$, then $b = a^{1/c}$ and $c = \log a \div \log b$
$a^{1/n} = \sqrt[n]{a}$
$a^{-n} = 1 \div a^n$
$a^b \times a^c = a^{(b+c)}$ and $a^b \div a^c = a^{(b-c)}$
$(a^b)^c = a^{(b \times c)}$
$a \times b = \text{antilog}(\log a + \log b)$

Presenting calculations in assessed
work – always show the steps in your
calculations; most markers will only
penalise a mistake once, and part marks
will be given if the remaining operations
are performed correctly. This can only be
done if those operations are visible.

Steps in tackling a numerical problem

The step-by-step approach outlined below may not be the fastest method of arriving at an answer, but most mistakes occur where steps are missing, combined or not made obvious, so a logical approach is often better.

Have the right tools ready

Scientific calculators (p. 126) greatly simplify the numerical part of problem-solving. However, the seeming infallibility of the calculator may lead you to accept an absurd result that could have arisen because of faulty key-pressing or faulty logic. Make sure you know how to use all the features on your calculator, especially how the memory works; how to introduce a constant multiplier or divider; and how to obtain an exponent (note that the 'exp' button on most calculators gives you 10^x, not 1^x or y^x; so 1×10^6 would be entered as $\boxed{1}\,\boxed{\text{exp}}\,\boxed{6}$, *not* $\boxed{10}\,\boxed{\text{exp}}\,\boxed{6}$).

Approach the problem thoughtfully

If the individual steps have been laid out on a worksheet, the 'tactics' will already have been decided. It is more difficult when you have to adopt a strategy on your own, especially if the problem is presented as a story and it is not obvious which equations or rules need to be applied.

- **Read the problem carefully** as the text may give clues as to how it should be tackled. Be certain of what is required as an answer before starting.
- **Analyse what kind of problem it is**, which effectively means deciding which equation(s) or approach will be applicable. If this is not obvious, consider the dimensions/units of the information available and think how they could be fitted to a relevant formula. In examinations, a favourite ploy of examiners is to present a problem such that the familiar form of an equation must be rearranged (see Table 74.2 and Box 74.1). Another is to make you use two or more equations in series (see Box 74.2). If you are unsure whether a recalled formula is correct, a dimensional analysis can help: write in all the units for the variables and make sure that they cancel out to give the expected answer.
- **Check that you have, or can derive, all of the information required** to use your chosen equation(s). It is unusual but not unknown for examiners to supply redundant information. So, if you decide not to use some of the information given, know why you do not require it.
- **Decide in what format and units the answer should be presented.** This is sometimes suggested to you. If the problem requires many changes in the prefixes to units, it is a good idea to convert all data to base SI units (multiplied by a power of 10) at the outset.
- **If a problem appears complex, break it down into component parts.**

Present your answer clearly

The way you present your answer obviously needs to fit the individual problem. The example shown in Box 74.2 has been chosen to illustrate several important points, but this format would not fit all situations. Guidelines for presenting an answer include:

- **Make your assumptions explicit.** Most mathematical models of biological phenomena require that certain criteria are met before they

> **Box 74.1** **Example of using the algebraic rules of Table 74.2**
>
> Problem: if $a = (b - c) \div (d + e^n)$, find e
>
> 1. Multiply both sides by $(d + e^n)$; formula becomes:
> $$a(d + e^n) = (b - c)$$
>
> 2. Divide both sides by a; formula becomes:
> $$d + e^n = \frac{b - c}{a}$$
>
> 3. Subtract d from both sides; formula becomes:
> $$e^n = \frac{b - c}{a} - d$$
>
> 4. Raise each side to the power $1/n$; formula becomes:
> $$e = \left(\frac{b - c}{a} - d\right)^{1/n}$$

can be legitimately applied (e.g. 'assuming the tissue is homogeneous …'), while some approaches involve approximations that should be clearly stated (e.g. 'to estimate the mouse's skin area, its body was approximated to a cylinder with radius x and height y …').

- **Explain your strategy for answering**, perhaps giving the applicable formula or definitions that suit the approach to be taken. Give details of what the symbols mean (and their units) at this point.

- **Rearrange the formula to the required form** with the desired unknown on the left-hand side (see Table 74.2).

- **Substitute the relevant values into the right-hand side of the formula**, using the units and prefixes as given (it may be convenient to convert values to SI beforehand). Convert prefixes to appropriate powers of 10 as soon as possible.

- **Convert to the desired units step by step, i.e. taking each variable in turn.**

- **When you have the answer in the desired units, rewrite the left-hand side and <u>underline the answer</u> for emphasis.** Make sure that the result is presented to an appropriate number of significant figures (see below).

Check your answer

Having written out your answer, you should check it methodically, answering the following questions:

- **Is the answer realistic?** You should be alerted to an error if a number is absurdly large or small. In repeated calculations, a result standing out from others in the same series should be double-checked.

- **Do the units make sense and match up with the answer required?** Do not, for example, present a volume in units of m^2.

- **Do you get the same answer if you recalculate in a different way?** If you have time, recalculate the answer using a different 'route', entering the numbers into your calculator in a different form and/or carrying out the operations in a different order.

> Units – never write any answer without its unit(s) unless it is truly dimensionless.

> Rounding off to a specific number of significant figures – do not round off numbers until you arrive at the final answer or you will introduce errors into the calculation.

Rounding: decimal places and significant figures

In many instances, the answer you produce as a result of a calculation will include more figures than is justified by the accuracy and precision of the original data. Sometimes you will be asked to produce an answer to a specified number of decimal places or significant figures and other times you will be expected to decide for yourself what would be appropriate.

> ### Box 74.2 Model answer to a typical biological problem
>
> **Problem**
>
> Estimate the total length and surface area of the fibrous roots on a maize seedling from measurements of their total fresh weight and mean diameter. Give your answers in m and cm² respectively to four significant figures.
>
> **Measurements**
>
> Fresh weight[a] = 5.00 g, mean diameter[b] = 0.5 mm.
>
> **Answer**
>
> Assumptions: (1) the roots are cylinders with constant radius[c] and the 'ends' have negligible area; (2) the root system has a density of 1000 kg m⁻³ (i.e. that of water[d]).
>
> Strategy: from assumption (1), the applicable equations are those concerned with the volume and surface area of a cylinder (Table 74.3), namely:
>
> $$V = \pi r^2 h \qquad\qquad [74.1]$$
> $$A = 2\pi rh \,(\text{ignoring ends}) \qquad\qquad [74.2]$$
>
> where V is volume (m³), A is surface area (m²), $\pi \approx 3.14159$, h is height (m) and r is radius (m). The total length of the root system is given by h and its surface area by A. We can find h by rearranging eqn [74.1] and then substitute its value in eqn [74.2] to get A.
>
> To calculate total root length: rearranging eqn [74.1], we have $h = V/\pi r^2$. From measurements[e], $r = 0.25$ mm $= 0.25 \times 10^{-3}$ m.
>
> From density = weight/volume,
>
> $$V = \text{fresh weight/density}$$
> $$= 5\,\text{g}/1000\,\text{kg m}^{-3}$$
> $$= 0.005\,\text{kg}/1000\,\text{kg m}^{-3}$$
> $$= 5 \times 10^{-6}\,\text{m}^3$$
>
> Total root length,
>
> $$h = V/\pi r^2$$
> $$5 \times 10^{-6}\text{m}^3/3.14159 \times (0.25 \times 10^{-3}\text{m})^2$$
>
> $\therefore$ Total root length = 25.46 m
>
> To calculate surface area of roots: substituting value for h obtained above into eqn [74.2], we have:
>
> Root surface area
>
> $$= 2 \times 3.14159 \times 0.25 \times 10^{-3}\,\text{m} \times 25.46\,\text{m}$$
> $$= 0.04\,\text{m}^2$$
> $$= 0.04 \times 10^4\,\text{cm}^2$$
> (there being $100 \times 100 = 10^4$ cm² per m²)
>
> $\therefore$ Root surface area = 400.0 cm²
>
> **Notes**
>
> - The fresh weight of roots would normally be obtained by washing the roots free of soil, blotting them dry and weighing.
> - In a real answer you might show the replicate measurements giving rise to the mean diameter.
> - In reality, the roots will differ considerably in diameter and each root will not have a constant diameter throughout its length.
> - This will not be wildly inaccurate as about 95%, of the fresh weight will be water, but the volume could also be estimated from water displacement measurements.
> - Note conversion of measurements into base SI units at this stage and on line 3 of the root volume calculation. Forgetting to halve diameter measurements where radii are required is a common error.

> **KEY POINT** Do not simply accept the numerical answer from a calculator or spreadsheet, without considering whether you need to modify this to give an appropriate number of significant figures or decimal places.

Rounding to n decimal places

This is relatively easy to do.

1. **Look at the number to the right of the *n*th decimal place.**
2. **If this is less than five, simply 'cut off' all numbers to the right of the nth decimal place** to produce the answer (i.e. round down).
3. **If the number is greater than five, 'cut off' all numbers to the right of the *n*th decimal place and add one to the *n*th decimal place** to produce the answer (i.e. round up).

Examples

The number 4.123 correct to two decimal places is 4.12

The number 4.126 correct to two decimal places is 4.13

The number 4.1251 correct to two decimal places is 4.13

The number 4.1250 correct to two decimal places is 4.12

The number 4.1350 correct to two decimal places is 4.14

The number 99.99 correct to one decimal place is 100.0.

Examples

The number of significant figures in 194 is 3

The number of significant figures in 2305 is 4

The number of significant figures in 0.003482 is 4

The number of significant figures in 210×10^8 is 3 (21×10^9 would be 2).

Examples

The number of significant figures in 3051.93 is 6

To five significant figures, this number is 3051.9

To four significant figures, this number is 3052

To three significant figures, this number is 3050

To two significant figures, this number is 3100

To one significant figure, this number is 3000

3051.93 to the nearest 10 is 3050

3051.93 to the nearest 100 is 3100

Note that in this last case you must include the zeros before the decimal point to indicate the scale of the number (even if the decimal point is not shown). For a number less than 1, the same would apply to the zeros before the decimal point. For example, 0.00305193 to three significant figures is 0.00305. Alternatively, use scientific notation (in this case, 3.05×10^{-3}).

4. **If the number is 5, then look at further numbers to the right to determine whether to round up or not.**

5. **If the number is *exactly* 5 and there are no further numbers to the right, then round to the nearest even number.** *Note:* When considering a large number of calculations, this procedure will not affect the overall mean value. Some rounding systems do the opposite to this (i.e. round to the nearest odd number), while others always round up where the number is exactly 5 (which *will* affect the mean). Take advice from your tutor and stick to one system throughout a series of calculations.

Whenever you see any number quoted, you should assume that the last digit has been rounded. For example, in the number 22.4, the '.4' is assumed to be rounded and the calculated value may have been between 22.35 and 22.45.

Quoting to n significant figures

The number of significant figures indicates the degree of approximation in the number. For most cases, it is given by counting all the figures except zeros that occur at the beginning or end of the number. Zeros *within* the number are always counted as significant. The number of significant figures in a number like 200 is ambiguous and could be one, two or three; if you wish to specify clearly, then quote as e.g. 2×10^2 (one significant figure), 2.0×10^2 (two significant figures), etc. to avoid spurious accuracy (pp. 175, 489). When quoting a number to a specified number of significant figures, use the same rules as for rounding to a specified number of decimal places, but do not forget to keep zeros before or after the decimal point. The same principle is used if you are asked to quote a number to the 'nearest 10', 'nearest 100', etc.

When deciding for yourself how many significant figures to use, adopt the following rules of thumb:

- **Always round *after* you have done a calculation.** Use *all* significant figures available in the measured data during a calculation.

- **If adding or subtracting with measured data, then quote the answer to the number of decimal places in the data value with the least number of decimal places** (e.g. $32.1 - 45.67 + 35.6201 = 22.1$, because 32.1 has one decimal place).

- **If multiplying or dividing with measured data, keep as many significant figures as are in the number with the least number of significant places** (e.g. $34901 \div 3445 \times 1.3410344 = 13.59$, because 3445 has four significant figures).

- **For the purposes of significant figures, assume 'constants' have an infinite number of significant figures** (e.g. number of mm in a metre).

Some reminders of basic mathematics

Errors in calculations sometimes appear because of faults in mathematics rather than computational errors. For reference purposes, Tables 74.1–74.3 give some basic mathematical principles that may be useful. Eason *et al.* (1992) or Stephenson (2010) should be consulted for more advanced/specific needs.

Table 74.3 Geometry and trigonometry – analysing shapes

Shape/object	Diagram	Perimeter	Area
Two-dimensional shapes			
Square		$4x$	x^2
Rectangle		$2(x+y)$	xy
Circle		$2\pi r$	πr^2
Ellipse		$\pi[1.5(a + b) - \sqrt{a*b}]$ (approx.)	πab
Triangle (general)		$x+y+z$	$0.5zh$
(right-angled)		$x + y + r$ $\sin\theta = y/r, \cos\theta = x/r,$ $\tan\theta = y/x; r^2 = x^2 + y^2$	$0.5xy$

Shape/object	Diagram	Surface area	Volume
Three-dimensional shapes			
Cube		$6x^2$	x^3
Cuboid		$2xy+2xz+2yz$	xyz
Sphere		$4\pi r^2$	$4\pi r^3/3$
Ellipsoid		no simple formula	πrab
Cylinder		$2\pi rh + 2\pi r^2 h$	$\pi r^2 h$
Cone and pyramid		$0.5PL+B$	$BL/3$

Key: x, y, z = sides a, b = half minimum and maximum axes; r = radius or hypotenuse; h = height; B = base area; L = perpendicular height; P = perimeter of base.

> **Examples**
>
> 1/8 as a percentage is $1 \div 8 \times 100 = 100 \div 8 = 12.5\%$.
>
> 0.602 as a percentage is $0.602 \times 100 = 60.2\%$.

> **Examples**
>
> 190% as a decimal fraction is $190 \div 100 = 1.9$
>
> 5/2 as a percentage is $5 \div 2 \times 100 = 250\%$.

> **Example** A population falls from 4 million to 3.85 million. What is the percentage change? The decrease in numbers is $4 - 3.85 = -0.15$ million.
>
> The fractional decrease is $-0.15 \div 4 = -0.0375$ and we multiply by 100 to get the percentage change = minus 3.75%.

> **Example** $2^3 = 2 \times 2 \times 2 = 8$.

> **Example** Avogadro's number, $\approx$ 602 352 000 000 000 000 000 000, is more conveniently expressed as 6.02352×10^{23}.

Percentages and proportions

A percentage is just a fraction expressed in terms of hundredths, indicated by putting the percentage sign (%) after the number of hundredths. So 35% simply means 35 hundredths. To convert a fraction to a percentage, just multiply the fraction by 100. When the fraction is in decimal form, multiplying by 100 to obtain a percentage is easily achieved just by moving the decimal point two places to the right.

To convert a percentage to a fraction, just remember that, since a percentage is a fraction multiplied by 100, the fraction is the percentage divided by 100. For example: $42\% = 42/100 = 0.42$. In this example, since we are dealing with a decimal fraction, the division by 100 is just a matter of moving the decimal point two places to the left (42% could be written as 42.0%). Percentages greater than 100% represent fractions greater than 1. Percentages less than 1 may cause confusion. For example, 0.5% means half of one per cent (0.005) and must not be confused with 50% (which is the decimal fraction 0.5).

To find a percentage of a given number, just express the percentage as a decimal fraction and multiply the given number. For example: 35% of 500 is given by $0.35 \times 500 = 175$. To find the percentage change in a quantity, work out the difference (= value 'after' – value 'before'), and divide this difference by the original value to give the fractional change, then multiply by 100.

Exponents

Exponential notation is an alternative way of expressing numbers in the form a^n ('a to the power n'), where a is multiplied by itself n times. The number a is called the base and the number n the exponent (or power or index). The exponent need not be a whole number, and it can be negative if the number being expressed is less than 1. See Table 74.2 for other mathematical relationships involving exponents.

Scientific notation

In scientific notation, also known as 'standard form', the base is 10 and the exponent a whole number. To express numbers that are not whole powers of 10, the form $c \times 10^n$ is used, where the coefficient c is normally between 1 and 10. Scientific notation is valuable when you are using very large numbers and wish to avoid suggesting spurious accuracy. Thus if you write 123 000, this may suggest that you know the number to ± 0.5, whereas 1.23×10^5 might give a truer indication of measurement accuracy (i.e. implied to be ± 500 in this case). Engineering notation is similar, but treats numbers as powers of 10 in groups of 3, i. e. $c \times 10^0$, 10^3, 10^6, 10^9, etc. This corresponds to the SI system of prefixes (p. 177).

A useful property of powers when expressed to the same base is that when multiplying two numbers together, you simply add the powers, while if dividing, you subtract the powers. Thus, suppose you counted 8 bacteria in a known volume of a 10^{-7} dilution (see p. 149), there would be 8×10^7 in the same volume of undiluted solution; if you now dilute this 500-fold (5×10^2), then the number present in the same volume would be $8/5 \times 10^{(7-2)} = 1.6 \times 10^5 = 160\,000$.

> **Examples** The logarithm to the base 10 ($\log_{10}$) of 1000 is 3, since $10^3 = 1000$. The logarithm to the base e ($\log_e$ or ln) of 1000 is 6.907755 (to six decimal places).

> **Examples** (use to check the correct use of your own calculator)
> 102963 as a $\log_{10}$ = 5.012681 (to six decimal places)
> $10^{5.012681} = 102962.96$
> (Note loss of accuracy due to loss of decimal places.)
> 102963 as a natural logarithm (ln) = 11.542125 (to six decimal places), thus $2.718282^{11.542125} = 102963$.

Logarithms

When a number is expressed as a logarithm, this refers to the power n that the base number a must be raised to give that number. Any base could be used, but the two most common are 10, when the power is referred to as $\log_{10}$ or simply log, and the constant e (2.718282), used for mathematical convenience in certain situations, when the power is referred to as $\log_e$ or ln (natural logarithm). Note that (a) logs need not be whole numbers; (b) there is no log value for the number zero; and (c) that $\log_{10} = 0$ for the number 1.

To obtain logs, you will need to use the log key on your calculator, or special log tables (now largely redundant). To convert back (antilog), use

- the $\boxed{10^x}$ key, with x = log value;
- the $\boxed{\text{inverse}}$ then the $\boxed{\log}$ key; or
- the $\boxed{y^x}$ key, with $y = 10$ and x = log value.

If you have used log tables, you will find complementary antilogarithm tables to do this.

There are many uses of logarithms in biology, including pH (= $-\log[H^+]$), where $[H^+]$ is expressed in moll^{-1} (see p. 163); the exponential growth of micro-organisms, where if log(cell number) is plotted against time, a straight-line relationship is obtained (see p. 291); and allometric studies of growth and development, where if data are plotted on log axes, a series of straight-line relationships may be found (see p. 314).

Linear functions and straight lines

One of the most straightforward and widely used relationships between two variables x and y is that represented by a straight-line graph, where the corresponding mathematical function is known as the equation of a straight line, where:

$$y = a + bx \qquad [74.3]$$

In this relationship, a represents the intercept of the line on the y (vertical) axis, i.e. where $x = 0$, and therefore $bx = 0$, while b is equivalent to the slope (gradient) of the line, i.e. the change in y for a change in x of 1. The constants a and b are sometimes given alternative symbols, but the mathematics remains unchanged, e.g. in the equivalent expression for the slope of a straight line, $y = mx + c$. Figure 74.1 shows what happens when these two constants are changed, in terms of the resultant straight lines.

The two main applications of the straight-line relationship are:

1. **Function fitting.** Here, you determine the mathematical form of the function, i.e. you estimate the constants a and b from a data set for x and y, either by drawing a straight line by eye (p. 479) and then working out the slope and y intercept, or by using linear regression (p. 523) to obtain the most probable values for both constants. When putting a straight line of best fit by eye on a hand-drawn graph, note the following:

- Always use a *transparent* ruler, so you can see data points on either side of the line.
- For a data series where the points do not fit a perfect straight line, try to have an equal number of points on either side of the line, as

(a)

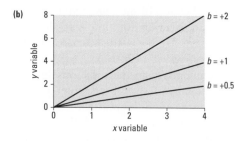

(b)

Fig. 74.1 Straight-line relationships ($y = a + bx$), showing the effects of (a) changing the intercept at constant slope, and (b) changing the slope at constant intercept.

in (Fig. 74.2(a), and try to minimise the average distance of these points from the line.

- Once you have drawn the line of best fit use this line, rather than your data values, in all subsequent procedures (e.g. in a calibration curve, Chapter 53).
- Tangents drawn to a curve give the slope (gradient) at a particular point, e.g. in an enzyme reaction progress curve (Fig. 60.3). These are best drawn by bringing your ruler up to the curve at the exact point where you wish to estimate the slope and then trying to make the two angles immediately on either side of this point approximately the same, by eye (Fig. 74.2(b)).
- Once you have drawn the straight line or tangent, choose two points reasonably far apart at either end of your line and then draw construction lines to represent the change in y and the change in x between these two points: make sure that your construction lines are perpendicular to each other. Determine the slope as the change in y divided by the change in x (Fig. 74.2).

2. **Prediction.** Where a and b are known, or have been estimated, you can use eqn [74.3] to predict any value of y for a specified value of x, e.g. during exponential growth of a cell culture (p. 291), where $\log_{10}$ cell number (y) increases as a linear function of time (x): note that in this example the dependent variable has been transformed to give a linear relationship. You will need to rearrange eqn [74.3] in cases where a prediction of x is required for a particular value of y (e.g. in calibration curves, p. 326, or bioassays, p. 315), as follows:

$$x = (y - a) \div b \qquad [74.4]$$

This equation can also be used to determine the intercept on the x (horizontal) axis, i.e. where $y = 0$.

Hints for some typical problems

Calculations involving proportions or ratios

The 'unitary method' is a useful way of approaching calculations involving proportions or ratios, such as those required when making up solutions from stocks (see also Chapter 25) or as a subsidiary part of longer calculations.

1. **If given a value for a multiple, work out the corresponding value for a single item or unit.**

2. **Use this 'unitary value' to calculate the required new value.**

Calculations involving series

Series (used in e.g. dilutions, see also p. 148) can be of three main forms:

1. **Arithmetic,** where the *difference* between two successive numbers in the series is a constant, e.g. 2, 4, 6, 8, 10,. . .

2. **Geometric,** where the *ratio* between two successive numbers in the series is a constant, e.g. 1, 10, 100, 1000, 10 000,. . .

3. **Harmonic,** where the values are reciprocals of successive whole numbers, e.g. $1, \frac{1}{2}, \frac{1}{3}, \frac{1}{4}, \ldots$

Note that the logs of the numbers in a geometric series will form an arithmetic series (e.g. 0, 1, 2, 3, 4,. . . in the above case). Thus, if a quantity

---

Examples Using eqn [74.3], the predicted value for y for a linear function where $a = 2$ and $b = 0.5$, where $x = 8$ is:
$y = 2 + (0.5 \times 8) = 6$.
Using eqn [74.4], the predicted value for x for a linear function where
$a = 1.5$ and $b = 2.5$, where $y = -8.5$ is:
$x = (-8.5 - 1.5) \div 2.5 = -4$.
Using eqn [74.4] the predicted x intercept for a linear function where $a = 0.8$ and $b = 3.2$ is: $x = (0 - 0.8) \div 3.2 = -0.25$.

Example A lab schedule states that 5 g of a compound with a relative molecular mass of 220 are dissolved in 400 mL of solvent. For writing up your Materials and Methods, you wish to express this as mol L^{-1}.

1. If there are 5 g in 400 mL, then there are $5 \div 400$ g in 1 mL.

2. Hence, 1000 mL will contain $5 \div 400 \times 1000$ g = 12.5 g.

3. 12.5 g = $12.5 \div 220$ mol = 0.0568 mol, so [solution] = 56.8 mmol L^{-1} (= 56.8 mol m^{-3}).

Examples For a geometric dilution series involving ten-fold dilution steps, calculation of concentrations is straightforward, e.g. two serial decimal dilutions (= 100-fold dilution) of a solution of NaCl of 250 mmol L^{-1} will produce a dilute solution of $250 \div 100 = 2.5$ mmol L^{-1}. Similarly, for an arithmetic dilution series, divide by the overall dilution to give the final concentration, e.g. a sixteen-fold dilution of a solution of NaCl of 200 mg mL^{-1} will produce a dilute solution of $200 \div 16 = 12.5$ mg mL^{-1}.

(a)

(b)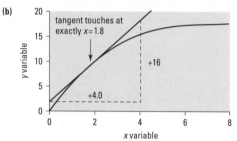

Fig. 74.2 Drawing straight lines. (a) Simple linear relationship, giving a straight line with an intersect of 2.3 and a slope of $-1.6 \div 4.0 = 0.4$. (b) Tangent drawn to a curve at $x = 1.8$, giving a slope of $16 \div 4 = 4$.

y varies with a quantity x such that the rate of change in y is proportional to the value of y (i.e. it varies in an exponential manner), a semi-log plot of such data will form a straight line. This form of relationship is relevant for exponentially growing cell cultures (p. 291) and radioactive decay (p. 416).

Statistical calculations

The need for long, complex calculations in statistics has largely been removed because of the widespread use of spreadsheets with statistical functions (Chapter 12) and specialised programs such as *SPSS* and *Minitab*. It is, however, important to understand the principles behind what you are trying to do (see Chapters 75 and 76) and interpret the program's output correctly, either using the 'help' function or a reference manual.

Problems in Mendelian genetics

These cause difficulties for many students. The key is to recognise the different types of problems and to practise so you are familiar with the techniques for solving them. Chapter 61 deals with the different types of crosses you will come across and methods of analysing them, including the use of the χ^2 (Chi^2) test.

Text references

Eason, G., Coles, C.W. and Gettinby, G. (1992) *Mathematics and Statistics for the Bio-Sciences*. Ellis Horwood, Chichester.

Sources for further study

Anon. *S.O.S. Mathematics*. Available: http://www.sosmath.com/
Last accessed 19/07/15.
[A basic Web-based guide with very wide coverage.]

Britton, N.F. (2003) *Essential Mathematical Biology*. Springer-Verlag. New York.
[Describes a range of mathematical applications in population dynamics, epidemiology, genetics, biochemistry and medicine.]

Cann, A.J. (2002) *Maths from Scratch for Biologists*. Wiley, Chichester.
[Deals with basic manipulations, formulae, units, molarity, logs and exponents as well as basic statistical procedures.]

Causton, D.R. (1992) *A Biologist's Basic Mathematics*. Cambridge University Press, Cambridge.

Forster, P.C. (1999) *Easy Mathematics for Biologists*. Harwood, Amsterdam.

Stephenson, F.H. (2010) *Calculations in Molecular Biology and Biotechnology*, 2nd edn. Academic Press, London.
[Covers basic principles, units, logarithms, ratios and proportions, concentrations, equations, rates and graphs.]

Harris, M., Taylor, G. and Taylor, J. (2005) *Catch Up Maths and Stats for the Life and Medical Sciences*. Scion, Bloxham.
[Covers a range of basic mathematical operations and statistical procedures.]

Koehler, K.R. *College Physics for Students of Biology and Chemistry*. Available: http:/kias.dyndns.org/biophys.2ed/contents.html.
Last accessed 19/07/15.
[A 'hypertextbook' written for first-year undergraduates. Assumes that you have a working knowledge of algebra.]

Lawler, G. (2011) *Understanding Maths. Basic Mathematics Explained*, 4th edn. Aber Publishing, Abergele.

STUDY EXERCISES

74.1 Rearrange the following formulae.

  (a) If $y = ax + b$, find b
  (b) If $y = ax + b$, find x
  (c) If $x = y^3$, find y
  (d) If $x = 3^y$, find y
  (e) If $x = (1 - y)(z^p + 3)$, find z
  (f) If $x = (y - z)^{1/n}/pq$, find n

74.2 Work with decimal places or significant figures.
Give the following numbers to the accuracy indicated:

  (a) 214.51 to three significant figures
  (b) 107 029 to three significant figures
  (c) 0.0450 to one significant figure
  (d) 99.817 to two decimal places
  (e) 99.897 to two decimal places
  (f) 99.997 to two decimal places
  (g) 6255 to the nearest 10
  (h) 134 903 to the nearest ten thousand

State the following:

  (i) the number of significant figures in 3400
  (ii) the number of significant figures in 3400.3
  (iii) the number of significant figures in 0.00167
  (iv) the number of significant figures in 1.00167
  (v) the number of decimal places in 34.46
  (vi) the number of decimal places in 0.00167

74.3 Carry out calculations involving percentages.
Answer the following questions, giving your answers to two decimal places:

  (a) What is 6/35ths expressed as a percentage?

  (b) What is 0.0755 expressed as a percentage?
  (c) What is 4.35% of 322?
  (d) A rat's weight increased from 55.23 g to 75.02 g. What is the percentage increase in its weight?

74.4 Practise using exponents and logarithms.

  (a) Write out (i) the charge on an electron and (ii) the speed of light *in vacuo* in longhand (i.e. without using powers of 10). See Table 28.4 for values given in scientific notation.

  (b) Compute the following values: 10^{3624}; $\log(6.37)$; $e^{-2.32}$; $\ln(1123)$; $6^{3.2}$. A calculator should be used, but round the output to give five significant figures.

74.5 Practise using geometric formulae. Select appropriate formulae from Table 74.3, and use them to compare the surface-area-to-volume ratios of the following:

  (a) a spherical unicellular alga of diameter 30 μm;
  (b) a cylindrical root of radius 0.5 mm (ignoring ends).

74.6 Practise working with linear functions (note also that Chapter 53 includes study exercises based on linear functions and plotting straight lines). Assuming a linear relationship between x and y, calculate the following (give your answers to three significant figures):

  (a) x, where $y = 7.0$, $a = 4.5$ and $b = 0.02$;
  (b) x, where $y = 15.2$, $a = -2.6$ and $b = -4.46$;
  (c) y, where $x = 10.5$, $a = 0.2$ and $b = -0.63$;
  (d) y, where $x = 4.5$, $a = -1.8$ and $b = 4.1$.

Answers to these study exercises are available at www.pearsoned.co.uk/practicalskills

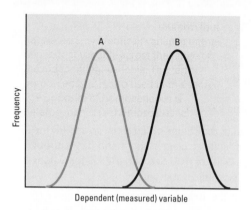

Fig. 75.1 Two distributions with different locations but the same dispersion. The data set labelled B could have been obtained by adding a constant to each datum in the data set labelled A.

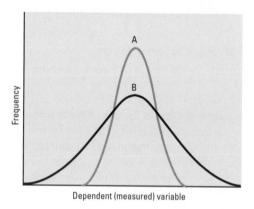

Fig. 75.2 Two distributions with different dispersions but the same location. The data set labelled A covers a relatively narrow range of values of the dependent (measured) variable, while that labelled B covers a wider range.

Example Box 75.1 shows a set of data and the calculated values of the measures of location, dispersion and shape for which methods of calculation are outlined here. Check your understanding by calculating the statistics yourself and confirming that you arrive at the same answers.

Whether obtained from observation or experimentation, most data in biology exhibit variability. This can be displayed as a frequency distribution (e.g. (Fig. 72.7). Descriptive (or summary) statistics quantify aspects of the frequency distribution of a sample. You can use them to:

- condense a large data set for presentation in figures or tables;
- provide estimates of parameters of the frequency distribution of the population being sampled (pp. 516–19).

The three important features of a frequency distribution that can be summarised by descriptive statistics are:

- the sample's location, i.e. its position along a given dimension representing the dependent (measured) variable (Fig. 75.1);
- the dispersion of the data, i.e. how spread out the values are (Fig. 75.2);
- the shape of the distribution, i.e. whether symmetrical, skewed, U-shaped, etc. (Fig. 75.3).

The appropriate statistics to use depend on both the type of data (i.e. whether quantitative, ranked or qualitative, p. 173) and the nature of the underlying frequency distribution. The latter may be assumed from theoretical considerations, but if not, it can be determined graphically from a representative sample (preferably with $n > 100$) or via statistical tests (pp. 511).

The methods used to calculate descriptive statistics also depend on whether the data have been grouped into classes. You should use the original data set if at all possible, because grouping loses information and reduces accuracy. However, large data sets require unwieldy calculations and are best handled using computer programs.

> **KEY POINT** Although their use is commonplace, care is required in the selection and calculation of descriptive statistics otherwise they may convey a misleading representation of your results.

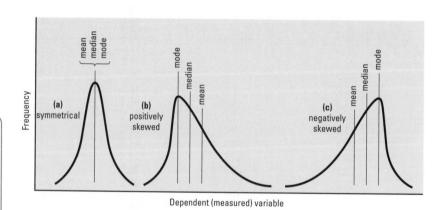

Fig. 75.3 Symmetrical and skewed frequency distributions, showing relative positions of mean, median and mode.

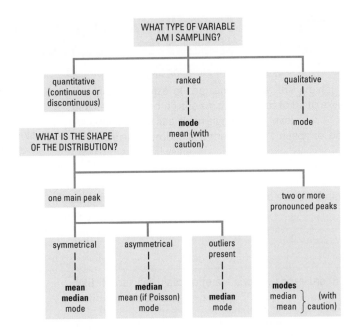

Fig. 75.4 Choosing a statistic for characterising a distribution's location. Statistics written in bold are the preferred option(s).

Measuring location

Here, the objective is to pinpoint the 'centre' of the frequency distribution, i.e. the value about which most of the data are grouped. The chief measures of location are the mean, median and mode. Figure 75.4 shows how to choose among these for a given data set.

Mean

The mean (denoted $\overline{Y}$ and also referred to as the arithmetic mean) is the average value of the data. It is obtained from the sum of all the data values divided by the number of observations (in symbolic terms, $\sum Y/n$). The mean is a good measure of the centre of symmetrical frequency distributions. It uses all of the numerical values of the sample and therefore incorporates all of the information content of the data. However, the value of a mean is greatly affected by the presence of outliers (extreme values). The arithmetic mean is a widely used statistic in biology, but there are situations when you should be careful about using it (see Box 75.2 for examples).

Median

The median is the mid-point of the observations when ranked in increasing order. For odd-sized samples, the median is the middle observation; for even-sized samples it is the mean of the middle pair of observations. Where data are grouped into classes, the median can only be estimated. This is most simply done from a graph of the cumulative frequency distribution, but can also be worked out by assuming the data to be evenly spread within the class. The median may represent the location of the main body of data better than the mean when the distribution is asymmetric or when there are outliers in the sample.

Mode

The mode is the most common value in the sample. The mode is easily found from a tabulated frequency distribution as the most frequent value. If data

Use of symbols – *Y* is used in Chapters 75 and 76 to signify the dependent variable in statistical calculations (following the example of Sokal and Rohlf, 2012, Heath, 1995 and Wardlaw, 2000). Note, however, that some authors use *X* or *x* in analogous formulae and many calculators refer to e.g. $\bar{x}$, $\sum x^2$, etc., for their statistical functions.

Definition

Rank – the position of a data value when all the data are placed in order of ascending magnitude. If ties occur, an average rank of the tied variates is used. Thus, the rank of the datum 6 in the sequence 1, 3, 5, 6, 8, 8, 10 is 4; the rank of each datum with value 8 is 5.5.

Definition

An outlier – any datum that has a value much smaller or bigger than most of the data.

Box 75.1 Descriptive statistics for an illustrative sample of data

Value (Y)	Frequency (f)	Cumulative frequency	fY	fY²
1	0	0	0	0
2	1	1	2	4
3	2	3	6	18
4	3	6	12	48
5	8	14	40	200
6	5	19	30	180
7	2	21	14	98
8	0	21	0	0
Totals	$21 = \sum f(=n)$		$104 = \sum fY$	$548 = \sum fY^2$

In this example, for simplicity and ease of calculation, integer values of Y are used. In many practical exercises, where continuous variables are measured to several significant figures and where the number of data values is small, giving frequencies of 1 for most of the values of Y, it may be simpler to omit the column dealing with frequency and list all the individual values of Y and Y² in the appropriate columns.

To gauge the underlying frequency distribution of such data sets, you would need to group individual data into broader classes (e.g. all values between 1.0 and 1.9, all values between 2.0 and 2.9, etc.) and then draw a histogram (p. 483). Calculation of certain statistics for data sets that have been grouped in this way (e.g. median, quartiles, extremes) can be tricky and a statistical text should be consulted.

Statistic	Value*	How calculated
Mean	4.95	$\sum fY/n$, i.e. 104/21
Median	5	Value of the $(n + 1)/2$ variate, i.e. the value ranked $(21 + 1)/2 = 11$th (obtained from the cumulative frequency column)
Mode	5	The most common value (Y value with highest frequency)
Upper quartile	6	The upper quartile is between the 16th and 17th values, i.e. the value exceeded by 25% of the data values
Lower quartile	4	The lower quartile is between the 5th and 6th values, i.e. the value exceeded by 75% of the data values
Semi-interquartile range	1.0	Half the difference between the upper and lower quartiles, i.e. $(6 - 4)/2$
Upper extreme	7	Highest Y value in data set
Lower extreme	2	Lowest Y value in data set
Range	5	Difference between upper and lower extremes
Variance (s²)	1.65	$s^2 = \dfrac{\sum fY^2 - [(\sum fY)^2/n]}{n - 1}$ $= \dfrac{548 - (104)^2/21}{20}$
Standard deviation (s)	1.28	$\sqrt{s^2}$
Standard error (SE)	0.280	$s/\sqrt{n}$
95% confidence limits	4.36 – 5.54	$\bar{Y} \mp t_{0.05}[20] \times SE$, (where $t_{0.05}[20] = 2.09$, Table 76.2)
Coefficient of variation (CoV)	25.9%	$100s/\bar{Y}$

*Rounded to three significant figures (see p. 497), except when it is an exact number.

Describing the location of qualitative data – the mode is the only statistic that is suitable for this task. For example, 'the modal (most frequent) eye colour was crimson'.

have been grouped into classes then the term modal class is used for the class containing most values. The mode provides a rapidly and easily found estimate of sample location and is unaffected by outliers. However, the mode is affected by chance variation in the shape of a sample's distribution and it may lie distant from the obvious centre of the distribution.

The mean, median and mode have the same units as the variable under discussion. However, whether these statistics of location have the same or similar values for a given frequency distribution depends on the symmetry and shape of the distribution. If it is near-symmetrical with a single peak, all three will be very similar; if it is skewed or has more than one peak, their values will differ to a greater degree (see Fig. 75.3).

Measuring dispersion

Here, the objective is to quantify the spread of the data about the centre of the distribution. Figure 75.5 indicates how to decide which measure of dispersion to use.

Range

The range is the difference between the largest and smallest data values in the sample (the extremes) and has the same units as the measured variable. The range is easy to determine, but is greatly affected by outliers. Its value may also depend on sample size: in general, the larger this is, the greater will be the range. These features make the range a poor measure of dispersion for many practical purposes.

Example In a sample of data with values 3, 7, 15, 8, 5, 10 and 4, the range is 12 (i.e. the difference between the highest value, 15, and the lowest value, 3).

Semi-interquartile range

The semi-interquartile range is an appropriate measure of dispersion when a median is the appropriate statistic to describe location. For this, you need

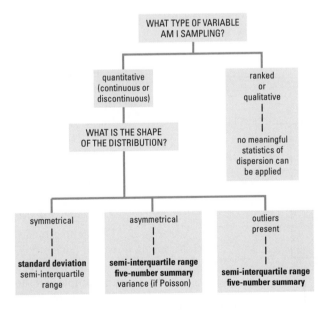

Fig. 75.5 Choosing a statistic for characterising a distribution's dispersion. Statistics written in bold are the preferred option(s). Note that you should match statistics describing dispersion with those you have used to describe location, i.e. standard deviation with mean, semi-interquartile range with median.

Fig. 75.6 Illustration of median, quartiles, range and semi-interquartile range.

to determine the first and third quartiles, i.e. the medians for those data values ranked below and above the median of the whole data set (see (Fig. 75.6). To calculate a semi-interquartile range for a data set:

1. **Rank the observations in ascending order.**
2. **Find the values of the first and third quartiles.**
3. **Subtract the value of the first quartile from the value of the third.**
4. **Halve this number.**

For data grouped in classes, the semi-interquartile range can only be estimated. Another disadvantage is that it takes no account of the shape of the distribution at its edges. This objection can be countered by using the so-called 'five-number summary' of a data set, which consists of the three quartiles and the two extreme values; this can be presented on graphs as a box-and-whisker plot (see (Fig. 75.7) and is particularly useful for summarising skewed frequency distributions. The corresponding 'six-number summary' includes the sample's size.

Variance and standard deviation

For symmetrical frequency distributions, an ideal measure of dispersion would take into account each value's deviation from the mean and provide a measure of the average deviation from the mean. Two such statistics are the sample variance, which is the sum of squares ($\sum (Y - \overline{Y})^2$) divided by $n - 1$ (where n is the sample size), and the sample standard deviation, which is the positive square root of the sample variance.

The variance (s^2) has units that are the square of the original units, while the standard deviation (s, or SD) is expressed in the original units, one reason s is often preferred as a measure of dispersion. Calculating s or s^2

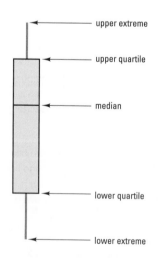

Fig. 75.7 A box-and-whisker plot, showing the 'five-number summary' of a sample as it might be used on a graph.

longhand is a tedious job and is best done with the help of a calculator or computer. If you do not have a calculator that calculates s for you, an alternative formula that simplifies calculations is:

$$s = +\sqrt{\frac{\sum Y^2 - (\sum Y)^2/n}{n-1}} \qquad [75.1]$$

To calculate s using a calculator:

1. **Obtain $\sum Y$, square it, divide by n and store in memory.**
2. **Square Y values, obtain $\sum Y^2$, subtract memory value from this.**
3. **Divide this answer by $n - 1$.**
4. **Take the positive square root of this value.**

Take care to retain significant figures, or errors in the final value of s will result. If continuous data have been grouped into classes, the class mid-values or their squares must be multiplied by the appropriate frequencies before summation (see example in Box 75.1). When data values are large, longhand calculations can be simplified by coding the data, e.g. by subtracting a constant from each datum, and decoding when the simplified calculations are complete (see Sokal and Rohlf, 2012).

Coefficient of variation

The coefficient of variation (CoV) is a dimensionless measure of variability relative to location that expresses the sample standard deviation, usually as a percentage of the sample mean, i.e.

$$\text{CoV} = 100s/\overline{Y}(\%) \qquad [75.2]$$

This statistic is useful when comparing the relative dispersion of data sets with widely differing means or where different units have been used for the same or similar quantities.

A useful application of the CoV is to compare different analytical methods or procedures, so that you can decide which involves the least proportional error – create a standard stock solution, then base your comparison on the results from several subsamples analysed by each method. You may find it useful to use the CoV to compare the precision of your own results with those of a manufacturer, e.g. for an autopipettor (pp. 140–1). The smaller the CoV, the more precise (repeatable) is the apparatus or technique (note: this does not mean that it is necessarily more *accurate*, see p. 175).

Measuring the precision of the sample mean as an estimate of the true value using the standard error

Most practical exercises are based on a limited number of individual data values (a sample) that are used to make inferences about the population from which they were drawn. For example, the haemoglobin content might be measured in blood samples from 100 adult females and used as an estimate of the adult female haemoglobin content, with the sample mean ($\overline{Y}$) and sample standard deviation (s) providing estimates of the true values of the underlying population mean (μ) and the population standard deviation (σ). The reliability of the sample mean as an estimate of the true (population) mean can be assessed by calculating a statistic termed the standard error of the sample mean (often abbreviated to standard error or SE), from:

$$\text{SE} = s/\sqrt{n} \qquad [75.3]$$

Using a calculator for statistics – make sure you understand how to enter individual data values and which keys will give the sample mean (usually shown as $\overline{X}$ or $\bar{x}$) and sample standard deviation (often shown as σ_{n-1}). In general, you should not use the population standard deviation (usually shown as σ_n).

Example Consider two methods of bio-assay for a toxin in fresh water. For a given standard, Method A gives a mean result of = 50 'response units' with $s = 8$, while Method B gives a mean result of = 160 'response units' with $s = 18$. Which bioassay gives the more reproducible results? The answer can be found by calculating the CoV values, which are 16% and 11.25% respectively. Hence, Method B is the more precise, even though the absolute value of s is larger.

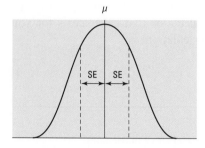

Fig. 75.8 Frequency distribution of sample means around the population mean (μ). Note that SE is equivalent to the standard deviation of the sample means, for sample size = n.

Strictly, the standard error is an estimate of the dispersion of repeated sample means around the true (population) value: if several samples were taken, each with the same number of data values (n), then their means would cluster around the population mean (μ) with a standard deviation equal to SE, as shown in Fig. 75.8. Therefore, the *smaller* the SE, the more reliable the sample mean is likely to be as an estimate of the true value, since the underlying frequency distribution would be more tightly clustered around μ. At a practical level, eqn [75.3] shows that SE is directly affected by the dispersion of individual data values within the sample, as represented by the sample standard deviation (s). Perhaps more importantly, SE is inversely related to the *square root* of the number of data values (n). Therefore, if you wanted to increase the precision of a sample mean by a factor of 2 (i.e. to reduce SE by half), you would have to increase n by a factor of 2^2 (i.e. four-fold).

Summary descriptive statistics for the sample mean are often quoted as $\overline{Y} \pm SE(n)$, with the SE being given to one significant figure more than the mean. For example, summary statistics for the sample mean and standard error for the data shown in Box 75.1 would be quoted as 4.95 ± 0.280 ($n = 21$). You can use such information to carry out a t-test between two sample means (Box 76.1); the SE is also useful because it allows calculation of confidence limits for the sample mean (p. 523).

Describing the 'shape' of frequency distributions

Frequency distributions may differ in the following characteristics:

- **number of peaks;**
- **skewness (asymmetry);**
- **kurtosis (pointedness).**

The shape of a frequency distribution of a small sample is affected by chance variation and may not be a fair reflection of the underlying population frequency distribution: check this by comparing repeated samples from the same population or by increasing the sample size. If the original shape were due to random events, it should not appear consistently in repeated samples and should become less obvious as sample size increases.

Genuinely bimodal or polymodal distributions may result from the combination of two or more unimodal distributions, indicating that more than one underlying population is being sampled (Fig. 75.9). An example of a bimodal distribution is the height of adult humans (females and males combined).

Figure 75.9 Frequency distributions with different numbers of peaks. A unimodal distribution (a) may be symmetrical or asymmetrical. The dotted lines in (b) indicate how a bimodal distribution could arise from a combination of two underlying unimodal distributions. Note here how the term 'bimodal' is applied to any distribution with two major peaks – their frequencies do not have to be exactly the same.

Box 75.2	Three examples where simple arithmetic means are inappropriate

Mean	n
6	4
7	7
8	1

1. **If means of samples are themselves meaned, an error can arise if the samples are of different size.** For example, the arithmetic mean of the means in the table shown left is 7, but this does not take account of the different 'reliabilities' of each mean due to their sample sizes. The correct weighted mean is obtained by multiplying each mean by its sample size (n) (a 'weight') and dividing the sum of these values by the total number of observations, i.e. in the case shown, $(24 + 49 + 8)/12 = 6.75$.

2. **When making a mean of ratios (e.g. percentages) for several groups of different sizes, the ratio for the combined total of all the groups is not the mean of the proportions for the individual groups.** For example, if 20 rats from a batch of 50 are male, this implies 40% are male. If 60 rats from a batch of 120 are male, this implies 50% are male. The mean percentage of males $(50 + 40)/2 = 45\%$ is not the percentage of males in the two groups combined, because there are $20 + 60 = 80$ males in a total of 170 rats $= 47.1$ approx.

pH value	$[H^+]$ (mol L^{-1})
6	1×10^{-6}
7	1×10^{-7}
8	1×10^{-8}
mean	3.7×10^{-7}
$-\log_{10}$ mean	6.43

3. **If the measurement scale is not linear, arithmetic means may give a false value.** For example, if three media had pH values 6, 7 and 8, the appropriate mean pH is not 7 because the pH scale is logarithmic. The definition of pH is $-\log_{10}[H^+]$, where $[H^+]$ is expressed in mol L^{-1} ('molar'); therefore, to obtain the true mean, convert data into $[H^+]$ values (i.e. put them on a linear scale) by calculating $10^{(-\text{pH value})}$ as in the table shown left. Now calculate the mean of these values and convert the answer back into pH units. Thus, the appropriate answer is pH 6.43 rather than 7. Note that a similar procedure is necessary when calculating statistics of dispersion in such cases, so you will find these almost certainly asymmetric about the mean.

Mean values of log-transformed data are often termed geometric means – they are sometimes used in microbiology and in cell culture studies, where log-transformed values for cell density counts are averaged and plotted (p. 291), rather than using the raw data values. The use of geometric means in such circumstances serves to reduce the effects of outliers on the mean.

A distribution is skewed if it is not symmetrical, a symptom being that the mean, median and mode are not equal (Fig. 75.3). Positive skewness is where the longer 'tail' of the distribution occurs for higher values of the measured variable; negative skewness where the longer tail occurs for lower values. Some biological examples of characteristics distributed in a skewed fashion are volumes of plant protoplasts, insulin levels in human plasma and bacterial colony counts.

Kurtosis is the name given to the 'pointedness' of a frequency distribution. A platykurtic frequency distribution is one with a flattened peak, while a leptokurtic frequency distribution is one with a pointed peak (Fig. 75.10). While descriptive terms can be used, based on visual observation of the shape and direction of skew, the degree of skewness and kurtosis can be quantified and statistical tests exist to test the 'significance' of observed values (see Sokal and Rohlf, 2012), but the calculations required are complex and best done with the aid of a computer.

Fig. 75.10 Examples of the two types of kurtosis.

Using computers to calculate descriptive statistics

There are many specialist statistical packages (e.g. *SPSS*) that can be used to simplify the process of calculation of statistics. Note that correct interpretation of the output requires an understanding of the terminology used and the underlying process of calculation, and this may best be obtained by working through one or more examples by hand before using these tools. Spreadsheets offer increasingly sophisticated statistical analysis functions, some examples of which are provided in Box 75.3.

Box 75.3 How to use a spreadsheet to calculate descriptive statistics

Method 1: Using spreadsheet functions to generate the required statistics

Suppose you had obtained the following set of data, stored within an array (block of columns and rows) of cells (A2:L6) within a spreadsheet:

	A	B	C	D	E	F	G	H	I	J	K	L
1	**My data set**											
2	4	4	3	3	5	4	3	7	7	3	5	3
3	6	2	9	7	3	4	5	6	6	9	4	8
4	5	3	2	5	4	5	7	2	8	3	6	3
5	11	3	5	2	4	3	7	8	4	4	4	3
6	3	6	8	5	6	4	3	4	3	6	10	5

Taking Microsoft *Excel* as an example, the following functions could be used to extract information from the data set. Note that the precise syntax required may depend on the spreadsheet version used.

Descriptive statistic	Example of use of function[a, b]	Result for the above data set
Sample size n	=COUNT((A2:L6)	60
Mean	=AVERAGE(A2:L6)[c]	4.9
Median	=MEDIAN(A2:L6)	4.0
Mode	=MODE(A2:L6)	3
Upper quartile	=QUARTILE(A2:L6,3)[d]	6.0
Lower quartile	=QUARTILE(A2:L6,1)	3.0
Semi-interquartile range	=QUARTILE(A2:L6,3)-QUARTILE(A2:L6,1)	3.0
Upper extreme	=QUARTILE(A2:L6,4) *or* =MAX(A2:L6)	11
Lower extreme	=QUARTILE(A2:L6,0) *or* =MIN(A2:L6)	2
Range	=MAX(A2:L6)-MIN(A2:L6)[e]	9.0
Variance	=VAR(A2:L6)	4.464
Standard deviation	=STDEV(A2:L6)	2.113
Standard error	=STDEV(A2:L6)/(SQRT(COUNT(A2:L6)))[f]	0.273
Coefficient of variation	=100*STDEV(A2:L6)/AVERAGE(A2:L6)	43.12%

Notes:

[a]Typically, for *Excel*, in an appropriate cell, you would, for example, select *Formulas > More Functions > Statistical > COUNT*, then select the input range and press return.

[b]Other descriptive statistics can be calculated – these mirror those shown in Box 75.1, but for this specific data set.

[c]There is no 'MEAN' function in Microsoft *Excel*.

[d]The first argument within the brackets relates to the array of data, the second relates to the quartile required (consult the *Help on this function* feature for further information).

[e]There is no direct 'RANGE' function in Microsoft *Excel*.

[f]There is no direct 'STANDARD ERROR' function in Microsoft *Excel*. The SQRT function returns a square root and the COUNT function determines the number of filled data cells in the array.

(continued)

Box 75.3 (continued)

Method 2: Using the *Tools* > *Data Analysis* option in *Excel*

This can automatically generate a table of descriptive statistics for the data array selected, although the data must be presented as a single row or column. This option might need to be installed for your network or personal computer before it is available to you (see Box 72.2, p. 482 for instructions). Having entered or rearranged your data into a row or column, the steps involved are as follows:

1. Select *Data* > *Data Analysis*.

2. From the *Data Analysis* box, select *Descriptive Statistics*.

3. Input your data location into the *Input Range* (left-click and hold down to highlight the column of data).

4. From the menu options, select *Summary Statistics* and *Confidence Level for Mean: 95%*.

5. When you click *OK* you should get a new worksheet, with descriptive statistics and confidence limits shown. Alternatively, at step 3, you can select an area of your current worksheet as a data output range (select an area away from any existing content as these cells would otherwise be overwritten by the descriptive statistics output table).

6. Change the format of the cells to show each number to an appropriate number of decimal places. You may also wish to make the columns wider so you can read their content.

7. For the data set shown above, the final output table should look as shown in Table 75.1.

Table 75.1 Descriptive statistics for a data set

Column1[a, b]	
Mean	4.9
Standard error	0.27
Median	4.0
Mode	3
Standard deviation	2.113
Sample variance	4.464
Kurtosis	0.22
Skewness	0.86
Range	9.00
Minimum	2.0
Maximum	11.0
Sum	294
Count	60
Confidence level (95.0%)	0.55

Notes:
[a]These descriptive statistics are specified (and are automatically presented in this order) – any others required can be generated using Method 1.
[b]A more descriptive heading can be added if desired - this is the default.

Notes: Spreadsheet commands illustrated here may vary among the various versions of Microsoft *Office* programmes and for different spreadsheet products. Check the precise functions and syntax using the insert function menu option (f_x button in some versions of *Excel*).

Text references

Heath, D. (1995) *An Introduction to Experimental Design and Statistics for Biology.* UCL Press, London.

Sokal, R.R. and Rohlf, F.J. (2012) *Biometry,* 4th edn. W.H. Freeman, San Francisco.

Wardlaw, A.C. (2000) *Practical Statistics for Experimental Biologists,* 2nd edn. Wiley, New York.

Source for further study

Schmuller, J. (2013) *Statistical Analysis with Excel for Dummies,* 3rd edn. Wiley, Hoboken, NJ.

STUDY EXERCISES

75.1 Choose appropriate descriptive statistics. For the following data and distribution types, state what statistics of location and dispersion (Figs 75.4 and 75.5) you should choose.

a) Colour of each egg in nests as judged against a colour chart.

b) Clutch size (number of eggs) in bird nests visited.

c) Beak lengths for chicks hatched in nests, where there is a pronounced difference in beak length between the sexes.

75.2 Practise calculating descriptive statistics. Using the data set given below, calculate the following statistics:

a) range

b) variance

c) standard deviation

d) coefficient of variation

e) standard error.

Answers (b) to (e) should be given to four significant figures.

Set of data

9	6	7	5	7	7	8	6	5	5	7	8
5	8	7	7	6	7	8	6	5	7	7	6
3	6	8	9	9	6	7	8	5	6	5	5
8	8	7	5	6	5	8	6	7	5	7	6
5	6	7	8	7	6	7	7	8	8	9	4

75.3 Calculate and interpret standard errors. Two samples, A and B, gave the following descriptive statistics (measured in the same units): Sample A, mean = 16.2, standard deviation = 12.7, number of data values = 12; Sample B, mean = 13.2, standard deviation 14.4, number of data values = 20. Which has the lower standard error in absolute terms and in proportion to the sample mean? (Express answers to three significant figures.)

75.4 Compute a mean value correctly. A researcher finds that the mean diameter of limpets on three seashore stones designated A, B and C is 3.0, 2.5 and 2.0 mm respectively. He computes the mean limpet diameter as 2.5 mm, but forgets that the sample sizes were 24, 37 and 6 respectively. What is the true mean diameter of the limpets in this sample? (Answer to three significant figures.)

Answers to these study exercises are available at www.pearsoned.co.uk/practicalskills

This chapter outlines the philosophy of hypothesis-testing statistics, indicates the steps to be taken when choosing a test, and discusses features and assumptions of some important tests. For details of the mechanics of tests, appropriate texts should be consulted (e.g. Heath, 1995; Wardlaw, 2000; Sokal and Rohlf, 2012). Most tests are now available in statistical packages for computers (see p. 82) and many in spreadsheets (pp. 87–8).

To carry out a statistical test:

1. **Decide what it is you wish to test** (create a null hypothesis and its alternative).

2. **Determine whether your data fit a standard distribution pattern.**

3. **Select a test and apply it to your data.**

Setting up a null hypothesis

Hypothesis-testing statistics are used to compare the properties of samples either with other samples or with some theory about them. For instance, you may be interested in whether two samples can be regarded as having different means, whether the counts of an organism in different quadrats can be regarded as randomly distributed, or whether property A of an organism is linearly related to property B.

> **KEY POINT** You cannot use statistics to *prove* any hypothesis, but they can be used to assess *how likely* it is to be wrong.

Statistical testing operates in what at first seems a rather perverse manner. Suppose you think a treatment has an effect. The theory you actually test is that it has no effect; the test tells you how improbable your data would be if this theory were true. This 'no effect' theory is the null hypothesis (NH). If your data are very improbable under the NH, then you may suppose it to be wrong, and this would support your original idea (the 'alternative hypothesis'). The concept can be illustrated by an example. Suppose two groups of subjects were treated in different ways, and you observed a difference in the mean value of the measured variable for the two groups. Can this be regarded as a 'true' difference? As Fig. 76.1 shows, it could have arisen in two ways:

1. **Because of the way the subjects were allocated to treatments,** i.e. all the subjects liable to have high values might, by chance, have been assigned to one group and those with low values to the other (Fig. 76.1(a)).

2. **Because of a genuine effect of the treatments,** i.e. each group came from a distinct frequency distribution (Fig. 76.1(b)).

A statistical test will indicate the probabilities of these options. The NH states that the two groups come from the same population (i.e. the treatment effects are negligible in the context of random variation). To test this, you calculate a test statistic from the data, and compare it with tabulated critical values giving the probability of obtaining the observed or a more extreme result by chance (see Boxes 76.1 and 76.2). This probability is sometimes called the significance of the test.

Note that you must take into account the degrees of freedom (d.f.) when looking up critical values of most test statistics. The d.f. is related to the

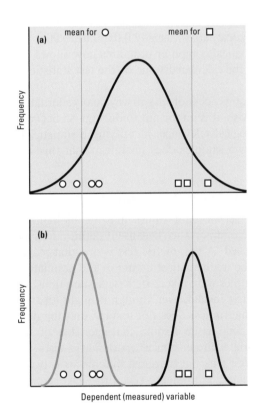

Fig. 76.1 Two explanations for the difference between two means. In case (a) the two samples happen by chance to have come from opposite ends of the same frequency distribution, i.e. there is no true difference between the samples. In case (b) the two samples come from different frequency distributions, i.e. there is a true difference between the samples. In both cases, the means of the two samples are the same.

Quoting significance – the convention for quoting significance levels in text, tables and figures is as follows:

$P > 0.05$ = 'not significant' (or NS)

$P \leq 0.05$ = 'significant' (or *)

$P \leq 0.01$ = 'highly significant' (or **)

$P \leq 0.001$ = 'very highly significant' (or ***)

Thus, you might refer to a difference in means as being 'highly significant $(P \leq 0.01)$'. For this reason, the word 'significant' in its everyday meaning of 'important' or 'notable' should be used with care in scientific writing. The asterisk coding is most often used as a short form in tables and should be explained in the table legend.

Choosing between parametric and non-parametric tests – always plot your data graphically when determining whether they are suitable for parametric tests as this may save a lot of unnecessary effort later.

size(s) of the samples studied; formulae for calculating it depend on the test being used. Biologists normally use two-tailed tests, i.e. we have no expectation beforehand that the treatment will have a positive or negative effect compared with the control (in a one-tailed test we expect one particular treatment to be bigger than the other). Be sure to use critical values for the correct type of test.

By convention, the critical probability for rejecting the NH is 5% (i.e. $P = 0.05$). This means we reject the NH if the observed result would have come up by chance a maximum of one time in twenty. If the modulus of the test statistic is less than or equal to the tabulated critical value for $P = 0.05$, then we accept the NH and the result is said to be 'not significant' (NS for short). If the modulus of the test statistic is greater than the tabulated value for $P = 0.05$, then we reject the NH in favour of the alternative hypothesis that the treatments had different effects and the result is 'statistically significant'.

Two types of error are possible when making a conclusion on the basis of a statistical test. The first occurs if you reject the NH when it is true and the second if you accept the NH when it is false. To limit the chance of the first type of error, choose a lower probability, e.g. $P = 0.01$, but note that the critical value of the test statistic increases when you do this and results in the probability of the second type of error increasing. The conventional significance levels given in statistical tables (usually 0.05, 0.01, 0.001) are arbitrary. Increasing use of statistical computer programs now allows the actual probability of obtaining the calculated value of the test statistic to be quoted (e.g. $P = 0.037$).

Note that if the NH is rejected, this does not tell you why: many alternative explanations may be possible. Also, it is important to distinguish between statistical significance and biological relevance: identifying a statistically significant difference between two samples does not mean that this will carry any biological importance.

Comparing data with parametric distributions

The distribution pattern of a set of data values may be biologically relevant, but it is also of practical importance because it defines the type of statistical tests that can be used. A parametric test is one that makes particular assumptions about the mathematical nature of the population distribution from which the samples were taken. If these assumptions are not true, then the test is obviously invalid, even though it might give the answer we expect. A non-parametric test does not assume that the data fit a particular pattern, but it may assume some characteristics of the distributions. Used in appropriate circumstances, parametric tests are better able to distinguish between true but marginal differences between samples than their non-parametric equivalents (i.e. they have greater 'power').

The properties of the main distribution types found in biology are given below with both rules of thumb and more rigorous tests for deciding whether data fit these distributions.

Binomial distributions

These apply to samples of any size from populations when data values occur independently in only two mutually exclusive classes (e.g. type A or type B). They describe the probability of finding the different possible combinations of the attribute for a specified sample size k

Fig. 76.2 Examples of binomial frequency distributions with different probabilities. The distributions show the expected frequency of obtaining *n* individuals of type A in a sample of 5. Here *P* is the probability of an individual being type A rather than type B.

(e.g. out of 10 specimens, what is the chance of 8 being type A?). If *p* is the probability of the attribute being of type A and *q* the probability of it being type B, then the expected mean sample number of type A is *kp* and the standard deviation is √*kpq*. Expected frequencies can be calculated using mathematical expressions (see Sokal and Rohlf, 2012). Examples of the shapes of some binomial distributions are shown in Fig. 76.2. Note that they are symmetrical in shape for the special case *p* = *q* = 0.5 and the greater the disparity between *p* and *q*, the more skewed the distribution.

Some biological examples of data likely to be distributed in binomial fashion are: possession of two alleles for seed-coat morphology (e.g. smooth and wrinkly); whether an organism is infected with a microbe or not; whether an animal is male or female. Binomial distributions are particularly useful for predicting gene segregation in Mendelian genetics and can be used for testing whether combinations of events have occurred more frequently than predicted (e.g. more siblings being of the same sex than expected). To establish whether a set of data is distributed in binomial fashion: calculate expected frequencies from probability values obtained from theory or observation, then test against observed frequencies using a χ^2 test (p. 330) or a *G* test (see Wardlaw, 2000).

Poisson distributions

These apply to discrete characteristics that can assume low whole number values, such as counts of events occurring in area, volume or time. The events should be 'rare' in that the mean number observed should be a small proportion of the total that could possibly be found. Also, finding one count should not influence the probability of finding another. The shape of Poisson distributions is described by only one parameter, the mean number of events observed, and has the special characteristic that the variance is equal to the mean. The shape has a pronounced positive skewness at low mean counts, but becomes more and more symmetrical as the mean number of counts increases (Fig. 76.3).

Some examples of characteristics distributed in a Poisson fashion are: number of plants in a quadrat; number of microbes per unit volume of medium; number of animals parasitised per unit time; number of radioactive disintegrations per unit time. One of the main uses for the

Tendency towards the normal distribution – under certain conditions, binomial and Poisson distributions can be treated as normally distributed:

- where samples from a binomial distribution are large (i.e. > 15) and *p* and *q* are close to 0.5;
- for Poisson distributions, if the number of counts recorded in each outcome is greater than about 15.

Fig. 76.3 Examples of Poisson frequency distributions differing in mean. The distributions are shown as line charts because the independent variable (events per sample) is discrete.

Definition

Coefficient of dispersion = $s^2/\overline{Y}$. This is an alternative measure of dispersion to the coefficient of variation (p. 509).

Poisson distribution is to quantify errors in count data such as estimates of cell densities in dilute suspensions (see p. 294). To decide whether data are Poisson distributed:

- **Use the rule of thumb that if the coefficient of dispersion ≈1, the distribution is likely to be Poisson.**
- **Calculate 'expected' frequencies from the equation for the Poisson distribution and compare with actual values using a χ^2 test or a *G*-test.**

It is sometimes of interest to show that data are *not* distributed in a Poisson fashion, e.g. the distribution of parasite larvae in hosts. If $s^2/\overline{Y} > 1$, the data are 'clumped' and occur together more than would be expected by chance; if $s^2/\overline{Y} < 1$, the data are 'repulsed' and occur together less frequently than would be expected by chance.

Normal distributions (Gaussian distributions)

These occur when random events act to produce variability in a continuous characteristic (quantitative variable). This situation occurs frequently in biology, so normal distributions are very useful and much used. The bell-like shape of normal distributions is specified by the population mean and standard deviation (Fig. 76.4): it is symmetrical and configured such that 68.27% of the data will lie within ±1 standard deviation of the mean, 95.45% within ±2 standard deviations of the mean, and 99.73% within ±3 standard deviations of the mean.

Some biological examples of data likely to be distributed in a normal fashion are: fresh weight of plants of the same age; linear dimensions of bacterial cells; height of either adult female or male humans. To check whether data come from a normal distribution, you can:

- **Use the rule of thumb that the distribution should be symmetrical** and that nearly all the data should fall within ±3*s* of the mean and about two-thirds within ±1*s* of the mean.
- **Plot the distribution on normal probability graph paper.** If the distribution is normal, the data will tend to follow a straight line (see Fig. 76.5). Deviations from linearity reveal skewness and/or kurtosis

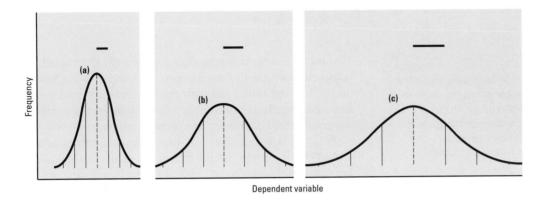

Fig. 76.4 Examples of normal frequency distributions differing in mean and standard deviation. The horizontal bars represent population standard deviations for the curves, increasing from (a) to (c). Vertical dashed lines are population means, while vertical solid lines show positions of values ± 1, 2 and 3 standard deviations from the means.

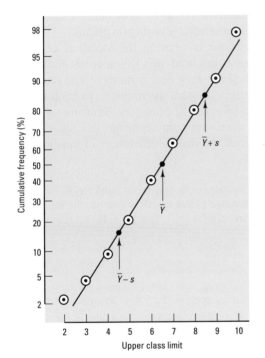

Fig. 76.5 Example of a normal probability plot. The plotted points are from a small data set where the mean $\overline{Y} = 6.93$ and the standard deviation $s = 1.895$. Note that values corresponding to 0% and 100% cumulative frequency cannot be used. The straight line is that predicted for a normal distribution with $\overline{Y} = 6.93$ and $s = 1.895$. This is plotted by calculating the expected positions of points for $\overline{Y} \pm s$. Since 68.3% of the distribution falls within these bounds, the relevant points on the cumulative frequency scale are $50 \pm 34.15\%$; thus this line was drawn using the points (4.495, 15.85) and (8.285, 84.15) as indicated on the plot.

(see p. 510), the significance of which can be tested statistically (see Sokal and Rohlf, 2012).

- **Use a suitable statistical computer program to generate predicted normal curves** from the $\overline{Y}$ and s values of your sample(s). These can be compared visually with the actual distribution of data and can be used to give 'expected' values for a χ^2 test or a G-test.

The wide availability of tests based on the normal distribution and their relative simplicity means you may wish to transform your data to make them more like a normal distribution. Table 76.1 provides transformations that might be applied (see also Fig. 71.3). The transformed data should be tested for normality as described above before proceeding – do not forget that you may need to check that transformed variances are homogeneous for certain tests (see below).

A very important theorem in statistics, the Central Limit Theorem, states that as sample size increases, the distribution of a series of means from any frequency distribution will become normally distributed. This fact can be used to devise an experimental or sampling strategy that ensures that data are normally distributed, i.e. using means of samples as if they were primary data.

Table 76.1 Suggested transformations altering different types of frequency distributions to the normal type. To use, modify data by the formula shown; then examine effects with the tests described on pp. 518–19

Type of data; distribution suspected	Suggested transformation(s)
Proportions (including percentages); binomial	arcsine $\sqrt{x}$ (also called the angular transformation)
Scores; Poisson	$\sqrt{x}$ or $\sqrt{(x + 1/2)}$ if zero values present
Measurements; negatively skewed	x^2, x^3, x^4, etc. (in order of increasing strength)
Measurements; positively skewed	$1/\sqrt{x}$, $\sqrt{x}$, ln X, $1/x$ (in order of increasing strength)

Understanding 'degrees of freedom' – this depends on the number of values in the data set analysed, and the method of calculation depends on the statistical test being used. It relates to the number of observations that are free to vary before the remaining quantities for a data set can be determined.

Checking the assumptions of a test – always acquaint yourself with the assumptions of a test. If necessary, test them before using the test.

Choosing a suitable statistical test

Comparing location (e.g. means)

If you can assume that your data are normally distributed, the main test for comparing two means from independent samples is Student's t-test (see Boxes 76.1 – 76.3, and Table 76.2). This assumes that the variances of the data sets are homogeneous. Tests based on the t-distribution are also available for comparing means of paired data or for comparing a sample mean with a chosen value.

When comparing means of two or more samples, analysis of variance (ANOVA) is a very useful technique. This method also assumes data are normally distributed and that the variances of the samples are homogeneous. The samples must also be independent (e.g. not subsamples). The test statistic calculated is denoted F and it has two different degrees of freedom related to the number of means tested and the pooled number of replicates per mean (Box 76.1). The nested types of ANOVA are useful for letting you know the relative importance of different sources of variability in your data. Two-way and multi-way ANOVAs are useful for studying interactions between treatments.

For data satisfying the ANOVA requirements, the least significant difference (LSD) is useful for making planned comparisons among several means (see Sokal and Rohlf, 2012). Any two means that differ by more than the LSD will be significantly different. The LSD is useful for showing on graphs.

The chief non-parametric tests for comparing the locations of two samples are the Mann–Whitney U-test and the Kolmogorov–Smirnov test. The former assumes that the frequency distributions of the samples are similar, whereas the latter makes no such assumption. In both cases the sample's size must be ≥ 4 and for the Kolmogorov–Smirnov test the samples must have equal sizes. In the Kolmogorov–Smirnov test, significant differences found with the test could be due to differences in location or shape of the distribution, or both.

Table 76.2 Critical values of Student's t statistic (for two-tailed tests). Reject the null hypothesis at probability P if your calculated t value equals or exceeds the value shown for the appropriate degree of freedom $= (n_1 - 1) + (n_2 - 1)$

Degrees of freedom	Critical values for $P = 0.05$	Critical values for $P = 0.01$	Critical values for $P = 0.001$
1	12.71	63.66	636.62
2	4.30	9.92	31.60
3	3.18	5.84	12.94
4	2.78	4.60	8.61
5	2.57	4.03	6.86
6	2.45	3.71	5.96
7	2.36	3.50	5.40
8	2.31	3.36	5.04
9	2.26	3.25	4.78
10	2.23	3.17	4.59
12	2.18	3.06	4.32
14	2.14	2.98	4.14
16	2.12	2.92	4.02
20	2.09	2.85	3.85
25	2.06	2.79	3.72
30	2.04	2.75	3.65
40	2.02	2.70	3.55
60	2.00	2.66	3.46
120	1.98	2.62	3.37
∞	1.96	2.58	3.29

Box 76.1 How to carry out a *t*-test

The *t*-test was devised by a statistician who used the pen name 'Student', so you may see it referred to as Student's *t*-test. It is used when you wish to decide whether two samples come from the same population or from different ones (Fig. 76.1). The samples might have been obtained by selective observation (Chapter 29) or by applying two different treatments to an originally homogeneous population (Chapter 33).

The null hypothesis (NH) is that the two groups can be represented as samples from the same overlying population (Fig. 76.1(a)). If, as a result of the test, you accept this hypothesis, you can say that there is no significant difference between the group means.

The alternative hypothesis is that the two groups come from different populations (Fig. 76.1(b)). By rejecting the NH as a result of the test, you can accept the alternative hypothesis and say that there is a significant difference between the sample means, or, if an experiment were carried out, that the two treatments affected the samples differently.

How can you decide between these two hypotheses? On the basis of certain assumptions (see below), and some relatively simple calculations, you can work out the probability that the samples came from the same population. If this probability is very low, then you can reasonably reject the NH in favour of the alternative hypothesis, and if it is high, you will accept the NH.

To find out the probability that the observed difference between sample means arose by chance, you must first calculate a '*t* value' for the two samples in question. Some computer programs (e.g. *Minitab*) provide this probability as part of the output, otherwise you can look up statistical tables (e.g. Table 76.2). These tables show 'critical values' – the borders between probability levels. If your value of *t* equals or exceeds the critical value for probability *P*, you can reject the NH at this probability ('level of significance').

Note that:

- for a given difference in the means of the two samples, the value of *t* will get larger the smaller the scatter within each data set; and

- for a given scatter of the data, the value of *t* will get larger the greater the difference between the means.

So, at what probability should you reject the NH? Normally, the threshold is arbitrarily set at 5% – you quite often see descriptions like 'the sample means were significantly different ($P < 0.05$)'. At this 'significance level' there is still up to a 5% chance of the *t* value arising by chance, so about one in twenty times, on average, the conclusion will be wrong. If *P* turns out to be lower, then this kind of error is much less likely.

Tabulated probability levels are generally given for 5, 1 and 0.1% significance levels (see Table 76.2). Note that this table is designed for 'two-tailed' tests, i.e. where the treatment or sampling strategy could have resulted in either an increase or a decrease in the measured values. These are the most likely situations you will deal with in biology.

Examine Table 76.2 and note the following:

- The larger the size of the samples (i.e. the greater the 'degrees of freedom'), the smaller *t* needs to be to exceed the critical value at a given significance level.

- The lower the probability, the greater *t* needs to be to exceed the critical value.

The mechanics of the test

A calculator that can work out means and standard deviations is helpful.

1. Work out the sample means $\bar{Y}_1$ and $\bar{Y}_2$ and calculate the difference between them.

2. Work out the sample standard deviations s_1 and s_2. (NB if your calculator offers a choice, chose the '$n - 1$' option for calculating s – see pp. 508–9).

3. Work out the sample standard errors $SE_1 = s_1/\sqrt{n_1}$ and $SE_2 = s_2/\sqrt{n_2}$; now square each, add the squares together, then take the positive square root of this (n_1 and n_2 are the respective sample sizes, which may, or may not, be equal).

4. Calculate *t* from the formula:

$$t = \frac{\bar{Y}_1 - \bar{Y}_2}{\sqrt{[(SE_1)^2] + [(SE_2)^2]}} \qquad [76.1]$$

The value of *t* can be negative or positive, depending on the values of the means; this does not matter and you should compare the modulus (absolute value) of *t* with the values in tables.

5. Work out the degrees of freedom $= (n_1 - 1) + (n_2 - 1)$.

6. Compare the *t* value with the appropriate critical value (see e.g. Table 76.2) and decide on the significance of your findings (see pp. 515–16).

Boxes 76.2 and 76.3 provide worked examples – use these to check that you understand the above procedures.

Assumptions that must be met before using the test

The most important assumptions are:

- The two samples are independent and randomly drawn (or if not, drawn in a way that does not create bias). The test assumes that the samples are quite large.

- The underlying distribution of each sample is normal. This can be tested with a special statistical test, but a rule of thumb is that a frequency distribution of the data should be (a) symmetrical about the mean and (b) nearly all of the data should be within 3 standard deviations of the mean and about two-thirds within 1 standard deviation of the mean (see p. 518).

- The two samples should have uniform variances. This again can be tested (by an *F*-test), but may be obvious from inspection of the two standard deviations.

Box 76.2 **Worked example of a *t*-test**

Suppose the following data were obtained in an experiment (the units are not relevant):

Control: 6.6, 5.5, 6.8, 5.8, 6.1, 5.9
Treatment: 6.3, 7.2, 6.5, 7.1, 7.5, 7.3

Using the steps outlined in Box 76.1, the following values are obtained (denoting control with subscript 1, treatment with subscript 2):

1. $\bar{Y}_1 = 6.1167$; $\bar{Y}_2 = 6.9833$; difference between means $= \bar{Y}_1 - \bar{Y}_2 = -0.8666$

2. $s_1 = 0.49565$; $s_2 = 0.47504$

3. $SE_1 = 0.49565/2.44949 = 0.202348$
   $SE_2 = 0.47504/2.44949 = 0.193934$

4. $t = \dfrac{-08666}{\sqrt{(0.202348^2 + 0.193934^2)}} = \dfrac{-0.8666}{0.280277} = -3.09$

5. d.f. $= (5 + 5) = 10$

6. Looking at Table 76.2, we see that the modulus of this *t* value exceeds the tabulated value for $P = 0.05$ at 10 degrees of freedom ($= 2.23$). We therefore reject the NH, and conclude that the means are different at the 5% level of significance. If the modulus of *t* had been ≤ 2.23, we would have accepted the NH. If the modulus of *t* had been > 3.17, we could have concluded that the means are different at the 1% level of significance.

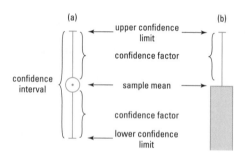

Fig. 76.6 Graphical representation of confidence limits as 'error bars' for (a) a sample mean in a plotted curve, where both upper and lower limits are shown; and (b) a sample mean in a histogram, where, by convention, only the upper value is shown. For data that are assumed to be symmetrically distributed, such representations are often used in preference to the 'box and whisker' plot shown in Fig. 75.7. Note that SE is an alternative way of representing sample imprecision/error (e.g. Fig. 72.1).

Confidence limits for statistics other than the mean – consult an advanced statistical text (e.g. Sokal and Rohlf, 2012) if you wish to indicate the reliability of estimates of e.g. population variances.

Suitable non-parametric comparisons of location for paired data (sample size ≥ 6) include Wilcoxon's signed rank test, which is used for quantitative data and assumes that the distributions have similar shape. Dixon and Mood's sign test can be used for paired data scores where one variable is recorded as 'greater than' or 'better than' the other.

Non-parametric comparisons of location for three or more samples include the Kruskal–Wallis *H*–test. Here, the number of samples is without limit and they can be unequal in size, but again the underlying distributions are assumed to be similar. The Friedman *S*-test operates with a maximum of five samples and data must conform to a randomised block design. The underlying distributions of the samples are assumed to be similar.

Comparing dispersions (e.g. variances)

If you wish to compare the variances of two sets of data that are normally distributed, use the *F*-test. For comparing more than two samples, it may be sufficient to use the F_{max}-test, on the highest and lowest variances. The Scheffé–Box (log-ANOVA) test is recommended for testing the significance of differences between several variances. Non-parametric tests exist but are not widely available: you may need to transform the data and use a test based on the normal distribution.

Determining whether frequency observations fit theoretical expectation

The χ^2 test (Box 61.2) is useful for tests of 'goodness of fit', e.g. comparing expected and observed progeny frequencies in genetical experiments or comparing observed frequency distributions with some theoretical function. One limitation is that simple formulae for calculating χ^2 assume that no expected number is less than 5. The *G*-test ($2I$-test) is used in similar circumstances.

Comparing proportion data

When comparing proportions between two small groups (e.g. whether 3/10 is significantly different from 5/10), you can use probability tables such as those of Finney *et al.* (1963) or calculate probabilities from formulae; however, this can be tedious for large sample sizes. Certain proportions can be transformed so that their distribution becomes normal.

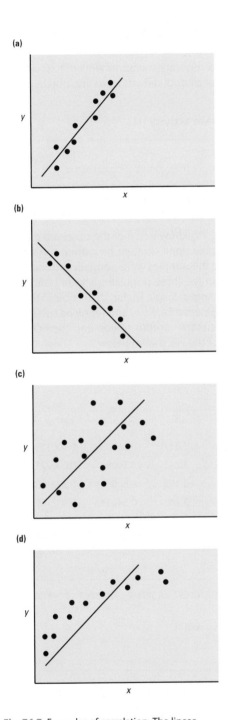

Fig. 76.7 Examples of correlation. The linear regression line is shown. In (a) and (b), the correlation between x and y is good: for (a) there is a positive correlation and the correlation coefficient, r, would be close to 1; for (b) there is a negative correlation and the correlation coefficient would be close to -1. In (c) there is a weak positive correlation and r would be close to 0. In (d) the correlation coefficient may be quite large, but the choice of linear regression is clearly inappropriate.

Placing confidence limits on an estimate of a population parameter

On many occasions, a sample statistic is used to provide an estimate of a population parameter, and it is often useful to indicate the reliability of such an estimate. This can be done by putting confidence limits on the sample statistic, i.e. by specifying an interval around the statistic within which you are confident that the true value (the population parameter) is likely to fall, at a specified level of probability. The most common application is to place confidence limits on the mean of a sample taken from a population of normally distributed data values. In practice, you determine a confidence factor for a particular level of probability that is added to and subtracted from the sample mean ($\overline{Y}$) to give the upper confidence limit and lower confidence limit respectively. These are calculated as:

$$\overline{Y} + (t_{P[n-1]} \times SE) \text{ for the upper limit and} \qquad [76.2]$$
$$\overline{Y} - (t_{P[n-1]} \times SE) \text{ for the lower limit}$$

where $t_{P[n-1]}$ is the tabulated critical value of Student's t statistic for a two-tailed test with $n - 1$ degrees of freedom at a specified probability level (P) and SE is the standard error of the sample mean (pp. 509–10). The 95% confidence limits (i.e. $P = 0.05$) tell you that, on average, 95 times out of 100 the interval between the upper and lower limits will contain the true (population) value. Confidence limits are often shown as 'error bars' for individual sample means plotted in graphical form. Figure 76.6 illustrates how this is applied to plotted curves and histograms (note that this can be carried out for data series within a Microsoft *Excel* graph (chart) using the *Layout > Error Bars* commands).

Correlation and regression

These methods are used when testing the relationship between data values for two variables. Correlation is used to measure the extent to which changes in the two sets of data values occur together in a linear manner. If one variable can be assumed to be dependent upon the other (i.e. a change in X causes a particular change in Y), then regression techniques can be used to provide a mathematical description of the underlying relationship between the variables, e.g. to find a line of best fit for a data series. If there is no a priori reason to assume dependency, then correlation methods alone are appropriate.

A correlation coefficient measures the strength of the linear relationship between two variables, but does not describe the relationship. The coefficient is expressed as a number between -1 and $+1$: a positive coefficient indicates a direct relationship, where the two variables change in the same direction, while a negative coefficient indicates an inverse relationship, where one variable decreases as the other increases (Fig. 76.7). The nearer the coefficient is to -1 or $+1$, the stronger the linear relationship between the variables, i.e. the less 'scatter' there would be about a straight line of best fit (note that this does *not* necessarily imply that one variable is dependent upon the other). A coefficient of 0 implies that the two variables show no linear association and therefore the closer the correlation coefficient is to zero, the weaker the linear relationship. The importance of graphing data is shown by the case illustrated in Fig. 76.7(d).

Pearson's product moment correlation coefficient (r) is the most commonly used statistic for testing correlations. The test is valid only if

Box 76.3 Using a spreadsheet to calculate hypothesis-testing statistics

Presented below are three examples of the use of Microsoft *Excel* to investigate hypotheses about specific data sets. In each case, there is a brief description of the problem; a table showing the data analysed; an outline of the commands (for Microsoft *Excel*) used to carry out the analysis and an annotated table of results from the spreadsheet.

Example 1: A *t*-test

As part of a project, a student applied a chemical treatment to a series of flasks containing fungal cultures with nutrient solution. An otherwise similar set of control flasks received no chemical treatment. After three weeks' growth, she measured the wet mass of the filtered cultures:

Wet mass of samples (g)

Replicate	1	2	3	4	5	6	7	8	Mean	Variance
Treated with ZH52	2.342	2.256	2.521	2.523	2.943	2.481	2.601	2.449	2.515	0.042
Control	2.658	2.791	2.731	2.402	3.041	2.668	2.823	2.509	2.703	0.038

The student proposed the null hypothesis that there was no difference between the two means and tested this using a *t*-test, as she had evidence from other studies that the fungal masses of replicate flasks were normally distributed. She also established, by calculation, that the assumption that the populations had homogeneous variances was likely to be valid. She selected the option *t-Test: Two-Sample Assuming Equal Variances*, then entered data ranges as above and then 0 for *Hypothesized Mean Difference* and 0.05 for *Alpha* (= *P*). She selected a suitable output option and then *OK*. This returned a table of data as shown below.

t-test: Two-sample assuming equal variances

	Variable 1	Variable 2
Mean	2.515	2.703
Variance	0.042	0.038
Observations	8	8
Pooled variance	0.040	
Hypothesised mean difference	0	
d.f.	14	
t Stat	-1.881	
$P(T <= t)$ one-tail	0.040	
t Critical one-tail	1.761	
$P(T <= t)$ two-tail	0.081	
t Critical two-tail	2.145	

The value of *t* obtained was -1.881 (row 7, '*t* stat') and the probability of obtaining this value for a two-tailed test (row 10) was 0.081 (or 8.1%), so the student was able to accept the null hypothesis and conclude that ZH52 had no significant effect on fungal growth in these circumstances.

Example 2: An ANOVA test

A biochemist made six replicate measurements of four different batches (A–D) of alcohol dehydrogenanse, obtaining the following data:

Alcohol dehydrogenase activity (U L^{-1})

Batch/ Replicate	1	2	3	4	5	6	Mean	Variance
A	0.562	0.541	0.576	0.545	0.542	0.551	0.552833	0.000189
B	0.531	0.557	0.537	0.521	0.559	0.538	0.540500	0.000221
C	0.572	0.568	0.551	0.549	0.564	0.559	0.560500	0.000085
D	0.532	0.548	0.541	0.538	0.547	0.536	0.540333	0.000039

The biochemist wanted to know whether the observed differences were statistically significant, so he carried out an ANOVA test, assuming the samples were normally distributed and the variances in the three populations were homogeneous. He selected *Anova: Single Factor*, then entered the data ranges, selected *Grouped By: Rows* and set *Alpha* to 0.05 (= *P*). He selected a suitable output option and then *OK*. This returned a table of data as shown below.

ANOVA: Single Factor

SUMMARY

Groups	Count	Sum	Average	Variance
A	6	3.317	0.552833	0.000189
B	6	3.243	0.5405	0.000221
C	6	3.363	0.5605	8.51E-05
D	6	3.242	0.540333	3.95E-05

ANOVA

Source of variation	SS	d.f.	MS	F	P-value	F crit
Between groups	0.001761	3	0.000587	4.397856	0.015669	3.098391
Within groups	0.002669	20	0.000133			
Total	0.00443	23				

The *F* value calculated was 4.397856. This comfortably exceeds the stated critical value (F_{crit}) of 3.098391, and the probability of obtaining this result by chance (*P*-value) was calculated as 0.015669 (1.57% to three significant figures); hence the biochemist was able to reject the null hypothesis and conclude that there was a significant difference in average enzyme activity between the four batches, since $P < 0.05$. Such a finding might lead on to an investigation into why there was batch variation, e.g. had they been stored differently?

(continued)

Box 76.3 (continued)

Example 3: Testing the significance of a correlation

A researcher wanted to know whether his observations of earthworm casts on the surface of closely mown grass were related to how wet the soil was. He took weekly measurements of precipitation using a rain gauge and counted the mean number of casts per m² taking mean results from nine quadrats per weekly observation:

Observation	Precipitation in previous week (mm)	Mean density of casts (m⁻²)
1	11	4.4
2	1	3.1
3	0	2.3
4	5	4.6
5	8	4.5
6	2	3.3
7	4	3.5
8	15	6.4

The researcher used the Microsoft *Excel* function PEARSON(array1, array2) to obtain a value of +0.927857674 for Pearson's product moment correlation coefficient r, specifying the precipitation data as array1 and the cast density as array2. He then used a spreadsheet to calculate the t statistic for this r value, using eqn [76.3]. The value of t was 7.037, with six degrees of freedom. The critical value from tables (e.g. Table 76.2) at $P = 0.001$ is 5.96, so he concluded that there was a very highly significant positive correlation between his two sets of observations. The investigator moved on from this observation and next investigated the effect of artificial hosepipe rainfall in a sheltered grass plot, to test whether there was a causal relationship involved.

Spreadsheet commands illustrated here may vary among the various versions of Microsoft *Office* programs and for different spreadsheet products. Check the precise functions and syntax using the insert function menu option (f_x button in some versions of *Excel*).

both variables are normally distributed. Statistical tests can be used to decide whether the correlation is significant, e.g. using a one-sample t-test to see whether r is significantly different from zero, based on the equation:

$$t = r \div \sqrt{[(1 - r^2) \div (n - 2)]} \text{ at } n - 2 \text{ degrees of freedom,} \quad [76.3]$$

where n is the number of paired observations. If one or both variables are not normally distributed, then you should calculate an alternative non-parametric coeffcient, e.g. Spearman's coefficient of rank correlation (r_s) or Kendall's coefficient of rank correlation (τ). These require the two sets of data to be ranked separately, and the calculation can be complex if there are tied (equal) ranks. Spearman's coefficient is said to be better if there is any uncertainty about the reliability of closely ranked data values.

If underlying theory or empirical graphical analysis indicate a linear relationship between a dependent and an independent variable, then linear regression can be used to estimate the mathematical equation that links the two variables. Model I linear regression is the standard approach, and is available within general-purpose software programs such as Microsoft *Excel* (Box 76.3), and on some scientific calculators. It is suitable for experiments where a dependent variable Y varies with an *error-free* independent variable X in accordance with the relationship $Y = a + bX + e_Y$, where e_Y represents the residual (error) variability in the Y variable. For example, this relationship might apply in a laboratory procedure where you have carefully controlled the independent variable and the X values can be assumed to have zero error (e.g. in a calibration curve, see Chapter 53, or in a time course experiment where measurements are made at exact time points). The regression analysis gives estimates for a and b (equivalent to the slope and intercept of the line of best fit, p. 500): computer-based programs usually provide additional features, e.g. residual values for Y (e_Y), estimated errors for a and b, predicted values of Y along with graphical

Using more advanced types of regression – these include:

- Model II linear regression, which applies to situations where a dependent variable Y varies with an independent variable X, and where both variables may have error terms associated with them.
- Multiple regression, which applies when there is a relationship between a dependent variable and two or more independent variables.
- Non-linear regression, which extends the principles of linear regression to a wide range of functions. Technically, this method is more appropriate than transforming data to allow linear regression.

Advanced statistics books should be consulted for details of these methods, which may be offered by some statistical computer programs.

plots of the line of best fit (the trend line) and the residual values. For the model to be valid, the residual (error) values should be normally distributed around the trend line and their variance should be uniform (homogeneous), i.e. there should be a similar scatter of data points around the trend line along the x-axis (independent variable).

If the relationship is not linear, try a transformation (see pp. 472–3). For example, this is commonly done in analysis of enzyme kinetics (see Fig. 60.4). However, you should be aware that the transformation of data to give a straight line can lead to errors when carrying out linear regression analysis: take care to ensure that (a) the assumptions listed in the previous paragraph are valid for the transformed data set and (b) the data points are evenly distributed throughout the range of the independent variable. If these criteria cannot be met, non-linear regression may be a better approach, but for this you will require a suitable computer program, e.g. *GraphPad Prism*.

The strength of the relationship between Y and X in Model I linear regression is best estimated by the coefficient of determination (r^2 or R^2), which is equivalent to the square of the Pearson correlation coefficient. The coefficient of determination varies between 0 and +1 and provides a measure of the goodness of fit of the Y data to the regression line: the closer the value is to 1, the better the fit. In effect, r^2 represents the fraction of the variance in Y that can be accounted for by the regression equation. Conversely, if you subtract this value from 1, you will obtain the residual (error) component, i.e. the fraction of the variance in Y that cannot be explained by the line of best fit. Multiplying the values by 100 allows you to express these fractions in percentage terms.

> **Example** If a regression analysis gives a value for r^2 of 0.75 (i.e. $r = 0.84$), then 75% of the variance in Y can be explained by the trend line, with $1 - r^2 = 0.25\,(25\%)$ remaining as unexplained (residual) variation.

Using computers to calculate hypothesis-testing statistics

As with the calculation of descriptive statistics (Chapter 75), specialist statistical packages such as *SPSS* and *MINITAB* can be used to simplify the calculation of hypothesis-testing statistics. The correct use of the software and interpretation of the output requires an understanding of relevant terminology and of the fundamental principles governing the test, which is probably best obtained by working through one or more examples by hand before using these tools (e.g. Box 61.2, Box 76.2). Spreadsheets offer increasingly sophisticated statistical analysis functions, three examples of which are provided in Box 76.3.

Text references

Finney, D.J., Latscha, R., Bennett, B.M. and Hsu, P. (1963) *Tables for Testing Significance in a 2 × 2 Table.* Cambridge University Press, Cambridge.

Heath, D. (1995) *An Introduction to Experimental Design and Statistics for Biology.* UCL Press, London.

Sokal, R.R. and Rohlf, F.J. (2012) *Biometry,* 4th edn. W.H. Freeman, San Francisco.

Wardlaw, A.C. (2000) *Practical Statistics for Experimental Biologists,* 3rd edn. Wiley, New York.

Source for further study

Schmuller, J. (2013) *Statistical Analysis with Excel for Dummies,* 3rd edn. Wiley, Hoboken, New Jersey.

STUDY EXERCISES

76.1 Calculate 95% confidence limits. What are the 95% confidence limits of a sample with a mean = 24.7, standard deviation = 6.8 and number of data values = 16? (Express your answer to three significant figures.)

76.2 Use the Poisson distribution. In a sample of 15 snails, a researcher finds the following number of parasite larvae per snail: 0, 0, 0, 0, 1, 1, 1, 3, 4, 5, 5, 7, 7, 9, 9. Using the rule of thumb on p. 518, decide whether the parasites are 'clumped' or 'repulsed' in distribution on their host. What might this mean in biological terms?

76.3 Practise using a t-test. A biology student examines the effect of adding a plant hormone to pea plants. She dissolves an appropriate amount of the compound in ethanol and applies 25 µL of this to the uppermost stipules of the treated plants. With the controls, she applies the same amount of pure ethanol. After three days, she measures the distance between the 2nd and 3rd internodes of the plants and obtains the results shown at the top of the next column. Carry out a t-test on the data and draw appropriate conclusions.

Internode distance in cm

Control	7.5	8.1	7.6	6.2	7.5	7.8	8.9
Treatment	5.6	7.5	8.2	6.7	3.5	6.5	5.9

76.4 Interpret the output from Excel linear regression analysis. The output in the two tables below represents a regression analysis for an experiment measuring the uptake of an amino acid by a cell suspension (in pmol cell^{-1}) against time (in minutes). These were obtained as output using the 'Regression' tool within the *Data > Data Analysis* menu in *Excel*. Note that 'R' here is the same as r, Pearson's produce moment correlation coefficient (p. 523). Based on this output, what is the form and strength of the underlying linear relationship? (Express the coefficients to three significant figures.)

Summary output from Microsoft *Excel* spreadsheet linear regression analysis

Regression statistics	
Multiple R	0.985335951
R square	0.970886937
Adjusted R square	0.963608672
Standard error	2.133876419
Observations	6

ANOVA					
	d.f.	SS	MS	F	Significance F
Regression	1	607.4062857	607.4063	133.3954	0.000320975
Residual	4	18.21371429	4.553429		
Total	5	625.62			

	Coefficients	Standard error	t Stat	P-value	Lower 95%	Upper 95%	Lower 95.0%	Upper 95.0%
Intercept	1.171428571	1.544386367	0.758507	0.490383	–3.11648428	5.459341423	–3.11648428	5.459341423
X variable 1	2.945714286	0.255047014	11.54969	0.000321	2.237588784	3.653839787	2.237588784	3.653839787

Answers to these study exercises are available at www.pearsoned.co.uk/practicalskills

Index

absolute humidity, defined 367
absorbance, defined 447
Abstracts, defined 55
abundance, defined 335
accuracy
    defined 320
    in measurement 175
    in observations 183
acetic acid 225
    pK$_a$ value 168
acid rain, measurement of 370
acids, disassociation of 163
activity coefficient 157
activity in solution chemistry 157–8
    equivalent mass 158
    normality 158
    osmolality 158
    osmolarity 158
adjustable pipettor 140, 141
Adobe Dreamweaver 83
Adobe Illustrator 83
Adobe Photoshop 83
adsorption chromatography 457
aerosol, defined 369
aerotolerant anaerobe 233
affinity chromatography 458–9
agar
    molten, using 284
    plate count of bacteria in 295
    as sterile media 284
agar plate, harvesting bacteria 290
agarose gel electrophoresis 463
    separation of DNA 402–3, 464
agarose system in electrophoresis 462
age, measuring in organisms 341
    and size structure 340–1
agglutination tests 378
aggregated structures in soil
        minerals 360
air displacement pipettor 140, 141
air pressure, measurement of 366
air space volume of soils and
        sediments 358
Airey, P. 422
alfisols 355
algebra, rules 494–5
alkalis 163
    disassociation of 163–4
alleles in Mendelian genetics 394, 395
allometry in growth of organisms 314
alpha decay 416
altruism 130
Ames, B.N. 317
Ames test on mutagenicity 317
amino acids in Mendelian genetics 394
ammonium sulphate fractionation
        enzyme purification 390
amount of substance 180
ampholyte 163
amplification of DNA 403–4

analysis of variance (ANOVA) 520
    in Excel 524
analytical writing 108
andisols 355
anemometers 368
aneroid barometer 366
animal subjects
    cell culture systems 300–3
    collecting 221–3
        devices 222
        edge effects 222
        equipment for 221–3
        formal samples 222–3
        general strategies 223
        location, defined 222
        reasons for 221
        recording 223
    fixing and preserving 225–8
        agents 225–6
            selection of 226
        humane actions 225
        narcotisation 225–6
        preservation 227–8
            dry 228
            wet 227–8
        storage 228
    identifying 240–4
        comparing specimens with
            descriptions 244
        guides for 240–1
        holotypes 244
        keys for 241–3
            bracketed keys 241
            computerised keys 243
            flowchart keys 242
            indented keys 241–2
            multi-access keys 242–3
            using keys 243–4
        rare specimens, dealing with 240
        species distribution 242
    marking in ecological analyses 338
    organs, working with 298–9
    parts and sections through body 279
    research involving 134
    tissues and cells, working
        with 298–305
        sterile technique 302
Animal Welfare Act (UK, 2006) 134
animalia in taxonomy of organisms 235
Animals Scientific Procedures Acts (UK,
        1976, 1986) 134, 298, 377
anode, defined 462
ANOVA (analysis of variance) 520
    in Excel 524
anthropomorphism, defined 315
antibiotics
    in cell culture 301
    in microbe isolation 233
antibodies
    defined 377

production of 377–8
    monoclonal 378
    polyclonal 377
    structure of 377
antigens 377
    typing in identifying microbes 251
antishake, defined 306
apoplasm, defined 299
apparatus, diagrams of 186, 187
aquatic environments 345–52
    bottom-dwelling biological
        materials (benthos) 351
    dissolved oxygen, biological oxygen
        demand, chemical oxygen de-
        mand 352
    field probes in 346
    pH 346
    river discharge
        measurements 348–9
    salinity, conductivity and
        temperature 345
    suspended biological materials
        nekton 351
        plankton 349, 351
    water flow velocity 347–8
    water transparency 346–7
        optical instruments 347
        Secchi disk 346–7
aqueous solutions, storing 142
area, measuring 151–2
aridisols 355
artefacts
    in microscopy 280
        eliminating 277
    in quality of observations 182
assays
    of DNA 404
    for enzymes 387–9
assertiveness in time management 10
assessment
    oral 114
    of poster display 97
    role of 30–1
atmosphere
    in microbe isolation 232
atmospheric conditions, con-
        trolling 152–3
    gas composition 152–3
    pressure 153
atmospheric variables,
        measurement of 364–70
    air pressure 366
    chemical analysis 369–70
        acid rain 370
        particulates 369–70
    cloud cover 368
    humidity 367
    precipitation 367
    sampling 364
    solar radiation 368